Pest Management: Biologically Based Technologies

EDITED BY

Robert D. Lumsden
Agricultural Research Service, U.S. Department of Agriculture

James L. Vaughn
Agricultural Research Service, U.S. Department of Agriculture

**Proceedings of Beltsville Symposium XVIII,
Agricultural Research Service, U.S. Department of Agriculture,
Beltsville, Maryland,
May 2–6, 1993**

American Chemical Society, Washington, DC 1993

Library of Congress Cataloging-in-Publication Data

Beltsville Symposium (XVIII: 1993: Beltsville, Md.)
Pest management: biologically based technologies: proceedings of the Beltsville Symposium XVIII, Agricultural Research Service, U.S. Department of Agriculture, Beltsville, Maryland, May 2–6, 1993 /   Robert D. Lumsden, James L. Vaughn, editors.

p.   cm.—(Conference proceedings series, ISSN 1054–7487)

Includes bibliographical references and index.

ISBN 0–8412–2726–8

1. Biological pest control agents—Congresses.   2. Agricultural pests—Biological control—Congresses.

I. Lumsden, R. D.   II. Vaughn, James L., 1934–   .   III. Beltsville Agricultural Research Center. IV. Title. V. Series: Conference proceedings series (American Chemical Society)

SB975.B45   1993
632'.96—dc20   93–26355
                                         CIP

1993 Advisory Board

ACS Symposium Series

M. Joan Comstock, *Series Editor*

Symposium Organizing Committee

Barbara A. Leonhardt, Co-chair
James L. Vaughn, Co-chair
James D. Anderson
David J. Chitwood
Dora K. Hayes
Robin N. Huettel
George W. Irving
Robert D. Lumsden
Julius J. Menn
J. David Warthen, Jr.
William P. Wergin

Special Thanks

The Symposium Organizing Committee thanks
the secretarial staff, who made a substantial
contribution to making this Symposium "work"
through their conscientious and diligent efforts:

Sandra L. Dusch
Gloria V. Gantt
Virginia L. Hupfer
Veronica A. Jameson
Kendra K. Trzupek

Sponsors of the Symposium

The Friends of Agricultural Research–Beltsville (FAR–B) are cosponsors of the Beltsville Symposium Series. FAR–B is a nonprofit group dedicated to supporting the research and educational programs at the Beltsville Agricultural Research Center. Membership is composed of former and current employees plus a growing number of industry supporters. The Symposium Organizing Committee thanks the members of FAR–B for their many contributions to the success of this meeting.

The organizers also thank the following sponsors of the Symposium for their generous financial contributions:

Patrons

Ciba-Geigy Corporation
Greensboro, North Carolina

Entotech, Inc.
Davis, California

ISK Biotech Corporation
Mentor, Ohio

Sandoz Agro, Inc.
Palo Alto, California

DuPont
Newark, Delaware

Gardens Alive
Lawrenceburg, Indiana

Pioneer Hi-Bred International, Inc.
Johnston, Iowa

O. M. Scott & Sons Company
Marysville, Ohio

Valent USA Corporation
Walnut Creek, California

Supporters

American Cyanamid Company
Princeton, New Jersey

EcoScience Corporation
Worcester, Massachusetts

Grace Sierra
Allentown, Pennsylvania

Contents

BIOCONTROL AGENTS FOR SUPPRESSION
OF PLANT PATHOGENS

IMPLEMENTATION: NEEDS, ISSUES, AND CHALLENGES

INDEX

Preface

THE BELTSVILLE AGRICULTUAL RESEARCH CENTER (BARC), Agricultural Research Service (ARS), U.S. Department of Agriculture (USDA) annually sponsors a symposium on agricultural research. These symposia provide an arena for BARC, ARS, and scientists from academe and industry to present the latest research findings and concepts in a variety of disciplines related to agricultural science and technology. Symposium V, "Biological Control in Crop Production", held in 1980, was the most recent related program.

The current Symposium XVIII provides new information on this rapidly evolving science and technology of biologically based pest-management systems. The goal is to harmonize integrated pest-management concepts, sustainable agriculture, and use of natural resources in developing crop protection technologies. Current methods rely less on synthetic chemical pesticides and integrate better with environmental and societal requirements.

Biologically based pest management encompasses a multicomponent approach to crop and livestock protection. These proceedings are based on 57 papers that were presented at the conference. Twelve papers were selected from 75 posters for inclusion in this volume. An introductory section deals with the definition, framework, and role of biologically based control, with emphasis on development of a holistic systems approach to lead the field well into the 21st century.

The following five sections include papers on biocontrol agents for suppression of insects, biocontrol agents for suppression of plant pathogens, biocontrol agents for suppression of weeds, natural compounds in pest management, and genetic manipulation of biocontrol agents. The last section deals with implementation of needs, issues, and challenges.

Many advances have taken place in biocontrol technology in the 13 years since the previous BARC Symposium on this topic. Impetus for development of biological control measures has come from societal demands and perceptions, environmental needs and regulations, concerns for food safety and water quality, pest resistance to chemical pesticides, and a changing regulatory climate.

Although biocontrol-based crop protection currently represents only a very small fraction of the chemical pesticide market (less than 1%), market studies indicate a growth rate of 10% per annum from the current base. Thus, by the turn of the century, biocontrol-based crop-protection technologies should capture a $2–3 billion market share of the $20–30 billion pesticide market at the producers level worldwide.

Biologically based pest management is an interdisciplinary effort involving biologists, biochemists, agronomists, molecular biologists, agricultural engineers, economists, and regulatory professionals, all working in concert. The papers presented in this volume reflect this diversity of professions and disciplines.

We acknowledge, with our sincere thanks, Friends of the Agricultural Research Center, Beltsville (FAR–B), the corporations, government agencies, and members of the Symposium Organizing Committee, and the secretarial staff (all names appear elsewhere in this volume) whose generous financial support, contributed time, and talents made this outstanding conference possible.

Finally, our special thanks are due to Anne Wilson and her colleagues in the American Chemical Society Books Department for their enthusiastic support and assistance in making this proceedings volume possible.

JULIUS J. MENN
Organizing Committee
Plant Sciences Institute
Agricultural Research Center, BARC–West
U.S. Department of Agriculture
Building 003, Room 232
Beltsville, MD 20705–2350

ROBERT D. LUMSDEN
Editor
Biocontrol of Plant Diseases Laboratory
Plant Sciences Institute
Agricultural Research Center, BARC–West
U.S. Department of Agriculture
Building 011A
Beltsville, MD 20705–2350

JAMES L. VAUGHN
Editor
Insect Biocontrol Laboratory
Plant Sciences Institute
Agricultural Research Center, BARC–West
U.S. Department of Agriculture
Building 011A
Beltsville, MD 20705–2350

June 11, 1993

INTRODUCTION

Biologically Based Pest Management in Managed Ecosystems: Increasing Its Acceptance

Ralph W.F. Hardy, Boyce Thompson Institute for Plant Research Inc., Tower Road, Ithaca, NY 14853-1801

Pests and disease—referred to collectively as pests—have been, are, and will be a continuing challenge for managed ecosystems. An extended time of change is projected in which we progress from the period of about 1940-2000, where synthetic chemical pesticides were the primary method used to control pests, to the 2000s, where an ecosystem focus for pest management with a holistic approach will be employed, utilizing the most goal-compatible biological, chemical, and physical/cultural knowledge. Considerable overlap occurs in the classification of methods as biological, chemical, or physical/cultural. The holistic approach will need to be interdisciplinary, interfunctional, and cooperative. Involvement of a broad spectrum of key decision makers and implementers is essential for the establishment of goals and for their broad acceptance. These suggested goals include meeting all pest-problem needs, minimum economic cost based on full-cost accounting, maximum human safety, minimum environmental impacts, maximum sustainability, and consideration of social impacts. Major operational challenges must be solved.

Pests and disease—referred to collectively as pests—have been, are, and will continue for the foreseeable future to be a demanding problem for managed ecosystems. The 1990s and early 2000s will be a period of substantial change in the methods used for pest management as were the 1940s and 1950s. This period of change will take time and occur by a series of steps to reach the goal of a much improved and more acceptable pest management system. We cannot move immediately from our present methods to new ones and continue to meet production needs of managed ecosystems. A view on how to facilitate this transition is presented in this paper with an integrative emphasis on biologically based methods and their acceptance. The experience of the past is examined and the transition to the future is projected.

Prior to 1940, pest management used biological, non-synthetic chemical and physical/cultural methods to control pests. The discovery of the utility of synthetic chemical pesticides in the 1940s introduced the era in which these products became the dominant purchased agents for pest control. This era produced many benefits but also significant limitations. The limitations of the synthetic chemical approach, coupled with a recognition

of the need for an ecosystem perspective and the expanding potential of biology, will drive changes and opportunities in the 1990s and early 2000s. Pest management, not pest eradication, will be the objective for managed ecosystems. This new era will be characterized by a holistic approach. Goals of pest management should be defined by a broad group of key decision makers and implementers so that there will be broad acceptance extending from society to adopters. The methods for pest management will be generated from biological, cultural/physical, and chemical knowledge. The relative importance of biological, cultural/physical, or chemical knowledge will be based on their compatibility with the defined goals. It is reasonable to predict that the role of biologically based approaches will increase while that of synthetic chemical pesticides will decline.

The XVIIIth Beltsville Symposium focused on the biologically based technologies. Many of the leaders of this biologically driven change reported the status of their science and technology. We are still early into this transition period. We need to expand greatly our knowledge base and couple that knowledge to development, regulatory approval, implementation, and adoption. If synthetic chemical pesticides were banned today, we would not have alternative products, processes, systems, or information to manage pests in most of our managed ecosystems. We must develop alternatives that are competitive with or superior to synthetic chemical pesticides on a full-cost accounting basis.

The Board on Agriculture of the United States National Research Council (NRC) has recognized the need to make this change happen. A study, titled "Pest and Pathogen Control through Management of Biological Control Agents and Enhanced Natural Cycles and Processes," was initiated in June 1992, and the report should issue in the first half of 1994. An interdisciplinary committee of fourteen members from academe, government, research institutions, industry, and public-interest organizations is developing a common conceptual framework, goals, and approaches and is identifying science and technology needs, defining an appropriate regulatory process, and suggesting how to facilitate implementation and adoption. As most in this field are aware, this is not an easy task. Probably the primary impediment to increasing the importance of biological control is fragmentation—fragmentation within disciplines, fragmentation between disciplines, and fragmentation within organizations. We need to integrate and be holistic in our approach: interdisciplinary, interfunctional, and cooperative.

Although I am chairing the above-mentioned NRC committee, this paper is not a product of the committee, as the deliberations of the committee are still in process and the outcome of these deliberations will not be available until the committee has decided on content and recommendations and until the report has passed the rigorous NRC report review process. Rather, the content of this paper is my personal view about an area which I believe must receive very high priority in the 1990s and early 2000s. My views have been shaped in part by participation in the 1987 National Academy of Sciences report on biological control in managed ecosystems (NAS, 1987); the 1991 Cornell/Boyce Thompson Institute Workshop on "Biological Control: Making It Work" (Chabot et al., 1991); and the 1991 report of the annual meeting of the National Agricultural Biotechnology Council, *Agricultural Biotechnology at the Crossroads: Biological, Social, and Institutional Concerns*, section on biological control (Granados et al., 1991). In addition, on a daily basis I am in contact with biologically based pest management research with about one-third of Boyce Thompson Institute's (BTI) research now focused on biological alternatives for pest management while twenty years ago a significant part of BTI's research was focused on the biological evaluation of synthetic chemical pesticide candidates.

Scope of Biologically Based Pest Management

An expansive, not restrictive, scope for biologically based pest management is necessary for its optimal development and acceptance. A restrictive scope is ill-advised and probably would constrain rather than enhance even those areas to which the scope is restricted. A win-win, not a win-lose, perspective is needed.

Products, processes, systems, or information based on biological materials or knowledge are biologically based. Biocontrol agents for pest suppression, natural compounds in pest management, and genetic manipulation—as were discussed in this symposium—are included in this scope, together with even broader methods as suggested below. Such a broad scope provides a common canopy for entomologists, nematologists, plant pathologists, and weed scientists. This broad scope includes classical biological control and natural cycles and processes. This broad scope has substantial overlap with cultural/physical and chemical approaches (Table 1). There is a continuum of methods and there are no clear boundaries—nor is there a need for or benefit from such clear boundaries. The objective is pest management and the use of those methods that are most compatible with goals. The establishment and acceptance of goals will guide the selection of methods. The range of biologically based approaches extends from biobased synthetic chemical pesticides—such as 2,4-D which is also chemically based—to crop rotation which is also cultural/physical-based. Undoubtedly, there will be new methods that are biologically based. For example, transgenic organisms, a technological advance of the 1980s, provide an additional method with high expectations for its utility.

Table 1. The Continuum of Pest Management Methods Based on Chemical, Biological, and Physical Approaches

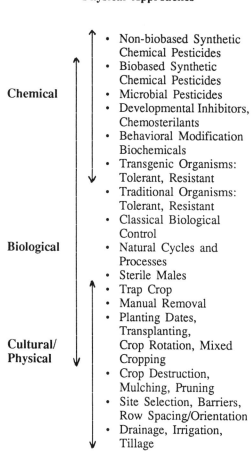

Chemical
- Non-biobased Synthetic Chemical Pesticides
- Biobased Synthetic Chemical Pesticides
- Microbial Pesticides
- Developmental Inhibitors, Chemosterilants
- Behavioral Modification Biochemicals
- Transgenic Organisms: Tolerant, Resistant
- Traditional Organisms: Tolerant, Resistant

Biological
- Classical Biological Control
- Natural Cycles and Processes
- Sterile Males
- Trap Crop
- Manual Removal

Cultural/ Physical
- Planting Dates, Transplanting, Crop Rotation, Mixed Cropping
- Crop Destruction, Mulching, Pruning
- Site Selection, Barriers, Row Spacing/Orientation
- Drainage, Irrigation, Tillage

Acceptance

My assigned title focuses on increasing the acceptance of biologically based pest management. Acceptance means favorable reception, approval, or meeting-of-the-minds. Other equivalents to acceptance are backing, endorsement, esteem, recognition, respect, sanction, sponsorship, and support. All of these words convey various objectives that are desirable for biologically based pest management. For example, biologically based pest management needs approval, backing, endowment, sanction, and support.

Whose acceptance are we seeking? I suggest it is the favorable reception, approval, backing, endorsement, esteem, recognition, respect, sanction, sponsorship, and support for biologically based pest management by key decision makers and implementers. These are (1) society, including public-interest groups that understand science and are primarily concerned with serving the public rather than with their individual motives such as raising donations for their organizations (Fumento, 1993); (2) consumers; (3) the input, processing and retail industry for private sector research and development, seeking regulatory approval, implementation, and, in some cases, adoption such as occurs with contract production; (4) government for appropriate legislation, funding public research and development, appropriate regulation, implementation, and in some cases adoption; (5) scientists, engineers, economists, and social scientists for planning and conducting research and development; (6) regulators for regulation (although they are part of government, they are identified separately because of their key role); (7) industrial field people, extension workers, and consultants for implementation; and (8) growers for adoption. The list of key decision makers and implementers is diverse and large, but it is this broadly heterogenous group that must be involved in the 1990s and beyond. There are many and varied voices that demand involvement, and it is critical that acceptance be obtained to the maximum extent possible from all these key decision makers and implementers. Such broad acceptance is equally important for the for-profit as well as for the not-for-profit sector. No one wants to develop pest management methods whose implementation and adoption will be met with opposition that produces extensive delays or even blockages.

Early and continuous dialogue is necessary so as to avoid prolonged delays in technology use such as we have seen with the case of bovine somatotropin (BST) for enhanced dairy productivity. Would industry have invested in developing BST if they had realized at an early stage that there would be excessive delays, not because there were concerns about efficacy or human health but because of socio-economic concerns. However, agricultural technologists in the 1970s, when the BST decision was made, did not recognize the need for broad acceptance. Productivity and human health were the key considerations. Now, in the 1990s, forward-looking agricultural technologists—whether in the for-profit or not-for-profit sectors—are coming to recognize the need for broad communication and acceptance.

An open forum is desirable where all these key groups can meet to speak, to listen, and to learn. The National Agricultural Biotechnology Council (NABC) was formed in 1988 to provide such a forum for agricultural biotechnology. The NABC (NABC, 1989, 1990, 1991, 1992) may provide the model for a process to establish goals and seek broad-based acceptance for biologically based pest management. As discussed later, goal definition and acceptance of the goals would be the objective of such an effort.

Pest Management Eras and the Case for Change to Increase Acceptability

Pest management can be divided into three time periods based on the dominant use of synthetic chemical pesticides. The pre-1940 era preceded this use of synthetic chemical pesticides. Biological, cultural/physical, and non-synthetic chemical methods were used with no one overall approach dominant. Pest control, for the most part, was less effective than in the following era.

The 1940-2000 era is the period during which the use of synthetic chemical pesticides became dominant. The insecticide DDT and the selective herbicide 2,4-D probably represent the seminal events for the synthetic chemical pesticide era (Menn & Henrick, 1985). The growth of this chemical era was driven, for the most part, by the expanding knowledge of synthetic organic chemistry. Inventive chemists synthesized novel chemical structures which biologists routinely tested for biological activity. Most of the chemical structures had no relationship to biological molecules. There are exceptions, such as 2,4-D whose synthesis and discovery of its potent biological activity by Boyce Thompson scientists in the late 1930s was based on the plant hormone indoleacetic acid and its promotion of root growth (Hardy, 1991). Most synthetic chemical pesticides are chemical-synthesis based, however, and there has been poor predictability of activity. Statements like "spray and pray" have been used to realistically describe the discovery process for synthetic chemical pesticides. In most cases it was impossible to predict whether a synthetic chemical would be a potential herbicide, insecticide, or fungicide or for that matter, have any potential pest control activity. For example, scientists at Boyce Thompson Institute proposed that the chemical which became the insecticide carbaryl would be a herbicide. A synthetic chemical with potential pest-control activity would be analogued extensively to identify the preferred chemical structures for development and the chemical scope for patent filing. Because of patents, synthetic chemical pesticides are highly proprietary, enabling pricing based on the value provided. Synthetic chemical pesticides usually produce consistent performance and broad geographic utility. Production and marketing are readily scaled. The most useful synthetic chemical pesticides have annual markets of about $1 billion and those with less than about $100 million are, for the most part, too small to justify the cost of research and development, regulatory approval, manufacture, and marketing.

There are many parallels between the agrichemical and pharmaceutical industry. Synthetic chemicals as pesticides and drugs dominate their respective industries. There are similar benefits and limitations. Synthetic chemical pesticides along with fertilizers and genetic improvements of crops are the major contributors to total world crop production and improved productivity. Without synthetic chemical pesticides, it is doubtful that world crop production would be able, at this time, to meet the food needs of over 5 billion people to the extent now achieved.

However, there are limitations from the dominant use of synthetic chemical pesticides and opportunities for more acceptable pest management in the future. There are several unmet needs and these may expand. There are not effective chemical pesticides for many soil-borne pests and insect-vectored pathogens. Minor crops often lack registered pesticides since the market size does not justify the cost of testing, regulatory approval, manufacture, and marketing. There will be losses of existing pesticides from industry's decision not to re-register or government's decision to ban. In addition, the nature of pests is that regularly there will be new pests for managed ecosystems such as the silverleaf whitefly (Klassen, 1993).

Successfully marketed pesticides have favorable direct cost economics; otherwise they would not survive in the market place. The world-wide purchase cost of pesticides used in managed ecosystems has grown to about $25 billion which probably provides a direct benefit to crop yield or equivalent of $100 billion or more. However, there is a developing recognition that there are additional costs associated with chemicals so that full-cost accounting is needed (Popoff and Beizzelli, 1993). This additional cost relates to the indirect cost associated with limitations such as human health effects, negative environmental impacts, decreased sustainability, and possibly social costs as outlined below. In addition, there is an increasing cost and decreasing rate of discovery and registration of new chemical pesticides. Today, syntheses of probably 10 to 20 times as many new chemicals as in the 1950s are required to discover a marketable product. The synthetic chemical pesticide industry is a maturing industry with consolidation; thus, most sales are concentrated within less than ten large agrichemical companies, thereby reducing competition and probably total research, development, and implementation novelty.

The agrichemical industry is only beginning to explore the design of agrichemicals based on the expanding knowledge of structural biochemistry, as the pharmaceutical industry has been successfully doing in recent years. This approach should replace "spray and pray" with design of synthetic chemicals for predictable activity. Synthesis chemists will be evaluated on the quality of their designs rather than on the number of new chemical structures made. This design approach will be driven by biological knowledge with chemistry used simply as a synthesis tool in much the way that biology was used as a tool in the old era.

There is a concern about the human safety of synthetic chemical pesticides. Applicator exposure is a significant concern. Many synthetic pesticides are chlorinated hydrocarbons about which there is growing health concern (Hileman, 1993). The public is more concerned about pesticide residues in food than justified by the real risk (Stevens, 1991).

There is increasing societal concern about synthetic chemical residues in soil, water, and air. This concern may be valid for a small number of pesticides, but the risk is probably smaller than that of fertilizer-derived nitrate in soil and water of intensively nitrogen-fertilized soils. There are also non-target effects of pesticides on wildlife such as fish and birds. Soil residues of slowly degrading pesticides decrease growers' options to rotate crops, e.g., impossibility of immediately following corn with soybeans when atrazine has been used as a herbicide for corn. Synthetic chemical pesticides favor monoculture, which is not viewed favorably from an environmental perspective.

Sustainability is increasingly identified as a goal for many of the activities conducted by humans. Manufacturing, mining, energy, and other industries are considering how to increase the sustainability of their activities (Schmidheiny, 1992). Agriculture is sensitive to the need to develop production systems that are more sustainable (NRC, 1991). Some of the major limitations of the dominant use of synthetic chemical pesticides relate to the issue of sustainability. Synthetic chemical pesticides are inadequately selective and often destroy beneficial organisms as well as pest organisms. Extensive kill creates opportunities for old or new pests to establish themselves. In addition, the dominant and extensive use of synthetic chemical pesticides creates pesticide-resistant pests, rendering the pesticide ineffective (NRC, 1986). The pharmaceutical industry has experienced the same problem with drugs for infectious diseases. Most scientists agree that pesticide resistance, especially in short life-cycle pests, is unavoidable, and a pest control agent must be managed to control the problem of resistance.

Finally, there is the question of social sensitivity. There is strong opposition by society to the use of synthetic chemical pesticides. This opposition exceeds the reality. Alar (daminozide) is an example of a synthetic chemical from which strong negative societal response to apparently misleading media reports led to the removal of the product by the manufacturers (Fumento, 1993). Legal actions have been taken by growers who suffered major economic loss because of the misleading information. However, perception is as important to the consumer as reality and as such, limits the dominant use of pesticides and the need for change in pest management.

The above limitations document the need for change. The 2000s era will be a post synthetic chemical pesticide one. There will be a decreased role of chemical-based and an increased role of biological-based methods. However, the chemical-based methods will not be eliminated but should be used consistent with their ability to meet the established goals.

Pest-management approaches and goals need to be established for the post synthetic chemical era. For purposes of acceptance it is essential to involve representatives from all major key decision makers and implementers. The groups of key decision makers and implementers are the same as those identified already for acceptance. There may be resistance to including public-interest groups, but failure to include them probably will lead to decreased acceptance and major delays in being able to implement and adopt methods. We must learn how to speak, to listen, and to learn with representatives from all groups. It will be an important and essential skill and key to seeking to maximize the acceptance of pest-management methods. These groups will not go away if we elect not to include them; rather, they will use political and other approaches to achieve their positions. Inclusion of all groups will lead to maximum learning by all groups; every group will learn. We should not be so arrogant to believe that we will not learn from such group interactions.

A Holistic Approach to Pest Management in Managed Ecosystems

We need to identify collectively the needs and opportunities for pest management. In my view, these may include (1) pest management for existing and new pest problems for all managed ecosystems independent of size, (2) minimizing economic cost based on full-cost accounting. (If we were able to measure the cost in economic terms for the needs listed below, we might need to consider only the above two needs. However, economists have not developed that capability.); (3) maximizing safety for workers, consumers, and users; (4) minimizing negative impact on the local and global environment; (5) maximizing sustainability; and (6) considering social questions. Change usually results in winners and losers, and such will be the case for changes in pest-management methods. These changes must be considered before development of new methods; the impact on losers from method change must be addressed. The above list of suggested needs and opportunities should be refined or modified to represent the agreement of key decision makers and then implemented.

A holistic approach should be used to seek methods that are maximally compatible with the above needs. All sources—biological, chemical, and physical/cultural—should be utilized. We should also use all available tools and knowledge. The new molecular tools of biology such as RFLPs (restruction fragment linked polymorphisms), PCR (polymerase chain reaction) and LCR (ligase chain reaction), gene cloning, sequencing, synthesis, transformation, and control elements provide powerful capabilities in both analysis and genetic modification. We should employ them as we do those of traditional biology because all methods will be needed to achieve our goals. We should not be constrained by definitions. Definitions are of use only if they help to generate highly compatible pest-management methods. The user, such as the grower or consumer, is interested in effective and acceptable pest-management methods; the user has little or no interest in definitions.

We will need to be interdisciplinary. Individuals with training in biochemistry, chemistry, crop science, economics, engineering, entomology, environmental science, forestry, horticulture, microbiology, molecular biology, plant pathology, social science, soil science, and weed science need to work together.

We also need to be interfunctional. Workers in research need to interact with others in development, regulation, implementation, and adoption. These interdisciplinary and interfunctional structures might be viewed as both horizontal and vertical integration. In addition, there needs to be cooperation, especially between the for-profit and not-for-profit sectors.

A holistic approach to pest management in managed ecosystems is synthesized in Figure 1. Step 1 is setting goals. Step 2 is generation of knowledge to enable the subsequent steps to meet the goals. Generation of knowledge is probably the primary limiting factor controlling the rate of change from the 1940-2000 era to the 2000s era. This knowledge will come from all areas, and areas will often overlap and interact. Step 3 is development of products, processes, and systems by not-for-profit and for-profit sectors. These products, processes, or systems may be in information, management, chemicals, diagnostics, seeds, microbes, macrobes, and/or equipment. Step 4 is appropriate regulation to

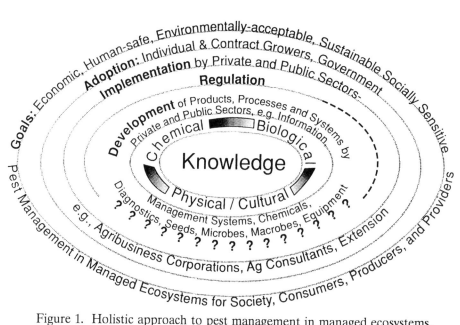

Figure 1. Holistic approach to pest management in managed ecosystems.

minimize total risks in relationship to total benefits. Step 5 is implementation which involves both for-profit and not-for-profit sectors, e.g. industrial field workers, extension staff, and consultants. Step 6 is adoption. Adoption will involve individual growers, decision makers for contract growers—a group which will probably expand greatly in the future—and in some cases government, when broad decision making and adoption is essential to the application of certain methods such as occurred with the sterile-male technique for control of the screw worm.

Operational Problems to be Solved

There are many operational problems to be solved for the successful use of the holistic approach. National and international leadership is necessary. Goals must be established, utilizing broad input for maximum acceptance. Economists must develop full-cost accounting procedures for pest-management methods so that meaningful direct economic comparisons can be made. The for-profit-sector corporations, for which engineering and the physical sciences are key, are establishing consortia to support research in the pre-commercial area. Such partnerships may be appropriate for the for-profit pest-control companies for pre-commercial knowledge

generation. We need to facilitate interdisciplinary relationships. Such interdisciplinary teams will need leadership, and these interdisciplinary teams will be facilitated by national and international interdisciplinary meetings and journals. Funding for interdisciplinary teams and interdisciplinary training will be needed. Establishment of real interdisciplinary teams will be difficult because professional societies and their meeting and publication activities and academic structure are on a strictly disciplinary basis. I am not suggesting that we abandon our disciplinary structure because such would be almost impossible to do and I believe that one needs a disciplinary base. What we do need is to make it easy for interdisciplinary teams to be formed, to dissolve, and to re-arrange as the need or opportunity occurs.

We will need to facilitate interfunctional activities so that each function communicates with the other functions and works as an interfunctional team.

Risk investment by both the public and private sectors will be needed. Managers usually feel more comfortable investing in better-known, established technologies than in less well-known, potential future technologies. However, I believe the high-reward opportunities lie in the future technologies. We must invest if the potential pest-management methods of the 2000s are to become reality.

Information and systems will become increasingly important. The process of development and implementation will be different from that for products and processes.

Appropriate and efficacious regulations must be established for products, processes, or must be established for products, processes, or systems that are relevant for other than synthetic chemical pesticides.

Adoption by growers and others of some new methods other than synthetic chemical pesticides where there is good reliability initially may require incentives or, alternatively, taxes on synthetic chemical pesticides to reflect the full cost accounting of the synthetic chemical pesticides. Other creative ways may be needed to encourage adoption.

All of the above are operational problems that represent substantial challenges but are solvable. It probably will be more difficult to establish goals with broad input and acceptance than to solve the above operational problems. The opportunities are large in this period of change, and the benefits that will result will also be large. The creative will capitalize.

References

Chabot, B. F. et al. "Biological Control: Making It Work." In *Agricultural Biotechnology at the Crossroads: Biological, Social, and Institutional Concerns;* Fessenden MacDonald, J., Ed.; NABC Report 3; National Agricultural Biotechnology Council (NABC): Ithaca, NY, 1991, pp. 199-250.

Fumento, M. *Science under Siege: Balancing Technology and the Environment;* William Morrow & Co., Inc.: New York, NY, 1993; 448 pp.

Granados, R. R.; Hinkle, M. K.; Swinnen, J. F. "Biological Control of Pests." In *Agricultural Biotechnology at the Crossroads: Biological, Social, and Institutional Concerns;* Fessenden MacDonald, J., Ed.; NABC Report 3; NABC: Ithaca, NY, 1991, pp. 32-37.

Hardy, R. W. F. "Biotechnology and New Directions for Agrochemicals." In *Regulation of Agrochemicals: A Driving Force in Their Evolution*; Marco, G. J.; Hollingworth, R. M.; Plimmer, J. R., Eds.; American Chemical Society: Washington, DC, 1991; pp. 131-144.

Hileman, B. "Concerns Broaden over Chlorine and Chlorinated Hydrocarbons." *Chem. Engr. News* **1993**, *71(#16)*, pp. 8-10.

Klassen, P. "Whitefly Update." *Ag Consultant* **1993**, March; p. 20.

Menn, J. J.; Henrick, C.A. "Newer Chemicals to Insect Control." In *Agricultural Chemicals of the Future*; Hilton, J.L., Ed.; BARC Symp. VIII; Rowman & Allanheld: Totowa, NJ, 1985; pp. 247-265.

National Academy of Sciences (NAS). *Research Briefings 1987: Report of the Research Briefing Panel on Biological Control in Managed Ecosystems.* Committee on Science, Engineering, and Public Policy. National Academy Press: Washington, DC, 1987.

National Agricultural Biotechnology Council (NABC). *Biotechnology and Sustainable Agriculture: Policy Alternatives*; Fessenden MacDonald, J., Ed.; NABC Report 1. NABC: Ithaca, NY, 1989; 221 pp.

National Agricultural Biotechnology Council. *Agricultural Biotechnology, Food Safety and Nutritional Quality for the Consumer;* Fessenden MacDonald, J., Ed.; NABC Report 2. NABC: Ithaca, NY, 1990; 222 pp.

National Agricultural Biotechnology Council. *Agricultural Biotechnology at the Crossroads: Biological, Social, and Institutional Concerns;* Fessenden MacDonald, J., Ed.; NABC Report 3. NABC: Ithaca, NY, 1991; 307 pp.

National Agricultural Biotechnology Council. *Animal Biotechnology: Opportunities & Challenges;* Fessenden MacDonald, J., Ed.; NABC Report 4. NABC: Ithaca, NY, 1992; 181 pp.

National Research Council. *Pesticide Resistance;* National Academy Press: Washington, DC, 1986; 471 pp.

National Research Council. *Toward Sustainability;* National Academy Press: Washington, DC, 1991; 145 pp.

Popoff, F. P.; Beizzelli, D. T. "Full-Cost Accounting." *Chem. Engr. News* **1993**, *71(#2)*, pp . 8-10.

Schmidheiny, S. *Changing Course: A Global Business Perspective on Development and the Environment;* MIT Press; Cambridge, MA, 1992; 374 pp.

Stevens, W. K. "What Really Threatens the Environment." *The New York Times* **1991,** Jan. 29; p. C4.

The Role of Biological Control in Pest Management in the 21st Century

R. James Cook, USDA-ARS, Root Disease and Biological Control Unit, Washington State University, Pullman, Washington 99164-6430

Biological control, including through the use of genes transferred from natural enemies and antagonists to crop plants, has the potential to become the major component of pest management of the 21st century, with no significant compromise in the quantity or quality of plant or animal products, the environment, or the expectations of society. Full employment of biological control will require the management of populations or genes of thousands if not tens of thousands of site- and pest- or disease-specific biological control agents. It will also require changes in the institutional infrastructure, scientific resources, and socioeconomic resolve to develop this still largely untapped biological and genetic resource in time to make this vision a reality in the 21st century.

Biological control has played a major role in pest management since the beginning of agriculture. Farmers learn pest management practices through careful observation, and by trial and error that takes advantage of natural biological control even though they may not recognize its contributions to the well-being of their crops and livestock. Even plants and animals in the wild benefit from natural biological control of their pests and diseases, having evolved interdependencies with other organisms in their natural environment that science is only beginning to unravel.

Looking ahead to the 21st century, and recognizing that food production must double in the next 30-50 years with less land, declining natural resources, and less dependency on chemical pesticides, agriculture may well come "full circle" by again depending on biological control as the primary, if not dominant, component of pest management. However, there will be two major differences: Biological control in the 21st century will be delivered or achieved, increasingly, based on scientific understanding and modern technologies; and biological control of the future will be needed to achieve optimal if not maximal productivity of crops and livestock heretofore associated mainly with chemical control.

Every technology from the bow and arrow to modern nuclear energy can be characterized by a sigmoid-type curve of a) lag phase during the earliest stages of its development or while the idea catches on, b) a rapid growth phase as the technology becomes widely adopted, and finally c) a leveling out or even a declining phase as the technology matures or is replaced by a newer and better technology. Pest control with synthetic chemical pesticides is a maturing technology, because of the increasingly greater difficulty of finding or producing new synthetic chemicals, the increasing cost of developing new synthetic chemical pesticides, and the evolution of pest populations resistant to these chemicals. In contrast, in spite of being the oldest component of pest management, biological control is still very much a developing technology, possibly near or still in the lag phase, especially now considering the opportunities to customize, improve, and deploy it using recombinant DNA techniques.

An Expanding Concept of Biological Control for the 21st Century

It is to be expected that the scientific concepts underpinning biological control must and will expand as new information becomes available and we gain more experience with mechanisms and organism interactions. Already, the once seemingly clear conceptual boundaries between the various methods of biological control have become blurred. Consider, for example, control of herbivorous insects by endophytic fungi that produce substances such as alkaloids in the leaf tissues of their plant hosts (Siegel et al., 1987). These endophytes provide biological control of insects, but they do not fall into any of the earlier categories of natural enemies of insects, referred to as "predators, parasites, and pathogens" (DeBach, 1964). They fit more logically into the plant pathology concept of "antagonist." Furthermore, if the plant rather than the endophytic symbiont produced these same substances, earlier concepts would have classified this mechanism as host-plant "antibiosis." Taking this one step further, Freeman and Rodriguez (1993) have reported that a single-gene mutation in the cucurbit anthracnose fungus, *Colletotrichum magna*, converts the fungus into a "nonpathogenic, endophytic mutualist" with ability to protect cucurbits against the pathogenic wild-type. They have proposed "the use of such mutants as biocontrol agents." This is one of many emerging examples of the potential for "pathogen-derived biocontrol agents." Taking this still another step, genes from a wide range of plant viruses, when expressed as a transgenes in otherwise susceptible plants, provide host-plant resistance to those specific viruses (Beachy et al., 1991). A similar approach is now being tested as a means to control certain virus diseases of animals (Clements et al., submitted). This approach for plants exemplifies what is now being called "pathogen-derived resistance".

Examples such as these undermine completely any scientifically based rationale for excluding the plant or animal that benefits or the target pest or pathogen as agents of biological control. The 21st century will bring even more examples of this kind.

A report from the National Academy of Sciences (NRC, 1987) defines biological control as "the use of natural or modified organisms, genes, or gene products to reduce the effects of undesirable organisms (pests), and to favor desirable organisms such as crops, trees, animals, and beneficial insects and microorganisms." This NRC report includes, in addition to more traditional natural enemies and antagonists, both pathogen- or pest-derived strains or biotypes and the host plant or animal that benefits as components (agents) of biological control. The report places the many strategies of biological control into three general groups: 1) reduce the pest population; 2) prevent, repel, or deter pest attack (as with endophytes that produce substances toxic to insects); and 3) turn on systems of self-defense in the host-plant or animal (including through use of pathogen-derived host-plant resistance).

Garrett (1965), in a keynote address at the first international symposium on ecology of soilborne plant pathogens held at Berkeley in 1963, stated that biological control was "any condition under which, or practice whereby survival or activity of a pathogen is reduced through the agency of any other living organism (except man himself), with the result that there is a reduction in the incidence of disease caused by the pathogen." Garrett considered that "Biological control cannot be separated from the whole subject of disease control, which involves eventually a complete knowledge of the biology and epidemiology of a disease, and of the ecology of the crop plant." Beirne (1967) similarly regarded biological control as an integral part of pest management rather than a separate discipline and stated that "Any living organisms that can be manipulated by man for pest control purposes are biological control agents."

Gabriel and Cook (1990), in a call for "a new scientific framework," proposed that the many methods of pest and disease control be divided simply into biological, physical, and chemical (Fig. 1). Physical controls include heat, drying, solar and UV radiation, light waves, microwaves, irradiation, mowing, plowing, and physically separating crops or animals from their pests by choice of planting dates, seed indexing, meristem culture, separate penning, quarantines, etc. Physical controls are a mixture of developing and maturing technologies.

We must also rethink whether an agent formulated as a chemical, such as products formulated from the delta endotoxins of *Bacillus thuringiensis* (Bt), fit even with the broadest concept of biological control. As a kind of litmus test, if the organism only works when alive, it is biological control, but if it works just as well dead as alive, surely this would be more logically considered under chemical control. Gene products delivered by the organism would be biological control, but gene products extracted from organisms and applied directly would fit more logically under chemical control. It seems to be a general principle that organisms can deliver the pest-control products of their genes and biosynthetic pathways far more efficiently and safety than we can deliver them as extracted, formulated chemicals.

On the other hand, the opportunities in chemical control with natural products from plants, microorganisms, and insects, is still very much a developing technology and unquestionably will play an increasingly more important role in pest management for the 21st century.

My concept of integrated *pest management is the use of the most effective, economical, safest and sustainable combination of physical, chemical, and biological methods to limit the effects of pests and improve the yield and quality of plant and animal products.*

A Projection of Biological Control as a Maturing Technology

Biological control, including "the use of natural or modified organisms, genes, or gene products" (delivered by organisms), has the potential to become the major component or center-piece of integrated pest management in the United States in the 21st century. However, having stated where I think this technology can evolve, as a "maturing" technology itself, I must add *provided that ways are found, ultimately, to apply and/or manage populations or genes of thousands if not tens of thousands of site-, and pest-, or disease-specific biological control agents.*

Success stories such as control of cottony-cushion scale insect by the vedalia beetle (DeBach, 1964), crown gall by *Agrobacterium radiobacter* strain K-84 (Kerr, 1980), and wheat stem rust in the Great Plains by gene deployment in response to knowledge of emerging races of the pathogen (Roelfs, 1988) show that pests and diseases can be brought into a permanent state of suppression, and with no significant economical, ecological, or social compromises, by use of biological control methods. The real issue with my stated destination for biological control as a maturing technology at sometime in the future is whether there is the institutional infrastructure, scientific resources, and socioeconomic resolve to find, understand, and use populations or genes of

thousands if not tens of thousands of site-, pest-, and disease-specific biological control agents, or whether this can be accomplished in time to make this vision a reality in the 21st century.

Rationale for the Multiple-Agent Approach to Biological Control

There are several lines of evidence or experiences to support my thesis that biological control, to become fully employed, will depend on our ability to apply and/or manage a great diversity of species, biotypes, and strains of biological control agents or their genes. Some examples are given below.

1. Biological control is widely recognized both scientifically and based on empirical experience as highly pest- or disease-specific. The biological control of crown gall cited above works against biovars 1 and 2 of the crown gall pathogen but not against biovar 3 pathogenic on grape (Farrand, 1990). Moreover, rarely if ever are biological control agents perfectly adapted to the full ecological range of the target pest, except, for example, pathogen-derived biocontrol agents such as ice-minus bacteria produced by single-gene deletion from the genome of the target ice-nucleation active (INA) bacteria (Lindow, 1985). Obviously, there are and will be widely-adapted agents, but in general, different biological control agents are and will be needed to control the same pest in different environments.

2. A multi-cultivar approach has long been the basis for success of traditional plant breeding. Most of the approximately 150

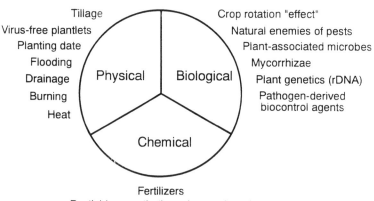

Fig. 1. Diagrammatic representation of components of pest management.

12

economically important crops grown in the United States are represented, in turn, by tens and even hundreds of cultivars or hybrids bred and selected for their adaptation to specific environments or to solve specific problems such as pest damage, disease, or environmental stress effects. As examples, some 350 cultivars of wheat and more than 500 cultivars of soybeans were grown in 1989 in the United States just for seed (McLaughlin and Jones, 1989). Breeders aspire to produce broadly adapted cultivars, but typically these do not perform as well in any given environment compared with cultivars developed specifically for that environment. Even with intensive management of the environment of crop plants, different cultivars are needed in different geographic regions or with different soils to maximize biological productivity.

3. The multi-agent approach is also the basis for success of vaccines for control of human and animal diseases. Even different strains of the same virus group, e.g., the influenza virus group, each may require a different vaccine.

4. An approach to biological control that uses several agents for each pest or disease is consistent with the principles of evolution. Natural selection is towards species, subspecies, and strains ever-better adapted to specific habitats, including food sources and physical environments. Even apparently widely adapted species can be represented by genetically and phenotypically different subpopulations or individuals, thereby allowing the population to exploit and compete in different environments and take advantage of new evolutionary opportunities.

5. Perhaps the best argument for the large number of different biological control agents is simple arithmetic. Considering the number of pests and diseases on the approximately 150 different kinds of crops grown in the United States, for example, and that multiple agents will likely be needed for optimal control of most of these pests and diseases, how many biological control agents will be required to reach full employment of this technology for all of these crops?

6. Finally, there is strength in diversity, both in the biological control agents used to control a pest or disease and the mechanisms used by any one biological control agent. As rephrased by Marston Bates from C. S. Elton, "The greater the complexity of the biological community, the greater its stability," (Baker and Cook, 1974).

The good news is that biological control

agents and mechanisms represent an enormous and still largely untapped natural biological and genetic resource (Cook, 1991). Ways must be found to make significantly greater use of this resource.

Biological Control as a Developing Technology

Biological control agents can be applied or managed as a) indigenous (resident) populations or communities by cultural practices, b) as individual species or subspecies introduced occasionally or repeatedly when and where needed, or c) as a source of genes to improve the level or consistency of biological control by natural enemies, antagonists, pathogen-derived biocontrol agents, crop plants, and possibly now also livestock with transgenes from their pathogens (Fig. 2).

Management of Natural Communities of Beneficial Organisms

Management of natural communities of beneficial organisms suppressive to pests is the basis for the biological control concept of *conservation*. One of the best examples is less use of pesticides that, because of their nontarget effects, reduce populations of naturally occurring (resident or native) beneficial organisms. Another example might be the use of signals or food sources associated with the plant to attract or enhance timely populations of naturally occurring (native) beneficial insects with ability to subsequently feed on a pest of that crop (Coll and Bottrell, 1991). Management (conservation) can also play an important role in maintenance of biological control by exotic natural enemies once they are released.

In the terminology of *alternative agriculture* (NRC, 1989a), this approach includes the management of natural cycles of beneficial organisms without resorting to purchased inputs other than as cultural practices necessary to conserve, enhance, or take advantage of these beneficial communities and biological control cycles. Garrett (1965) described this as bringing about a change in environmental conditions so as to favor the multiplication and activity of one or more biocontrol agents already present. His perspective was greatly influenced by experiences on biological control of soilborne plant pathogens, going back to the 1920s, which showed the great difficulty of achieving

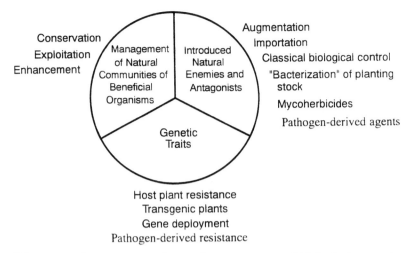

Fig. 2. Diagrammatic representation of components of biological control.

biological control with introduced microorganisms. To quote Garrett: "...the flora and fauna of a habitat will be selected by the environmental conditions from amongst the species currently available. The balance of the flora and fauna can be upset only temporarily by augmenting artificially the population of a species already present, ...most species of microorganisms that could thrive in a particular soil are there already." An example is greater use of crop rotation to permit more time between susceptible crops for the naturally occurring, sanitizing microorganisms in the soil to reduce the inoculum concentration of pathogens below some threshold level (Cook and Baker, 1983). Other examples are the soils that, because of their microbiological properties, are naturally suppressive to certain soilborne plant pathogens (Baker and Cook, 1974).

Introduced Native and Exotic Organisms

Introduced native or exotic organisms is the basis for the biological control concepts of *augmentation* and *classical biological control*. This also includes the use of occasional or repeated applications of microorganisms referred to by the unfortunate labels of "microbial pesticide" or "biopesticide." Those who might shudder at the thought of lumping classical biological control with biopesticides may not realize that the U.S. Environmental Protection Agency (EPA) considers all microorganisms intended for pest control as "microbial pesticides," whether indigenous or exotic and whether applied as an

inoculative or inundative release (EPA,1993).

The best examples of biological control with introduced organisms are the examples of classical biological control--the control of immigrant naturalized arthropod pests and weeds with natural enemies introduced from the original home of the arthropod pest or weed. An estimated 750-800 natural enemies have been imported into the United States during the past 100+ years for control of immigrant naturalized pests of U.S. crops and farm animals (J.R. Coulson, personal communication). Of these releases, an estimated 250 have established.

For weeds, an estimated 65 arthropods have been released in the United States, of which about 48 have established. In addition, one nematode has been released for biological control of weeds, which may or may not have established, and three fungal pathogens have been introduced, of which two may have established but the third, *Colletotricum gloesporoides* f.sp. *aeschymones* (Collego) for biological control of northern joint vetch in rice and soybean fields of Arkansas, must be applied again and again (Templeton and Heiny, 1990).

Considering the amazingly high rate of success, and also that only a relatively few of the hundreds of arthropod pests and weeds have thus far been targeted for this approach (in spite of these large numbers), the trends would suggest that classical biological control, like the other approaches to biologically based pest-control technologies, is still far from a maturing technology.

In contrast to biological control with arthropods introduced as the natural enemies,

14

there has been very little progress in the use of introduced microorganisms (including antagonists of plant pathogens, nematodes, and insect pests) for biological control. According to EPA records, only 24 microorganisms have been registered for control of all categories of pests since 1948 (EPA, 1993) when the first product consisting of the insect pathogen *Bacillus popilliae* for the control of Japanese beetle larvae was registered under statutes of the federal insecticide, fungicide, and rodenticide act (FIFRA). Today about one-third of the 24 microbial agents/products registered under FIFRA are strains of *B. thuringiensis* (Bt).

Possibly as many as 10 species of microorganisms have been released as classical biological control agents of insect pests or weeds, such as the exotic rust fungus, *Puccinia chondrilla*, for biological control of rush skeleton-weed in the Pacific Northwest (Emge and Kingsolver, 1977). The introduction of dsRNA-carrying *Cryptonectria parasitica* into North America as naturally infected cultures from Europe for biological control of chestnut blight can be included in this group--as a classical biological control of a plant disease (Pfeifer and Van Alfen, 1992). No microorganism is currently used or registered in the United States for biological control of nematodes, although a strain of *Paecilomyces lilacinus* is used in the Philippines for nematode control in soil (Davide, 1991).

Most successes with introduced microorganisms as biological control agents have been with native or indigenous strains used in augmentative biological control, either as an inoculative or more commonly an inundative release. This is because microorganisms tend not to persist in the environment at populations greater than occurs naturally, unless a) supported by a specific substrate and b) they have a competitive advantage over other microorganisms for that substrate. Both of these requirements are met for pathogens of weeds and insect pests; a specific substrate is provided in the form of the susceptible plant or insect, and the necessary competitive advantage for that substrate over other microorganisms is provided by the abilities of the biological control agents as pathogens. With saprophytic microorganisms, on the other hand, such as antagonists of plant pathogens, provision of both a unique substrate and a competitive advantage in the environment is much more difficult.

A major breakthrough for biological control using microorganisms is the discovery that on every plant or within every plant population, there resides microorganisms with the ability as antagonists, if their populations were higher, to protect that plant or plant population against disease or insect pests (NRC, 1989b). Typical of many biological controls, any given population of antagonistic microorganisms rarely controls more than one pathogen. Moreover, populations of these beneficial microorganisms (or communities of microorganisms) tend to build-up in response to (rather than in advance of) the disease and, therefore, too late or slow to control the disease. However, the benefits of these microorganisms can be increased significantly, depending on the case, by an "inoculative release" with the seed or other planting material at the time of planting (Cook, 1993).

Agrobacterium radiobacter K84, having the ability to protect plants against crown gall caused by *A. tumefaciens* (Kerr, 1980), and the antibiotic-producing strains of fluorescent *Pseudomonas* spp. suppressive to take-all of wheat (Cook and Weller, 1987) both fit the model of a) naturally occurring strains of microorganisms on the plant (in the rhizosphere) that b) increase in numbers in response to the disease caused by the pathogens they inhibit and c) provide biological control when introduced in sufficient numbers in advance of infections by inoculation of the planting material.

Nonpathogenic bacteria of the type identified as *A. radiobacter* K84 presumably co-evolved with the pathogenic types. While they lack pathogenicity, they are equipped to compete with the pathogen for special amino acids (opines) produced by crown galls, and they have the ability to produce a bacteriocin (agrocin 84) inhibitory to the pathogen (Farrand, 1990). Thus, the two requirements are met of a) a specific substrate (opines) for their support and b) a competitive advantage over other microorganisms (their unique ability to catabolize opines *and* produce bacteriocins). However, even in the case of *A. radiobacter* K84, which was brought to the United States from Australia, it is necessary to inoculate all the roots of each transplant seedling of peach, apple, rose, or other crown-gall susceptible plant to obtain biological control of the crown gall pathogen on those plants. There is no evidence of useful spread of the biological control agent from plant to plant or even root to root.

Plant-associated microorganisms also have potential for biological control of arthropod pests. The endophytic fungi discussed above are shown to provide protection of a wide range of grasses against insect pests (Siegel et al., 1987) are examples of plant-associated biological control agents. Some of the best studied endophyte-plant systems are *Acremonium*-infected ryegrasses and fescues, where the endophytes have been shown to produce alkaloids and other chemicals toxic to aphids and insect herbivores. Unfortunately, two diseases of cattle known as fescue toxicosis and ryegrass staggers result from the chemicals produced by endophytic fungi. Through further research, including genetic manipulation of the fungal endophytes, it may be possible to develop strains toxic to insects but not to cattle.

With plant-associated microorganisms, useful genes can be delivered with microorganisms on or in the seed. Some, such as the grass endophytes, protect by direct inhibition of the pest or pathogen, while others protect through induced resistance in the plant to the pest or pathogen (Kuć, 1987). Clearly, this and other applications of microorganisms are only in the very earliest stages of development as a technology.

Transfer of Useful Genetic Traits from or to Biological Control Organisms

There is virtually no limit to the possibilities for making greater use of genetic traits for biological control. Microorganisms, in particular, lend themselves to genetic improvement as agents of biological control, either as classical biological control agents or antagonists, including as plant-associated biological control agents (Cook, 1993).

The examples of host plant resistance conferred against insect pests by transfer of *Bt* genes for production of the delta endotoxins of strains of *Bacillus thuringiensis* further illustrate the enormous potential of this approach (Vaeck et al., 1987), now nearing practical use. Microorganisms in particular, but also arthropods and other kinds of organisms with traits inhibitory to target pests, can now be considered as part of the total genetic resource available for plant improvement through increased resistance to pests and diseases. In these cases, the plant becomes the biological control agent.

One flaw in this approach, of course, is the risk of overuse of single-gene resistance; plant breeding experience has shown that this will favor evolution within the target pest towards ability to attack plants with that gene (Browning

et al., 1977). Nevertheless, there are many strategies for extending the utility of genes for host-plant resistance to pests and diseases (Roelfs, 1988). It may also be preferable, if not essential, to use tissue-specific promoters to insure that the genes are only expressed in tissues where and when needed for both maximum efficiency and safety.

Implications of the Multi-Agent Approach to Biological Control

The projection of such a large number and diversity of biological control agents delivered into practice to reach the full potential of this approach to pest management has implications for a) science and education, b) public and private investments, c) governmental guidelines and regulations, d) societal concerns for safety, and e) acceptance by users.

Progress towards full employment of biological control can only be as fast as allowed by the weakest link. Since all five areas are critical, any one or all of these five areas are or could become weak links.

Science and Education

Science has already revealed an enormous number of candidate biological control agents. In plant pathology, for example, it is almost axiomatic that virtually every one of the hundreds of scientists working on microbial biocontrol agents worldwide is working with a different species or strain of antagonist (Tjamos et al., 1992). This is a reflection of the range of untapped microbial germplasm, and the ease with which effective antagonists of plant pathogens can be found. Insect pathologists have identified well over 1000 strains of insect viruses (Payne, 1988). As pointed out above, relatively few of the several hundred arthropod pests important in agriculture and forestry have been targeted to date for classical biological control. I am confident that there are tens of thousands of excellent natural enemies, including thousands of microbial biocontrol agents already discovered or soon to be discovered in nature.

An unusual problem in plant pathology relates to the fact that new interesting microbial biocontrol agents with novel mechanisms of control turn up so frequently. Until recently, the tendency was to collect or screen 10, 50, or at most 100 candidate organisms, pick one or two, and then focus for the next several years on the interesting mechanisms of biological control used by these organisms. This has been valuable in building a science but has limited our progress toward use because the strains have performed inconsistently. On average, we can expect better results working with the best of a thousand than with the best of only 10 to 100 candidate strains.

There has also been too little attention paid to the ecological range of biological control agents, especially microbial biocontrol agents (Deacon, 1991). This information is needed not only to know the limits of different biological control agents, but also to develop management practices to provide the environment needed by introduced organisms. While the interest or goal naturally will be in favor of organisms with an ecological range at least as broad as the target pest agent, we should not reject agents on the basis that their range of effectiveness may be too narrow. Consider, for example, that Washington State University developed the soft white winter wheat cultivar 'Sprague' (Bruehl et al., 1978) to control one disease (snow mold) in only one area of the state, but this cultivar probably saved the wheat industry in that area.

Finally, and perhaps most importantly, there is a need for renewal and expansion of research on the ecology of pests and disease agents for the purpose of developing or fine-tuning cultural controls that take advantage of natural biological control. Herein lies an equally great need for education since these kinds of biological controls are achieved with cultural or management practices.

Public vs Private Investments

The rate-limiting step towards reaching the full potential of biological control most likely will be funds available for the required research and development. If state and federal budgetary constraints of this current decade are any prediction for the medium (next decade) and long-range (next 20-30 years), public investments will fall far short of that required to develop biological control much beyond its current stage as a technology.

On the other hand, it seems obvious because of the nature of this technology, that the great majority of biological control agents and systems will represent little or no market potential no matter how well they work. Classical biological control agents traditionally have represented little or no market potential because these agents characteristically are introduced only once or occasionally. Biological controls achieved through conservation (management practices) likewise represent little market potential other than as information possibly conveyed by consultants. Any given antagonist of a soilborne plant pathogen, not surprisingly, tends to be limited in ecological range to certain soils, and therefore, conceivably, some antagonists will be effective on less than 100,000 acres. Such a market size could be attractive to a local entrepreneur but probably not to a company with any size. Without new and more efficient ways for companies to deal with multiple niche-market products, the future of biological control as a major component of pest management will depend, as in the past, on public-supported research and extension, which may not be available.

One possible solution to this funding paradox is more efforts and incentives to develop public-private consortia. State and federal laboratories should have the means and encouragement to formalize relationships with local entrepreneurs or agribusinesses, local farmer cooperatives, and regional or state commodity groups as well as national or multinational seed companies and agrichemical corporations. The U.S. model for these approaches is the current relationships between public-supported plant breeding programs, the state crop improvement associations, and the private seed producers and seed companies. This has the additional positive aspect of creating new employment and business opportunities at the local level.

Another possible solution to the funding paradox is to broaden the host ranges and/or ecological ranges of natural enemies and antagonists as agents through genetic manipulations. This would expand the market size and increase the economic attractiveness/viability of these biological control agents. Classical biological control agents that establish more or less permanently might be more useful with a wider ecological range but would still represent little or no market opportunity. Wider ecological range could also be counter-productive for classical biological control because of safety concerns for those agents with potential to establish in the environment.

The third possibility to increase economic attractiveness of biological control agents for private investments is through identification of traits in the agents that can be transferred to crop plants. Such traits could be used to develop elite lines of crop plants for use in either public or private plant breeding programs. The means to take advantage of new genetic variability in plants for defense against diseases and insect pests is well developed within the United States and world-wide. Moreover, proprietary rights are or can be protected by the Plant Variety Protection Act, breeder's rights, and trade secrets. In the case of "novel, nonobvious, and useful" new genes, genetic constructs, or delivery systems, proprietary protection is provided as intellectual property rights through patenting.

Regulatory Oversight

Regulatory oversight of biological control must become significantly more efficient to facilitate and ideally accelerate the use of more biological control agents as a component of pest management. Towards this end, both the EPA and the USDA's Animal and Plant Health Inspection Service (APHIS) are considering new guidelines intended to expedite the review and approval procedures for biological control agents. Without an efficient, low-cost approval procedure, regulatory oversight could effectively filter out many biological control agents, not because they would be of too little benefit or unsafe to a user, processor, or consumer, but because the cost of providing the necessary data for approval would preclude further consideration of the agent.

In the interest of streamlining federal regulatory oversight, consideration should be given to having one (rather than the current two or more) federal agency provide oversight for organisms deliberately released into the environment. The USDA with its long history of experience in the release and management of organisms (crops, livestock, and beneficial insects and microorganisms) could provide this oversight.

Societal Concerns for Safety

Societal concerns for safety of biological control might increase in response to a trend towards more natural enemies and antagonists as biological control agents released into the environment. The greatest concern at present seems to be for genetically manipulated biological control agents, especially for genetically manipulated microbial biocontrol agents, although transgenic plants have also been a source of public concern.

Herein exists another obvious paradox; while overall environmental safety increases with the number and diversity of site-, and pest- or disease-specific biological control agents, and while breeding of organisms to fit specific uses, environments, or management strategies has tended to make organisms less rather than more likely to develop as feral populations, the number of citizens likely to question releases in their back yard can also be expected to increase. Society no longer questions the need for hundreds or possibly thousands of species and strains of fungi to make all the cheeses in the world (among other commercial applications). Potentially this kind of familiarity will also emerge with experience and public education on biological control, including biological control developed through the use of the tools of rDNA technology.

Without public support and confidence, regulatory agencies will be forced to move more cautiously and will require more data for approvals. And without both public confidence and efficient regulatory oversight, private investments will become more tentative. Moving past these weak links will require more science and education.

Users of Biological Control Technology

Major changes will also be required among the users of pest management technology, including farmers, foresters, agribusinesses, homeowners, golf course superintendents, and managers of parks, roadways, waterways, and other managed ecosystems. More and better information will be needed on best management practices to enhance, optimize, or at least conserve the benefits of biological control provided by indigenous natural enemies and antagonists or prior introductions, as examples. More and better ways also will be needed through education to deliver and process this information, which could be increasingly more complex and sophisticated. Cooperation and coordination will be more important than ever among users, researchers, extension personnel, regulators, and advisors or agribusiness company representatives.

In general, the users of biological control should be one of the strongest links (together with science and education) in the several fronts that must advance simultaneously for full employment of this technology. The major requirements for most users are a) that it works, b) is economical, and c) for an increasing percentage of users, that it is ecologically sound and sustainable. Unfortunately, there is still the unknown of public acceptance since political actions such as boycotts have stopped users from adopting some new technologies.

Thus, again we return to the issue of public support and confidence in this technology. With public support and confidence, we move forward with expanding science and education, increased public and private funding, efficient regulatory oversight, and rapid adoption by users; without it, we move forward slowly or not at all.

References

Baker, K.F.; Cook, R.J. *Biological Control of Plant Pathogens*; W.H. Freeman, San Francisco, CA, 1974; 433 pp. (Book reprinted in 1982, Am. Phytopathol. Soc., St. Paul, MN).

Browning, J.A.; Simons, M.D.; Torres, E. "Managing host genes: Epidemiologic and genetic concepts." In *Plant Disease: An Advanced Treatise*; Horsfall, J.G. and Cowling, E.B., Eds.;1977, Vol. 1; pp 192-213.

Bruehl, G.W. et. al. "Registration of Sprague wheat." *Crop Science* **1978**, 18:695-696.

Coll, M.; Bottrell, D.G. "Microhabitat and resource selection of the European corn borer (*Lepidoptera: Pyralidae*) and its natural enemies in field corn." *Environ. Entomol.* **1991**, 20:526-533.

Cook, R.J. "Making greater use of introduced microorganisms for biological control of plant pathogens." *Annu. Rev. Phytopathol.* **1993**, 31: In Press.

Cook, R.J. "Challenges and rewards of sustainable agriculture research and development." In *Sustainable Agriculture Research and Education in the Field, a Proceedings*; National Research Council; National Academy Press: Washington, D.C., 1991, pp. 32-76.

Cook, R.J.; Baker, K.F. *The Nature and Practice of Biological Control of Plant Pathogens*; American Phytopathological Society, St. Paul, MN, 1983.

Cook, R.J.; Weller, D.M. "Management of take-all in consecutive crops of wheat or barley." In *Innovative Approaches to Plant Disease Control*; Chet, I., Ed.; John Wiley and Sons: New York, NY, 1987, pp. 41-76.

Davide, R.G. "Biological control of plant diseases in the Philippines." In *The Biological Control of Plant Diseases*; Bay-Peterson, J., Ed.; Food and Fertilizer Technology Center Book Series No. 42. 1991, pp. 186-191.

Deacon, J.W. "Significance of ecology in the development of biocontrol agents against soil-borne plant pathogens." *Biocontrol Sci. Technol.*, **1991**, 1:5-20.

DeBach, P., Ed.; *Biological Control of Insect Pests and Weeds*; Reinhold: New York, NY; 1964

Emge, R.G.; Kingsolver, C.H. "Biological control of rush skeletonweed with *Puccinia chondrilla*." *Proc. Am. Phytopathol. Soc.* **1977**, 4:215.

Environmental Protection Agency. "Microbial Pesticides; experimental use permits, 40 CFR part 172, proposed rules." *Federal Register* **Jan. 22, 1993**, 58:5878-5902.

Farrand, S.K. "*Agrobacterium radiobacter* strain K84: A model biocontrol system." In *New Directions in Biological Control: Alternatives for Suppressing Agricultural Pests and Diseases*; Baker, R.R.; Dunn, P.E., Eds.; Alan R. Liss, Inc., New York, NY, 1990, pp. 679-691.

Freeman, S.; Rodriquez, R.J. "Genetic conversion of a fungal plant pathogen to a nonpathogenic, endophytic mutualist." *Science.* **1993**, 260:75-78.

Gabriel, C.J.; Cook, R.J. "Biological control--the need for a new scientific framework." *BioScience.* **1990**, 40, 204-206.

Kerr, A. "Biological control of crown gall through production of agrocin 84." *Plant Dis.*. **1980**, 64:25-30.

Kué, J. "Plant immunization and its applicability for disease control." In *Innovative Approaches to Plant Disease Control*; Chet, I, Ed.; John Wiley and Sons, New York, NY, 1987, pp. 255-274.

Lindow, S.E. "Strategies and practice of biological control of ice nucleation active bacteria on plants." In *Microbiology of the Phyllosphere*; Fokkema, M.N., Ed.; Cambridge Univ. Press: Cambridge UK, 1985, pp. 293-311.

McLaughlin, F.W.; Jones, L.L. "Report of acres applied for certification in 1989 by seed certification agencies." *Assoc. Official Seed Certifying Agencies*, 3709 Hillsborough St., Raleigh, NC, 1989

National Research Council. *Alternative Agriculture.* National Academy Press, Washington, D.C., 1989a.

National Research Council. *The Ecology of Plant-Associated Microorganisms.* National Academy Press, Washington, D.C., 1989b.

Payne, C.C. "Pathogens for the control of insects: where next?" *Phil. Trans. R. Soc. Lond. B* **1988**, 318:225-248.

Pfeiffer, P.; Van Alfen, N.K. "The genetic mechanism of hypovirulence in *Cryphonectria (Endothia) parasitica*." In *Biological Control of Plant Diseases*; Tjamos, E.S.; Papavizas, G.E; Cook, R.J., Eds.; Plenum Press, New York, NY, 1992, pp. 305-316.

Roelfs, A.P. "Genetic control of phenotypes in wheat stem rust." *Annu. Rev. Phytopathol.* **1988**, 26:351-367.

Siegel, M.R.; Latch, G.C.M.; Johnson, M.C. "Fungal endophytes of grasses." *Annu. Rev. Phytopathol.* **1987**, 25:293-315.

Templeton, G.E.; Heiny, D.K. "Mycoherbicides." In *New Directions in Biological Control: Alternatives for Suppressing Agricultural Pests and Diseases*; Baker, R.R.; Dunn, P.E., Eds.; Alan R. Liss, Inc., New York, NY, 1990, pp. 279-286.

Tjamos, E.C.; Papavizas, G.C.; Cook, R.J., Eds.; *Biological Control of Plant Diseases: Progress and Challenges for the Future*; Proceedings of a NATO Advanced Research Workshop, May 19-24, 1991, Cape Sounion, Athens, Greece; Plenum Press: New York, NY, 1992.

Vaeck, M; Reynaerts, A.; Hufte, H; Janseno, S.; DeBueckeleer, M.; Dean, C.; Zabeau, M.; van Montagu, M.; Leemans, J. "Transgenic plants protected from insect attack." *Nature.* **1987**, 328:33-39.

A Proposed Definition of Biological Control and its Relationship to Related Control Approaches

Pedro Barbosa and Susan Braxton, Department of Entomology, University of Maryland, College Park, MD 20740

The incorporation of biotechnology in the "biological control" of pests has resulted in some novel approaches for the control of pest species. The use of transgenic plants, the incorporation of endophytes into plants with lethal properties, and the alteration of microorganisms to carry or express novel compounds, are but a few examples. These new approaches, often termed by their developers as biological control have led some to suggest a redefinition of the term biological control. This sentiment has caused concern and significant controversy, and has the potential to divide important parts of the scientific community that would benefit from collaboration.

The definition set forth by the National Academy of Sciences report (NAS, 1987), proposed that biological control be defined as "the use of natural or modified organisms, genes, or gene products to reduce the effects of undesirable organisms (pests) and to favor desirable organisms such as crops, trees, animals and beneficial insects and microorganisms". This suggestion has stirred a great deal of concern and controversy (Gabriel and Cook, 1990; Garcia et al., 1988). For example, Garcia et al. (1988) contend that a redefinition such as that given by the NAS "has broadened the meaning " of biological control "to such an extent that it is vague, confusing, and incompatible with historical usage." Thus, the "NAS definition clouds the issue", in large part according to Garcia et al.(1988), because of the great differences between the approaches that would be grouped together under the redefinition. Garcia et al. (1988), emphasized that the self sustainability of traditional biological is one major difference between those approaches and recent ones. The counter argument has been that historical definitions were developed within a context of concern about **insect pests** and that currently biological control is viewed as a viable option against various types of pests (Gabriel et al. 1988). Gabriel et al. (1988) conclude that what some view as vague others view as

logically comprehensive, and what some view as historically accurate others view as exclusionary.

Underlying the debate, although rarely explicitly stated, is the concern that broadening of the definition of biological control (and thus the research considered to be biological control) will dilute the funding for "traditional" biological control. Others fear that funds would be disproportionately diverted to these novel forms of control. The adoption of a scheme and terminology which provides a clear distinction between biologically based approaches and other traditional biological control approaches can go a long way towards insuring appropriate and equitable support of all aspects of biological control against all types of pests.

Another reason for the inability to agree on definitions may stem from the apparent parallel but independent development of the use of biological control in Entomology and Plant Pathology. Researchers and practitioners tend to engage in "biological control" as circumscribed by the pest of concern. Thus, individuals rarely work across pest types but focus on the "biological control" of insects, or weeds, or plant pathogens, etc. Indeed, if one examines the definition of "biological control" one finds that an Entomologist might define it as "the utilization of natural enemies to reduce the damage caused by noxious

2726–8/93/0021$06.00/0 © 1993 American Chemical Society

organisms to tolerable levels" (Debach and Rosen, 1991), and a Plant Pathologist might define it as "the reduction of inoculum density or disease producing activity of a pathogen or parasite in its active or dormant state, by one or more organisms, accomplished naturally or through manipulation of the environment, host, or antagonist, or by mass introduction of one or more antagonists" (Baker and Cook, 1982). Or "any means of controlling disease or reducing the amount or effect of pathogens that relies on biological mechanisms or organisms other than man" (Campbell, 1989), including crop rotation, fertilizer manipulations which affect microbes, use of antagonists, plant breeding, etc.

Even if we were to restrict ourselves to the most common form of traditional biological control of insects there would be some degree of ambiguity. Historical usage of the term classical biological control has been less than consistent. The one example of classical biological control which is perhaps cited more often than any other is that of the importation of the vedalia beetle, *Rodolia cardinalis* (DeBach, 1974; Caltigarone and Doutt, 1989). This success was a milestone in the development of biological control in the U.S. and worldwide (Coppel and Mertins, 1977). It unequivocally and dramatically demonstrated the importance of natural enemies in population control (DeBach and Rosen, 1991), and marked the beginning of renewed widespread activity in biological control. To the degree that this control program was a milestone in the history of biological control in the U.S., the technique used to achieve this success has come to be considered "classical".

Classical biological control has been variously defined, often by implication; with definitions ranging from the highly restrictive to the wildly inclusive (Schroth and Hancock, 1985). Several authors have, more or less, rigidly defined classical biological control as natural enemy importation directed against exotic pests (Bernays, 1985; Drea and Hendrickson, 1986; Ehler and Andres, 1983; Hoy, 1988; Huffaker, 1985). A more flexible definition is that classical biological control is usually, but not necessarily, limited to exotic pests (Caltigarone, 1981; DeBach, 1974; Hull

and Beers, 1985; Simmonds et al., 1976; Van Den Bosch et al., 1982; Yeargan, 1985). In the latter circumstances the use of biological control against native species is mentioned but then subsequently ignored. Still others have treated classical biological control as completely synonymous with natural enemy importation. Classical biological control in a book index may refer readers to any discussion of natural enemy importation (Hoy and Herzog, 1985), or it may be cross-referenced with "natural enemies," or "importations of" (Huffaker and Messenger, 1976). In Coppel and Mertins (1977) classical biological control and natural enemy importation are used interchangeably throughout the book.

In some publications (USDA, 1978), classical biological control is asserted to involve the introduction of biological control agents with the expectation that they will become established but completely ignores the target pest origin entirely. Still others have explicitly proposed the potential for the effective use of classical biological control against native pests. This concept, sometimes referred to as "novel associations" is discussed and evaluated by Hall and Ehler, 1979; Hall et al., 1979; Hokkanen, 1985; Hokkanen and Pimentel, 1989; and Waage, 1990. Finally, some authors have included under classical biological control practices such as augmentation and conservation (Van Gundy, 1985), and the importation of pollinators, scavengers or competitors (Howarth, 1991).

The diversity of opinions as to what does and does not constitute classical biological control is illustrated by the variation in definitions found in a single edited text. Although there are various examples, one of the best recent texts on biological control can serve as an example. Examples illustrating the points made above can be found in "Biological Control in Agricultural IPM Systems" (Hoy and Herzog, 1985). More troublesome is that different definitions are occasionally implied by the same author. Not only is this evident in older publications (e.g., DeBach, 1974 and Huffaker and Messenger, 1976; vs. Huffaker, 1985) but apparent discrepancies still persist in other recent publications. Hoy (1988) states that "classical" biological control is "based on the importation of exotic natural enemies... and their long-term establishment in the new environment, a strategy which may then provide long-term control of the target *exotic* pest arthropod or weed." In another article, Hoy (1985) asserts that exotic natural enemies may be used against native pests, and that this is a compelling argument in favor of classical biological control. Ehler and Andres (1983) write that natural control of species in their native environment is the "basic premise" of classical biological control, which the authors define as the "intentional introduction of exotic natural enemies to control introduced or naturalized species." In earlier work, however, Ehler and colleagues discuss the effectiveness of classical biological control against native species (Hall and Ehler, 1979; Hall et al. 1980). Similarly, Indeed, Ehler (1990) recently noted that "whereas classical biological control has been applied extensively against introduced (exotic) pests, it is now recognized that native pests can be suitable targets as well."

A useful resolution of the problems outlined above may be a clearly delineated framework that encompasses and juxtaposes a large variety of commonly used approaches for the control not only of insects, but also, of weeds, pathogens, nematodes, and other pests.

The Coexistence of Different Perspectives: Discussion

Definitions not only enable researchers to communicate meaningfully but also provide a heuristic conceptual framework for conducting and evaluating research. It is in the interest of a better understanding of "biological control" and its evolution that a scheme is proposed here that defines and juxtaposes various forms of "biological control" and related tactics. It is not our intent to demonstrate, nor are we presumptuous enough to assume, that our scheme perfectly defines and juxtaposes all traditional views of biological control and recent so-called biologically based controls. However, our objective is to provide a basis

from which a dialogue can be initiated which will lead to a conceptual framework agreeable, not to all, but to most.

At the core of the scheme is the concept of natural biological control which encompasses all of the biological ways in which the survival and abundance of populations are regulated, unaided by human intervention (see Fig. 1). In contrast, when natural controls are directly and purposefully utilized for the control of a pest, or when our understanding of living organisms serves as the basis for a control tactic or strategy they are then either defined as Biological Control or as Parabiological (i.e., near Biological) Control (Fig. 1). The latter two forms of applied natural controls are distinguished from each other on the basis of the organism that is used or manipulated.

Thus, Biological Control encompasses the manipulation of natural enemies, the competitors (in the broad sense of the tern) of the pest and/or the resources (e.g., components of the habitat) of either natural enemies or competitors. Indeed, competitors of the pest can be viewed, broadly speaking, as "natural enemies" of the pest. In contrast, Parabiological Control comprises the manipulation of the pest or the pest's resources. In some circumstances there may be some overlap of the two types of biological control. Most obviously, the habitat of the pest and the natural enemy may be identical. The distinction lies in that a Parabiological Control involving habitat manipulation aims to affect the pest, whereas a Biological Control involving habitat manipulation aims to affect the natural enemy. A given tactic may, regardless of the intent influence both pest and natural enemy and in these circumstances the approach can be classified based on the original intent or simply be viewed as a rare example which fits either concept.

Similarly, it is obvious that some approaches listed under Parabiological Control are more similar to each other than they are to other tactics, e.g., the use of pheromones and hormones are more similar than the use of cultural controls. Thus, there may be logical subdivisions of Parabiological Control. However, for the sake of simplicity and to avoid distractions which are not

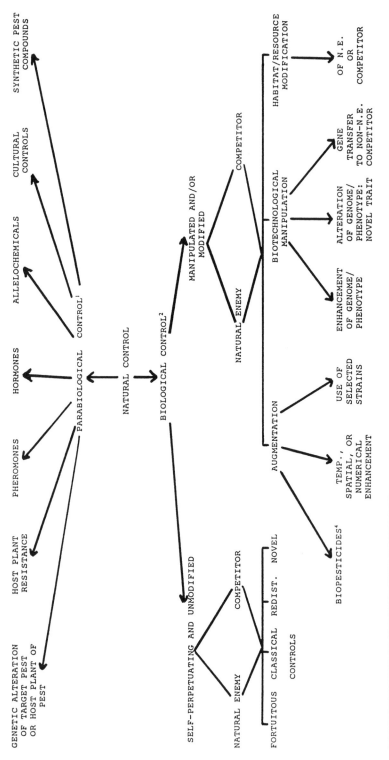

[1] An applied natural control involving the use and/or manipulation of the target pest, or its essential resources (e.g., one or more components of the habitat) (see Sailer, 1981).

[2] An applied natural control involving the use and/or manipulation of the pest's natural enemies and/or its competitors, or their essential resources (e.g., one or more components of the habitat).

[3] "Fortuitous" means biological control (BC) of a species other than the target by a released BC agent, or control of a target species by other than the released species (e.g., by a native species). "Classical" means importation BC, "Redist." mean redistribution of established, successful species, and "Novel" means novel associations.

[4] Use of pathogens which may e.g., may have incorporated protectants against environmental challenges, virulence enhancers synergists, etc. Typically involves the use of microbes or microbial products, formulated for use in a similar fashion to that of synthetic pesticides.

Fig. 1   Relationships and Components of Biological and Parabiological Control

germane to the main objective of this paper we have refrained from detailed subcategorization of Parabiological Control.

Figure 1 illustrates a variety of approaches to Parabiological Control. These include alteration of the target pest such as genetic alteration, either by traditional means (e.g., sterile male technique) or changes achieved through the use of biotechnology. An example is the use of altered host plants of insect pests, i.e., the use of transgenic plants which have incorporated within them the *Bacillus thuringiensis* gene regulating toxin production. The use of pest products such as sex attractants, aggregation pheromones, or hormones would also fall under this category as would the use of synthetic compounds based on an understanding of the pest's biology. For example, inhibitors of hormones, enzymes, pheromones, etc., would fall in the latter group. Similarly, the use of antibiotic or behavior modifying chemicals from the pest's food (or non-host plants) would comprise parabiological controls. Finally, manipulation of the pest's habitat (such as occurs when cultural controls are used) in ways which reduce the survival or fitness of the pest, or its potential to inflict damage, would also be considered Parabiological Control.

Biological Control is divided into two major sub-categories. One approach comprises the manipulation of unmodified organisms and results in self-perpetuating populations of natural enemies or of the competitors of the pest. These self-perpetuating populations can be the result of redistribution, fortuitous or classical biological controls, or novel associations. The second group comprises tactics which may or may not be self-perpetuating but all of which involve the manipulation and or modification of the natural enemy, the pest's competitors, or their resources. These involve traditional augmentation by ecological or numerical enhancement of populations, and alteration of genotypes or phenotypes. Another approach,termed conservation aims to modify essential resources (such as the habitat) as to enhance survival, fitness, and/or performance of the natural enemy or the pest's competitor. Conservation, in the broad sense, can also include enhancing the habitat or environment by avoidance of agrichemicals harmful to natural enemies or competitors, the use of selective agrichemicals (such as pesticides), or the appropriate timing of their use, in order to Still other tactics may involve the use of natural enemies or pest competitors which have been altered using biotechnological approaches.

The use of competitors, although not common in the biological control of insects, or weeds for that matter, is a form of antagonism and of great importance in the biological control of plant pathogens. Antagonism is the reduction in disease or pathogen inoculum as a result of the activities of specific or general microorganisms or other microfauna. This may involve the use of competitor microbes which utilize a limited resource more efficiently than the pest pathogen. In other circumstances, organisms produce some metabolic product which inhibits the growth or causes lysis of cells and tissues of pest pathogens. Plant Pathologists often refer to the latter as antibiotic organisms; a designation that fails to specify the adaptive value of target cell lysis by the antagonist. The limited resource required by both organisms may not, in fact, be well defined and thus the antibiotic organism might indeed be a competitor. Apparently, in neither of the first two types of antagonism are lysed tissues consumed, nor does the microbe replicate in tissues of the target organism. In contrast, predation and parasitism by microbes, as well as micro-invertebrates, does involve tissue consumption or microbe replication, and comprises the third form of antagonism.

Working Definitions

In the scheme presented here the ideal definition of *Biological Control* would be the "DIRECT AND PURPOSEFUL MANIPULATION OF NATURAL ENEMIES, PEST COMPETITORS (IN WHOLE OR IN PART), OR THE RESOURCES REQUIRED BY THESE ORGANISMS FOR THE REDUCTION OF NEGATIVE PEST EFFECTS, OR PEST SPECIES' DENSITY TO LEVELS AT OR BELOW THEIR ECONOMIC THRESHOLDS. Similarly,

Parabiological Control would be defined as "DIRECT AND PURPOSEFUL MANIPU-LATION OF PEST SPECIES OR THE PESTS' RESOURCES (IN WHOLE OR IN PART) FOR THE REDUCTION OF NEGATIVE PEST EFFECTS, OR PEST SPECIES' DENSITY TO LEVELS AT OR BELOW THEIR ECONOMIC THRES-HOLDS." However, the lack of economic (or aesthetic) thresholds for very many pest species requires the more practical definitions for biological and parabiological control of "DIRECT AND PURPOSEFUL MANIPU-LATION OF NATURAL ENEMIES, PEST COMPETITORS (IN WHOLE OR IN PART), OR THE RESOURCES RE-QUIRED BY THESE ORGANISMS FOR THE CONTROL OF PEST SPECIES OR THEIR NEGATIVE IMPACT" and "DIRECT AND PURPOSEFUL MANIPU-LATION OF PEST SPECIES OR THE PESTS' RESOURCES (IN WHOLE OR IN PART) FOR THE CONTROL OF PEST SPECIES OR THEIR NEGATIVE IMPACT," respectively. As we noted above, the scheme provided in this paper is not viewed as an endpoint but a beginning. We hope and anticipate that further input will improve what we present here. The develop-ment of new or modified terminology is not an exercise that should be undertaken hastily or frequently but we hope that we have presented a sufficiently compelling argument for our proposed changes and that these changes parallel the advances being made by biological control practitioners.

References

Baker, K. F.; Cook, R. J. "Biological Control of Plant Pathogens." *Amer. Phytopathol. Soc.*: St. Paul, MN, **1982.**

Bernays, E. A. In *Biological Control in Agricultural IPM Systems.* Editors, Hoy, M. A.; Herzog, D. C.; Academic Press: Orlando, FL, 1985, 373-388.

Caltigarone, L. E. "Landmark examples in class-ical biological control." *Annu. Rev. Ent.* **1981**, 26, 213-232.

Caltigarone, L. E.; Doutt, R. L. "The history of the vedalia beetle introduction to California and its impact on the development of biological control." *Annu. Rev. Ent.* **1989**, 34, 1-16.

Campbell, R. *Biological Control of Microbial Plant Pathogens.* Cambridge University Press: Cambridge, U.K., 1989.

Coppel, H. C.; Mertins, J. W. *Biological Insect Pest Suppression.* Springer-Verlag: Berlin, Germany, 1977.

DeBach, P. *Biological Control by Natural Enemies.* Cambridge University Press: Cambridge, U.K., 1974.

DeBach, P.; Rosen, D. *Biological Control of Natural Enemies.* Cambridge University Press: Cambridge, U.K. 1991, 2nd Ed.

Drea, J. J.; Hendrickson, R. M. "Analysis of a successful classical biological control project: the alfalfa blotch leafminer (Diptera: Agromyzidae) in the northeastern United States." *Environ. Ent.* **1986** 15, 448-455.

Ehler, L. E. In *Critical Issues In Biological Control.* Editors, Mackauer, M.; Ehler, L. E.; Roland, J.; Intercept Ltd.: Andover, U.K., 1990, 111-134.

Ehler, L. E.; Andres, L. A. In *Exotic Plant Pests and North American Agriculture.* Editors, Wilson, C. L.; Graham, C. L.; Academic Press: New York, NY, 1983, 396-418.

Gabriel, C. J.; Cook, R. J. "Biological control - the need for a new scientific framework." *Biosci.* **1990**, 40, 204-207.

Garcia, R.; Caltagirone, L. E.; Gutierrez, A. P. "Comments on a redefinition of biological control." *Biosci.* **1988**, 38, 692-694.

Hall, R. W.; Ehler, L. E. "Rate of establishment of natural enemies in classical biological control." *Bull. Ent. Soc. Amer.* **1979**, 25, 280-282.

Hall, R. W.; Ehler, L. E.; Bisabri-Ershadi, B. "Rate of success in classical biological control of arthropods." *Bull. Ent. Soc. Amer.* **1980**, 26, 111-114.

Hokkanen, H. M. T. "Success in classical biological control." *Critical Reviews in Plant Science* **1985**, 3, 35-72.

Hokkanen, H. M. T.; Pimentel, D. "New associa-tions in biological control: theory and practice." *Can. Ent.* **1989**, 121, 829-840.

Howarth, F. G. "Environmental impacts of classical biological control." *Annu. Rev. Ent.* **1991**, 36, 485-509.

Hoy, M. In *Biological Control in Agricultural IPM Systems.* Editors, Hoy, M. A.; Herzog, D. C.;

Academic Press: Orlando, FL, 1985, 151-166.

Hoy, M. A. "Biological control of arthropod pests: traditional and emerging technologies." *Amer. J. Alter. Agric.* **1988**, 3, 63-68.

Hoy, M. A.; Herzog, D. C. *Biological Control in Agricultural IPM Systems*. Academic Press: Orlando, FL, 1985.

Huffaker, C. B. In *Biological Control in Agricultural IPM Systems*. Editors, Hoy, M. A.; Herzog, D. C.; Academic Press: Orlando, FL, 1985, 13-23.

Huffaker, C. B.; Messenger, P. S. *Theory and Practice of Biological Control*. Academic Press: New York, NY, 1976.

Hull, L. A.; Beers, E. H. In *Biological Control in Agricultural IPM Systems*. Editors, Hoy, M. A.; Herzog, D. C.; Academic Press: Orlando, FL, 1985, 103-121.

National Academy of Sciences, *Report of the Research Briefing Panel on Biological Control in Managed Ecosystems*. National Academy Press: Washington, DC, 1987.

Sailer, R. In *CRC Handbook of Pest Management in Agriculture*. Editor, Pimentel, D.; CRC Press: Boca Raton, FL, 1981, Vol. II; 57-67.

Schroth, M. N.; Hancock, J. G. In *Biological Control in Agricultural IPM Systems*. Editors, Hoy, M. A.; Herzog, D. C.; Academic Press: Orlando, FL, 1985, 415-431.

Simmonds, F. J.; Franz, J. M.; Sailer, R. I. In *Theory and Practice of Biological Control*. Editors, Huffaker, C. B.; Messenger, P. S. Academic Press: New York. NY, 1976, 17-39.

U. S. Department of Agriculture. *Biological Agents for Pest Control. Status and Prospects*. U.S. Government Printing Office: Washington, DC, 1978.

Van den Bosch, R.; Messenger, P. S.; Gutierrez, A. P. *An Introduction to Biological Control*. Plenum Press: New York, NY, 1983.

Van Gundy, S. D. In *Biological Control in Agricultural IPM Systems*. Editors, Hoy, M. A.; Herzog, D. C.; Academic Press: Orlando, FL, 1985, 467-478.

Waage, J. K. In *Critical Issues In Biological Control*. Editors, Mackauer, M.; Ehler, L. E.; Roland, J.; Intercept Ltd.: Andover, U.K., 1990, 135-157.

Yeargan, K. V. In *Biological Control in Agricultural IPM Systems*. Editors, Hoy, M. A.; Herzog, D. C.; Academic Press: Orlando, FL, 1985, 521-536.

BIOCONTROL AGENTS FOR SUPPRESSION OF INSECTS

Viruses for Control of Arthropod Pests

Patrick V. Vail, Horticultural Crops Research Laboratory, USDA, ARS, 2021 S. Peach Ave., Fresno, CA 93727

Of the many viruses isolated from arthropods, the baculoviruses have been the most studied as potential microbial control agents. In many cases these DNA viruses are highly virulent and efficacious. Four have been registered by the U.S. Environmental Protection Agency for control of agricultural and forest pests. Others are candidates for registration. Development of baculoviruses to this state has depended on many disciplines including virology, production systems, formulation and application technology, and microbial control/pest management. Because of their lack of direct effects on other biological control agents they are good candidates for insect control in IPM systems. To completely utilize their potential, several areas of research need to be explored.

Entomogenous viruses of many types have been described from a multitude of insect orders, but primarily the Lepidoptera (moths and butterflies). One group of these viruses, the baculoviruses, has been studied more than any other and the lists of hosts now exceeds 400 species, many of them economic pests to agriculture and forestry throughout the world. Viruses have also been isolated from phytophagous mites and other arthropods. In addition to the baculoviruses other types of viruses have been isolated from many other insects (e.g., mosquitoes, tsetse fly) which may have significant impacts on their populations. Baculoviruses have been isolated from a number of major production and postharvest pests and shown to be efficacious as pest control agents. Several baculoviruses have been brought to large scale production and field testing. A number of these viruses have been approved for use in agriculture or forestry: the *Heliothis* sp. and the Douglas fir tussock moth (*Orgyia pseudotsugata*), gypsy moth (*Lymantria dispar*), and European pine sawfly (*Neodiprion sertifer*) nuclear polyhedrosis viruses (NPV).

Currently there is increasing pressure by regulatory agencies, environmentalists, and consumer groups to increase the safety of food products and reduce environmental impacts of pest control. Coupled with this are the increasing costs for registration of chemical pesticides and the potential liability associated with their use. These factors combined have reduced the incentives for the development of new chemical pesticides as documented by the reduction in availability of these materials. One of the potential alternatives to chemical pesticides and registration are entomogenous viruses.

Because of these concerns and regulatory restraints, there is a need to continue to develop integrated pest management (IPM) systems for economic insect/arthropod pests of agriculture, forestry, man and livestock. I will focus my discussion on agricultural and forest pests. IPM supports the minimal and timely use of pesticides to reduce pest species below the economic thresholds. Often, only a few key species are involved in the development of IPM programs and many of these species are susceptible to viruses that may have practical use. The advantages and limitations of baculoviruses as insect control agents are shown in Table 1.

Baculoviruses have been reported from most economic species of Lepidoptera. They are often genus or species specific. However, some have a relatively broad host range which provides the possibility of controlling several insect species on a crop or several crops with a single virus. Safety and environmental studies have been conducted on a number of baculoviruses with no detrimental effects being shown to date. Generally they have been shown

Table 1. Advantages and Disadvantages to the Use of Baculoviruses

Advantages	Disadvantages
Safety for non-target organisms (beneficials, vertebrates, etc.)	High specificity
	Low heat, UV stability
Persistence through marketing channels	Requires strict quality controls in production
No demonstrated resistance	Control not immediate
Comparable costs	
	Only *in vivo* production systems available (to date)
Registration procedures streamlined	
	Patent problems

Baculoviridae is divided into three major groups (genera) depending on the presence or absence of an inclusion body and the morphology of that inclusion body. Those most studied include the nuclear polyhedrosis and granulosis viruses. The former has many virions occluded in occlusion bodies (OB); while the latter has a single virion occluded in a structure termed a capsule (Fig. 1a & b). OBs are visible with the light microscope (Fig. 2). Historically these viruses were considered to be very specific (species or genus), however currently many examples exist of one virus infecting many members of different families of the Lepidoptera including a number of economic pests (Hostetter and Puttler 1991; Vail *et al.* 1971). The lack of specificity, within limits, is arguably advantageous to the development of a microbial control agent (Table 2).

Infection Process

Unlike chemical insecticides which often have contact, oral, and fumigant action, baculoviruses are only infective orally. In the normal route of infection, either OBs or capsules are ingested by

Table 2. Selected Hosts of Ac*M*NPV (8 families)[+]

Species
Autographa californica
Trichoplusia ni[+]
Pseudoplusia includens
Autographa biloba
Spodoptera exigua[+]
Spodoptera frugiperda[+]
Spodoptera praefica
Diparopsis watersi (African bollworm)
Helicoverpa zea
Heliothis virescens[+]
Galleria mellonella[+]
Bucculatrix thurberiella
Pectinophora gossypiella
Plutella xylostella[+]
Hemerocampa pseudotsugata
Ostrinia nubilalis
Cactoblastis cactorum

[+]Indicates susceptibility of homologous cell lines.

to have no direct (i.e., infectivity) effects on predators or parasitoids. Several baculoviruses have been registered and undergone extensive safety testing in the USA: *Heliothis*, gypsy moth, Douglas fir tussock moth, alfalfa looper (*Autographa californica*) and European pine sawfly NPVs, and codling moth (*Cydia pomonella*) granulosis virus (GV). Based on this information, the registration protocols for baculoviruses have been considerably simplified and registration costs significantly reduced. Thus, this family of viruses provides the potential to incorporate a non-disruptive insect control method into IPM systems that would provide more stability to these systems. In the postharvest realm they provide the potential for long term protection against storage pests. Since baculoviruses have been studied most intensively, I will restrict my comments to this group. I will discuss isolation, characterization, production, standardization, field testing, and commercialization and the problem areas unique to the utilization of these viruses.

Description and Pathology

Baculoviruses are DNA viruses that have a unique replication strategy. The family

way of a contaminated food source (Fig. 3). Upon entrance into the alkaline midgut of the insect, OBs dissolve and release the infectious virions which are adsorbed onto the cells of the midgut epithelium. The first replication of the virus may occur in the midgut epithelium and then spread to other tissues via the hemocoel. As shown in Table 3 once in the hemocoel many major organs and tissues of the insect may be infected although this is variable depending on the specific virus-host system. First signs of infection at the cellular level may be evident within 12 hours (nuclear hypertrophy and extra cellular virus (ECV) production) after inoculation. The entire process from infection to polyhedra production and cell lysis may occur within 72 hr. Depending on dose, gross pathology may become evident in 2–5 days. Time of mortality varies but may occur in as little as 3 days depending upon dose, species, insect age, and virus isolate.

In vitro systems utilizing insect cell lines have also been widely studied. OBs are not infectious *in vitro* as the alkaline requirements for dissolution are not present, therefore non-occluded forms of the virus must be used. The infection cycle may be very rapid as resistance mechanisms are diminished in some *in vivo* systems.

Virus Production Systems

All large scale baculovirus production for microbial control to date have been accomplished using *in vivo* systems. Many baculoviruses have been produced for small scale field testing. In early production procedures the host plant of the insect was used necessitating maintenance of large supplies of live host plants and provisions to exclude contamination (Elmore *et al.* 1961). In some cases infected field collected larvae are used. Generally these early preparations were held in water and the insects allowed to decompose. Little further purification was done, except to homogenize the preparation. These "lightly processed" preparations often contained excessive extraneous microbial contaminants. The presence of more than one baculovirus (Heimpel and Adams 1966) in field collected larvae used in some early viral preparations

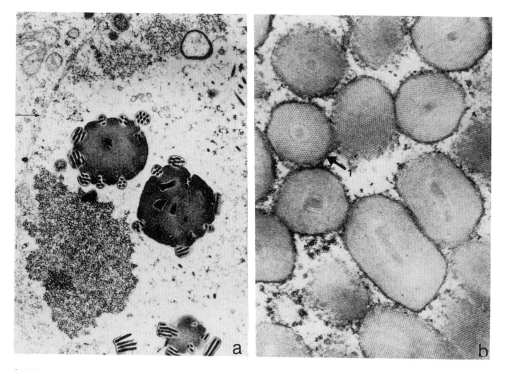

Fig. 1. Electron micrographs of inclusion bodies of nuclear polyhedrosis (a) and granulosis (b) viruses.

32

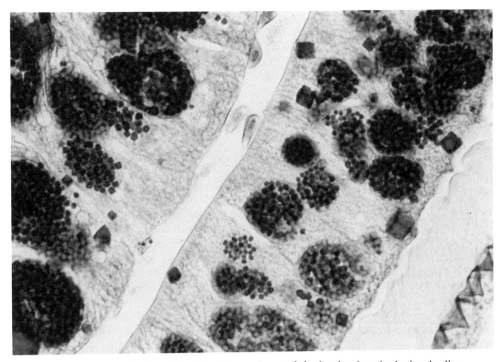

Fig. 2. Histological section showing nuclear polyhedrosis virus inclusion bodies (polyhedra) in nucleus of cells of susceptible tissues.

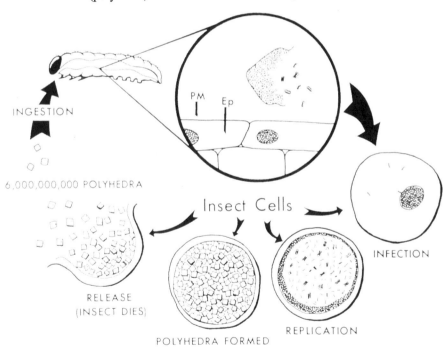

INGESTION

6,000,000,000 POLYHEDRA

PM   Ep

Insect Cells

INFECTION

REPLICATION

POLYHEDRA FORMED

RELEASE
(INSECT DIES)

Fig. 3. Life cycle of a baculovirus (NPV). Up to 6 billion polyhedra may be produced in one diseased larva. (From: Vail, P.V. "Microbial Pest Control. II. Baculoviruses." Calif. Dept. Food and Agric. Seminar Series #2; **1983**, 16 pp.)

complicated the issues of production and standardization of these types of preparations. The advent of semi-synthetic diets (Vanderzant *et al.* 1962) not only eliminated the need for host plants with their associated problems, but also provided the opportunity to truly mass rear hosts for large scale production of viruses and other beneficial organisms.

The first large scale production of baculoviruses was accomplished by Ignoffo (1964b, 1965) using baculoviruses isolated from cabbage looper and *Heliothis* sp. These methods were improved by International Minerals Corp. and later Sandoz Inc. who developed highly mechanized production systems for a commercial product Elcar®, based on the nuclear polyhedrosis virus isolated from *Heliothis* sp. (Ignoffo and Couch, 1981). Large scale production of many candidate baculoviruses could be accomplished with this technology. Viruses for which large scale production methods have been developed are shown in Table 4.

Some of these systems provide varying degrees of efficiency depending on their state of mechanization and the relative ease by which the appropriate host(s) can be reared. Baculoviruses such as those isolated from the alfalfa looper (Vail *et al.* 1971) and celery looper (Hostetter and Puttler 1991), demonstrated to have broad host ranges, provide the opportunity to select the host most amenable to mass production. As a working figure, each larva generally produces $1–6 \times 10^9$ OBs while application rates in the field generally range between 10^{11} and 10^{12} OBs/acre. Yields for rearing systems may vary considerably. Such factors as yield of OBs/larva, whether the larvae are gregarious or cannibalistic, ease and efficiency of rearing and virulence of the isolate impact on the efficiency and cost of production and the product. In some instances the larval production diet has been used to formulate the virus (Vail *et al.* 1991) although in most cases the larvae are removed from the diet prior to extraction of the OBs or capsules.

In vitro systems utilizing insect cell lines have also been developed for the study of entomogenous viruses. At the present time *in vitro* systems are used primarily for titration of ECV, basic virology, viral genetics and foreign gene expression. *In vitro* cell culture systems

Table 3. Tissues Infected by Nuclear Polyhedrosis and Granulosis Viruses

Fat body	Ganglia
Tracheal matrix	Foregut
Hypodermis	Midgut
Malpighian tubules	Hindgut
Muscle	Juvenile tissue
Hemocytes	Testes
	Silk glands

are also used for production of small quantities of inocula of various types (OBs and ECV forms of baculoviruses), and small OB productions for laboratory or limited field testing. More recently attempts have been made to develop either continuous flow or batch systems for large scale production of OBs for use as microbial control agents (Hink and Strauss 1980; Weiss *et al.* 1992). The feasibility of using *in vitro* systems in large scale production for microbial control purposes has increased with the development of low cost serum-free medium (Godwin *et al.* 1991). Although a GV has been produced in cell lines (Naser *et al.* 1984) yields have not been high.

Table 4. Baculoviruses produced on a large scale using *in vivo* production systems. Those produced in cell lines on a small scale are also indicated.

Baculovirus from:	Reference
Cabbage looper	Ignoffo 1964b
Alfalfa looper	Vail *et al.* 1973[+]
Codling moth	Benz 1981
. . . .	Brassel 1978
. . . . Falcon (personal communication)	
Corn earworm/tobacco	
budworm*	Ignoffo 1965[+]
Gypsy moth*	Shapiro *et al.* 1981[+]
Douglas fir tussock	
moth*	Martignoni 1979
European pine sawfly*	Lewis 1970
Indianmeal moth	
(*Plodia interpunctella*)	
. . . .	Cowan *et al.* 1986
. . . .	Vail *et al.* 1991

*Registered by the EPA.
[+]Also produced in insect cell lines.

The advantages to *in vitro* systems are (1) they are more easily manipulated and (2) the products are free of other microbial contaminants. The absence of contaminants can eliminate or considerably reduce harvesting and clean up procedures. To date, *in vitro* systems have not been developed that are competitive with *in vivo* systems for baculovirus based microbial control agents. Unfortunately few investigators are involved in this research because of its inherent high cost. Interestingly, insect cell lines and engineered baculoviruses are now routinely used to produce complex biologically active materials such as animal and human viral proteins (Summers 1991). In most cases the cell lines and viruses used in these systems were derived from the alfalfa looper NPV and cell lines, or variations thereof, developed by ARS (Vail *et al.* 1971; Vaughn *et al.* 1977). The availability of these virus-cell line systems has made the baculoviruses one of the most extensively studied groups among animal viruses. This is partially due to the relative simplicity of the virus-cell line systems but also due to the relative ease by which these viruses can be genetically modified. Research is now being conducted to increase the effectiveness of baculoviruses by genetic engineering including the introduction of known insect specific toxin genes into the viral genome (Summers 1991).

Standardization/Quality Control

Key components to the successful use of baculoviruses as control agents are standardization and quality control (Summers *et al.* 1975). One only has to study the early literature on the commercialization of *Bacillus thuringiensis* to understand the negative impacts that the lack of concern for these two factors can have on the predictable field performance of a microbial control agent. As mentioned above, many early studies were conducted with baculoviruses that were poorly standardized as to virus type and activity. Standardization and quality control are inherently important parts of the development process. Now even small scale field tests with baculoviruses are conducted with standardized materials to avoid the pitfalls of variability previously encountered.

With few exceptions, baculoviruses used for microbial control are assayed by various *per os* techniques. The most common method of assay is to layer different concentrations of OBs on the surface of artificial diets (Ignoffo 1964a). Cups containing diet are then infested with one larva of a known age. Mortality is observed at certain time intervals after infestation. Activity of the virus is usually determined by probit analysis which provides slopes, LC_{50} and LC_{95} values expressed as OBs/mm^2 of diet surface area. Another technique incorporates different concentrations of OBs into the warm liquid diet prior to solidification (Dulmage *et al.* 1976). Activity in these assays is expressed in terms of OBs/$\mu\ell$ or ml of diet. Assays have also been developed in which larvae are fed droplets containing OBs (Hughes *et al.* 1986) or virus contaminated diet or plant discs which are completely consumed. The advantage to these assays is that the number of OBs consumed by each larva is known. In some cases time/mortality curves have been used to express activity.

Factors Affecting Efficacy

Many baculoviruses are efficacious towards their hosts when applied in the field; however, numerous factors may affect efficacy. In the early days of baculovirus field experimentation, most formulations were in aqueous form. More recently, commercial type formulations have been dry. Generally this is accomplished by freeze or flash drying. The "technical" product is then diluted with substances generally accepted as safe (GRAS list) for production of the formulated product. Dry formulations are generally preferred also because they reduce the growth of microbial contaminants. For application, various materials such as surfactants, wetting agents, or feeding stimulants may be added to the formulation or the tank-mixes just prior to application. In addition, a number of chemicals to protect the applied virus from UV inactivation have been described (Shapiro and Robertson 1990; Shapiro *et al.* 1983). Several of these compounds also enhance baculovirus activity by several orders of magnitude in the laboratory (Hamm and Shapiro 1992; Shapiro 1992; Shapiro and

Robertson 1992). If similar effects can be demonstrated in the field, these chemicals could have a significant impact on the economics and consequent use of baculoviruses as insect control agents.

Since OBs and granulosis capsules are only infectious *per os*, timing of application for insect control becomes very critical in order to infect insects in the early stages of development. Coverage of foliage is also extremely important because of the sole *per os* route of inoculation. Recent developments in application technology will benefit the use of baculoviruses. The problems of coverage and mode of entry are further exacerbated by the relative short field life of OBs and capsules due to UV inactivation. Thus, the number of applications may have to be increased as compared to chemical insecticides which often have extended residual activity. The importance of maintaining field activity for longer periods (5–10 days) than is now possible becomes obvious. Clearly, baculoviruses can potentially control insects in diverse situations, however certain critical factors still need to be addressed and solved.

Safety and Registration

As noted above, four baculoviruses (*Heliothis*, gypsy moth, Douglas fir tussock moth and European pine sawfly NPVs) have been approved for use in the United States. Several, including the codling moth GV have also been registered in Europe. Safety protocols are rigorous and involve tests with non-target species, both vertebrate and invertebrate, and assessment of potential environmental impacts. Testing is arranged in three tiers. Tier I tests maximize the possibility of response of the test organism(s) to the virus. If Tier I tests are completed successfully and without implications as to positive responses, testing is stopped. To date all baculoviruses have been registered based on successful completion of Tier I testing. Because of the rigorous testing both *in vivo* and *in vitro* in a number of vertebrate and higher mammal systems, and the absence of responses, most experts in the field consider baculoviruses safe for use. In addition, they are naturally occurring organisms and generally the costs of

registration are significantly lower than that required for chemical insecticides.

Use Strategies

A number of diverse methods (strategies) have been employed for the successful use of baculoviruses. Perhaps the most simple are inoculative type releases in which the virus is introduced into the host populations and allowed to spread naturally and thus gradually lower the overall population density of the host. One of the best and earliest examples of the successful use of this method was for control of a rhinoceros beetle (*Oryctes rhinoceros*) attacking coconut palms in the South Pacific (Marshall 1970). Since the introductions, rhinoceros beetle populations and damage have been reduced significantly (Bedford 1981). A similar strategy has been proposed for the western grapeleaf skeletonizer (*Harrisina brillians*) a sporadic but continual pest of grapes which is infected by a highly virulent granulosis virus (Stern and Federici 1990). The virus is also readily transmitted from generation to generation. The reproductive biology and behavior of this pest also makes it especially amenable to this type of control. This type of introduction is analogous to classical biological control in which natural enemies are introduced and gradually reduce pest species populations over an extended period of time.

The most commonly used method for control is the application of large quantities of OBs or capsules for direct and immediate control of insect populations. Numerous examples of this use have been presented already in this chapter, but several need to be highlighted. The NPV isolated from *Heliothis* (Elcar) was developed for the immediate control of corn earworm and tobacco budworm populations infesting cotton and other crops in the United States. The previously mentioned NPVs of forest pests have also been developed for the immediate control of forest pests in the western and eastern United States. Several other viruses mentioned in this paper have been used in a similar way. The Indianmeal moth GV has been used to protect grains and dried fruits (Cowan *et al.* 1986). Because UV or high temperature are seldom problems in commodity storage, long term protection can be afforded after a single application of this virus. Because many agricultural crops and forests are infested by a complex of lepidopterous pests, the benefits of using a single virus having a broad host range has also been explored. The use of baculoviruses will become more widespread as the need for "softer" insecticides to integrate into pest management systems utilizing other beneficial species (predators and parasitoids) increases.

Conclusions

The use of baculoviruses as microbial insecticides started in the early 1950s. They provide the opportunity to develop pesticides that are not harmful to nontarget organisms or the environment and are nondestructive tools for the development of or incorporation into pest management systems. Since that time, research has intensified in this area with the registration of baculoviruses here and abroad. These viruses provide the opportunity to control production and postharvest agricultural and forest pests. A number of baculoviruses in addition to those already registered have been studied intensively and should be registered in the near future. Sophisticated *in vivo* production systems have been developed for their host(s) which allow large scale production of the candidate viruses. However efficient and cost effective *in vitro* techniques need to be developed. The increasing regulatory constraints and cost of registration for chemical insecticides makes the utilization of baculoviruses more appealing today than in the recent past. To reach our goals of utilizing these important agents in integrated systems a number of important problem areas need to be addressed:

1. Conduct more research to find highly virulent viruses infectious to production and postharvest coleopteran (beetle) pests;

2. Develop economical *in vitro* production systems for entomogenous viruses that are candidates as microbial control agents;

3. Develop an understanding of the biology and genetics of entomogenous viruses particularly as they relate to virulence and host range;

4. Define the limitations and environmental requirements of these organisms;

5. Develop the technology to improve entomogenous viruses by genetic engineering;

6. Significantly increase the environmental persistence (5–10 days) of entomogenous viruses in order to make them more efficacious and cost effective; and

7. Design formulations and application technologies for specific virus-insect-host plant systems.

References

Bedford, G.O. "Control of the Rhinoceros beetle *Oryctes rhinoceros* by Baculovirus Pests of Palms, Biological Control, Southeast Asia, Mauritius, Africa, South Pacific." In *Microbial Control of Pests and Plant Diseases*; Burges, H.D., Ed.; Academic Press, London, **1981**, pp. 409–426.

Benz, G. "Use of Viruses for Insect Suppression." *Bio. Control in Crop Prod. Symp. 5*; Papavisas G.C., Ed.; Allanheld, Osmun, Totowa, **1981**, pp. 259–272.

Brassel, J. "Development of Techniques for the Production of a Granulosis Virus Preparation for Microbial Control of the Codling Moth, *Laspeyresia pomonella* L. (Lepidoptera, Tortricidae), and Estimation of Production Costs [Fruit Pests]." *Mitt. Schweiz. Entomol. Ges.*, **1978**, *51*, 155–211.

Cowan, D.K.; Vail, P.V.; Kok-Yokomi, M.L.; Schreiber, F.E. "Formulation of a Granulosis Virus of *Plodia interpunctella* (Hübner) (Lepidoptera: Pyralidae): Efficacy, Persistence, and Influence on Oviposition and Larval Survival." *J. Econ. Entomol.*, **1986**, *79*, 1085–1090.

Dulmage, H.T.; Martinez, A.J.; Pena, T. "Bioassay of *Bacillus thuringiensis* (Berliner) δ-endotoxin using the tobacco budworm." USDA-ARS Tech. Bull. No. 1528, **1976**, 15 pp.

Elmore, J.C. "Control of the Cabbage Looper With a Nuclear Polyhedrosis Virus Disease." *J. Econ. Entomol.*, **1961**, *54*, 47–50.

Godwin, G.; Gorfien, S.; Tilkins, M.L.; Weiss, S. "Development of a Low Cost Serum-Free Medium for the Large-Scale Production of Viral Pesticides in Insect Cell

Culture;" In *Proc. 8th Intl. Confr. Invertebrate and Fish Tissue Culture with 1991 World Congr. on Cell and Tissue Culture,* Fraser, M.J. Jr., Ed.; **1991**, Anaheim, Calif., pp. 102–110.

Hamm, J.J.; Shapiro, M. "Infectivity of Fall Armyworm (Lepidoptera: Noctuidae) Nuclear Polyhedrosis Virus Enhanced by a Fluorescent Brightener." *J. Econ. Entomol.,* **1992**, *85*, 2149–2152.

Heimpel, A.M.; Adams, J.R. "A New Nuclear Polyhedrosis of the Cabbage Looper, *Trichoplusia ni.*" *J. Invertebr. Pathol.,* **1966**, 8, 340–346.

Hink, W.F.; Strauss, E.M. "Semi-Continuous Culture of the TN-368 Cell Line in Fermentors with Virus Production in Harvested Cells;" In *Invertebrate Systems in vitro*; Kurstak, E.; Maramorousch, K.; Dübendorfer, A., Eds.; Elsevier/North-Holland Biomedical Press, The Netherlands, **1980**, pp. 27–33.

Hostetter, D.L.; Puttler, B. "A New Broad Host Spectrum Nuclear Polyhedrosis Virus Isolated From a Celery Looper, *Anagrapha falcifera* (Kirby), (Lepidoptera: Noctuidae)." *Environ. Entomol.,* **1991**, *20*, 1480–1488.

Hughes, P.R.; van Beek, N.A.M.; Wood, H.A. "A Modified Droplet Feeding Method for Rapid Assay of *Bacillus thuringiensis* and Baculoviruses in Noctuid Larvae." *J. Invertebr. Pathol.,* **1986**, *48,* 187–192.

Ignoffo, C.M. "Bioassay Technique and Pathogenicity of a Nuclear-Polyhedrosis Virus of the Cabbage Looper, *Trichoplusia ni* (Hübner)." *J. Invertebr. Pathol.,* **1964a**, *6,* 237–245.

Ignoffo, C.M. "Production and Virulence of a Nuclear-Polyhedrons Virus from Larvae of *Trichoplusia ni* (Hübner) Reared on a Semi-Synthetic Diet." *J. Invertebr. Pathol.,* **1964b**, *6,* 318–326.

Ignoffo, C.M. "The Nuclear-Polyhedrosis Virus of *Heliothis zea* (Boddie) and *Heliothis virescens* (Fabricius)." I. Virus Propagation and its Virulence. *J. Invertebr. Pathol.,* **1965**, *7*, 209–216.

Ignoffo, C.M.; Couch, T.L. "The Nucleo-Polyhedrosis Virus of *Heliothis* Species as a Microbial Insecticide;" In *Microbial Control of Pests and Plant Diseases;* Burges, H.D.,

Ed.; Academic Press, London, **1981**, pp 329–362.

Lewis, F.B. "Mass Propagation of Insect Viruses with Specific Reference to Forest Insects;" *Proc. IV Intl. Coll. of Ins. Pathol.,* College Park, MD, **1970**, pp. 320–326.

Marshall, K.J. "Introduction of a New Virus Disease of the Coconut Rhinoceros Beetle in Western Samoa." *Nature* (London), **1970**, *225*, 288–298.

Martignoni, M.E. "The Douglas Fir Tussock Moth: A Synthesis;" Brookes, M.H.; Stark, R.W.; Campbell, R.W., Eds.; USDA Forest Service Tech. Bull. 1585, **1979**, pp. 140–147.

Naser, W.L.; Miltenburger, H.G.; Harvey, J.P.; Huber, J.; Huger, A.M. "In Vitro Replication of the *Cydia pomonella* (Codling Moth) Granulosis Virus." *FEMS Microbiol. Ltrs.,* **1984**, *24*, 117–121.

Shapiro, M.; Bell, R.A.; Owens, C.D. "In Vivo Mass Production of Gypsy Moth Nucleo-polyhedrosis Virus;" In *The Gypsy Moth: Research Toward Integrated Pest Management*; Doane, C.C.; McManus, M.L.; Eds.; USDA-FS Tech. Bull. 1584, **1981**, pp. 633–655.

Shapiro, M.; Agin, P.P.; Bell, R.A. "Ultraviolet Protectants of the Gypsy Moth (Lepidoptera: Lymantriidae) Nucleo-polyhedrosis Virus." *Environ. Entomol.,* **1983**, *12*, 982–985.

Shapiro, M.; Robertson, J.L. "Laboratory Evaluation of Dyes as Ultraviolet Screens for the Gypsy Moth (Lepidoptera: Lymantriidae) Nuclear Polyhedrosis Virus." *J. Econ. Entomol.,* **1990**, *83*, 168–172.

Shapiro, M. "Use of Optical Brighteners as Radiation Protectants for Gypsy Moth (Lepidoptera: Lymantriidae) Nuclear Polyhedrosis Virus." *J. Econ. Entomol.,* **1992**, *85*, 1682–1686.

Shapiro, M; Robertson, J.L. "Enhancement of Gypsy Moth (Lepidoptera: Lymantriidae) Baculovirus Activity by Optical Brighteners." *J. Econ. Entomol.,* **1992**, *85*, 1120–1124.

Stern, V.M.; Federici, B.A. "Granulosis Virus: Biological Control for Western Grapeleaf Skeletonizer." *Calif. Agr.,* **1990**, *44*, 21–22.

Summers, M.D.; Engler, R.; Falcon, L.A.;

Vail, P.V. Eds.; *Baculoviruses for Insect Control: Safety Considerations* EPA-USDA Working Symp., Amer. Soc. Microbiol., Washington, D.C.; **1975**.

Vail, P.V. "Standardization and Quantification: Insect Laboratory Studies;" In *Baculoviruses for Insect Control: Safety Considerations*; Summers, M.D.; Engler, R.; Falcon, L.A.; Vail, P.V. Eds.; EPA-USDA Working Symp., Amer. Soc. Microbiol., Washington, D.C.; **1975**, pp. 44–46.

Vail, P.V.; Jay, D.L.; Hunter, D.K. "Cross Infectivity of a Nuclear Polyhedrosis Virus Isolated from the Alfalfa Looper, *Autographa californica*." Proc. IVth Intl. Colloq. Insect Pathol., College Park, MD, **1971**, pp. 297–304.

Vail, P.V.; Anderson, S.J.; Jay, D.L. "New Procedures for Rearing Cabbage Loopers and Other Lepidopterous Larvae for Propagation of Nuclear Polyhedrosis Viruses." *Environ. Entomol.*, **1973**, *2*, 339–344.

Vail, P.V.; Morris, T.J.; Collier, S.S. "An RNA Virus in *Autographa californica* Nuclear Polyhedrosis Virus Preparations: Gross Pathology and Infectivity." *J. Invertebr. Pathol.*, **1983**, *41*, 179–183.

Vail, P.V.; Tebbets, J.S.; Cowan, D.C.; Jenner, K.E. "Efficacy and Persistence of a Granulosis Virus Against Infestations of *Plodia interpunctella* (Hübner) (Lepidoptera: Pyralidae) on Raisins." *J. Stored Prod. Res.*, **1991**, *27*, 103–107.

Vanderzant, E.S.; Richardson, C.D.; Fort, Jr., S.W. "Rearing of the Bollworm on Artificial Diet." *J. Econ. Entomol.*, **1962**, *55*, 140.

Vaughn, J.L.; Goodwin, R.H.; Tompkins, G.J.; McCawley, P. "The Establishment of Two Cell Lines from the Insect *Spodoptera frugiperda* (Lepidoptera: Noctuidae);" *In Vitro*, **1977**, *13*, 213–217.

Weiss, S.A.; Godwin, G.P.; Whitford, W.G.; Gorfien, S.F.; Dougherty, E.M. "Viral Pesticides: In Vitro Process Development." Proc. 10th Australian Biotechnol. Confr., **1992**, 67–71.

The Use of Fluorescent Brighteners as Activity Enhancers for Insect Pathogenic Viruses

Martin Shapiro and Edward M. Dougherty, USDA, ARS, Insect Biocontrol Laboratory, PSI, Beltsville, MD, 20705-2350

Recently, we demonstrated that certain optical brighteners (*i.e.,* selected stilbenes) act as activity enhancers for the gypsy moth NPV. For the last two years, collaborators have demonstrated significant enhancement under field conditions. Moreover, enhancement could also occur with several homologous and heterologous viruses against the gypsy moth, the fall armyworm, and the corn earworm. The use of brighteners as activity enhancers of insect viruses was awarded a U.S. patent on June 23, 1992, and was licensed to two large companies developing viruses as microbial control agents. The importance and impact of these materials is discussed. From a basic viewpoint, the brightener is a tool which will better enable us to understand the role of the host in the host-virus relationship, affecting both viral activity and viral host range. From a practical standpoint, cooperative research is continuing to determine which brighteners act as enhancers in different host-virus systems. The mode of action of these brighteners is discussed, on the basis of present knowledge.

During the past decade, the gypsy moth, *Lymantria dispar*, has demonstrated the capability of survival in many areas of the United States, including the Pacific Northwest and California (Miller *et al.* 1991). Control measures are varied, but effective biological control is needed in environmentally sensitive areas such as ports, residential areas, and municipal watersheds (Shapiro and Dougherty 1985).

A nuclear polyhedrosis virus (NPV) has been used as a microbial agent and was registered as Gypchek (Lewis *et al.* 1979 a,b). Although NPV has been used successfully to reduce gypsy moth populations (Rollinson *et al.* 1965, Injac and Vasiljevic 1978, Podgwaite 1989), the virus is slow acting and larvae may continue to feed for almost 2 weeks after exposure.

Three factors influencing the use and performance of the gypsy moth NPV and other entomopathogenic viruses are production, biological activity (=virulence) and persistence. During the past 25 years, efforts have been devoted to the enhancement of baculovirus efficacy by increasing host susceptibility with selected chemicals (Doane and Wallis 1964, Yadava 1971, Bell and Kanavel 1975) and by affecting inherent viral activity by selection of more virulent biotypes (Shapiro and Ignoffo 1970, Reichelderfer and Benton 1973, Vasiljevic and Injac 1973, Wood *et al.* 1981, Shapiro *et al.* 1992). During the past decade, we demonstrated that chemicals such as boric acid (Shapiro and Bell 1982), chitinase (Shapiro *et al.* 1987) and the dye Congo red (Shapiro unpublished data) reduced the $LC_{50}s$ and $LT_{50}s$ of gypsy moth NPV suspensions. In a recent study of optical brighteners as ultraviolet ([UV] screens), it was noted that several brighteners not only provided complete protection but also affected the lethal incubation period, as evidenced by $LT_{50}s$ (Shapiro 1992).

Optical Brighteners

Optical brighteners (*i.e.,* fluorescent brighteners) were discovered more than 50 years ago (Paine *et al.* 1937, Eggert and Wendt 1939) and are widely used in the detergent, paper, plastics, and organic coatings industries (Lanter 1966) and as fluorochromes for microorganisms (Darken 1962, Slifken and Cumbie 1988). The compounds readily absorb ultraviolet (UV) radiation and transmit light in the blue portion of the visible spectrum (Villaume 1958). Twenty three brighteners, belonging to several chemical classes (e.g., stilbene, oxazole, pyrazoline, naphthalic acid, lactone, coumarin), were tested as UV protectants for the gypsy moth NPV (LdNPV). While effective brighteners belong to each of these groups, the four superior brighteners (Leucophor BS and BSB, Phorwite AR, and Tinopal LPW), all belong to the stilbene group. These brighteners not only provided complete protection but also affected the lethal incubation, as evidenced by $LT_{50}s$. These compounds appear to be very promising as radiation protectants and are the focus of further research as adjuvants (Shapiro 1992).

The addition of stilbene brighteners (Leucophor BS, BSB; Phorwite AR, RKH; Tinopal LPW) to LdNPV reduced the average $LC_{50}s$ from $\approx$ 18,000 polyhedral inclusion bodies (PIB) to values between 10 and 44 PIB/ml (i.e., a reduction of $\approx$ 400-1,800-fold) (Table 1) (Shapiro and Robertson, 1992). The viral enhancement of these brighteners is very high, and significant reduction in the LC_{50} can occur with a concentration as low as 0.01%. $LT_{50}s$ were also greatly reduced by the addition of these optical brighteners to LdNPV. Moreover, reduction in $LC_{50}s$ and $LT_{50}s$ among mature larvae (fourth-fifth instar) were also significant, indicating that the

Table 1.   Effects of selected brighteners upon gypsy moth NPV activity: $LC_{50}s$

Treatment	LC_{50} (PIB/ml)	Relative Activity
NPV + H_2O	18,000	1.00
+ BS	19	~960
+ BSB	44	~418
+ AR	15	~1,225
+ RKH	10	~1840
+ LPW	11	~1671

Source: Adapted from Shapiro and Robertson, 1992

combination of virus and brighteners could also be effective for reducing late-instar populations if a second application of NPV would be desirable (Shapiro and Robertson, 1992).

Because of these unprecedented results, we determined whether a selected brightener (Phorwite AR) could enhance the activities of other entomopathogenic viruses (e.g., a cytoplasmic polyhedrosis [CPV] from the gypsy moth, an NPV from the noctuid, Autographa californica (AcNPV), and an entomopoxvirus [EPV] from the arctiid, Amsacta moorei, against the gypsy moth. The addition of Phorwite AR to L. dispar CPV enhanced viral activity ca 800-fold. AcNPV has a wide host spectrum (Gröner 1986), but does not undergo complete replication in gypsy moth cell lines (McClintock et al., 1980). In addition, the virus was not pathogenic to gypsy moth larvae. When Phorwite AR was added to the AcNPV suspension, the NPV was pathogenic to gypsy moth larvae and PIBs were observed in hemocytes and other tissues. While the LC_{50} was very high (i.e., > 10^6 PIB per ml) and the slope was low (i.e., 0.79), the virus was able to complete replication in an insect not considered a normal host for AcNPV (Groner, 1986).

The Amsacta EPV replicates in cell lines from the saltmarsh caterpillar (Arctiidae), the gypsy moth (Lymantriidae), the cotton bollworm (Noctuidae), and the silkworm (Bombycidae) (Granados 1981). While Amsacta EPV was not pathogenic for gypsy moth larvae, the virus underwent complete replication in a L. dispar cell line (IPLB-LD-652) (Goodwin et al. 1990). The addition of Phorwite AR to Amsacta EPV resulted in complete viral replication (i.e., formation of viral inclusion bodies in infected tissues and virus-caused larval mortality. As with AcNPV the LC_{50} value was very high (i.e., ca $7x10^5$ PIB per ml) and the slope was low (i.e., 0.84); larval mortality was obtained at concentrations as low as 10^4 PIB per ml per cup and a concentration-dependent response was obtained. Thus, as did AcNPV, the Amsacta EPV, in the presence of Phorwite AR, became pathogenic to a non-permissive host (L. dispar). With the addition of a selected

41

brightener, it was possible to increase the susceptibility of the gypsy moth and to expand the host range of these viruses. Thus, the brightener is an important tool in the study of host susceptibility and virus host range.

While it was very exciting to obtain viral enhancement with the addition of selected brighteners, was the gypsy moth unique in its response to the NPV-brightener combination? Subsequent cooperative research with John Hamm (ARS-Tifton, GA) demonstrated that selected stilbenes also enhanced the activities of the fall armyworm NPV against the fall armyworm, *Spodoptera frugiperda* (Hamm and Shapiro 1992), and the fall armyworm granulosis virus (GV) against *S. frugiperda*, (Shapiro *et al.* 1992).

At this time it would be most beneficial to summarize (1) what we know about the fluorescent brighteners as activity enhancers, (2) what we can infer, and (3) what we do not know. For the next several years, much effort will continue to determine the mode of action of these materials, since this knowledge will lead not only to better understanding of host response and viral host range but will enable us to make better use of insect virus-brightener combinations.

What We Know

(1) Active brighteners appear to be stilbenes.

The only brighteners that appear to act as viral enhancers (for LdNPV) are stilbenes. Previously, we demonstrated that several stilbenes (*e.g.,* Phorwite AR, RKH; Leucophor BS, Leucophor BSB; Tinopal LPW) reduced LC_{50} values of LdNPV from 400 to 1800-fold (Shapiro and Robertson 1992).

(2) Not all stilbenes are active.

Subsequent bioassays (Shapiro unpublished data) indicate that (a) of all the stilbenes tested (*e.g.,* both brighteners and non-brighteners), only brighteners act as enhancers for LdNPV, and (b) structure activity studies are underway and indicate that Tinopal LPW (= Calcofluor M_2R) and closely related compounds appear to be the most efficacious.

(3) Host susceptibility can be increased in certain virus-host systems by the addition of selected brighteners.

Although one normally speaks or refers to viral activity, it is quite apparent that a virus cannot act independently of its host. Moreover, a given virus can be active or inactive, depending upon the species of host insect challenged. Thus, AcNPV is inactive against the gypsy moth, has low activities against both the corn earworm and the fall armyworm, and is highly active against the cabbage looper. The addition of Calcofluor M_2R to AcNPV suspensions enhances activities against *H. zea*, *S. frugiperda* and *L. dispar*, but does not affect activity against *T. ni*. It would appear that host defenses of the gypsy moth, corn earworm and fall armyworm were compromised by the brightener.

(4) The brightener must be ingested. Feeding and injection experiments have clearly demonstrated that the virus and the brightener must be ingested together for the enhancement to occur. The brightener may even be fed to larvae within 24 hours of virus challenge and no enhancement will occur. If either the brightener or the virus is injected or both are injected, no enhancement occurs.

(5) Larvae stop feeding within 48 hours. In a typical bioassay employing either second or fourth-stage larvae, insects exposed to both virus and brightener are much smaller than virus infected larvae (*e.g.,* less than 1/2 the size, depending upon virus concentration). The brightener, in combination with (LdNPV) causes a cessation in larval feeding within 2-3 days.

(6) Midguts are clear at 48 hours, the peritrophic membrane is abnormal (in texture) within 48 hours; and frass production has been greatly reduced.

(7) Gut pH reduction occurs within 48 hours. While many biochemical changes have been observed during nuclear polyhedrosis ingestion, midgut and hemolymph pH values remain unchanged (Sheppard and Shapiro, unpublished data). With the addition of Calcofluor M_2R to LdNPV, however, midgut pH values are reduced by 1 or 2 pH units within 48 hours, depending on NPV concentration. At the end of 72 hours, pH values are further decreased, but hemolymph pH values remain unchanged. During infection with LdCPV, midgut pH values are reduced (possibly indicating viral replication?), and this phenomenon is well known with cytoplasmic

polyhedrosis virus infection (Sheppard and Shapiro, manuscripts in preparation). The addition of the brightener causes a further decrease in midgut pH. However, the pH reduction with NPV is unprecedented, and may enable us to better understand the mode of action of these brighteners.

(8) The effect of the brightener occurs in the midgut. The virus exposed to brightener is not altered by the brightener, as measured by biological activity of progeny virus (produced in larval exposed to NPV + brightener and by DNA analysis. We know that the virus and brightener must be ingested and that the site of action is the larval midgut. In the case of the gypsy moth, the brightener allows the virus to replicate in a nonpermissive tissue (*e.g.,* columnar cells of the midgut), which causes cessation of larval feeding within 48 hours, and changes in midgut pH. Immuno-chemistry, histopathology, and electron microscopy of midgut are currently being undertaken.

(9) Using radioactive virus as inoculum, higher levels of virions were found in the hemolymph among gypsy moth larvae exposed to LdNPV+ brightener than among larval exposed to brightener alone. More research is needed to pinpoint the initial occurrence of virions in hemolymph, however.

(10) The host spectrum of baculoviruses can be expanded. From a basic viewpoint, the brightener is a tool which will enable us to better understand the role of the host in the host-virus relationship. For example, why is the gypsy moth susceptible to a given NPV and not another? What changes occur in virus attachment, uptake, and replication with the addition of the brightener? The brightener is also able to change host susceptibility (gypsy moth), so that other viruses now cause lethal infection (alfalfa looper NPV, *Amsacta,* EPV. Moreover, enhancement could also occur with such agriculturally important insects as the corn earworm, fall armyworm, soybean looper, and velvet bean caterpillar with several homologous and heterologous baculoviruses.

What We Can Infer About The Mode of Action

(1) Effects upon peritrophic membrane. At present, the mode of action of these brighteners is not known, but some clues do exist. Several brighteners are known to interfere with cellulose (Haigler *et al.* 1980, Quader 1981, Roberts *et al.* 1981, Itoh *et al.* 1984) and chitin (Herth 1980 Elorza et al 1983, Herth and Hausser 1984, Selitrennikoff 1984) fibrillogenesis. In insects, the peritrophic membrane (pm) lines the midgut and is composed of chitin microfibrils. The pm may serve as a barrier for the invasion of microorganisms, including insect viruses (Brandt *et al.* 1978). Selected optical brighteners may inhibit or alter the chitinous pm, creating gaps in the lining. Electron microscope studies have revealed that an initial cycle of viral replication occurs in midgut columnar cells in the virus + brightener combination treatment, followed by budding from the basement membrane into the hemocoel. However, there may also be an effect of the brightener on the membranes of the gap portions between the columnar and goblet cells, allowing for passage of virions directly into the hemocoel. In the case of NPV (and CPV) greater uptake of virus into the midgut may occur in the presence of brightener.

(2) In at least one case, the dissolution of viral inclusion bodies (fall armyworm NPV) was faster when brightener was added to an alkaline dissolution solution (= carbonate, chloride). Is the increase in activity of *S. frugiperda* NPV at least partially due to faster release of virions?

(3) The maintainence of a high gut pH depends upon the ability of the midgut goblet cells to regulate the flow of $K+$ and $CO_3=$ from the hemolymph to the gut. Since the midgut pH decreases, it may be inferred that transport of these ions is inhibited. Disulfonic acids are known to affect ion transport in mammalian systems. For example, DIDS (= 4,4'-diisothiocyandstilbene-2,2'-disulfonic acid) and SICS (=4-acetamido-4'-isothiocyanostilbene-2,2'-disulfonic acid) are known to inhibit Ca-Mg ATP-ase of human red blood cells. Moreover, they also have a direct action on the CA^{2+} and inhibit CA^{2+} transport (or have an indirect action on the Ca^{2+} by inhibiting anion transport (=chloride, biocarbonate, sulfate) in human red blood cells [BAND III Protein] Dix et al. 1986, Romero and Ortiz, 1988.

43

What We Do Not Know

At this point, we have barely "scratched the surface" of the virus-host brightener relationship. When the story is complete, we have a much better understanding of the virus-host relationship and how to manipulate it for more efficious insect control.

(1) Is the increase in activity due to increased attachment, uptake, invasion, replication?

(2) What is the significance of pH changes to ultimate activity? Primary? Secondary?

(3) What allows LdNPV to replicate in the midgut?

(4) Is the replication in the midgut sufficient to lead to the large increase in activity? How much is enough?

(5) Is increased viremia sufficient to lead to the large increase in activity? How much is enough?

(6) Does the brightener act like the viral enhancer described by Tanada (1971) and later studied and characterized by Derksen and Granados (1988)?

(7) What causes a given virus to infect a given host or not? Conversely, what causes a given host to become susceptible to a given virus?

(8) What factors are responsible for viral specificity?

This research will give us a better understanding of the infection process. In addition, we will investigate systems where the brightener is a virus enhancer (gypsy moth and gypsy moth NPV) and where the effect is slight (corn earworm and corn earworm NPV) to gain better insights into host susceptibility and virus activity. This knowledge will enable us to eventually utilize the most effective combinations of virus(es) and brighteners to control pest insects.

References

Bell, R.A.; Kanavel, R.F. Potential of bait formulations to increase infectiveness of nuclear polyhedrosis virus against the pink bollworm. *J. Econ. Entomol.* **1975**; *68*, 389-391.

Brandt, C.R.; Adang, M.J.; Spence, K.D.S. The peritrophic membrane: unltra- structural analysis and function as a mechanical barrier to microbial infection in *Orgyia pseudotsugata*. *J. Invertebr. Pathol.* **1978**; *32*, 12-24.

Darken, M.A. Absorption and transport of fluorescent brighteners in biological techniques. *Science.* **1962**; *10*, 387-393.

Derksen, A.C.G.; Granados, R.R. Alteration of a lepidopteran peritrophic membrane by baculoviruses and enhancement of viral activity. *Virology.* **1988**; *167*, 242-250.

Doane, C.C., Wallis, R.C. Enhancement of the action of *Bacillus thuringiensis* var. *thuringiensis* Berliner on *Porthetria dispar* (Linnaeus) in laboratory tests. *J. Insect Pathol.* **1964**; *6*, 423-429.

Eggert, J.; Wendt, B. Wrapping materials. U.S. Patent 2,171,427; **1939**; 6 pp.

Elorza, M.V.; Rico, H.; Sentandreu, R. Calcofluor white alters the assembly of chitin fibrils in *Saccharomyces cerevisiae* and *Candida albicans* cells. *J. Gen. Microbiol.* **1983**; *129*, 1577- 1582.

Goodwin, R.H.; Adams, J.R.; Shapiro, M. Replication of the entomopoxvirus from *Amsacta moorei* in serum-free cultures of a gypsy moth cell line. *J. Invertebr. Pathol.* **1990**; *56*, 190- 205.

Granados, R.R. In *Pathogenesis of invertebrate diseases*; Davidson, E. W. Ed.; Allanheld: Totowa, NJ, **1981**; pp. 101- 126.

Gröner, A. In *The biology of baculoviruses;* Granados, R. R.; Federici, B. A., Eds.; CRC Press: Boca Raton, FL; **1986**, Vol. 1; pp. 177-202.

Haigler, C.H.; Brown, Jr.; R.M., Benziman, M. Calcofluor white ST alters the in vivo assembly of cellulose microfibrils. *Science.* **1980**; *210*, 903-906.

Hamm, J.J.; Shapiro, M. Infectivity of fall armyworm (Lepidoptera: Noctuidae) nuclear polyhedrosis virus enhanced by a fluorescent brightener. *J. Econ. Entomol.* **1992**; *854*, 2149-2152.

Herth, W. Calcofluor white and congo red inhibit chitin microfibril assembly in *Poteriochromonas:* evidence for a gap between polymerization and microfibril formation. *J. Cell. Biol.* **1980**; *87*, 442-450.

Herth, W.; Hausser J. *In Structure, function, and biosynthesis of plant cell walls;* Dugger, W.M., Bartnicki-Garcia, S. Eds.; Am. Soc. Plant Physiol., Rockville, MD; **1984**, pp. 89-119.

Injac, M.; Vasiljevic, L. Lutte contre le *Bombyx disparte (Lymantria dispar,* L.) par les virus de polyedrose nucleaire (Baculovirus) a l'aide dee l'avion. *Plant Protect.* **1978**; *143-144*, 43-56.

Itoh, T.; O'Neill, R.M.; Brown, Jr.; R.M. Interference of cell wall regeneration of *Boergesenia forbesii* protoplasts by Tinopal LPW, a fluorescent brightening agent. *Protoplasma.* **1984**; *123*, 174-183.

Lanter, J. Properties and evaluation of fluorescent brightening agents. *J. Soc. Dyes & Colourists.* **1966**; *82*, 125-132.

Lewis, F.B.; McManus, M.L.; Schneeberger, N.F. Guidelines for the use of Gypchek to control the gypsy moth. *U.S. For. Serv. Res. Pap.* **1979a**; **NE-441**, 9 pp.

Lewis, F.B.; Reardon, R.C.; Munson, A.S.; Hubbard, Jr., .B.; Scneeberger, N.F.; White, W.B. Observations on the use of Gypchek. *U.S. For. Serv. Res. Pap.* **1979b**; *NE-447*, 8 pp.

McClintock, J.T.; Dougherty, E.M.; Weiner, R.M. Semipermissive replication of a nuclear polyhedrosis virus of *Autographa californica* in a gypsy moth cell line. *J. Virol.* **1986**; *57*, 197-204.

Miller, J.C.; Hanson, P.E.; Kimberling, D.N. Development of the gypsy moth (Lepidoptera: Lymantriidae) on Douglas-fir foliage. *J. Econ. Entomol.* **1991**; *84*, 461-465.

Paine, C.; Radley, J.A.; Rendell, L.P. Fabrics fluorescent to ultraviolet light. U.S. Patent 2,089,413. **1937**; 2 pp.

Podgwaite, J.D. Gypchek... when you care enough to kill the very best! *Gypsy Moth News*, August **1989**, 6-7.

Quader, M. Interruption of cellulose micro-fibril crystallization. *Naturwissenschaften.* **1981**; *68*, 428-430.

Reichelderfer, C.F., Benton, C.F. The effect of 3-methylcholanthrene treatment on the virulence of nuclear polyhedrosis virus

of *Spodoptera frugiperda.* J. Invertebr. Pathol. 1973; 3: 38-41.

Roberts, E.; Haigler, C.; Brown, Jr., R.M. A fluorescent brightener affects wall morphology and cell development in the green alga *Oocystis apiculata. J. Cell. Biol.* **1981**; *91*, 155a.

Robinson, W.D.; Lenis, F.B.; Waters, W.E. The successful use of a nuclear polyhedrosis virus against the gypsy moth. *J. Invertebr. Pathol.* **1965**; *7*, 515-517.

Roncero,C.; Duran, A. Effect of calcofluor white and congo red on fungal cell wall morphogenesis: *in vivo* activation of chitin polymerization. *J. Bacteriol.* **1985**; *163*, 1180-1185.

Selitrennikoff, C.P. Calcofluor white inhibits *Neurospora* chitin synthetase activity. *Exp. Mycol.* **1984**; *8*, 269-272.

Shapiro, M. Use of optical brighteners as radiation protectants for the gypsy moth (Lepidoptera: Lymantriidae) nuclear polyhedrosis virus. *J. Econ. Entomol.* **1992**; *85*, 1682-1686.

Shapiro, M., Bell, R.A. Enhanced effectiveness of *Lymantria dispar* (Lepidoptera: Lymantriidae) nucleopolyhedrosis virus formulaed with boric acid. *Ann. Entomol. Soc. Am.* **1982**; *5*, 346-349.

Shapiro, M.; Dougherty, E. In *Microbial control of spruce budworms and gypsy moths;* Grimble, D. G., Lewis, F. B. Eds.; *U.S. For. Serv: Washington, DC,* **1985**; pp. 115-122.

Shapiro, M., Ignoffo, C.M. Nucleopolyhedrosis of *Heliothis zea. J. Invertebr. Pathol.* **1970**; *16,* 107-111.

Shapiro, M., Preisler, H.K., Robertson, J.L. Enhancement of baculovirus activity on gypsy moth (Lepidoptera: Lymantriidae) by chitinase. *J. Econ. Entomol.* **1987**; *80,* 1113-1116.

Shapiro, M.; Robertson, J.L. Enhancement of gypsy moth (Lepidoptera: Lymantriidae) baculovirus activity by optical brighteners. *J. Econ. Entomol.* **1992**; *85*, 1120-1124.

Shapiro, M.; Hamm, J.J.; Dougherty, E.M. Compositions and methods for biocontrol using fluorescent brighteners. U.S. Patent 5,124,149; **1992**. 16 pp.

Shapiro, M.; Lynn, D.E.; Dougherty, E.M. More virulent biotype isolated from wild-type virus. U.S. Patent 5,132,220; **1992**.

Slifkin, M. Cumbie, R. Congo red as a fluorochrome for the rapid detection of fungi. *J. Clin. Microbiol.* **1988**; *26*, 827-830.

Tanada, Y.; Hukuhara, T. Enhanced infection of a nuclear-polyhedrosis virus in larvae of the armyworm, *Pseudaletia unipuncta*, by a factor in the capsule of a granulosis virus. *J. Invertebr. Pathol.* **1971**, *17*, 116-126.

Vasiljevic, L., Injac, M. A study of gypsy moth viruses originating from different geographical regions. *Plant Prot.* **1973**: *24*, (124-125): 169-186.

Villaume, F.G. Optical bleaches in soaps and detergents. *J.Am.Oil Chemists Soc.* **1958**, *35*, 558-566.

Wood, A.H, Hughes, P.R., Johnston, L.B., Langridge, W.H.R. Increased virulence of *Autographa californica* nuclear polyhedrosis virus by mutagens. *J. Invertebr. Pathol.* **1981**; *38*, 236-241.

Yadava, R.L. On the chemical stressors of nuclear polyhedrosis virus of gypsy moth, *Lymantria dispar. Z. Angew. Entomol.* **1971**; *69*, 303-311.

Bacterial Control of Flies in Livestock Operations

E. T. Schmidtmann, Livestock Insects Laboratory, Beltsville Agricultural Research Center, Beltsville, Maryland 20705
D. W. Watson, Department of Entomology, Cornell University, Ithaca, New York, 14850
P. A. W. Martin, Insect Biocontrol Labororatory, Beltsville Agricultural Research Center, Beltsville, Maryland 20705

The association of immature muscoid flies and bacteria is reviewed relative to the development of a biologically-based management strategy for suppressing muscoid fly populations on dairy farms. Bedding calf hutches with materials alternative to straw, such as inorganic sand and gravel or microbe-limiting ground corncob and sawdust, suppressed house fly and stable fly larval density from 50 to 99 percent. We also report on the presence of spore-forming bacteria, *Bacillus* spp., including *B.thuringiensis,* in calf hutch bedding. Bioassay of this natural germplasm indicates that at least several isolates have activity against larvae of the house fly, hence potential for use in suppressing muscoid fly populations.

Flies of the family Muscidae are intimately associated with bacteria through the exploitation by immature stages of decomposing organic materials; these substrates, laden with microbes (Spiller 1964), provide essential nutrients for the growth and development of muscoid fly larvae (Brookes & Fraenkel 1958). That bacteria and the products of bacterial metabolism are essential to muscoid flies is indicated by the presence in larval stages of pharyngeal ridges that concentrate bacteria from liquified media (Dowding 1967), as well as the observation that immature house flies do not develop in media that has been heat-sterilized (autoclaved), but complete development if microbial activity is allowed to proceed for 48 h before sterilization (Greenberg 1954). Moreover, Levinson (1960) reported the growth of house fly larvae in autoclaved medium after addition of dessicated *Escherichia coli* cells, and other studies (Glaser 1924, Gerberic 1948, Silverman & Silverman 1953, Schmidtmann and Martin 1992) have likewise reported a positive relationship between bacteria and house fly maggot growth. Bacteria also are important in the growth of closely related species, the face fly, *Musca autumnalis* and horn fly, *Haematobia irritans* in cattle manure (Hollis et al. 1985 and Temeyer 1990a).

The intimate association between muscoid fly larvae and bacteria also represents opportunity for intervention or manipulation to suppress maggot density. Because suppressing the density of muscoid fly larvae represents a fundamental approach to reducing the abundance of adult muscoid flies (Anderson, 1966), we have pursued the development of methods that suppress the density of muscoid fly larvae by 1) the use of bedding materials that inherently limit microbial activity and therefore suppresss the growth of larval stages, and 2) developing the potential for using spore-forming bacteria, particularly insect-pathogenic *Bacillus* spp., to control muscoid fly larvae in livestock bedding. This approach, either individually or collectively as an integrated system, represents a biologically-based technology that can help limit muscoid fly abundance. The presence of adult muscoid flies on dairy farms compromises farm sanitation, bothers animals and farm workers (Bruce & Decker 1947) and, perhaps most importantly, represents a potential for dispersal into adjacent residental areas where flies are viewed as a public health concern (Miller 1993).

Bedding Materials

Our research with bedding materials has

focused on dairy outdoor calf hutches. Dairy-breed calves have historically been reared in pens or hutches bedded with straw. When fresh, straw is absorbent and effective insulation against low ground temperature and moisture. However, straw bedding readily soils with calf feces, urine and rainfall, and thereafter is highly suitable as a medium for the growth of immature house and stable flies (Thomsen & Hammer 1936, Pickens et al. 1967). Indeed, calf hutches on Maryland dairies supported average densities of house and stable fly larvae that represent a potential for producing 25,000 to 40,000 adult flies per hutch per summer (Schmidtmann et al. 1988).

Field trials in which sand and gravel, ground corncob, pine shavings and sawdust were evaluated for effect on suppressing the density of muscoid fly larvae when used as bedding in calf hutches were conducted at the Beltsville Agricultural Research Center dairy farm. Calf hutches were randomly bedded with either straw (= control) or experimental bedding materials, five hutches per type of bedding. Newborn calves were introduced individually into hutches at the start of six-week trials; house and stable fly densities were determined from samples of bedding taken weekly.

Selected results of these studies (Schmidtmann et al. 1989, Schmidtmann 1991) are summarized in Figure 1. Lowest house fly and stable fly densities, 85 to 99 percent reductions relative to straw, were observed in ground corncob (6.4 mm particle size), sand and gravel bedding substrates. The strong suppression of immature muscoid fly density in these substrates, due presumably to their inorganic composition or non-biodegradability, reflects their unsuitability as a growth medium for immature muscoid flies. Immature house and stable fly density in sawdust bedding was suppressed 84 and 64 percent, respectively.

Because of the general availability of sawdust and its suitability as a medium that both limits the density of muscoid fly larvae and provides acceptable sanitation (Schmidtmann 1991), we further compared different types (= tree species) of sawdust to assess possible variation in effect on larval density. The results show that sawdust bedding suppressed the density of house fly larvae by ca. 50 to 85 percent relative to straw (Fig. 2). House fly density in pine and tulip poplar sawdust was significantly less than straw, but did not differ statistically among types of sawdust. Reduction of stable fly density ranged from 43 to 66 percent; all stable fly densities in sawdust differed significantly from straw, but there was no difference among types of sawdust. This study further establishes the limiting effect of sawdust bedding on the density of muscoid fly larvae in calf hutches, and indicates that the type of sawdust (of those tested) is of less importance than use of sawdust relative to straw.

The suppression of immature musoid fly density in sawdust bedding can in part be attributed in part to the high carbon to nitrogen ratio of decomposing sawdust that inhibits microbial activity (Bollen & Glennie 1961). Because bacteria and the products of bacterial metabolism serve as food for immature stages of the house fly and stable fly, substrates that limit microbial activity, such as sand, gravel, corncob and sawdust, presumably limit their growth and development. The use of bedding unsuitable for immature muscoid flies represents a biologically rational and economically acceptable management strategy for controlling muscoid flies on dairy farms.

Spore forming bacteria

Using a blood agar assay system, Schmidtmann and Martin (1992) examined the effect of bacteria isolated from house fly rearing media (CSMA) on the growth and development of immature house flies. Pure cultures of bacteria representing gram-negative, gram-positive, coccoid and micrococcoid cell types each supported growth of larvae; *Escherichia coli*, used as a positive control, also supported larval growth. In contrast, virtually no growth occurred on blood agar in the absence of bacteria (= negative control). These findings further emphasize the importance of bacteria in the growth of immature house flies.

This study also reported inhibition of immature house fly growth in the presence of an isolate of *Bacillus cereus*. This culture, unlike a second strain of *B. cereus* that supported maggot growth, hemolyzed blood agar during colony growth. *Bacillus cereus* is recognized as a facultative pathogen that traumatizes the midgut of insects (Krieg 1987), and strains of *B. cereus*

Bedding types

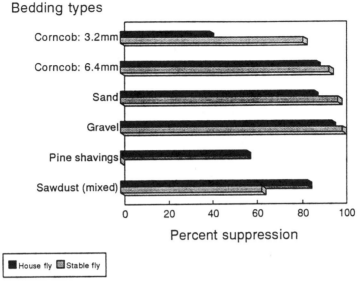

Fig. 1. Suppression of immature house and stable flies in experimental materials, as compared with straw, when used as bedding in calf hutches.

(Adapted from Schmidtmann et al. 1989, Schmidtmann 1991).

Bedding types

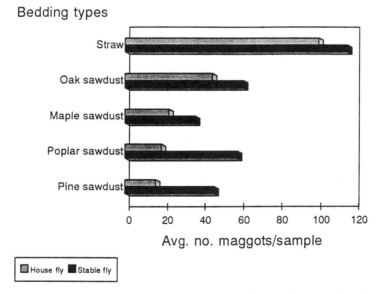

Fig. 2. Immature house and stable fly density in straw and several types of sawdust used as bedding in calf hutches.

are pathogenic for Lepidoptera and Hymenoptera (Heimpel and Angus 1963). Phospholipases and other extracellular enzymes produced by *B. cereus* may contribute to traumatization of the insect midgut (Heimpel 1955, Rahmet-Alla & Rowley, 1989). The hemolysis of blood agar by the strain of *B. cereus* lethal to immature stages of the house fly may reflect a similar effect on Diptera (Schmidtmann & Martin 1992).

In a follow-up study (unpublished data) we have characterized the abundance of spore-forming bacteria, *Bacillus* spp., associated with straw and sawdust bedding in calf hutches. Previous efforts to develop *Bacillus thuringiensis* for control of muscoid flies have focused largely on the beta-exotoxin described by Briggs (1960). Beta-exotoxin producing bacteria have been evaluated primarily as an oral feed-through larvicide against immature face flies, house flies and horn flies in cattle manure (Dunn 1960, Hower & Cheng 1968, Miller et al. 1971, Gingrich & Eschle 1971; Haufler & Kunz 1985). Although effective in killing muscoid fly larvae, commercial production of beta-exotoxin bacteria has not been aggressively pursued in the United States because of vertebrate teratogenicity (Sebesta et al. 1981). Scandinavian researchers have continued investigation of beta-exotoxin for control of insecticide resistant house flies (Holmberg et al. 1980, Jesperson & Keiding 1990).

The delta-endotoxin of *B. thuringiensis* has not been extensively tested against muscoid flies, although Singh et al. (1986) and Indraseth et al. (1992) report endotoxin activity in house flies. Mortality to purified crystals also has been reported for the closely related horn fly, *Hematobia irritans* (Temeyer 1990b). Most isolates of *B. thuringiensis* have originated from soil (Martin & Travers 1989), although isolates also have recently been recovered from the surface of plants (Smith & Couche 1991). The exploration of pest-specific habitats (e.g. livestock bedding) may therefore provide new germplasm, possible with activity for Diptera. It is further important, as a regulatory issue, to define the bacterial flora of a given habitat as a prerequisite for the release of a microbial pathogen. We therefore examined calf hutch bedding for spore-forming bacteria using the acetate-selection technique of Travers et al. (1987). We also conducted bioassays on selected isolates using the blood agar test system to assess activity against house fly larvae.

Samples of straw and four types of sawdust bedding were taken bi-weekly from inside and outside areas of calf hutches through the course of six-week periods. After acetate selection, selected colonies were transfered to a medium to induce sporulation. Isolates were then identified using a biochemical test procedure (Martin et al. 1985). Three species of *Bacillus* were identified. These isolates included 1,182 *B. megaterium*, 1,390 *B. sphaericus*, 4,623 *B. thuringiensis* and 3,783 unidentified bacilli. Of these, 292 isolates of *B. thuringiensis* were tested for activity against house fly maggots in the blood agar assay system. One hundred and one cultures demonstrated activity, expressed as mortality in house fly larvae (Table 1). These data illustrate the numerical abundance, as well as the wealth and variability of *Bacillus* spp. germplasm, that exist in calf hutch bedding.

The blood agar assay used to screen the *Bacillus* spp. isolates does not differentiate between mortality attributable to toxicity (e.g. exotoxin or endotoxin) versus bacteria-induced pathogenesis. To elucidate endotoxin activity, crystals were separated from vegetative cells and spores and then bioassayed. Ten *B. thuringiensis* isolates that demonstrated activity in previous bioassay were evaluated. The presence of parasporal bodies in each isolate was confirmed by microscopic examination after 72 - 96 h of growth. These cultures, tentatively identified as *B. thuringiensis israelensis*, *B. t. morrisoni* and several un-named isolates, were compared with *B. t. israelensis* HD522. The HD522 subtype is active against larval and pupal stages of the horn fly (Temeyer 1984, Temeyer 1990b). Respective cultures were grown in L-broth until cell lysis (72h) and centrifuged (Martin et al. 1985), and the supernatant discarded. The crystal fraction was enriched by froth flotation (Sharpe et al. 1981). Renografin gradients were used to separate the pellet into three fractions; crystals, spores and cellular debris (Milne et al. 1977, Martin et al. 1981). Each fraction was then bioassayed.

Immature house fly mortality in respective fractions was highly variable (Table 2). One *B. t. israelensis* isolate killed all immature house flies and was as toxic as the *B. t. israelensis*

Table 1. Number of *B. thuringiensis* isolates with activity against immature house flies

B. t. subtype	Isolates with activity	Total examined
morrisoni	5	50
israelensis	49	100
un-named	47	142

Table 2. Percent mortality in immature house flies exposed to *B. thuringiensis*

	Fraction		
B. t. subtype	Crystal	Spore	Debris
1 *morrisoni*	56	50	33
2 *israelensis*	100	100	100
3 *israelensis*	66	50	66
4 *israelensis*	33	50	66
5 *israelensis*	66	33	17
6 Unknown	100	66	84
7 Unknown	66	33	0
8 Unknown	12	17	17
9 Unknown	12	50	50
10 Unknown	12	50	50
HD522	100	100	84
Control	0	0	0

strain HD 522 standard. One un-named *B. thuringiensis* isolate also killed 100 percent of immature house flies; the other isolates, including *B. t. morrisoni,* showed activity, but were less lethal.

In characterizing the abundance of *B. thuringiensis* associated with hutch bedding, this study calls attention to the need for further developing the use of *B. thuringiensis* for controlling muscoid flies. To further this process we envision evaluating selected isolates by introducing preparations of bacteria or cell fractions into fly rearing medium seeded with immature house flies. Log-dose challenges will be used to further define and compare diptericidal activity in components of selected cultures. The potential for manipulation of selected isolates through modification of genetic components will be pursued contingent upon opportunities and funding. In an ecological context, possible increases in the density of *B. thuringiensis* through growth in house fly rearing medium or amplification in larval cadavers, a relationship postulated to explain the presence of *B. thuringiensis* in nature (Federici 1993), need investigation. It is anticipated that spore-forming bacteria can be developed, and used with bedding substrates that limit the density of muscoid fly larvae, as an integrated, biologically-based technology tailored to suppressing muscoid fly populations on dairy farms.

REFERENCES

Anderson, J. R. "Recent developments in the control of some arthropods of public health and veterinary importance. Muscoid flies." *Bull. Entomol. Soc. Am.,* **1966**, 12, 342-348.

Bollen, W. B.; Glennie, D.W. "Sawdust, bark and other wood wastes for soil conditioning and mulching." *Forest Prod. J.,* **1961**, 11, 38-46.

Briggs, J. D. "Reduction of adult house fly emergence by the effects of *Bacillus* spp. on the development of immature forms. *J. Insect. Pathol.,* **1960**, 2, 418-432.

Brookes, M. J.; Fraenkel, G. "The nutrition of the larva of the housefly, *Musca domestica* L." *Pysiol. Zool.,* **1958**. 31. 208-223.

Bruce, W. N.; Decker, G. C. "Fly control and milk flow." *J. Econ. Entomol.,* **1947**, 40, 530-536.

Dowding, V. M. "The function and ecological significance of the pharyngeal ridges occuring in the larvae of some cyclorrhaphous Diptera." *Parasitol.,* **1967**, 57, 371-388.

Federici, B. A. "Insecticidal bacterial proteins identify the midgut epithelium as a source of novel target sites for insect control." *Arch. Insect Biochem. Physiol.,* **1993**, 22, 357-372.

Gerberich, J. B. "Rearing house-flies on common bacteriological media." *J. Econ. Entomol.,* **1948**, 141, 125-126.

Gingrich, R. E.; Eschle, J. L. "Susceptibilty of immature horn flies to toxins of *Bacillus thuringiensis*." *J. Econ. Entomol.,* **1971**, 64, 1183-1188.

Glaser, R. W. "The relation of microorganisms to the development and longevity of flies." *Am. J. Trop. Med.,,* **1924**, 4, 85-107.

Greenberg, B. "A method for the sterile culture of housefly larvae, Musca domestica L." *Can. Entomol.,* **1954**, 86, 527-528.

Heimpel, A. M. "Investigation of the mode of action of strains of *Bacillus cereus* Fr. and Fr. pathogenic for the larch sawfly, *Pristiophora erichsonii* (Htg.)." *Can. J. Zool.,* **1955**, 33, 311- 326.

Heimpel, A. M.; Angus, T. A. "Diseases caused by certain sporeforming bacteria." In *Insect Pathology, an Advanced Treatise, Vol. 2.* Steinhaus, E. A., Ed., Academic Press: New York. 1963, pp. 21-73

Hollis, J.H,; Knapp, F. W.; Dawson, K. A. "Influence of bacteria within bovine feces on the development of the face fly (Diptera: Muscidae)." *Environ. Entomol.,* **1985**, 14, 568-571.

Hower, A. A.; Cheng, T. H. "Inhibitive effect of *Bacillus thuringiensis* on the development of the face fly in cow manure." *J. Econ. Entomol.,* **1968**, 61, 26-31.

Indraseth, L. S.; Suzuki, N.; Ogiwara, K.; Asano, S.; Hori, H. "Activated insecticidal crystal proteins from *Bacillus thuringiensis* serovars killed adult house flies." *Letter Appld. Microbiol.,* **1992**, 14, 174-177.

Jesperson, J. B.; Keiding, J. "The effect of *Bacillus thuringiensis var. thuringienis* on *Musca domestica* L. larvae resistant to insecticides." In Biocontrol of Arthropods Affecting Livestock and Poultry. Rutz, D. A.; Patterson, R. S. Ed., Westview Press: Boulder, CO. 1990, pp. 215-229.

Krieg, A. "Diseases caused by bacteria and other prokaryotes." *In Epizootiology of Insect Diseases.* Fuxa, J. R.; Tanada, Y. Ed., John Wiley and Sons: New York. 1987.

Levinson, Z. H. "Food of house fly larvae." *Nature*, **1960**, 188, 427.

Martin, P. A. W.; Dugan, R. R.; Tovinen, O. H. "Differentiation of acidophilic thiobacilli by cell density in renografin gradients." *Current Microbiol.*, **1981**, 6, 81-84.

Martin, P. A. W.; Haransky, E. B.; Travers, R. F.; Reichelderfer, C. F. "Rapid biochemical testing of large numbers of *Bacillus thuringiensis* isolates using agar dots." *BioTechniques*, **1985**, 3, 386-392.

Martin, P. A. W.; Travers, R. "Worldwide abundance and distribution of *Bacillus thuringiensis* isolates." *Appl. Environ. Microbiol.*, **1989**, 55, 2437-2442.

Miller, R. W.; Pickens, L. G.; Gordon, C. H. "Effect of *Bacillus thuringiensis* in cattle manure on house fly larvae." *J. Econ. Entomol.*, **1971**, 6, 902-903.

Miller, R. W. "The influence of dairy operations on the urban fly problem." In *Rural Flies in the Urban Enviroment.* Thomas, G. B.; Skoda, S. R. Ed.; Univ. Nebraska, Res. Bull. no. 317, **1993**, pp. 25-33.

Milne, R; Murphy, D; Fast, P. G. "*Bacillus thuringiensis* delta-endotoxin: an improved technique for separation of crystals from spores." *J. Invert. Pathol.* 1977, 29, 230-231.

Rahmet-Allah, M.; Rowley, A. F. "Studies on the pathogenicity of different strains of *Bacillus cereus* for the cockroach, *Leucophaea maderae.*" J. Invert. Pathol., **1989**, 53, 190-196.

Schmidtmann, E. T. "Exploitation of bedding in dairy outdoor calf hutches by immature house flies and stable flies (Diptera: Muscidae)." *J. Med. Entomol.*, **1988**, 25, 484-488.

Schmidtmann, E. T.; Miller, R. W.; Muller, R. "Effect of experimental bedding treatments on Muscidae the density of immature *Musca domestica* and *Stomoxys calcitrans* (Diptera: Muscidae) in outdoor calf hutches." *J. Econ. Entomol.*, **1989**, 82, 1134-1139.

Schmidtmann, E. T. "Suppressing immature house and stable flies in outdoor calf hutches with sand, gravel and sawdust bedding." *J. Dairy Sci.*, **1991**, 74, 3956-3960.

Schmidtmann, E. T.; Martin, P. A. W. "Relationship between selected bacteria and the growth of immature house flies, *Musca domestica,* in an axenic test system." *J. Med. Entomol.*, **1992**, 29, 232-235.

Sebesta, K.; Farkas, J; Horska, K; Vankova, K. "Thuringiensin, the beta-exotoxin of *Bacillus thuringiensis.*" In *Microbial Control of Pest and Plant Diseases 1970-1980*; Burges, B., Ed.; Academic Press: New York 1981, pp. 249-282.

Sharpe, E. S.; Herman, A. I.; Toolan, S. C. "Foam flotation process for separating *Bacillus thuringiensis* sporulation products." U.S. Patent document 4,247,644, **1981**.

Silverman, P. H.; Silverman, L. "Growth measurements on *Musca vicina* (Macq.) reared with a known bacterial flora." *Riv. Parassitol.*, **1953**, 14, 89-95.

Singh, G. J. P.; Schoest, L. P., Jr.; Gill, S. "Action of *Bacillus thuringiensis* subsp. *israelensis* endotoxin on the ultrastructure of the house fly larva neuromuscular system in vitro." *J. Invert. Pathol.* 1986, 47, 155-166.

Smith, R. A.; Couche, G. A. "The phylloplane as a source of *Bacillus thuringiensis* variants." *Appl. Environ. Microbiol.*, **1991**, 57, 311-315.

Spiller, D. "Nutrition and diet of muscoid flies." *Bull.Wld. Hlth. Org.*, **1964**, 31, 551-554.

Temeyer, K. B. "Larvicidal activity of *Bacillus thuringiensis* subsp. *israelensis* in the dipteran *Haematobia irritans.*" Appl. Environ. Microbiol., **1984**, 47, 952-955.

Temeyer, K. B. "Fecal supplementation with carbohydrate reduces surival of horn fly larvae - a cautionary note on bioassays and diet development." *Southwest. Entomol.*, **1990**a, 15, 447-452.

Temeyer, K. B. "Potential of *Bacillus thuringiensis* for fly control." In *The Vth International Colloquium on Invertebrate Pathology and Microbial Control - Proceedings and Abstracts -20-24 August, 1990 - Adelaide, Australia;* Pinnock, B. E., Ed.; Society for Invertebrate Pathology, **1990**b, pp. 352-356.

Thomsen, M,; Hammer, O. "The breeding media of some common flies." *Bull. Entomol. Res.*, **1936**, 27, 559-587.

Travers, R.; Martin, P. A. W.; Reichelderfer, C. F. "Selective processs for efficient isolation of soil *Bacillus spp.*" *Appl. Environ. Microbiol.*, **1987**, 53, 1263-1266.

New Options For Insect Control Using Fungi

Ann E. Hajek, Boyce Thompson Institute for Plant Research, Tower Road, Ithaca, NY 14853-1801

Epizootics in insect populations provide conspicuous displays of the control potential of fungal pathogens. Experience has demonstrated that simple inundative fungal releases do not always provide control; therefore, innovative methods to promote infection are being developed. Efforts are still focussed primarily on augmentative releases of fungi although classical biological control and environmental manipulation can be very successful. Innovations toward improving control include characterizing fungal strain diversity and exploring the potential for manipulation of pathogenicity, evaluating system specific methods for use of fungi for control, and gaining an understanding of the biotic and abiotic interactions resulting in epizootic development. Entomopathogenic fungi being developed for control are not infective or toxic to vertebrates.

Fungal infections of insects can be very noticeable and abundant; it is most certainly for this reason that fungi were the first recorded pathogens of insects. In fact, dramatic fungal epizootics provide a constant reminder of the potential of these pathogens for insect control. Fungi causing epizootics are known from a diversity of hosts in a wide range of habitats, including fresh water, soil, and aerial sites. Fungi are unique among insect pathogens because they infect by penetrating the external cuticle instead of the gut. For this reason, fungi are particularly important for control of sucking insects. In addition, many fungal pathogens infect Coleoptera, a group with few known bacterial and viral pathogens.

A great diversity of fungal species have evolved the ability to infect insects. Based on the occurrence of entomopathogenicity throughout the fungi, the pathogenic mode of life has most probably evolved independently many times. Generalities about fungal entomopathogens are difficult to make because basic life cycle strategies vary among species. Hyphomycetes are not obligate pathogens in nature and many have broad host ranges. By contrast, another major taxonomic group of entomopathogens, the Entomophthorales, are naturally obligate pathogens of generally limited host range, although many can be grown *in vitro* to some extent.

Of the over 700 fungal species known to be pathogenic to insects, only 8 species have been, or are presently being, commercially developed for insect control (Table 1). Our knowledge of fungal entomopathogens, from the biochemical/ molecular level to the level of epizootiology, is based on relatively few systems and many of the fungal species known to cause epizootics have not been investigated for control purposes. In addition, those fungi that do not cause epizootics in nature should not be ignored as potential control agents. This generality is based on the bacterial insect pathogen, *Bacillus thuringiensis*, which is not known to cause epizootics in nature, but is the entomopathogen most extensively used for control.

Insect control using fungi was first attempted in 1888 by mass producing and applying *Metarhizium anisopliae* for control of the sugar beet weevil, *Cleonus punctiventris* (Tanada & Kaya 1993). This type of augmentative strategy for fungal release is still used today for many host/pathogen systems. However, research has determined that simplified versions of this type of release strategy (augmentation) are not always efficacious (e.g., Hajek et al. 1987; Wilding et al. 1986a), especially in systems with refractory pests. As a response, innovative approaches to use of fungal pathogens have met with success. At present, entomopathogenic fungi are being registered for application for at least 4 new host/pathogen systems (Keller, 1992a; Kerwin, 1992; Rath, 1992; J. C. Lord, pers. comm.). This paper will specifically focus on recent trends in development of fungal pathogens for control, including safety. The breadth of control strategies being used or developed in this growing field will be discussed as well as major concerns in developing these agents for control. A number of recent reviews provide perspective on control using entomopathogenic fungi (e.g., Fuxa, 1987; Wraight & Roberts, 1987; McCoy et al., 1988; Samson et al., 1988; Roberts, 1989; Roberts & Hajek, 1992).

Strategies for Control

Standard strategies for manipulation of fungal entomopathogens include permanent introduction (classical biological control), environmental manipulation (conservation), and augmentative release (covering a continuum from inoculative to inundative releases). Fungi are ubiquitous so these latter releases are generally considered as augmenting naturally occurring populations.

Permanent Introduction. This control strategy has historically been used against introduced insects. Primary emphasis has focussed on introduction of predators and parasitoids and fungal pathogens have been successfully introduced in only 20 host/pathogen systems in 27 locations (Roberts & Hajek, 1992; T. Poprawski, pers. comm.). Due to the introduced status of many pest insects and the relative ease and low cost of classical biological control programs, this control strategy holds great potential for further application with fungal entomopathogens.

As with parasites and predators, introducing biotypes of pathogens adapted to the climate of the area of introduction can be crucial to successful establishment. To control spotted alfalfa aphid populations introduced to Australia, numerous strains of *Zoophthora radicans* were evaluated to find a climatically adapted isolate for introduction (Milner et al., 1982). The Israeli strain chosen for introduction became successfully established, spread across alfalfa-growing regions, and now is part of a suite of natural enemies holding alfalfa aphid populations in check (Carruthers & Hural, 1990).

In 1989-1990, epizootics of a Japanese fungal pathogen, *Entomophaga maimaiga*, infecting introduced gypsy moths occurred in 10 northeastern U.S. states. This fungus may originally have been successfully introduced to North America during biological control introductions to the Boston area in 1910-1911; however, circumstantial evidence also supports a hypothesis of more recent accidental introduction (AEH, unpubl. data). During 1989-1990, this pathogen did not occur in areas more recently colonized by spreading gypsy moth populations (Hajek et al., 1990; Elkinton

Table 1. Fungal pathogens produced for insect control or in the process of registration

Country of Production	Fungal Pathogen	Insect Hosts
USSR	*Aschersonia aleyrodis*	Whiteflies
USA, PRC, USSR	*Beauveria bassiana*	Caterpillars, beetles, grasshoppers
Switzerland	*Beauveria brongniartii*	Scarab grubs
USA	*Hirsutella thompsonii*	Planthoppers, mites
USA	*Lagenidium giganteum*	Mosquitoes
Brazil, USA, Tasmania	*Metarhizium anisopliae*	Spittlebugs, cockroaches, scarab grubs
Philippines	*Paecilomyces lilacinus*	Planthoppers
Netherlands, Denmark	*Verticillium lecanii*	Whiteflies, aphids, thrips

et al. 1991). Small scale studies suggested that introduction of *E. maimaiga* was most efficient when overwintering resting spores were released (Hajek & Roberts, 1991). This poorly understood spore stage has been considered for control use in few entomophthoralean systems. *E. maimaiga* resting spores released at 39 of 41 locations became established and epizootics developed at many release sites (AEH & J.S. Elkinton, unpubl. data).

More recently, some permanent introduction programs have been based on the "new associations" theory, which suggests that natural enemies that do not evolve with hosts will exert greater impacts (Hokkanen & Pimentel, 1984). Using this concept, classical biological control is not limited to introduced insects but includes foreign introductions against endemic pests. However, releases of an Australian member of the *Entomophaga grylli* species complex to control North American grasshoppers have met with controversy. Public concern over the potential impact of this *E. grylli* strain on non-target species has led to extensive debate regarding the cost of potential risks and the future regulation of such releases (Lockwood, 1992; Carruthers & Onsager, 1993). This issue must be fully evaluated if permanent establishment is to continue being used for insect control.

At present, microbial agents to be introduced for classical biological control (therefore, establishment) are regulated as genetically engineered microorganisms. For further use of this valuable control strategy, separate regulatory guidelines specific to the introduction of pathogenic microorganisms for permanent establishment must be developed and implemented instead (Maddox, 1991).

Augmentative Releases. Augmentative releases of fungal pathogens generally are based on the ability of the fungal pathogen to repeatedly cycle through the host population after introduction. The degree and speed with which fungi are able to increase in prevalence as well as the damage thresholds for the crop determine whether releases are inoculative or inundative, the latter being frequently described as use of mycoinsectides. Among the major strategies for control, augmentative release has been adopted most extensively, especially among the Hyphomycetes. Hyphomycetes are the fungi most readily exploited for augmentative releases because of the stability

of spores and the relative ease with which they can be grown in large quantities.

The largest programs using fungi for insect control are found in China and Brazil in systems where some insect damage can be tolerated. In Brazil, *M. anisopliae* is grown by small companies or grower cooperatives for control of spittlebugs on sugarcane and pasture lands (Moscardi, 1989). For $\geq$ 10 years, extensive applications have resulted in increased sugar content in sugarcane without harming parasitoids attacking lepidopterous pests. In China, *B. bassiana* has been applied to control pine moth larvae on $\geq$ 1 million hectares (Xu, 1988). This fungus is produced locally on inexpensive substrates and applications are only necessary approximately every 3 years.

Successful augmentative releases of unformulated fungi have generally been limited to confined, moist habitats, such as greenhouses, where high humidities can be maintained (e.g., Ravensberg, 1990). Spores of many fungi cannot survive extended exposure to UV radiation, high temperatures, or desiccation. Therefore, formulation of fungal spores with protectants can be critical to spore survival. Recent investigations of oil-based emulsions for formulation of Hyphomycetes have yielded excellent control of grasshoppers (C. Prior, pers. comm.). Certain oils cause spores to stick to target insects, allow germination in a dry atmosphere, protect spores from UV, as well as causing some mortality when applied alone. In addition, spores of *M. anisopliae* stored in oil can survive for 3 months. Alginates have also been investigated for spore protection from UV and enhanced spore survival during storage (e.g., Pereira & Roberts, 1991).

One innovative approach for augmentative releases has been to release both asexual and sexual (long-lived) spores of *Lagenidium giganteum* for mosquito control (Kerwin & Washino, 1988). Initial control of several mosquitoes was high due to infections initiated by zoospores, but the pathogen also continued to cycle through the mosquito populations, presumably due, in part, to oospore germination. Development of epizootics in populations of floodwater mosquitoes occupying transient habitats, suggests that the environmentally resistant oospores can germinate synchronously under favorable conditions.

Fungi formulated with baits have been successfully applied in several host systems. Spraying fungi on ants in the field has been largely unsuccessful, although preliminary studies now demonstrate that ants will take fungus into colonies when it is mixed with a bait (Blowers, 1992). Baits containing *M. anisopliae* have also been successful in control of scarab grubs (D. E. Pinnock, pers. comm.). In particular, a bait containing a strain of *M. anisopliae* active at cold temperatures shows promise for control of grubs in Tasmanian pastures (Rath et al., 1990).

In Switzerland, *Beauveria brongniartii* grown on barley kernels is drilled into the ground for control of European chafer grubs (Keller, 1992b). This insect is a difficult pest to control because long-lived larvae in the soil feed on roots and adults are present only once every three years. Fungal treatments resulted in control levels superior to chemical insecticides. A second method of fungal application was evaluated for comparison. *B. brongniartii* was applied to adults swarming at the borders of forests that subsequently spread spores to the field while laying eggs. Although this second application method was successful, *B. brongniartii* applied in this way acted more slowly and may have been more dependent on host density.

Conservation. This control strategy has seldom been utilized, possibly because the epizootiology of host/pathogen systems is too poorly understood to determine likely applications. However, the sensitivity of fungi to environmental conditions suggests that system-specific manipulation of the environment could increase infection levels.

A successful fungus conservation strategy was applied to alfalfa weevils as a result of simulation modeling. Simulation models followed by field trials were used to evaluate the influence of cultural practices on prevalence of *Erynia* sp. infections of alfalfa weevils (Brown, 1987). When timing of alfalfa harvest was delayed and alfalfa remained in fields, the high densities of weevils within humid windrows of alfalfa allowed increased disease transmission. In conjunction, pesticide usage was altered so as to have minimal impact on *Erynia* sp.

In another attempt to promote infection by raising humidity, Wilding et al. (1986b) irrigated crops. This manipulation yielded increased infection of 2 species of fungi

infecting pea aphid, but the prevalence of 2 other fungal pathogens was not affected. These latter 2 fungi may not be as sensitive to humidity; more detailed knowledge of their epizootiology would help to determine critical factors necessary for development of epizootics of these species.

Use of pesticides can impact fungal pathogens by altering host density as well as survival and development of fungal pathogens. A multitude of tests have been conducted to evaluate pesticide/fungus interactions for many different fungal species and pesticides, with emphasis on fungicides (McCoy et al., 1988). For many fungal entomopathogens, fungicides can be inhibitory, although each pathogen differs in specific sensitivities. As in the alfalfa weevil/*Erynia* sp. system, knowledge of interactions can be used to minimize impact of pesticide treatments on fungal entomopathogens.

Safety of Entomopathogenic Fungi

Development of entomopathogenic fungi for control of insects is only feasible if these materials are safe. "Safe" has historically been defined as lacking any effect on mammals, especially humans, and domesticated and beneficial animals. However, due to growing concerns about our environment, safety is now being defined more broadly to include effects on all animals and plants with a view of resulting ecosystem-level effects. In fact, regulations based on safety considerations are presently impacting classical biological control programs and development of mycoinsecticides, as well as genetic engineering of fungi.

Effects on Vertebrates. Extensive safety testing regarding potential effects of entomopathogenic fungi on humans has been conducted with *M. anisopliae* and *L. giganteum* (Siegel & Shadduck, 1990). Based on studies with laboratory animals, neither species is considered either infective or toxic to humans.

A variety of studies exposing different vertebrates to entomopathogens have demonstrated that *M. anisopliae, Culicinomyces* spp., *L. giganteum, Entomophthora* spp., *Nomuraea rileyi, Paecilomyces* spp., and *Hirsutella thompsonii* have no deleterious effects (Saik et al. 1990). Due to conflicting reports of effects caused by *Beauveria* spp., additional testing is necessary with this species.

However, it is doubtful whether *Beauveria* can have systemic effects because it cannot grow at the mammalian body temperature, a characteristic of several other entomopathogens. The only entomopathogens that are not being considered for development due to safety considerations are *Conidiobolus coronatus* and *Aspergillus* spp. *C. coronatus* can cause mammalian infections while *Aspergillus* spp. produce toxins deleterious to mammals.

Fungi can also exert more subtle effects with prolonged exposure of individuals working in fungal production or application, including allergic or dermatological symptoms. *B. bassiana* has been cited occasionally causing moderate to severe allergic reactions with repeated exposure. *Metarhizium* and *Paecilomyces* have also been listed as associated with allergic responses in humans (Latgé & Paris, 1991).

Effects on Invertebrates. In general, safety research has focussed on the effects of entomopathogenic fungi on beneficial insects, including honeybees, pollinators, silkworms, and insect predators and parasitoids. In general, pathogens with broad host ranges (frequently Hyphomycete species) have a greater impact on non-target hosts than more specific pathogens, e.g., many Entomophthorales (Goettel et al., 1990). Entomopathogenic fungi are known to be able to infect insect predators; parasitoids within hosts are generally not infected by fungi although larvae of endoparasitoids can compete with fungi for host tissue. In many parasitoids, early instar parasitoids will not survive in infected larvae, while late instar parasitoids can complete development (see Goettel et al., 1990).

The ecosystem level impacts of fungal entomopathogens have seldom been addressed. Entomopathogens may have indirect effects on ecosystems, e.g., high levels of fungal infection could impact densities of non-susceptible predators and parasitoids due to lack of hosts. However, direct effects of fungi on non-target hosts would be more immediate and are simpler to evaluate. Effects of *B. brongniartii* on non-target organisms have been documented when this fungus was applied against scarab adults aggregating at the forest edge (Baltensweiler & Cerutti, 1986). An overall infection rate of 1.1% was reported among non-target invertebrates, although 9.0% of spiders were infected. We are interested in the

impact on non-target invertebrates as the gypsy moth pathogen *E. maimaiga* spreads throughout the gypsy moth distribution (see above). *E. maimaiga* is specific to Lepidoptera (Soper et al., 1988). During 1992, laboratory bioassays were conducted using native Lepidoptera from northeastern West Virginia where *E. maimaiga* is not yet established. Under laboratory conditions, very low levels of infection were found in only some of the species of Noctuidae, Arctiidae, Notodontidae, Geometridae, Lasiocampidae, and Saturniidae that were tested; only species in the Lymantriidae consistently demonstrated significant levels of infection. However, infection in the laboratory does not always correspond to infection levels found in the field. Our studies will be continued to evaluate the impact of *E. maimaiga* on non-target Lepidoptera using field-based experimentation as well as population sampling.

Areas for Future Development

Several aspects of the use of fungi for insect control deserve discussion because they hold promise for future advancement (improving fungal pathogens and methods for utilizing them) or because increased knowledge in these areas would accelerate progress (fungal formulation/production and understandings of the epizootiology of diseases).

Selecting and Improving Fungal Pathogens. The potential use of many species of fungal entomopathogens for control has not been evaluated. Among those species under investigation, characteristics determining pathogenicity and ecological adaptation show considerable intraspecific variation. This strain diversity is being described at the biochemical level within some cosmopolitan species such as *B. bassiana* and *M. anisopliae* (St. Leger et al., 1992a, c) as well as within species complexes (Hajek et al., 1990). In conjunction, biological and ecological studies will be necessary to correlate genetic diversity and fungal strain attributes.

With the advent of increasing knowledge of fungal pathogens at the molecular level, the prospect arises that determinants of pathogenicity can be improved. To this end, the genes encoding proteases of *M. anisopliae* that are used for cuticular penetration are being characterized (St. Leger et al., 1992b). Genetic transformation has already been accomplished using *M. anisopliae*; a gene for benomyl resistance from another fungus was inserted into the *M. anisopliae* genome (Goettel et al., 1989).

Improving Strategies for Use of Fungal Pathogens. Development of mycoinsecticides has been emphasized for control using entomopathogenic fungi, while both classical biological control and conservation strategies are seldom utilized. However, in most systems mycoinsectides do not provide adequate control. It has become evident that specific circumstances in each host/pathogen system must be understood and considered in developing fungi for control. In many instances, this system-specific attention results in integration of several different control techniques. For example, parasitoids, predatory mites, and *Verticillium lecanii* can be successfully applied together to control greenhouse pests (see Goettel et al., 1990). In China, control techniques are used in mushroom hothouses where cadavers of gnats dying from *Erynia ithacensis* infections are collected and distributed (Huang et al., 1992). Areas with cadavers are sprayed daily with water to create conditions for epizootic development.

Delivery of fungi is being improved through development of a trap for cockroaches that delivers doses of *M. anisopliae*. Pheromone traps for diamondback moth have been adapted to transfer spores of *Zoophthora radicans* to adults. Adults then fly to fields where they deliver the fungus to larval populations (Pell et al., 1993).

In one fascinating biological system, *B. bassiana* applied to corn plants can become endophytic. Thereafter, European corn borer larvae infesting plants become infected (e.g., Bing & Lewis, 1991). This novel interaction is thus far known only between this specific host and *B. bassiana*, a species of fungus that readily grows saprophytically.

A recurring subject has been the *in vitro* production by some entomopathogenic fungi of low molecular weight secondary metabolites. These toxins can interfere with insect physiology but in most instances have not been identified from moribund insects (Gillespie & Claydon, 1989). In the future, these toxins may be integrated in control programs through fungal transformation to heighten fungal aggressiveness by increased toxin production.

Fungal Production. For inoculative augmentative releases and some classical biological control programs, fungi must be produced *in vitro*; mass production is required for inundative releases of Hyphomycetes as mycoinsecticides. Reviews discussing fungal production (e.g., McCoy et al., 1988; Samson et al., 1988; Roberts, 1989) document industrial level mass production, but in cases where entomopathogenic fungi are part of established control systems, e.g., China, Brazil, production is accomplished by a labor-intensive "cottage industry."

Although many species of Entomophthorales naturally cause dramatic epizootics in nature, production of this group *in vitro* has been more difficult. For some species, mycelium can be grown and then carefully dried. This dry mycelium applied in the field subsequently rehydrates and produces spores. Entomophthoralean fungi grown and applied in this way have caused subsequent epizootics (Wraight et al., 1986).

Constant concerns regarding fungal production include quality control and maintenance of virulence. For mycoinsecticides, continuing improvements in formulation are critical to survival of fungi during storage and after application. Recent successes with oil formulations provide an exciting new avenue for mycoinsectide development (see above).

Epizootiology of Fungal Diseases. Improving our understanding of disease epizootiology is crucial to increased use of fungal entomopathogens for control. Detailed study and application of simulation models can help to determine those stages in disease transmission that can potentially be manipulated to increase infection levels (e.g., see Brown, 1987 described above). A simulation model of the *E. maimaiga*/gypsy moth system helped to evaluate timing of gypsy moth larval infection due to resting spores (AEH & T.S. Larkin, unpubl. data).

For classical biological control, studies of disease epizootiology can aid in evaluation of failures in fungal establishment, so that future efforts will succeed.

References

Baltensweiler, W.; Cerutti, F. "A study of the possible side effects of using the fungus

Beauveria brongniartii to control the May beetle on the fauna of the forest edge." *Mitt. Schweiz. Entomol. Ges.*, **1986**, *59*, 267-274.

Bing, L.A.; Lewis, L.C. "Suppression of *Ostrinia nubilalis* (Hübner) (Lepidoptera: Pyralidae) by endophytic *Beauveria bassiana* (Balsamo) Vuillemin." *Environ. Entomol.*, **1991**, *20*, 1207-11.

Blowers, M.; Jackson, C.W.; Knapp, J.J. Effect of composition of alginate granules on their potential as carriers of microbial control agents against the leaf-cutting ant *Atta sexdens*. In *Biology and Evolution of Social Insects*; Billen, J., Ed.; Leuven Univ. Press: Belgium, **1992**, pp 145-151.

Brown, G.C. Modeling, In *Epizootiology of Insect Diseases*; Fuxa, J.R.; Tanada, Y., Eds.; Wiley: New York, **1987**, pp 43-68.

Carruthers, R.I.; Hural, K. Fungi as naturally occurring entomopathogens, In *New Directions in Biological Control: Alternatives for Suppressing Agricultural Pests and Diseases*; Baker, R.R.; Dunn, P.E., Eds.; Liss: New York, **1990**, pp 115-138.

Carruthers, R.I.; Onsager, J.A. "A perspective on the use of exotic natural enemies for biological control of pest grasshoppers." *Environ. Entomol.*, **1993**, (In press).

Elkinton, J.S.; Hajek, A.E.; Boettner, G.H.; Simons, E.E. "Distribution and apparent spread of *Entomophaga maimaiga* (Zygomycetes: Entomophthorales) in gypsy moth (Lepidoptera: Lymantriidae) populations in North America." *Environ. Entomol.*, **1991**, *20*, 1601-1605.

Fuxa, J.R. "Ecological considerations for the use of entomopathogens in IPM." *Ann. Rev. Entomol.*, **1987**, *32*, 225-251.

Gillespie, A.T.; Claydon, N. "The use of entomogenous fungi for pest control and the role of toxins in pathogenesis." *Pestic. Sci.*, **1989**, *27*, 203-215.

Goettel, M.S.; Poprawski, T.J.; Vandenberg, J.D.; Li, Z.; Roberts, D.W. Safety to nontarget invertebrates of fungal biocontrol agents. In *Safety of Microbial Insecticides*; Laird, M.; Lacey, L.A.; Davidson, E.W., Eds.; CRC: Boca Raton, **1990**, pp 209-231.

Goettel, M.S.; St. Leger, R.J.; Bhairi, S.; Roberts, D.W.; Staples, R.C. "Transformation of the entomopathogenic fungus *Metarhizium anisopliae* using the benA3 gene from *Aspergillus nidulans*." *Curr. Genet.*, **1989**, *17*, 129-132.

Hajek, A.E.; Humber, R.A.; Elkinton, J.S.;

May, B.; Walsh, S.R.A.; Silver, J.C. "Allozyme and RFLP analyses confirm *Entomophaga maimaiga* responsible for 1989 epizootics in North American gypsy moth populations." *Proc. Natl. Acad. Sci. USA*, **1990**, *87*, 6979-82.

Hajek, A.E.; Roberts, D.W. "Pathogen reservoirs as a biological control resource: Introduction of *Entomophaga maimaiga* to North American gypsy moth, *Lymantria dispar*, populations." *Biol. Contr.*, **1991**, *1*, 29-34.

Hajek, A.E.; Soper, R.S.; Roberts, D.W.; Anderson, T.E.; Biever, K.D.; Ferro, D.N.; LeBrun, R.A.; Storch, R.H. "Foliar application of *Beauveria bassiana* (Balsamo) Vuillemin for control of the Colorado potato beetle, *Leptinotarsa decemlineata* (Say) (Coleoptera: Chrysomelidae): An overview of pilot test results from the northern United States." *Can. Entomol.*, **1987**, *119*, 959-974.

Hokkanen, H.; Pimentel, D. "New approach for selecting biological control agents." *Can. Entomol.*, **1984**, *116*, 1109-1121.

Huang, Y.; Zhen, B.; Li, Z. "Natural and induced epizootics of *Erynia ithacensis* in mushroom hothouse populations of yellow-legged fungus gnats." *J. Invertebr. Pathol.*, **1992**, *60*, 254-258.

Keller, S. "Control of *Melolontha melolontha* with *Beauveria brongniartii*: Comparison of two methods." *XXV Ann. Mtg. Soc. Invertebr. Pathol.*, **1992a**, p. 175 (Abstr.).

Keller, S. The *Beauveria-Melolontha* project: experiences with regard to locust and grasshopper control. In *Biological Control of Locusts and Grasshoppers*; Lomer, C.J.; Prior, C., Eds.; CAB Internat.: Wallingford, UK, **1992b**, pp 279-86.

Kerwin, J.L. "EPA registers *Lagenidium giganteum* for mosquito control." *Soc. Invertebr. Pathol. Newsl.*, **1992**, *24*, 8-9.

Kerwin, J.L.; Washino, R.K. "Field evaluation of *Lagenidium giganteum* (Oomycetes: Lagenidiales) and description of a natural epizootic involving a new isolate of the fungus." *J. Med. Entomol.*, **1988**, *25*, 452-460.

Latgé, J.-P.; Paris, S. The fungal spore: Reservoir of allergens. In *The Fungal Spore and Disease Initiation in Plants and Animals*; Cole, G.T.; Hoch, H.C., Ed.; Plenum: New York, **1991**, pp 379-401.

Lockwood, J.A. "Neoclassical biological control: a double-edged sword." *Soc. Invertebr. Pathol. Newsl.*, **1992**, *24*, 6-8.

Maddox, J.V. The effects of regulations on the use of insect pathogens as biological control agents. In *Regulations and Guidelines: Critical Issues in Biological Control*; Inst. Food & Agric., Univ. Florida, **1991**.

McCoy, C.W.; Samson, R.A.; Boucias, D.G. Entomogenous fungi. In *CRC Handbook of Natural Pesticides, Vol. 5. Part A, Microbial Insecticides*; C.M. Ignoffo, Ed.; CRC: Boca Raton, **1988**, pp 151-236.

Milner, R.J.; Soper, R.S.; Lutton, G.G. "Field release of an Australian strain of the fungus *Zoophthora radicans* (Brefeld) Batko for biological control of *Therioaphis trifolii* (Monell) f. *maculata*." *J. Austral. Entomol. Soc.*, **1982**, *21*, 113-118.

Moscardi, F. Production and use of entomopathogens in Brazil. In *Proc. Conf. Biotechnology, Biological Pesticides and Novel Plant-Pest Resistance for Insect Pest Management*, Boyce Thompson Inst., NY, **1989**, pp 53-60.

Pell, J.K.; Macaulay, E.D.M.; Wilding, N. "A pheromone trap for dispersal of the pathogen *Zoophthora radicans* Brefeld (Zygomycetes: Entomophthorales) amongst populations of the diamondback moth, *Plutella xylostella* L. (Lepidoptera: Yponomeutidae)." *Biocontrol Sci. Technol.*, **1993**, (In press).

Pereira, R.M.; Roberts, D.W. "Alginate and cornstarch mycelial formulations of entomopathogenic fungi, *Beauveria bassiana* and *Metarhizium anisopliae*." *J. Econ. Entomol.*, **1991**, 84, 1657-1661.

Rath, A.C., "The long-term control of the scarab, *Adoryphorus couloni*, with *Metarhizium anisopliae*." XXV *Ann. Mtg. Soc. Invertebr. Pathol.*, **1992**, p 180 (Abstr.).

Rath, A.C.; Koen, T.B.; Worladge, D.; Anderson, G.C. "Control of the subterranean pasture pest *Adoryphorus couloni* (Coleoptera: Scarabaeidae) with *Metarhizium anisopliae* isolate DAT F-001." *Proc. Vth Internat. Colloq. Invertebr. Pathol.*, **1990**, p 35 (Abstr.)

Ravensberg, W.J.; Malais, M.; Van der Schaaf, D.A. "*Verticillium lecanii* as a microbial insecticide against glasshouse whitefly." *Brighton Crop Protection Conf. Pests & Dis.*, **1990**, pp 265-268.

Roberts, D.W. "World picture of biological control of insects by fungi." *Mem. Inst. Oswaldo Cruz, Rio de Janeiro*, **1989**, *84*, Suppl. III, 89-100.

Roberts, D.W.; Hajek, A.E. Entomopathogenic fungi as bioinsecticides. In *Frontiers in Industrial Mycology*; Leatham, G.F., Ed.; Chapman & Hall: New York, **1992**, pp 144-159.

Saik, J.E.; Lacey, L.A.; Lacey, C.M. Safety of microbial insecticides to vertebrates--Domestic animals and wildlife. In *Safety of Microbial Insecticides*; Laird, M.; Lacey, L.A.; Davidson, E.W., Eds.; CRC: Boca Raton, **1990**, pp 115-132.

St. Leger, R.J.; Allee, L.L.; May, B.; Staples, R.C.; Roberts, D.W. "World-wide distribution of genetic variation among isolates of *Beauveria* spp." *Mycol. Res.*, **1992a**, *96*, 1007-1015.

St. Leger, R.J.; Frank, D.C.; Roberts, D.W.; Staples, R.C. "Molecular cloning and regulatory analysis of the cuticle-degrading protease structural gene for the entomopathogenic fungus *Metarhizium anisopliae*." *Eur. J. Biochem.*, **1992b**, *204*, 991-1001.

St. Leger, R.J.; May, B.; Allee, L.L.; Frank, D.C.; Roberts, D.W. "Genetic differences in allozymes and in formation of infection structures among isolates of the entomopathogenic fungus, *Metarhizium anisopliae*." *J. Invertebr. Pathol.*, **1992c**, *60*, 89-101.

Samson, R.A.; Evans, H.C.; Latgé, J.-P. *Atlas of Entomopathogenic Fungi*. Springer-Verlag, Berlin, 1988.

Siegel, J.P.; Shadduck, J.A. Safety of microbial insecticides to vertebrates--Humans. In *Safety of Microbial Insecticides*; Laird, M.; Lacey, L.A.; Davidson, E.W., Eds.; CRC: Boca Raton, **1990**, pp 101-113.

Soper, R.S.; Shimazu, M.; Humber, R.A.; Ramos, M.E.; Hajek, A.E. "Isolation and characterization of *Entomophaga maimaiga* sp. nov., a fungal pathogen of gypsy moth, *Lymantria dispar*, from Japan." *J. Invertebr. Pathol.*, **1988**, *51*, 229-241.

Tanada, Y.; Kaya, H.K. *Insect Pathology*, Academic Press, San Diego, CA, 1993.

Wilding, N.; Latteur, G.; Dedryver, C.A. Evaluation of Entomophthorales for aphid control: Laboratory and field data. In *Fundamental and Applied Aspects of Invertebrate Pathology*; Samson, R.A.; Vlak, J.M.; Peters, D., Eds.; Found. 4th Intern. Colloq. Invertebr. Pathol., **1986a**, pp 159-162.

Wilding, N.; Mardell, S.K.; Brobyn, P.J. "Introducing *Erynia neoaphidis* into a field population of *Aphis fabae*: form of the inoculum and effect of irrigation." *Ann. Appl. Biol.*, **1986b**, *108*, 373-385.

Wraight, S.P.; Galaini-Wraight, S.; Carruthers, R.I.; Roberts, D.W. Field transmission of *Erynia radicans* to *Empoasca* leafhoppers in alfalfa following application of a dry, mycelial preparation. In *Fundamental and Applied Aspects of Invertebrate Pathology*, Samson, R.A.; Vlak, J.M.; Peters, D., Eds.; Found. 4th Intern. Colloq. Invertebr. Pathol., **1986**, p 233.

Wraight, S.P.; Roberts, D.W. "Insect control efforts with fungi." *Dev. Indus. Microbiol.*, **1987**, *28*, 77-87.

Xu, Q. Some problems about study and application of *Beauveria bassiana* against agricultural and forest pests in China. In *Study and Application of Entomogenous Fungi in China*, Vol. 1, Academic Periodical Press, Beijing, PRC, **1988**, pp. 1-9.

The Past, Present and Future of Insect Parasitic Rhabditids for the Control of Arthropod Pests

George O. Poinar, Jr., Department of Entomological Sciences, University of California, Berkeley, CA 94720

Since the discovery of the first steinernematid in 1929, and the feasibility of commercially producing steinernematids and heterorhabditids for the biological control of soil insects, this discipline has grown explosively within the last decade. There still awaits a period of testing the many species and strains that are now available throughout the world in order to obtain their biological and physical parameters. With this information programmed on a computer database, strains can be selected which fit a particular host niche as well as specific field conditions.

The following discussion will be restricted to two families of insect parasitic nematodes, the Steinernematidae and the Heterorhabditidae since the great majority of all present cases of insect control by nematodes is involved with representatives of these two families. This account will be separated into sections covering the past, present and future.

The Past

Just when the insect-parasitic rhabditids evolved from their free-living, microbotrophic ancestors is not known, although a mid-Paleozoic origin for both Steinernematidae and Heterorhabditidae has been postulated (Poinar, 1983). Evidence has been presented indicating that these two families of nematodes evolved independently from one another and that their similar features (life cycles, acquisition of bacterial symbiotes, infective stage morphology) are a result of evolutionary convergence. Indeed, it is likely that *Heterorhabditis* evolved from representatives of the genus *Pellioditis* in a completely separate habitat from *Steinernema* which evolved from a *Rhabditonema* type of ancestor (Poinar, 1993).

Nematode fossils are rare and since the nature of the infective process with heterorhabditids and steinernematids results in the rapid breakdown of the host tissue, chances of discovering nematode infected cadavers as fossils are slight. We do know from fossils that other groups of insect parasitic nematodes (Mermithidae, Allantonematidae and Iotonchiidae) existed throughout the mid-Tertiary, and mermithids even in the Early Cretaceous (some 125 million years ago).

The first described *Steinernema* was *S. glaseri* (Steiner, 1929) which was discovered infecting the then newly introduced Japanese beetle (*Popillia japonica*) in New Jersey. The pioneering studies of R. W. Glaser on this nematode have already been discussed (Stoll, 1948; Poinar, 1992A). Essentially, Glaser's group were the first to culture *S. glaseri* on artificial medium (veal infusion dextrose agar covered with baker's yeast) and obtain sufficient numbers for the first field release of a rhabditid nematode against an insect pest (in this case, the Japanese beetle) (Fig. 1). Glaser's findings prompted others to search for similar nematodes and the second steinernematid was discovered in Europe in 1934 by I. Filipjev and named *S. feltiae*. Both of these steinernematid species play an important role today in the biological control of insect pests.

The genus *Heterorhabditis* was established in 1976 with the species *H. bacteriophora* Poinar from noctuid larvae in Australia. However, it is now clear that members of this genus were isolated earlier but placed in other genera without recognizing their unique morphological characters. The earliest of these descriptions was by Pereira in 1937 when the species *Heterorhabditis* (then *Rhabditis*) *hambletoni* was discovered infecting the cotton boring weevil (*Eutinobothrus brasiliensis* Hambleton) in Brazil.

The Present

Today there are 12 recognized species of *Steinernema* (Table 1) and four of *Heterorhabditis* (Table 2) as well as hundreds of isolates of these species and certainly many undescribed species throughout the world. The large number of new strains recorded in recent years corresponds to a dramatic increase in research workers now found in institutes,

Fig. 1. Historical photo showing William Rudolph Glaser (standing on right) and his assistants working on the cultivation of *Steinernema glaseri* on artificial media in the White Horse Laboratory in Trenton, New Jersey in 1939.

universities and commercial laboratories. Most countries have at least one person investigating these nematodes and also some type of cultivation in progress, either "in vitro" or "in vivo" and often commercial.

Table 1. Recognized species of *Steinernema* in chronological order (modified from Poinar, 1990A).

1. *glaseri* (Steiner, 1929)
2. *feltiae* (Filipjev, 1934)
3. *affinis* (Bovien, 1937)
4. *carpocapsae* (Weiser, 1955)
5. *intermedia* (Poinar, 1985)
6. *rara* (Doucet, 1986)
7. *kushidai* Mamiya, 1988
8. *scapterisci* Nguyen and Smart, 1990
9. *longicaudum* Shen, 1991
10. *neocurtilis* Nguyen and Smart, 1992
11. *serratum* Liu, 1992
12. *riobravis* Cabanillas, Poinar and Raulston 1993

Table 2. Recognized species of *Heterorhabditis* in chronological order (modified from Poinar, 1990A).

1. *bacteriophora* Poinar, 1976
2. *megidis* Poinar, Jackson & Klein, 1987
3. *zealandica* Poinar, 1990
4. *indicus* Poinar, Karunakar and David 1992

One present day theme is for research workers to concentrate on using nematodes against a range of pest insects in a particular geographical region. Investigations along these lines have appeared in China (Wang & Li, 1987), Japan (Ishibashi, 1987; 1990) and Latin America and the Caribbean (Pavis and Kermarrec, 1991) (Georgis & Hom, 1992). In addition, emphasis has been placed on determining which species or strain of nematodes is most effective on specific insect groups such as scarab beetles (Jackson & Glare, 1992), turf and ornamental insects (Georgis & Poinar, 1989) and social insects (Poinar and Georgis, 1989). Some idea of the development of this field can be obtained by examining the recently published bibliography on heterorhabditid and steinernematid nematodes which up to and including the year 1991, cites 1413 references pertaining to various aspects of these nematodes (Smith, Miller and Simser, 1992). Current and prospective markets for these nematodes have been summarized by Georgis (1992) and Kaya and Gaugler (1993) have recently reviewed general aspects of steinernematid and heterorhabditid nematodes. Summaries of field applications of these nematodes were published by Poinar (1986) (trials up to 1984) and Poinar (1990) (from 1984 to 1990).

An interest in evaluating the potential of steinernematids and heterorhabditids against medically important arthropods in terrestrial environments has resulted in laboratory studies showing susceptibility to these nematodes by stable flies (Poinar & Boxler, 1984), fleas (Mrácek & Weiser, 1983) (Silverman et al., 1982), ticks (Samish & Glazer, 1992), lice (Weiss et al., 1993) and phlebotomine flies (Poinar et al., 1993).

Attributes of these nematodes such as their wide host range, ability to kill the host within 48 hours, ease of cultivation on artificial media, presence of an infective stage capable of storage under shipment and distribution for agro-eco systems, lack of inducement of insect immunity and environmental safety, have made them ideal commercial objects throughout the world. In fact, it is staggering to realize that since the first commercialization of the nematodes in California in 1981 (Poinar, 1986), there are now 44 commercial suppliers of heterorhabditids and steinernematids in North America alone! (Hunter, 1992). At the present time, heterorhabditid and steinernematid nematodes are exempt from registration in the United States by the Environmental Protection Agency (Gorsuch, 1982).

Commercialization of heterorhabditids and steinernematids was certainly accelerated by the development of large scale "in vitro" means of cultivation. The two basic types include a solid media process (Bedding, 1984) and a liquid formulation (Freedman, 1990). There are two potential dangers associated with the mass production of any organism. The first is the necessity to provide adequate amounts of nutrients to produce vigorous infective stages. Fortunately, with all known in vitro methods, the symbiotic bacteria (of the genus *Xenorhabdus*) are present, and provide essential nutrients. In some aspects, successful cultivation of these nematodes is associated with the successful production of the symbiotic bacteria. The second problem is somewhat more nebulous and represents a general decline in vigor of the nematodes over time. It is difficult to define just why this occurs but it may be necessary to periodically re-introduce or start again with "wild" inocula.

Recent studies have shown some interesting variations in the typical life cycle of these nematodes. Until recently, the larvae of holometabolous insects have been the natural hosts of these nematodes (occasionally the adult stage as well). However, the discovery of two steinernematid species adapted to infecting developmental stages of insects with gradual metamorphosis (Orthoptera) has been an interesting variation of the typical pattern (Nguyen & Smart, 1993). Another variation occurs with a strain of *S. feltiae* in California

that naturally only infects the adult stages of a holometabolous insect (in this case, a mycetophilid fly) (Poinar, 1992B). These examples show that at least the steinernematids possess a wide degree of variability in their genome and can adapt to variations in host groups and stages.

The Future

There are now numerous strains of steinernematids and heterorhabditids known from throughout the world (Poinar, 1990) and it is time to become sophisticated in this computer age. Each strain should be tested under different laboratory conditions in regards its 1) temperature preference and range for survival and infectivity, 2) humidity tolerance, 3) movement and survival under different soil types, 4) preferred host group, 5) preferred location in soil of preferred host, 6) ability to tolerate various agricultural chemicals, 7) ability to tolerate natural enemies in various environments, 8) ability to be transported by adult hosts and 9) ability to locate different host stages. Additional characters can be added but the above are most important regarding the interaction between the nematode, the host and the environment. Of course, practical considerations such as cost competitive methods of cultivation, storage and shipping are also important but these are factors that can be modified by man. The point here is that there exists a tremendous amount of genetic variation within the various strains of entomogenous rhabditids that occur in nature and it is a challenge for us to recognize that, understand it and utilize it to our advantage. Once the above data is acquired for even a portion of the available strains, it can be incorporated into a computer database so that the nematode species or strains most likely to be successful against a specific insect pest in a specific habitat can be determined. Also entered into the data matrix would be specific requirements or modifications of the diet in regards artificial cultivation and special considerations for storage and shipping.

How can genetic engineering be helpful in enhancing heterorhabditid and steinernematid nematodes in pest control? One has to fall back on a knowledge of nematodes and their habitats in general as well as common sense to answer this question. These nematodes have evolved in a terrestrial soil environment and are most effective in this location. Specific cryptic habitats are also suitable but there are serious problems with their use in aquatic or above ground habitats. First, it is very unlikely that a desired trait can be genetically engineered into a nematode that would enable it to survive in a habitat in which no nematodes occur at present. Of a list of characters (see Poinar, 1991A) that were considered desirable for steinernematids and heterorhabditis, one of these was thermo-adaptation. There are many situations where these nematodes could be used in the tropics or on exposed surfaces except that temperatures often range beyond the tolerance level of the infective juveniles. In general, temperatures above 30°C inhibit nematode development in a host and temperatures above 35°C are detrimental to infective juveniles (Kaya, 1990). However, a review of nematodes that survive in hot water springs throughout the world (Poinar, 1991A) show that at least 9 species can complete their development in water ranging from 35-60°C. In such cases, it may be possible to introduce the genes responsible for this heat tolerance into the genome of heterorhabditis or steinernematids.

Increased pathogenicity and increased host finding in specific strains or foundation (formed by mixed mating of a certain number of naturally-occurring strains) strains have been reported and analyzed (Gaugler, 1987; Poinar, 1991A) and suggest that both of these characteristics can be experimentally enhanced through genetic selection.

Geneticists interested in strain enhancement can best follow the development of various experimental procedures being performed by a host of workers with *Caenorhabditis elegans*. Such techniques as microinjection (injecting DNA directly into the nematode germ line with a micromanipulator) and electric discharge (propelling DNA coated microprojectibles into a nematode by an electric discharge) have been used to introduce foreign DNA into germ lines of *C. elegans* and would be applicable for steinernematids and heterorhabditis.

There is still much work to be done regarding the understanding of the unique characteristics of various species and strains. The tools at our disposal in the rapidly developing fields of molecular biology and computer science can be indispensable for reaching some of these sophisticated goals. However, the greatest success can be expected when these high-tech methods are combined with basic nematological knowledge related to heterorhabditis and steinernematids and their present day relatives, several of which are at the evolutionary stage with invertebrates that steinernematids and heterorhabditis were some millions of years ago (Poinar, 1993).

References

Bedding, R.A. "Large scale production, storage and transport of the insect-parasitic nematodes *Neoaplectana* spp. and *Heterorhabditis* spp." *Ann. Appl. Biol.,* **1984,** 104, 117-120.

Filipjev, I. "Eine neue Art der Gattung *Neoaplectana* Steiner nebst Bemerkungen über die systematische Stellung der letzteren." *Tr. Parazitol. lab. zool. Inst. Akad. Nauk. SSSR,* **1934,** 4, 229-240.

Friedman, M.J. In *Commercial production and development;* Gaugler, R.; Kaya, H.K. Eds.; Entomopathogenic Nematodes in Biological Control; CRC Press: Boca Raton, FL, **1990;** pp. 153-172.

Gaugler, R. In *Entomogenous nematodes and their prospects for genetic improvement;* Maramorosch, K. Ed.; Biotechnology in Invertebrate Pathology and Cell Culture; Academic Press, NY, **1987;** pp. 457-484.

Georgis, R. "Present and future prospects for entomopathogenic nematode products." *Biocontrol Sci. Techn.,* **1992,** 2, 83-99.

Georgis, R.; Hom, A. "Introduction of entomopathogenic nematode products into Latin America and the Caribbean." *Nematropica,* **1992,** 22, 81-98.

Georgis, R.; Poinar, G.O. Jr. In *Field effectiveness of entomophilic nematodes Neoaplectana and Heterorhabditis;* Leslie, A.; Metcalf, R. Eds.; Integrated Pest Management in Turfgrass and Ornamentals; U.S. Environ. Prot. Agency, Govt. Printing Office: Washington, D.C,. **1989;** pp. 213-224.

Gorsuch, A.M. "Regulations for the enforcement of the Federal Insecticide, Fungicide and Rodenticide Act exemption from regulation of certain biological control agents." *Fed. Regist.,* **1982,** 47, 23928-23930.

Hunter, C.D. *Suppliers of beneficial organisms in North America.* California Environ. Prot. Agency: Sacramento, CA, **1992;** 31 pp.

Ishibashi, N. Ed. *Development of biological integrated control of agricultural pests by beneficial nematodes.* Ministry of Education, Culture and Science: Saga, Japan, **1990;** 159 pp.

Ishibashi, N. *Recent advances in biological control of insect pests by entomogenous nematodes in Japan.* Ministry of Education, Culture and Science: Saga, Japan, **1987;** 179 pp.

Jackson, T.A.; Glare, T.R. *Use of pathogens in scarab pest management.* Intercept, Ltd.: Andover Hampshire, England, **1992;** 298 pp.

Kaya, H.K.; Gaugler, R. "Entomopathogenic nematodes." *Ann. Rev. Entomol.,* **1993,** 38, 181-206.

Kaya, H.K. In *Soil Ecology.* Gaugler, R.; Kaya, H.K. Eds.; Entomopathogenic Nematodes in

Biological Control; CRC Press: Boca Raton, FL, **1990;** pp. 93-115.

Mrácek, Z.; Weiser, J. "Pathogenicity of *Neoaplectana carpocapsae* (Nematoda) for the flea, *Xenopsylla cheopis.*" *J. Invert. Path.*, **1983,** 42, 133-134.

Nguyen, K.B.; Smart, Jr. G.C. "*Steinernema neocurtillis* n.sp. (Rhabditida: Steinernematidae) and a key to species of the genus *Steinernema.*" *J. Nemat.*, **1992,** 24, 463-477.

Pavis, C.; Kermarrec, A., Eds. *Caribbean meetings on biological control.* Institut National de la Recherche Agronomique: Paris, **1991;** 569 pp.

Pereira, C. "*Rhabditis hambletoni* n.sp., nema apparentemente semiparasito da "broca do algodoeiro" *Gasterocercodes brasiliensis.*" *Arch. Inst. Biol. (São Paulo),* **1937,** 8, 215-230.

Poinar, Jr. G.O. *The Natural History of Nematodes;* Prentice Hall, Inc.; Englewood Cliffs, New Jersey; **1983**.

Poinar, Jr. G.O. "Entomophagous nematodes." *Fortschr. Zool.*, **1986,** 32, 95-121.

Poinar, Jr. G.O. In *Successful control of pests with entomopathogenic nematodes;* Deseo, K.V., Ed.; Alternative Systems in Plant Protection; Commune di Cesena: Italy (In Italian), **1990;** pp. 67-82.

Poinar, Jr. G.O. In *Taxonomy and Biology of Steinernematidae and Heterorhabditidae;* Gaugler, R; Kaya, H.K. Eds.; Entomopathogenic Nematodes in Biological Control; CRC Press: Boca Raton, FL, **1990;** pp. 23-61.

Poinar, Jr. G.O. In *Genetic engineering of nematodes for pest control;* Maramorosch, K., Ed.; Biotechnology for Biological Control of Pests and Vectors; CRC Press: Boca Raton, FL, **1991A;** pp. 5-7.

Poinar, Jr. G.O. In *Nematoda and Nematomorpha*; Thorp, J.H.; Covich, A.P., Eds.; Ecology and Classification of North American Freshwater Invertebrates; Academic Press, Inc.: New York, NY, **1991B;** pp. 249-283.

Poinar, Jr. G.O. "Rudolph W. Glaser (1888-1947) - a pioneer of steinernematid nematodes." *J. Invert. Path.* **1992A,** 60, 1-4.

Poinar, Jr. G.O. "*Steinernema feltiae* (Steinernematidae: Rhabditida) parasitizing adult fungus gnats (Mycetophilidae: Diptera) in California." *Fundam. appl. Nematol.*, **1992B,** 15, 427-430.

Poinar, Jr. G.O. "Origins and phylogenetic relationships of the entomophilic rhabditids, *Heterorhabditis* and *Steinernema.*" *Fund. Appl. Nemat.*, **1993,** In Press.

Poinar, Jr. G.O.; Boxler, D.J. "Infection of *Stomoxys calcitrans* (Diptera) by neoaplectanid nematodes." *IRCS Med. Sci.*, **1984,** 12, 481.

Poinar, Jr. G.O.; Georgis, R. In *Biological control of social insects with nematodes.* Leslie, A.R.; Metcalf, R.L., Eds.; Integrated Pest Management for Turfgrass and Ornamentals; U.S. Environmental Protection Agency: Washington, DC, **1989;** pp. 213-224.

Poinar, Jr. G.O.; Ferro, C.; Morales, A.; Tesh, R.B. "*Anandranema phlebotophaga* n.gen., n.sp. (Allantonematidae: Tylenchida), a new nematode parasite of phlebotomine sand flies (Psychodidae: Diptera) with notes on experimental infections of these insects with parasitic rhabditoids." *Fundam. appl. Nematol.*, **1993,** 16, 11-16.

Samish, M.; Glazer, I. "Infectivity of entomopathogenic nematodes (Steinernematidae and Heterorhabditidae) to female ticks of *Boophilus annulatus* (Arachnida: Ixodidae)." *J. Med. Entomol.*, **1992,** 29, 614-618.

Silverman, J.; Platzer, E.G.; Rust, M.K. "Infection of the cat flea, *Ctenocephalides felis* (Bouche) by *Neoaplectana carpocapsae* Weiser." *J. Nematol.*, **1982,** 14, 394-397.

Smith, K.A.; Miller, R.W.; Simser, D.H. "Entomopathogenic nematode bibliography: Heterorhabditid and Steinernematid nematodes." *Arkansas Agr. Expt. Sta. Bull.*, **1992,** 370, 1-81.

Steiner, G. "*Neoaplectana glaseri* n.g., n.sp. (Oxyuridae) a new nemic parasite of the Japanese beetle (*Popillia japonica* Newm.)." *J. Wash. Acad. Sci.*, **1929,** 19, 436-440.

Stoll, N.R. "In memoriam Rudolf W. Glaser (1888-1947)." *Science*, **1948,** 107, 131-132.

Wang, J.X.; Li, L.Y. "Entomogenous nematode research in China." *Rev. Nematol.*, **1987,** 10, 483-489.

Weiss, M.; Glazer, I.; Mumcuoglu, K.Y.; Elkind, Y.; Galun, R. "Infectivity of steinernematid and heterorhabditid nematodes for the human body louse *Pediculus humanus humanus* (Anoplura: Pediculidae)." *Fund. Appl. Nematol.*, **1993,** In Press.

Parasites and Predators Play a Paramount Role in Insect Pest Management

J.C. van Lenteren, Laboratory of Entomology, Wageningen Agricultural University
P.O. Box 8031, 6700 EH Wageningen, The Netherlands

Parasites and predators have been recognized as important control agents of pest insects for centuries. It is estimated that more than 95% of the potential pest organisms are kept under control by natural enemies. In this paper different ways for use of biological control as a reliable, environmentally safe means of pest reduction will be illustrated. Present day application of biological control is reviewed and a case study to illustrate recent developments is presented. Examples are given of research that may lead to increased application of biological control and its future is addressed.

Parasites and Predators: the Very Heart of Biological Control

In order to determine what the role of parasites and predators is in the control of pests, first several questions have to be answered, such as: how many potential pest organisms do occur on earth, why are they not all developing to pest status, and what kind of situations do lead to pest problems. The way in which a pest is caused determines the possibilities for and the type of biological control.

During the past 100 years of modern biological control, parasites and predators have played a key role. Until 1990 5500 introductions of natural enemies have been made to new areas worldwide. In 1200 cases the natural enemies have become established and this has led to successful control in 420 instances. Of these successful natural enemies, 340 are species of parasites, 74 are predators and 6 are pathogens. It is clear that during the first era of biological control, parasites and predators make up 99% of the successful cases.

The outstanding successes already obtained have, in general, not resulted in much respect for and research input in biological control by organizations funding agricultural research both in the developed and developing world.

In this paper I hope to make clear that biological control deserves more attention as an environmentally safe, long lasting, sustainable method of control. I have structured the paper by formulating a number of statements.

Only a Few Plant Eaters are Pests

The number of animal species is estimated to be between 20 and 50 million, out of which about 2 million have been identified and named. Eighty percent of all animal species consists of insects. Among the insects, plant eaters are our main rivals. They comprise about half of the insect world, i.e. several millions of species. The other half consists of insects which live off other insects (predators and parasites), or other animals, and insects which feed on dead organic material.

On the 300,000 species of higher plants some 6-12 million herbivorous species occur. They are potential pest species, but as man does not use all plant species, direct competition with herbivores is limited to those plants that both insect and man are interested in. Ninety percent of our food is made up of only 300 plant species. On plants used by man more than 200,000 herbivorous arthropod species occur.

Pests are organisms that cause significant economic injury or are a nuisance. Pests in the widest sense (animals, plant pathogens and weeds) lead to world crop losses of 35%, under the present application of pesticides. About 12% is due to insect and mite pests, 12% to plant pathogens, 10% to weeds and 1% to mammals and birds. Yield reductions are in the order of 400 billion US$ annually (Pimentel, 1986). Additional losses of some 10% occur after harverst.

Estimates of the number of animal pest

species vary between 350 and 10,000 species worldwide (Table 1). The last figure includes harmful insects which attack man or cattle, which damage man's dwellings or are generally a nuisance, and many minor pests. Thus, although the number of plant-eating insects is very large, the number of species regarded as pests is much smaller. Of the *described* insect species maximally 2 % may create problems to man (DeBach & Rosen, 1991). Even the highest estimate of 10,000 species makes it clear that very few insects are problematic.

Pest species are generally attacked by an array of natural enemies, from viruses to predatory mammals. Quantitatively insects form the largest group of natural enemies.

Pests are caused by Man

In nature, only 1 percent of the plant biomass is eaten by insects and other herbivores, which is in strong contrast to the potentially 20 - 100 percent consumption of food crops by insects. What has gone wrong? Comparison of population dynamics of potential pest species in natural ecosystems and agro-ecosystems points towards several general explanations with regard to the difference in degree of consumption:

- Wild plants possess resistance against a variety of insects, which results in no or slow population development.
- In natural ecosystems, host plants specific for certain insect species are not easily discovered, leading to high insect mortality during dispersal. Only a few insects are able to locate a host plant and reproduce.
- In natural ecosystems, a myriad of natural enemy species maintain plant-eating insects at low population densities and, even in agroecosystems, the number of potential pests is held at non-damaging levels by enemies which occur naturally (for examples see DeBach & Rosen, 1991).

These combined factors result in low population densities of plant-eating insects in nature. Agriculture brought the creation of monocultures of genetically identical crop plants which are maintained in an excellent condition and are often grown in the same geographical area over many years. This has led to the occurrence of maximal survival and reproduction rates for pest insects, whereas many species of their natural enemies have problems of surviving in monocultures because of lack of essential resources, like alternative hosts, pollen or nectar. Thus, man has created first-rate favourable environments for plant-eating insects through the provision of more

Table 1. Estimates of numbers of pest species

No. of pests	Category	Reference
350	key insect pests world wide	Simmonds & Greathead 1977
600	key arthropod pests in USA	Schwartz & Klassen 1981
1,000	key arthropod pests in agriculture and forestry world wide	Hill 1987
5,000	pests and diseases (arthropods, snails, weeds, plant pathogens) world wide	Waterhouse 1992
5,000	arthropod pests world wide	Huffaker & Messenger 1976
10,000	arthropod pests world wide	Ridgway & Vinson 1977
10,000	arthropod pests world wide	Schwartz & Klassen 1981

accessible and/or more nutritious food or lush growth, as well as by the decimation of their natural enemies (Huffaker et al., 1976).

In addition to the creation of monocultures, transport of plants -and other material - all over the world has led to the accidental importation of hundreds of pest species in areas where they previously did not occur (Sailer, 1983; van Lenteren, 1992).

An important third cause for the creation of pests is chemical control. The past 40 years have shown repeatedly how easy it is to foster pests by the use of pesticides. Both resurgence of old pests and appearance of new pests can be explained by the pest organism becoming resistant to the pesticide and the natural enemies being exterminated by the pesticide (for examples see DeBach & Rosen, 1991).

Parasites and Predators have been functioning for Millions of Years as Important Pest Suppressors

Biological control is the use of living organisms (natural enemies) to reduce damage caused by noxious organisms (pests, plant pathogens and weeds). This definition implies activity by man. It is distinguished from *natural control* where natural enemies reduce pest populations without any specific actions by man.

Extensive studies have been made to find out how agro-ecosystems influence pest population dynamics and how these situations can be changed to profit in a better way from the pest control mechanisms such as natural enemies which nature provides freely (for a review see van Lenteren, 1987). There are no accurate estimates of the numbers of natural enemies of insects and mites, but their numbers will be in the order of millions of species. Most pest insects carry an array of natural enemies. Some pest species have well over 100 recorded, rather specific enemies. Polyphagous predators such as e.g. spiders, true predatory bugs, lady beetles and frogs are not included in this figure.

In many cases of natural control it remains to be unravelled which of the natural enemies play the mayor role in reduction of the pest. For more than 400 cases of successful biological control it appears that insect parasites are applied most often. The main reason is that parasites are generally much more pest-species-specific than predators, so the potential negative effects of introducing them to new areas is smaller and, also because of this specifity, they are expected to be more effective in finding and killing the pest organisms. Pathogens (viruses, bacteria and fungi) are used to a much lesser extent because studies to evaluate their value started later than for predators and parasites, and registration procedures for pathogens are often cumbersome.

Parasites, predators and pathogens can be used in different types of biological control programs:

Inoculative biological control. Beneficial organisms are collected in an exploration area and introduced in the area where the pest occurs. Only a limited number of beneficials is released. The aim is long-term suppression of pest populations. This method has been used most frequently against introduced pests, which are presumed to have arrived in a new area without their natural enemies. These are then sought in the area of origin of the pest. As it was the first type of biological control practised extensively it is also called "classical" biological control. The method is particularly popular in North America with its many imported pests. The control of cottony cushion scale (*Icerya purchasi*) obtained 100 years ago with the predatory beetle *Rodolia cardinalis* is the oldest example of this approach. A recent, very successful example concerns biological control of the cassava mealybug, *Phenococcus manihoti*, which was accidentally introduced into Africa in the early 1970s and is now spread over almost the entire cassava belt in Sub-Saharan Africa, with the main exceptions of Uganda and Madagascar (Herren & Neuenschwander, 1991). In vast areas it is under control of the parasite *Epidinocarsus lopezi* (Table 2). Both pest and natural enemy originate from Latin America.

Inundative biological control. Indigenous beneficials are mass reared in the laboratory and periodically released in large numbers to obtain an immediate control effect of pests with one or two generations (i.e. use as biotic insecticide), with no anticipation of effects on subsequent generations. An example of this approach is the application of the parasitic

70

wasp *Trichogramma* against a variety of Lepidopteran pests (Table 2).

Seasonal inoculative biological control. Native or exotic natural enemies are mass reared and periodically released in short-term crops (6-9 months, e.g. in greenhouses) against pests with several generations. A large number of natural enemies is released to obtain both an immediate control effect and also a build-up of a natural enemy population for control later during the same season. An illustration of this technique is the control of *Trialeurodes vaporariorum* with the parasitic wasp *Encarsia formosa* in protected crops (for details see case, section below).

Conservation. Conservation is an indirect method where measures are taken to conserve natural enemies. Conservation may result in a richer diversity of beneficial species as well as in larger populations of each species, with better control of pests as an effect. An example of conservation concerns the realization that predators were important control agents of an insecticide created pest (brown plant hopper, *Nilaparvata lugens*) in rice in Asia, followed by a considerable reduction in chemical pest control and reestablishment of natural control of the pest. The green revolution of the 1960s encouraged cultivation of large areas of high producing varieties of rice in Asia. This stimulated massive pesticide inputs, resulting in recurring pest outbreaks caused by pesticide-induced resurgence which, in turn, caused serious crop losses. After this was discovered, an innovative program to conserve predators and parasites was implemented in many Asian countries (Lim, 1991). Another example is the reintroduction of predatory mites for the control of fruit tree red spider mite with a concurrent change in use from broad spectrum to selective pesticides in European apple orchards.

The cardinal idea behind theoretical explanations for successful biological control has long been that efficient natural enemies operate by creating a stable pest-enemy

Table 2. Areas on which natural enemies are controlling pests

Type of control	Pest and crop	Area under control
natural control	Tens of thousands of (potential) pests	surface of the world's ecosystems multiplied by the number of key natural enemies (thousands)
inoculative control	> 160 pest species e.g. *Phenococcus manihoti*, cassava e.g. *Icerya purchasi*, citrus e.g. *Eriosoma lanigerum*, apple	vast, no reliable estimates Sub-Saharan Africa, 3 million ha 55 countries worldwide 42 countries worldwide
inundative control	Lepidopteran pests, various Lepidopteran pests, various Lepidopteran pests, various Lepidopteran pests, various *Ostrinia nubilalis*, corn	Russia, 20 million ha China, 2 million ha South-East Asia, 300,000 ha South + North America, 800,000 ha Europe, 40,000 ha
seasonal inoculative	20 arthropod pests, greenhouses insect pests, citrus	Europe + North America, 20,000 ha North America, 20,000 ha
conservation	insect pests, rice arthropod pests, orchards	South and Southeast Asia, vast Europe, 150,000 ha

equilibrium at low densities and that aggregation of natural enemies at patches with high host or prey densities is the critical feature that results in stability (Waage & Hassell, 1982). The central role of a low, stable pest equilibrium has recently been challenged and different explanations have been put forward (see e.g. Murdoch, 1990). Murdoch and coworkers examined the role of natural enemy aggregation independently of pest distribution and looked at field examples of successful biological control which resulted in the following conclusions:

- Successful control does not require a control agent to aggregate at areas of high host density, nor does it require aggregation indepently of host density, or other mechanisms that create "partial refuges." Therefore, random movement by parasites is capable of producing excellent control.
- Control appears possible in the absence of conditions leading to local stable equilibria.
- In model systems, control is compatible with local extinction of the pest and polyphagy in the natural enemy.

It is still possible, of course, that the aggregate result of unstable local dynamics, when combined with movement, may produce stable dynamics over large areas.

For a number of years my research group has been developing criteria for pre-introductory evaluation of natural enemies for the different types of biological control as mentioned above (van Lenteren, 1980, 1983, 1986b; van Lenteren & Woets, 1988). The criteria for natural enemies that are supposed to determine their efficiency are strongly dependent on the theoretical framework one uses to explain how biological control works. It also has important consequences for the biological control practitioner whose aim is to obtain a pest density below an economic threshold all or most of the time following introduction of a natural enemy. At the moment we cannot conclude more than that different theories produce different, or even conflicting, advice on selection criteria and about the functioning of biological control. The practitioner is not only confused by this, but may also delay application of biological control measures while he is determining the wrong criteria. We presently apply an evaluation scheme based on a number

of unequivocal criteria and leave out dubious criteria. For an extensive discussion of this topic see van Lenteren, 1986b.

Biological Control is a Realistic Long-Lasting, Environmentally Safe Solution for managing Pests

Biological control has an excellent record of successes. In the age of modern biological control, empirical evidence and a large number of case studies (e.g. Huffaker & Messenger, 1976; Bellows, 1993) have shown that biological control practiced by experts is a very cost effective method of pest control. Besides its good pest reduction capacity, biological control has highly positive social, ecological and economic benefits. Additionally, once a good natural enemy has been identified, it can be used indefinitely as development of resistance to natural enemies comparable to pesticide resistance does not occur. Therefore, it should often be the first pest control tactic to be explored.

The most recent listing of benefit/cost analyses for 30 inoculative biological control programs shows that benefits are between 18 and 1500 to one, better than for any other pest control method (Bellows, 1993). Even though no concerted efforts have been made to measure the global economic benefits of natural and biological control, the total benefits must be reckoned in billions of dollars (Tisdell, 1990). The most thorough cost/benefit analyses for different pest control methods have been made for Australia and were summarized by Tisdell (1990). He presents data from four biological control and 9 chemical control projects. Both non-biological and biological control projects were considered a success because benefits were A$ 350.4 and costs A$ 32.93, so the overall benefit/cost ratio was 10.6:1. However, both the absolute economic impact and the relative returns from resources used by the Division of Entomology on biological pest control were much greater than for projects on non-biological means of control, and the benefits keep accumulating as no further activities are needed to keep pests under biological control. For Australia alone they exceed A$ 1 billion.

72

The benefit/cost ratio for the Africa-wide cassava mealybug biological control program is 149:1 (Herren & Neuenschwander, 1991). The benefits of the FAO Intercountry Rice IPC Programme in Asia are, for the national treasuries, the reduction or elimination of pesticide subsidies resulting in savings to national economies varying from US$ 5-60 million depending on the country. For the farmers implementation of IPM meant the same or higher rice yields, making greater profits, less risk because of better yield stability and less risks for their health (Lim, 1991).

From a recent review of commercially applied European IPM programs (van Lenteren et al., 1992a) it became clear that in all the major IPM programs natural enemies play a key role in reducing pests. In several programs other control methods are completely focussed around natural control (e.g. apple, pear, peach and cherry orchards, vineyards and soil systems), in others, biological control by inoculative, seasonal inoculative and inundative releases is central (e.g. orchards, major greenhousevegetable crops, and corn).

The area and number of pests under natural control is immense, and the extent to which natural control works can only be grasped from situations where man interferes with nature in a manner that natural control is upset. Abundant evidence for the role of natural control comes from experiences with chemical control: chemical control can strongly reduce or locally eradicate natural enemies and, therefore, lead to resurgence of the target pests and to the creation of new pests.

The area under inoculative or classical biological control is difficult to estimate. Results obtained with this type of biological control are forgotten fast, because no repeated, weekly to yearly action is needed to keep pest numbers down. The important and continuous role the introduced ladybird beetle *Rodolia cardinalis* has played since 1888 in California, and later in many countries all over the world, to keep the cottony cushion scale, *Icerya purchasi*, below damage levels, was stressed by resurgence of this pest in 1947 as a result of the widespread use of DDT in citrus (Stern et al., 1959). Cottony cushion scale is now controlled free of charge by *Rodolia* in 55 countries following its initial success in California in 1889. Woolly apple aphid (*Eriosoma lanigerum*) is under control by *Aphelinus mali* in 42 of the 51 countries after it was introduced initially in New Zealand at the start of this century. Both the pest and its natural enemy originated from western United States. These two examples show the enormous potentials of biological control with parasites and predators in inoculative programs.

It would be to the benefit of biological control if areas under inoculative biological control become better known. Specification of such data with (economic) benefits will command the necessary respect for biological control. During the past century inoculative biological control was aimed at 416 arthropod pests, of which 75 were brought under complete control (permanent reduction under economic threshold) and another 89 pests are under substantial or partial control (with chemical control reduced by at least 50%). As biological control is permanent, this is a considerable accomplishment. Up to 1988 4226 natural enemies of arthropod pests were imported worldwide (including repetitions) which lead to establishment of 1251 (29.6%) natural enemies in the new area (DeBach & Rosen, 1991); 932 cases are still being followed up and establishment could not be reliably judged; 2038 cases did not result in establishment.

It is easier to obtain data for those types of biological control where natural enemies have to be released regularly (Table 2). Inundative biological control is said to be extensively applied in the previous Soviet Union on 20 million ha for control of a number of lepidopterous pests, with *Trichogramma* as most used natural enemy. It is unclear, however, what the pest reduction effect of these releases is (see also King, this volume, for other comments). In China about 2 million ha are under inundative control with *Trichogramma* to manage a number of lepidopterous pests. In South East Asia, control of stemborers with a *Trichogramma* species takes place on circa 300,000 ha. In South and North America, control of Lepidoptera in cotton and sugar cane with *Trichogramma* species occurs on some 800,000 ha.

In apple and pear orchards and vineyards (some 150,000 ha) biological control of spider mites and several insect pests takes place in

Europe (van Lenteren et al. 1992). In citrus orchards, control of scale insects and mealybugs occurs with *Aphytis, Chrysoperla* and *Cryptolaemus* on about 20,000 ha in the USA. In greenhouses circa 20,000 ha are under seasonal inoculative and/or inundative biological control worldwide with a variety of natural enemies.

To summarize the important role of natural enemies in pest control I best quote Pimentel et al. (1992): "An estimated $20 billion is spent annually in the world for pesticides. Yet, predators and parasites existing in natural ecosystems are providing an estimated 5-10 times this amount of the pest control. Without the existence of natural enemies, crop losses by pests in agriculture would be catastrophic and costs of chemical control would escalate enormously."

Biological control compares very advantageously with chemical control. The data in the previous section illustrate the worldwide importance and applicability of biological control. Table 3 was compiled to compare characteristics of chemical with biological control. In all but one aspect - specificity - biological control comes out best, and even for that characteristic there is an important positive side. Natural enemies are usually so specific in their host or prey choice, that they do not attack more than a few, often closely related species. The positive side is that one does not have to be concerned with unwanted side effects like the killing of other beneficial or interesting insects. The problematic side of it is that one has to identify

a natural enemy species for each pest species. This means that when a complex of pests occurs in one crop, introduction of several natural enemies will be needed, whereas with chemical control spraying of one insecticide, or a cocktail of insecticides, may reduce several pests as a result of one action. That the release of a complex of natural enemies is not experienced as burdensome and uneconomic will be illustrated with the case study below.

Seasonal inoculative and inundative control in protected crops: a case study

It is often hard to judge the economic and biological success of inundative release programs because of the lack of essential information in publications. My own experience is mainly with seasonal inoculative and inundative types of biological control of greenhouse pests and I will illustrate some developments in this field to show that commercial successful biological control programs can be implemented and adapted fast. For detailed reviews of this work I refer to van Lenteren & Woets (1988) and van Lenteren et al. (1992b).

Biological pest control has been applied with commercial success for circa 25 years now in greenhouses. Although it is a new pest control method in this cropping system, the growers did readily accept and rely on it. Successful greenhouse production requires well-trained, intelligent growers who cannot afford to risk any

Table 3. Comparison of aspects related to the development and application of chemical and biological control (after van Lenteren 1986b, updated 1992)

	Chemical control	Biological control
Number of "ingredients" tested	> 1 million	5,500
Success ratio	1:30,000	1:20
Developmental costs	100 million US$	2 million US$
Developmental time	10 years	10 years
Benefit per unit of money invested	< 4[1]	30
Risk of resistance	large	nil/small
Specificity	small	large
Harmful side-effects	many	nil/few

[1] when social and indirect costs are included the benefits decrease to about 2

damage from insects for ideological reasons, e.g. that biological control may cause fewer negative side effects than chemical control. If chemical control works better and is cheaper they will certainly use it. In tomatoes, for example, pest control represents less than 2% of the total overall cost of production, thus the cost of chemical pest control is not a limiting factor.

Yet despite the serious constraint for implementation of biological control that chemical control is so easy and inexpensive to apply, acceptance has been remarkably fast. The main reason for developing biological control methods was the occurrence of resistance against pesticides of several key pests in greenhouses. A close relationship among researchers, extension workers, growers and producers of natural enemies has resulted in a rapid transfer and use of information on biological and integrated control. During the past 25 years 25 species of natural enemies have been introduced against more than 20 pest species (Table 4). Natural enemies in use are insect parasites (9 species), arthropod predators (9 species) and pathogens (4 species). Some 10 new species of natural enemies are in the process of being evaluated for future use. The greenhouse area on which biological control is applied has increased from 400 ha in 1970 to almost 20,000. Presently, biological control of the two key pests in greenhouses, whitefly (*Trialeurodes vaporariorum*) and spider mite (*Tetranychus urticae*), is applied in more than 20 countries out of a total of 35 countries having a greenhouse industry.

The fast evaluation and introduction of a number of natural enemies in situations where chemical control was either insufficient or impossible, has taught crop protection specialists that biological control, within IPM programs, is a powerful option in pest control and an economically profitable endeavour.

Biological Control Research can provide Many More Pest Control Options

In this section I will illustrate with two examples how research can contribute to intensified application of biological control.

First, *evaluation and selection of natural enemies can be drastically improved*. Most natural enemies were found through a trial-and-error procedure until now, and biological control can therefore often be characterized as a process whereby a diverse natural enemy complex is reduced to a few species for introduction. The selection process is highly arbitrary and not related to any aspect of an agent which might indicate its potential value. However, it is a fact that programs usually end before all promising agents have been introduced. Hence prioritizing agents on the basis of their likely efficiency would ensure that the best species are released. It would be much better for our profession if deliberate choices between possible candidates are made, particularly when this leads to a halt in importation of useless candidates. Further, if we intend to change biological control from an art into science, we should develop a basic understanding of how biological control works and be able to make predictions about the outcome of introduction programs. Ecological, genetic and behavioural theory might help to move the more effective agents to the front of the queue of species to be introduced. A lively debate on more efficient and scientifically satisfying evaluation methods has been going on for the past 20 years (see e.g. Mackauer et al., 1990).

Three approaches for the selection of natural enemies emerge from the literature: a. evaluation based on individual attributes of natural enemies, b. evaluation based on integration of individual attributes and c. evaluation based on ecosystem studies.

In the evaluation based on *individual attributes* of natural enemies (called the reductionist approach by Waage, 1990), agents are selected on the basis of particular biological attributes or life-history characteristics (e.g. duration of development, fecundity, searching efficiency). Theory dissects natural enemies into simple sets of characters, which can be viewed and compared independently. This approach is no longer popular, although it is still used.

In the evaluation based on *integration of individual attributes*, one comes up with a composite picture of the pest reduction potential of the natural enemy (e.g. van Lenteren, 1986b; van Lenteren & Woets, 1988). When carefully applied, this method has proved to be of value.

The evaluation based on *ecosystem studies* (named the holistic approach by Waage, 1990)

Table 4. Commercially produced natural enemies for control of greenhouse pests (after van Lenteren et al. 1992b, with additions)

Natuaral enemy	Target pest	In use since
Phytoseiulus persimilis	*Tetranychus urticae*	1968
Encarsia formosa	*Trialeurodes vaporariorum*	1970 (1926)
	Bemisia tabaci	1988
Opius pallipes	*Liriomyza bryoniae*	1980-1983[*]
Amblyseius barkeri	*Thrips tabaci*	1981-1990[*]
	Frankliniella occidentalis	1986-1990[*]
Dacnusa sibirica	*Liriomyza bryoniae*	1981
	Liriomyza trifolii	1981
	Liriomyza huidobrensis	1990
Diglyphus isaea	*Liriomyza bryoniae*	1984
	Liriomyza trifolii	1984
	Liriomyza huidobrensis	1990
Bacillus thuringiensis	Lepidoptera	1983
Heterorhabditis spp.	*Otiorrhynchus sulcatus*	1984
Steinernema spp.	Sciaridae	1984
Amblyseius cucumeris	*Thrips tabaci*	1985
	Frankliniella occidentalis	1986
Chrysoperla carnea	aphids	1987
Aphidoletes aphidimyza	aphids	1989
Aphidius matricariae	*Myzus persicae*	1990
Orius spp.	*Frankliniella occidentalis*	1991
Verticillium lecanii	aphids	1992
Aphidius colemnani	*Aphis gossypii*	1992
Aphidius colemani	*Aphis gossypii*	1992
Aphelinus abdominalis	*Macrosiphum euphorbiae*	1992
Trichogramma spp.	Lepidoptera	1992
Leptomastix dactylopii	*Planococcus citri*	1992
Cryptolaemus montrouzieri		
Amblys. cucumeris/degenerans	Thrips	1993
non-diapausing strains		
Eretmocerus californicus	*Bemisia tabaci*	1993
Metaseiulus occidentalis	*Tetranychus urticae*	1993
Hippodamia convergens	aphids	1993

[*] use terminated, other natural enemy available

proceeds from the theoretical notion of how natural enemies fit into the broad ecology of the pest and its other mortality factors. Here, community concepts predominate, expressed in arguments for density-specific agent complexes, multiple introductions and filling "empty" natural enemy niches. This approach is not often applied, but strongly supported by some biocontrol workers (e.g. Ehler, 1990). Although it is scientifically attractive, it is not applicable yet and it will take many more years before it may be of use, contrary to what Ehler (1990) suggests.

Currently, there are good evaluation criteria available to allow for a choice between useless and potentially promising natural enemies. Such a choice prevents research on and introduction of inefficient natural enemies. With a gradual improvement of evaluation criteria and a further integration of criteria, ranking among the promising natural enemies will be possible.

Downplaying of preintroductory evaluation based on sometimes confused reasoning, e.g. by Gonzalez and Gilstrap (1992) is not only unfair, but also unscientific. Their so-called information assimilation" approach is a mix of intuition and use of preintroduction evaluation criteria.

A sound preintroductory evaluation procedure takes some 18 person months per natural enemy or much shorter when the natural enemy shows very obvious inherent weaknesses. In that case no further money is spent on rearing, release and follow up studies of useless natural enemies. The data from this research are not only useful for selection, but also provide essential information for designing a mass production method, the type of releases (inundative, (seasonal) inoculative), the release program (timing, spacing and numbers to be released) and an extention program.

The second example of how research can provide more options for biological control concerns the **understanding of variability in natural enemy behaviour to enhance targeting of releases**. Recently, several papers have appeared on how to interpret and deal with variability in natural enemy behaviour (Lewis et al., 1990; Vet et al., 1990; Vet & Dicke, 1992). Most ecologists are aware that variability in natural enemy behaviour occurs abundantly, often to their despair. It is important to know how natural enemies function in agroecosystems, because such understanding may help in designing systems where natural enemies can play an even more important role in inundative and seasonal inoculative releases. In this section the sources of variability in behaviour are presented and the potential exploitation of this variability to improve biological control will be discussed.

The very core of natural enemy behaviour, host-habitat and host location behaviour shows great variability, and is repeatedly leading to inconsistent results in biological control. Most studies aimed at understanding variability have focused on extrinsic factors as causes for inconsistencies in foraging behaviour. Typically, however, foraging behaviour remained irregular when using precisely the same set of external stimuli. These irregularities are caused by intraspecific, interindividual variation in behaviour. In order to understand erratic behaviour and to be able to manipulate such variation, biological control researchers need to know the origins and width of variation.

Two types of adaptive variation are distinguished in the foraging behaviour of natural enemies (Lewis et al., 1990):

(1) *Genetically fixed differences* among individuals (fixed-behaviour; innate responses): e.g.

natural enemy strains with different capabilities for searching in different habitats, strains with different host acceptance patterns. Such variation is now used in selection of natural enemies. Genetically different strains of the same natural enemy species may react in very different ways to the same set of chemical stimuli being emitted by the host/plant complex. Knowledge of such inherited preferences for environments and matching of inherited preferences with stimuli in the environment is of vital importance in choosing correct natural enemy strains. For a population of natural enemies to be predictable and consistent in biological control, it must first of all have a proper blend of genetic traits appropriate to the target environment, and traits must occur sufficiently uniformly in the population. This statement has been recognized generally, but has been dealt with only on a gross level in applied programs (e.g. climate, habitat and host matching).

(2) *Phenotypic plasticity* (unfixed, learned, plastic behaviour): behaviour adapts as a result of experience for foraging more effectively in any one of a variety of circumstances that might be encountered. Preference develops for a habitat where suitable hosts were encountered. The response of a foraging natural enemy can be quite plastic, can be modified within the bounds of its genetic potential, and is dependent on the experience history of the individual. Modifications can be initiated during pre-imaginal stages and at eclosion, so the response of a "naive" adult will necessarily be altered as a routine consequence of rearing. Such alterations have seldom if ever been quantified, although changes in preference have been observed as a result of different hosts or host diets. Particularly for inundative and seasonal inoculative types of biological control, quantification of this variability is essential. An individual can often change its inherited response range, so it can develop an increased response for particular foraging environments as a result of experience with stimuli of these environments. Absence of reinforcement (i.e. absence of contact with host-related stimuli) will result in a waning of the level of that response and a reversion to the naïve preference. Natural enemies are plastic in their behaviour, but operate within genetically defined boundaries.

Only recently have we begun to appreciate the extent to which natural enemies can learn. Many parasite species are able to acquire by experience an increased preference for and ability to forage in a particular environmental situation (Vet et al., 1990; Vet & Dicke 1992). There is some indication for immature learning and abundant evidence for adult learning in natural enemies. Learning is mostly by association. Usually, close range, reliable, unconditional genetically fixed stimuli serve as associators and reinforcers for the longer range, more variable conditional stimuli. Foraging behaviour can continuously be modified according to the foraging circumstances encountered (Vet & Dicke, 1992).

Additionally, foraging behaviour can be strongly influenced by (3) the *physiological condition* of the natural enemy. Natural enemies face varying situations in meeting their food, mating, reproductive and safety requirements. Presence of strong chemical, visual or auditory cues, cues related to enemy presence, and (temporary) egg depletion can all reduce or disrupt the response to host-foraging cues. For example, hunger may result in increased foraging for food and decreased attention to hosts. In that case, the reaction to food and host cues will be different than when the natural enemy is well fed.

The sources of intrinsic variation in foraging behaviour (genetic, phenotypic and those related to the physiological state) are not mutually exclusive but overlap extensively, even within a singular individual. *The eventual foraging effectiveness of a natural enemy is determined by how well the natural enemy's net intrinsic condition is matched with the foraging environment in which it operates* (Lewis et al., 1990).

How can we manage variability in behaviour of natural enemies? In order to be efficient as a biological control agent, natural enemies must be able to (1) effectively locate and attack a host, and (2) stay in a host infested area until most/all hosts are attacked. (N.B. "efficient" as biological control agent in the anthropocentric view, which does not necessarily mean efficiency in a natural selection view.) Prediction of performance in efficiency is a product of proper matching of intrinsic conditions of the searching natural enemy with the target environments.

Management of the natural enemy component is particularly important when natural enemies are mass produced in the laboratory, sometimes in factitious hosts. In laboratory rearings the natural enemies are removed from the context of natural selection and are exposed to artificial selection for traits not appreciated in the field (van Lenteren, 1986a). In addition to the genetic component, associative learning may lead to many more changes in behavioural reactions. This, then, results in the need for quality control procedures in the establishment, maintenance and use of natural enemies. Quality control will have to include both genotypic and phenotypic aspects of behavioural traits. Currently, quality control is lacking completely or is minimal at the circa 50 mass production units in North America and Europe (van Lenteren, 1991).

Genetic qualities. Successful predation or parasitism of a target host in a confined situation does not guarantee that released individuals will be suitable for that host under field conditions. When selecting among strains of natural enemies, we need to ensure that the traits of the natural enemies are appropriately matched with the targeted use situations in the field.

Phenotypic qualities. Without care, insectary environments lead to weak or distorted responses. When we understand the sources and mechanism of learning, we can provide the appropriate level of experience before releasing the natural enemies. Also, prerelease exposure to important stimuli can help improve the responses of natural enemies through associated learning, leading to reduction in escape response and increased arrestment in target areas.

Physical and physiological qualities. Natural enemies should be released in a physiological state in which they are most responsive to herbivore or plant stimuli and not be hindered in their response by e.g. deprivation interfering with searching.

Biological Control with Parasites and Predators faces a Bright Future

Several current trends will stimulate the application of biological control. Fewer new insecticides are becoming available because of strongly increased costs for development and registration. Secondly, pests continue to develop resistance to pesticides, a particularly prevalent problem in developed countries where intensive management and repeated pesticide applications exert powerful selection on pest organisms. Thirdly, there is strong pressure by the general public (and in an increasing number of countries also by parliament) to reduce the use of chemical pesticides, resulting in a greater demand for other than currently used chemical control methods.

Particularly as a result of the last point, the future role of predators and parasites is expected to increase dramatically, as their positive role has been demonstrated extensively and registration costs are very low or nill. Both in tropical countries (e.g. Indonesia, China, Philippines, Sudan) and in temperate areas (e.g. Europe and Canada) integrated pest and disease management is strongly advocated as the official policy to realize short-term goals of obtaining a reduction in pesticicde use of 50%.

A strategy study of the Dutch Scientific Council for Government Policy (Rabbinge, 1992) shows that the area of cultivated land will be reduced to about half of its present size in the European Community. As a result, the anticipated reduction in the volume of pesticide used will be tremendous over the next two decades, i.e. in the order of 80-90% Europewide, because:

- The technology of pesticide application will be improved (better targeting) and the amount of active ingredient needed for effective control will decrease,
- The application of integrated pest and disease management systems will become standard for all crops,
- The abundant use of pesticides due to narrow crop rotations will be eliminated, which will particularly lead to a reduction of soil fumigants, and
- The area on which pesticides are used will be much smaller.

Such developments will strongly support the role that naturally occurring beneficial insects can play and also allow for increased use of manipulative types of biological control.

Biological control will not completely replace chemical control, but parasites and predators offer powerful control options and can be applied on a much larger scale than today. They should be used in combination with other pest control methods, including chemical control, in IPM programs. In this way mutual benefit will be harvested. For chemical control it may result in extended use of products because of slower development of resistance and a more positive perception of the role of the pesticide industry by laymen. In order to serve agriculture as well as the environment and human health, we should harvest the best from different methods to develop effective IPM methods. Designing such environmentally safer IPM programs is a challenge for our profession.

References

Bellows, T., Ed.;. *Principles and Application of Biological Control*. University of California Press, Berkeley, 1993 (in press).

DeBach, P.; Rosen, D. *Biological Control by Natural Enemies*, 2nd edition. Cambridge University Press, Cambridge, 1991: 440 pp.

Ehler, L.E. Introduction strategies in biological control of insects. In Mackauer, M.; Ehler, L.E.; Roland, J. Eds.;. "*Critical Issues in Biological Control*." Intercept, Andover, 1990: 111-134.

Gonzalez, D.; Gilstrap, F.E. Foreign exploration: assessing and prioritizing natural enemies and consequences of preintroduction studies. In Kauffman, W.; Nichols, J. *Selection criteria and ecological consequences of importing natural enemies*. Thomas Say Pub. in Entomology, ESA, Lankam, Md. 1992: 53-70.

Herren, H.R.; Neuenschwander, P. "Biological control of cassava pests in Africa." *Annual Review of Entomology* **1991**, *36*: 257-283.

Hill, D.S. *Agricultural Insect Pests of Temperate Regions and Their Control*. Cambridge University Press, London, 1987: 659 pp.

Huffaker, C.B.; Messenger, P.S., Eds.;. *Theory and Practice of Biological Control*. Academic Press, New York, 1976: 788 pp.

Huffaker, C.B.; Simmonds, F.J.; Laing, J.E. The theoretical and empirical basis of biological control. In Huffaker, C.B.; Messenger, P.S. Eds.; "*Theory and Practice of Biological Control*." Academic Press, New York, 1976: 42-80.

Lenteren, J.C. van. "Evaluation of control

capabilities of natural enemies: does art have to become science?" *Netherlands Journal of Zoology*, **1980**, *30*: 369-381.

Lenteren, J.C. van. "The potential of entomophagous parasites for pest control." *Agriculture, Ecosystems and Environment*, **1983**, *10*: 143-158.

Lenteren, J.C. van. Evaluation, mass production, quality control and release of entomophagous insects. In Franz, J.M. Ed.;. *"Biological Plant and Health Protection."* Fischer, Stuttgart, 1986a: 31-56.

Lenteren, J.C. van. Parasitoids in thegreen house: successes with seasonal inoculative release systems. In Waage, J.K.; Greathead, D.J. Eds.;. *"Insect Parasitoids."* Academic Press, London, 1986b: 342-374.

Lenteren, J.C. van. Environmental manipulation advantageious to natural enemies of pests. In Delucchi, V. Ed.;. *"IPM Quo Vadis."* Parasitis Symposium Book, Geneva, 1987:123-163.

Lenteren, J.C. van. Quality control of natural enemies: hope or illusion? Proceedings 5th workshop Global IOBC working group *"Quality control of mass reared organisms."* Wageningen, 25-26 March 1991: 1-14.

Lenteren, J.C. van. Insect invasions: origins and effects. In *"Ecological Effects of Genetically Modified Organisms."* Netherlands Ecological Society, Amsterdam, 1992: 59-80.

Lenteren, J.C. van; Woets, J. "Biological and integrated control in greenhouses." *Annual Review of Entomology*, **1988**, *33*: 239-269.

Lenteren, J.C. van, Minks, A.K., Ponti O.M.B. de eds *"Biological Control and Integrated Crop Protection: Towards Environmentally Safer Agriculture."* Pudoc, Wageningen, 1992a: 239 pp.

Lenteren, J.C. van; Benuzzi, M.; Nicoli, G.; Maini, M. Biological control in protected crops in Europe. In Lenteren, J.C. van; Minks, A.K.; Ponti, O.M.B. de, eds *"Biological Control and Integrated Crop Protection: Towards Environmentally Safer Agriculture."* Pudoc, Wageningen, 1992b: 77-89.

Lewis, W.J.; Vet, L.E.M.; Tumlinson, J.H.; Lenteren, J.C. van; Papaj, D.R. "Variations in parasitoid foraging behavior: essential element of a sound biological control theory." *Environmental Entomology*, **1990**, *19*: 1183-1193.

Lim, G.S. Integrated pest management in the Asia-Pacific context. In Ooi, P.A.C. Ed.; *"Integrated Pest Management in the Asia-Pacific Region."*, CAB International, Wallingford, UK, 1991: 1-11.

Mackauer, M.; Ehler, L.E.; Roland, J. Eds.;. *Critical Issues in Biological Control."* Intercept, Andover, 1990: 330 pp.

Murdoch, W.W. The relevance of pest-enemy models to biological control. In Mackauer, M.; Ehler, L.E.; Roland, J. Eds.; *"Critical Issues in Biological Control."* Intercept, Andover, 1990: 1-24.

Pimentel, D. Population dynamics and the importance of evolution in successful biological control. In Franz, J.M. Ed.; *"Biological Plant and Health Protection."* Fischer, Stuttgart, 1986: 3-18.

Pimentel, D.; Stachow, U.; Takacs, D.A.; Brubaker, H.W.; Dumas, A.R.; Meaney, J.J.; O'Neil, J.A.S.; Onsi, D.E.; Corzilius, D.B. "Conserving Biological Diversity in Agricultural/Forestry Systems." *BioScience*, **1992**, *42*, *5*: 354-362.

Rabbinge, R. Options for integrated agriculture in Europe. In Lenteren, J.C. van, Minks, A.K., Ponti, O.M.B. Eds.; *"Biological Control and Integrated Crop Protection: Towards Environmentally Safer Agriculture."* Pudoc, Wageningen, 1992: 211-218.

Ridgway, R.L.; Vinson, S.B. Eds.;. *"Biological Control by Augmentation of Natural Enemies. Insect and Mite Control with Parasites and Predators."* Plenum, New York, 1977: 480 pp.

Sailer, R.I. History of insect introductions. In Graham C.; Wilson, C. Eds.; *"Exotic Plant Pests and North American Agriculture."* Academic Press, New York, 1983: 15-38.

Schwartz, P.H.; Klassen, W. Estimate of losses caused by insects and mites to agricultural crops. In Pimentel, D. Ed.; *"CRC Handbook of Pest Management in Agriculture."* CRC Press, Boca Raton, Florida, 1981, 1: 15-77.

Simmonds, F.J.; Greathead, D. In Cherrett, J.M.; Sager, G.R. Eds.; *"Origins of Pest, Parasite, Disease and Weed Problems."* Blackwell, Oxford, 1977: 109-124.

Stern, V.M.; Smith, R.F.; Bosch, R. van den; Hagen, K.S. "The integrated control concept." *Hilgardia*, **1959**, *29(2)*: 81-101.

Tisdell, C.A. Economic impact of biological control of weeds and insects. In Mackauer, M.; Ehler, L.E.; Roland, J. Eds.; *"Critical Issues in Biological Control."* Intercept, Andover, 1990: 301-316.

Vet, L.E.M.; Dicke, M. "Ecology of infochemical use by natural enemies in a tritrophic context." *Annual Review of Entomology*, **1992**, *37*: 141-172.

Vet, L.E.M.; Lewis, W.J.; Papaj, D.R.; Lenteren, J.C. van. "A variable-response model for parasitoid foraging behaviour." *Journal of Insect Behavior*, **1990**, *3*: 471-490.

Waage, J.K. Biological theory and the selection of biological control agents. In Mackauer, M. Ehler, L.E., Roland, J. Eds.; *"Critical Issues in Biological Control."* Intercept, Andover, 1990: 135-157

Waage, J.K., Hassell, M.P. "Parasitoids as biological control agents - a fundamental approach." *Parasitology*, **1982**, *84*: 241-268.

Waterhouse, D.F. Biological control: a viable strategy for the tropics. In Ooi, P.A.C.; Lim, G.S.; Teng, P.S. Eds.; *"Biological Control: Issues in the Tropics."* Malaysian Plant Protection Organization Society, Kuala Lumpur, 1992: 1-13.

Introduction of Natural Enemies for Suppression of Arthropod Pests

T. S. Bellows, Department of Entomology, University of California, Riverside, CA 92521

The intentional use of natural enemies for suppression of arthropod populations has been practiced for several hundred years. Appreciation of the significance of arthropod natural enemies grew during the 19th century. In the last 100 years, there have been approximately 1200 introductions of natural enemies against arthropods around the world. About 60% of all efforts have resulted in successful control of the target pest. The risks associated with such introductions are generally outweighed by the benefits, which include permanent suppression of pest species, reduction in contamination of the physical environment by pesticides, and protection of an agricultural or other ecosystem which would otherwise be lost to a region.

Biological control through introductions of natural enemies has experienced a long history of travail and success, from early efforts in the 18th and 19th centuries until the present, where it is perhaps more widely practiced than at any previous time. During its development, and the compilation of ideas regarding theory, implementation, and a long list of case histories, the practice of biological control has undergone expansions and refinements to meet increased demands on its use and safety.

In this paper I trace the development of biological control from its early inception through its current use, to place in perspective contemporary policies regarding its use in agriculture and other areas of society. Current trends in its use are framed by this background, yet take advantage of modern technology for introducing and colonizing candidate beneficial species.

BACKGROUND

The significance of introductions of natural enemies and current issues surrounding their use to control insect pests must be considered within the context of the history of the development of this field. Initial discovery of the action of natural enemies is perhaps lost in antiquity. Early published accounts include those of Aldrovani in 1602 and Redi in 1668 who recorded the emergence of *Apanteles* sp. larvae from lepidopteran hosts, and Goedaert in 1662 who illustrated hymenopteran parasitoids emerging from what appears to be a nymphalid chrysalis (DeBach & Rosen, 1991). The intentional introduction of natural enemies of arthropods was practiced from an early date. Predaceous ants were introduced into agricultural systems by both the ancient Chinese and Yemenese date growers (DeBach & Rosen, 1991).

Knowledge of insect parasitoids and predators, and their life histories, expanded during the 18th century. A number of naturalists appreciated the role of arthropod natural enemies by this time, and credited them with population regulation of phytophagous groups. Kirby & Spence (1815, *Introduction to Entomology*, London) wrote: "From the observations hitherto made by entomologists, the great body of the Ichneumon tribe is principally employed in keeping within their proper limits the infinite host of lepidopterous larvae, destroying, however, many insects of other orders...."

Interest in augmentation of arthropod natural enemy populations was also significant during this period. Tests using insect predators collected and translocated in the field were reported in Europe in the 1840's. Professor Boisgiroud used the predatory carabid *Calosoma sycophanta* in 1840 in an attempt to control gypsy moth larvae in willows. An Italian Society for the Promotion of Arts and Crafts offered in 1843 a gold medal to anyone who conducted successful tests with artificial breeding of predatory insects for control of agricultural pests. The medal was awarded to Antovio Villa for his 1844 report to the society on control of lepidopteran pests, flower and ground-dwelling insects using predatory beetles.

Knowledge of insect diseases also grew substantially in the 18th century. Much of the early work on insect diseases was related to silkworm culture, although Pliny wrote about diseases of bees in A.D. 77, as had Aristotle. Louis Pasteur spent five years (1865-1870) studying silkworm diseases (DeBach & Rosen, 1991). During the 1800's, knowledge on the contagious and infectious nature of pathogens grew, and the first suggestions of introducing such pathogens into insect populations for suppression appeared. The Italian Agostino Bassi of Lodi suggested in 1836 the use of pathogen sprays against insect larvae. J. L. LeConte in 1873 and Pasteur in 1874 suggested the use of silkworm diseases against other lepidopteran pests, and Pasteur in 1882 suggested the use of diseases against the grape phylloxera. In 1879, E. Metchnikoff published a paper describing the culture of the fungal pathogen *Metarrhizium anisopliae* to produce spores artificially on sterilized beer mash for use in augmenting natural populations of the fungus against the wheat cockchafer *Anisoplia austriaca*. Production was initiated in Smela, Russia in 1884. Over 50 kg of spores were produced and tested against *Cleonus* larvae, causing 55-80% mortality. Hence the principal concepts for culture and application of insect pathogens were in place more than 100 years ago, and interest in the use of fungi and other pathogens accelerated subsequently. Biological control of rhinocerus beetle, *Oryctes rhinoceros*, in the south Pacific by the baculovirus *Rhabdionvirus oryctes* is significant in that it is the only project known where an imported insect pathogen has alone effected substantial biological control of an important pest (DeBach & Rosen, 1991).

Together with this growing awareness of the significance of arthropod natural enemies, appreciation of their potential use to suppress pest populations also grew. During the second half of the 19th century, North American scientists were aware that many accidentally introduced pests of European origin were considered to be held under control in Europe by natural enemies. This impression led to recommendations for the intentional intercontinental transfer of natural enemies for control of these pests. The earliest records are of the recommendations of Asa Fitch, State Entomologist for New York regarding the destructive wheat midge, *Sitodiplosis mosellana*. He noted that "we here are destitute of nature's appointed means for repressing and subduing this insect. Those other insects which have been created for the purpose of quelling this species and keeping it restrained within its appropriate sphere have never yet reached our shores." In May 1855 Fitch wrote to John Curtis, President of the London Entomological Society, requesting cooperation in obtaining shipments of parasitoids, although no shipments resulted. His efforts were continued by Benjamin Walsh of Illinois, who also stressed the difference between the damage done in the United States and that occurring in Britain. In 1866 he wrote, with perhaps uncanny foresight, the

following lines in the *Practical Entomologist* expressing his frustration that little had been done on behalf of the affected agricultural industries:

"The plain common sense remedy ... is, by artificial means to import European parasites, that in their own country prey upon the Wheat Midge, the Hessian Fly and other imported insects that afflict the North American Farmer. Accident has furnished us with the bane; science must furnish us with the remedy. ... The simplicity and comparative cheapness of the remedy, but more than anything else the ridicule which attaches in the popular mind, to the very names of 'Bugs' and 'Bug-hunters', are the principal obstacles to its adoption. Let a man profess to have discovered some new Patent powder pimperlimpimp, a single pinch of which being thrown into each corner of a field will kill every bug throughout its whole extent, and people will listen to him with attention and respect. But tell them of any simple commonsense plan, based upon correct scientific principles, to check and keep within reasonable bounds the insect foes of the Farmer, and they will laugh you to scorn."

A younger associate of Walsh's, C. V. Riley, also strongly championed the use of biological control during this period, as did many of the prominent entomologists of the day, including L. O. Howard and J. H. Comstock. Interest in and use of natural enemies continued to grow and, in 1873, Riley made the first recorded successful international transfer of an arthropod predator by sending the mite *Tyroglyphus phylloxerae*, a predator of grape phylloxera (*Daktulosphaira vitifolii*), to Planchon in France. The first intercontinental establishment of a predaceous insect resulted from shipments made of aphid predators from England to New Zealand in 1874 (DeBach & Rosen, 1991), when the coccinellid *Coccinella undecimpunctata* was reported to become established. In 1883, the first parasitoid species to be established following international shipment was *Cotesia glomerata*, a braconid parasitic on cabbage butterfly *Pieris rapae*. It was shipped from England to the United States, colonized in several locations and soon became widely distributed.

Perhaps the most visible early result of the intentional movement and use of natural enemies of arthropods was control of cottony cushion scale, *Icerya purchasi*, following the introduction into California of two natural enemies, the parasitic fly *Cryptochaetum iceryae* and the predaceous beetle *Rodolia cardinalis*. These natural enemies were collected, the fly initially by Frazer Crawford of Australia in 1888 and both species by Albert Koebele of the United States (an associate of Riley) in 1889, in Australia and shipped to California. The introduction of a small number of flies and 129 beetles, followed by colonization efforts over much of the affected region of California, devastated the scale populations and rescued the citrus industry from imminent disaster. The project "established the biological control method like a shot heard around the world" (DeBach & Rosen, 1991). It additionally resulted in the second recorded award for work on natural enemies: the citrus industry awarded a gold watch to Koebele and diamond earrings to his wife for this work.

There was a cascade of similar projects on many arthropod pests in the next several decades [see, for example, Debach & Rosen (1991) for a partial list]. In each of these and many later cases, the consequences of accidental introductions of the pest into a region were serious and immediate, causing widespread damage to a major cropping system. The intentional introduction of one or more natural enemies generally resulted in complete amelioration of the pest problem.

It is critical to recognize that the motivation during these programs was almost entirely the suppression of pest populations, with biological control being viewed as the alternative to complete loss of an affected agricultural industry.

Programs subsequent to the second world war generally included the use of air transport for travel by the explorer and for transport of collected material. Programs were conducted successfully against numerous pests in many different countries around the world (DeBach & Rosen, 1991; Gould et al., 1992; Bellows et al., 1992a).

An important issue in considering the history of biological control by introduced natural enemies is the objective of these programs. In many cases, especially in the early projects, the objective was simple: to preserve a particular agricultural system which was threatened with disaster. Biological control was rarely one of a suite of options for control; often there were no options. Only in recent decades have other options for management of pests been sufficiently efficacious that they could be considered, and the speed and determinacy with which management based on broad-spectrum toxins could be implemented often weighed heavily in favor of their use. Even today we find biological control used most frequently when no other option is available, rather than as a priority program among a suite of options. As we near the 21st century, biological control is still employed frequently when efforts at control by pesticides, the first choice, has failed. Estimates indicate that biological control through introduced natural enemies has been applied only to 5% of pest species on a world basis (Van Driesche & Ferro, 1987).

CURRENT TRENDS

Throughout the world the introduction of natural enemies is undertaken by agencies, typically government agencies, as publicly-funded, public-sector service projects. Although some early programs were undertaken directly by affected industries (DeBach, 1964), most today are conducted by public agencies, often with assistance from private sector or additional government funds. They are undertaken on behalf of a constituency, frequently an agricultural constituency, with the object of ameliorating a problem that the constituency has not been able to resolve on its own or with private resources.

These public-sector programs vary in scope and complexity. Some are as local and focused as a single individual working in an experiment station, working on a problem facing a narrow segment of agriculture or society. Slightly larger than these programs are those involving an informal confederacy of several workers in several experiment stations. Larger still are programs that have demanded national attention, where some portion of the research efforts are coordinated among several regions through a multi-state or federal agency. Finally, there are occasionally programs of such major significance to several neighboring countries that an effort is mounted by an international confederation, often employing the expertise of agencies capable of conducting international programs (Smith et al., 1964; Herren & Neuenschwander, 1991). In all of these cases, efforts toward natural enemy introduction often involve the cooperation and assistance of entomologists in many parts of the world.

Current programs in biological control take advantage of modern communication and air travel to an extent not possible in early programs. Individuals working in agencies and institutions throughout the world often cooperate in survey, collection, and shipment of natural enemies. This does not obviate the need for continued experience and expertise in foreign exploration or collection by principal investigators in a program, for much can be learned about a pest insect by examining its ecology in a native or undisturbed setting.

Indeed, knowledge by a single individual of the distribution and ecology of the world fauna of natural enemies can often be the key to making informed decisions which are critical in completing a biological control program, as was the case in introductions against the walnut aphid (van den Bosch et al., 1970).

While technical developments in travel and communication have enhanced the speed and efficiency with which we can procure natural enemies for introductions, work on colonization of these natural enemies in quarantine laboratories and in field settings is still very much a matter of husbandry over the colonies. Initial shipments often include very few individuals (although some programs have involved shipment of thousands of natural enemies, e.g., Smith et al., 1964). Maintaining these individuals and using them as stock for colonies for further experimentation and release continue to be critical areas of biological control. Some groups of natural enemies are so difficult to rear, or their alternative host associations or ontogenies so insufficiently known, that they have so far defied efforts to ship or colonize them in new areas, and thus they are not available for use in biological control.

One frequent result of current approaches is the availability of numerous stocks of natural enemy species and populations. Such availability has raised questions regarding how resources for the colonization, rearing, and establishment of new natural enemies should be allocated. Some authors have posited that studies should be conducted prior to colonization, or at least prior to release, to identify the more promising species and permit concentration of limited resources on them (Greathead, 1986). Other authors have argued that such studies, when conducted comprehensively, are much too resource-consuming (González & Gilstrap, 1992). There have been few studies where both laboratory evaluations and field assessments of efficacy have been conducted on the same suite of natural enemies so that a comparison of traits relating to efficacy could be made. In the few cases where either field studies in the native home of a pest (e.g. Varley & Gradwell, 1968, cf. Embree, 1966; Smith et al., 1964) or laboratory studies prior to field colonization were conducted (Mackauer, 1971, cf. González & Gilstrap, 1992), there has been no definitive relationship between the performance or effectiveness of natural enemies and their value assessed *a priori*. Thus there is little evidence that *a priori* assessments can identify an optimal candidate natural enemy. This perhaps indicates that this portion of the science of biological control is not sufficiently well-developed to permit general or even situation-specific conclusions on what traits will ensure which natural enemy will prove most effective, or even that a particular natural enemy will establish, in a particular system.

It is perhaps more likely that particular traits might be associated with inefficient natural enemies and that these natural enemies could be culled during the introduction process with little loss of efficiency if resources are limiting. Such traits might include association with the target pest only at high densities, evidence that the pest is used as a factitious host only under duress, or evidence that the candidate species is a general predator not closely tied to the target pest.

Evaluation of the impact of natural enemies following field colonization has received considerable attention in the last few decades. Early work relied on experimental manipulations of systems to provide information on pest density in the presence and in the absence of natural enemies (Greathead, 1986; Luck et al., 1988; DeBach & Rosen, 1991). These approaches are still used and are effective means of demonstrating the impact of an introduced or other natural enemy. In addition, analyses of such population-level phenomena as survival and reproduction have been developed for the evaluation of natural enemies. This latter approach permits quantitative estimates of the mortality or reduced reproductive rate attributable to any particular natural enemy operating in a system (e.g., Bellows et al., 1992b).

Finally, the overall success of many programs has been closely related to the number of personnel and degree of work invested in the program. Programs conducted with more resources tend to be more successful (Greathead, 1986).

EFFICACY AND ECONOMIC ANALYSIS

Two kinds of efficacy evaluations are relevant to biological control programs by introductions: (i) how many and how often programs are successful; and (ii) what quantitative levels of population suppression are associated with successful programs.

Many analyses have been addressed to the first of these questions, most recently by Hall & Ehler (1979), Hall et al. (1980), and Greathead (1986). The analyses are difficult because the documentation associated with different programs varies in degree of detail. Nonetheless, some general conclusions are possible. For example, the total fraction of predators and parasitoids established against exotic pests is 0.34. Successful control was reported in ca. 60% of all cases, and control was complete in ca. 17% of all cases.

The question of degree of suppression achieved in "successful" biological control programs against arthropods has been infrequently addressed quantitatively. Early experiences regularly demonstrated qualitative differences of such marked degree that little quantitative data was published. The outcome often was one where a previously devastating pest was relegated to a historical footnote following a successful biological control program (Dalgarno, 1935, Edwards, 1936, Byrne et al., 1990). Quantitative data were more readily available in terms of changes in harvestable product in the affected agricultural system, as in the case for cottony cushion scale where exportable product increased 250-fold in a single year following introduction of the natural enemies (DeBach, 1964).

More recently, quantitative data on population biology has been an objective in studies evaluating the efficacy of natural enemies. Such data have indicated that the typical degree of suppression of pest populations in successful biological control programs have been between 2-3 orders of magnitude, or 100- to 1,000-fold reductions in population density (Beddington et al., 1978, Bellows et al., 1992a). Although the quantitative data are limited to a few cases, records include such diverse target organisms as Lepidoptera (Embree, 1966), aphids (van den Bosch et al., 1970), diaspidid scales (Debach et al. 1971), sawflies (Ives, 1976), and whiteflies (Summy et al., 1983, Bellows et al., 1992a) (Figs. 1, 2).

In most of these programs, the declines in pest densities took place over a span of time which occupied between 6-10 generations of the pest (host) population. This appears to be a general phenomenon, and may be related to the time required for a natural enemy population to grow to a level where it can exert its influence on the overall population dynamics of the host or prey. This time, although perhaps brief for multivoltine arthropods, may be several years for univoltine or bivoltine pests. This is the time between the addition of an effective natural enemy to a system and the time that the natural enemy effects control or suppression, and does not include the time which may be required to locate, transport, and colonize the natural enemy in the system. Thus the total time expected for a project may weigh against investment in or implementation of a biological control program if substantial economic losses must be sustained during this period, particularly if other options providing more immediate relief are available.

Economic analyses of the results of biological control programs by introductions are uncommon, in part because the effects of an introduced pest before and after natural enemy introduction are often vastly different, and economic assessments based on equilibrium-point analyses or marginal value analyses do not apply. DeBach (1964) report that in general, the immediate return on resources invested in introduction programs for biological control is on the order of 30:1. Additional benefits accrue as time passes because of the absence of alternative control costs during each subsequent production season, and also for the amount of environmental contamination prevented by the obviation of the use of environmental toxins such as pesticides. These values do not include consideration of related economic benefit for the number of jobs retained which otherwise might have been lost should an agricultural industry fail in a particular region. Perhaps most significant is the economic value of a program which sustains the production of subsistence agriculture in areas where alternatives are unavailable, as in the case of cassava mealybug in Africa (Herren & Neuenschwander, 1991).

In spite of the obvious economic benefits of biological control programs, the financing of research and development for biological control by introductions remains in the public sector. This is largely because the value of a program is not in a product with repeated marketability, but rather in the permanence of its solution. One consequence of this is that development of biological control is subject to funding emphasis in the public sector, which may not recognize the significance of it or may group research and development of biological control together with competing areas of research. Decline of resources directed at biological control may indicate only a constant state for resources during a period of expanding horizons and opportunities in other areas, so that total resources are divided among more areas. One risk of such fiscal policies is that the long-term stability of introductory biological control programs may decline over a period of decades, and this may place future generations at a disadvantage by not permitting the maintenance of a committed program to this area vital to agricultural production and environmental protection. Training of future scientists may decline with time if successively fewer scientists are available to conduct research and teaching in this area.

RISKS AND BENEFITS

The benefits of biological control by

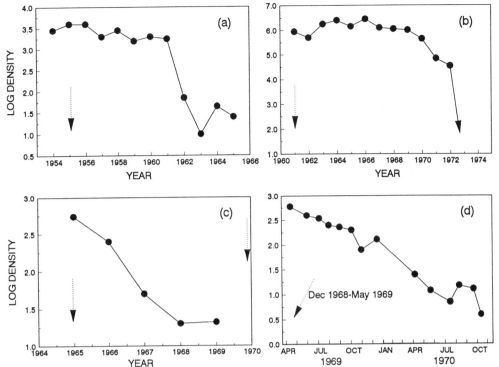

Fig. 1. Population densities of targeted species following introductions of natural enemies. (a) Decline in the univoltine winter moth (*Operophtera brumata*) following introductions of *Cyzenis albicans* (adapted from Embree 1966). (b) Decline in larch sawfly (*Pristiphora erichsonii*) following introductions of *Olesicampe benefactor* (adapted from Ives 1970). (c) Decline in California red scale (four generations per year) (*Aonidiella aurantii*) following introductions of *Aphytis melinus* in California (adapted from DeBach & Rosen 1971). (d) Decline in California red scale (*Aonidiella aurantii*) following introductions of *Aphytis melinus* in Australia (adapted from Campbell, 1976). Dotted arrows indicate approximate time natural enemy entered each system.

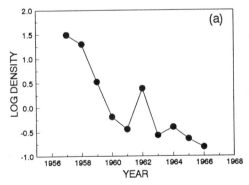

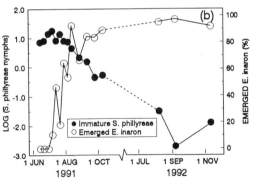

Fig. 2. Population densities of targeted species following the introduction of natural enemies. (a) Decline in olive scale (two or more generations per year) (*Parlatoria oleae*) following introductions of *Aphytis paramaculicornis* and *Coccophagoides utilis* (data from DeBach & Rosen 1971). (b) Decline in the whitefly *Siphoninus phillyreae* (approximately five generations per year) following introductions of *Encarsia inaron*; percentage of *S. phillyreae* bearing parasitoid larvae shown on right-hand ordinate (adapted from Bellows et al. 1992a).

introduced natural enemies have been profound (Howarth, 1991). Biological control has been crucial in the continued existence of many cultivation systems, and in the use of many ornamental plants, in countries around the world. For example, there is little doubt that citriculture in the western hemisphere would not have survived the onslaught of scales, mealybugs and whiteflies that have plagued it from Bermuda through the Caribbean, the Gulf States, Mexico, and California. Where alternative pest suppression methods are replaced by the presence of natural enemies, there may be additional benefit from reductions in environmental contamination by pesticides. Such benefits also may be profound, as reduction in pesticide usage presages reduction in ground water contamination, food-chain concentration of toxins, and other cascade effects. Thus the benefits of biological control by introduced natural enemies include both economic and environmental considerations.

The presence of benefits does not preclude the presence of risks, and potential risks associated with biological control have received attention. These perceived risks are usually associated with the biotic, rather than physical, environment, and are related to the effect of introduced natural enemies on non-target organisms. There is evidence for species extinctions following intentional introductions of natural enemies (Howarth, 1991), although much of the evidence reflects early practices, such as the introduction of vertebrates, that are no longer followed. However, risk to the environment appears real, and when considering effects of introductions into systems, the potential effects on related portions of the biotic community must be considered. It is unlikely that the complete impact of an introduction could ever be predicted entirely correctly, but some assessment of the risks involved can provide information to contrast with the potential benefits, and in this way considered decisions can be made which weigh the risks and benefits of this technology. While such additional considerations may not prevent the development of such programs, they may provide direction towards more risk-averse strategies in conducting introductions. In particular, polyphagous species are often considered as posing greater general risk than narrowly oligophagous species, and may be avoided in some programs.

experience in planning, structuring and executing such programs. This experience has led to some broad guidelines which aim at improved efficiency in the introduction process, consistent safety regarding the taxa involved, and efficacy of the intended solution. The guidelines often overlap in their area of impact. These guidelines are not absolute, but include the following ideas which are applicable in many current programs. (1) Vertebrates are rarely introduced against insect pests. (2) Insect parasitoids appear generally to be more effective natural enemies for use in introduction programs than either predators or diseases (Greathead, 1986). (3) Natural enemies with either a taxonomically or ecologically narrow host range are more likely to prove efficacious than generalists, and may have fewer unintended effects on the biotic environment. (4) Concern over harm to the physical environment is minimal following introductions, while increasing caution is used regarding harm to resident biota. This latter issue is rarely of definitive concern regarding the establishment or execution of a program, because programs are almost entirely motivated out of concern for public welfare; it may, however, affect the direction of a particular program, or decisions during the program regarding selection of species for introduction.

The general return on investments in successful biological control programs, when evaluated primarily on agricultural production, is of the order 30:1. Approximately 17% of programs are judged completely successful, and in these programs population levels of the target pest are typically reduced 100- to 1,000-fold, a remarkable achievement in any measure of pest suppression or management. At this rate of success (17%), the return on programs in general may be judged approximately 5:1. If return for partially successful programs (an additional 43% of projects) is included, this rate of return could be accounted higher. This is a superb return for public programs, and is evidence that resources placed into this area of endeavor are likely to return substantial benefits to society. Consequently, they are suitable areas for continued support for public research institutions, not only for crisis or regionally catastrophic pest problems, but also as a continuing basis for environmentally sound, biologically based technologies for the 21st century.

CONCLUSIONS

Current trends in the introduction of natural enemies take into consideration a century of

ACKNOWLEDGEMENTS

Thanks to R. D. Goeden and D. Headrick for reviews of the manuscript.

LITERATURE CITED

Beddington, J. R.; Free, C. A.; Lawton, J. H. "Characteristics of successful natural enemies in models of biological control of insect pests." Nature, **1978**, *273*, 513-519.

Bellows, T. S. Jr.; Paine, T. D.; Gould, J. R.; Bezark, L. G.; Ball, J. C.; Bentley, W.; Coviello, R.; Downer, J.; Elam, P.; Flaherty, D.; Gouveia, P.; Koehler, C.; Molinar, R.; O'Connell, N.; Perry, E.; Vogel, G. "Biological control of ash whitefly: a success in progress." California Agric, **1992a (January)**, *46(1)*, 24, 27-28.

Bellows, T. S. Jr.; Van Driesche, R. G.; Elkinton, J. S. "Life table construction and analysis in the evaluation of natural enemies." Annu. Rev. Entomol. **1992b**, *37*, 587-614.

Byrne, D. N.; Bellows, T. S. Jr.; Parrella, M. P. In *Whiteflies: their Bionomics, Pest Status and Management*; Gerling, D., Ed.; Intercept, Andover, Hants. (U.K.), 1990, pp. 227-262.

Campbell, M. M. "Colonization of *Aphytis melinus* Debach (Hymenoptera, Aphelinidae) in *Aonidiella aurantii* (Mask.) (Hemiptera, Coccidae) on citrus in South Australia." Bull. ent. Res. **1976**, *65*, 659-668.

Dalgarno, W. T. "Notes on the biological control of insect pests in the Bahamas." Trop. Agric. **1935**, *12*, 78.

DeBach, P.; Ed.; *Biological Control of Insect Pests and Weeds*; Chapman & Hall: London, 1964.

DeBach, P.; Rosen, D. *Biological Control by Natural Enemies*, *2nd edition*; Cambridge University Press: Cambridge, 1991.

DeBach, P.; Rosen, D.; Kennett, C. E. Biological control of coccids by introduced natural enemies. In *Biological Control*; Huffaker, C. B., Ed.; Plenum Press: New York, New York, 1971; pp 165-194.

Edwards, W. H. "Pests attacking citrus in Jamaica". Bull ent. Res. **1936**, *27*, 335-337.

Embree, D. G. "The role of introduced parasites in the control of the winter moth in Nova Scotia." Can. Ent. **1966**, *98*, 1159-1168.

González, D.; Gilstrap, F. E. "Foreign exploration: assessing and prioritizing natural enemies and consequences of preintroduction studies." In *Selection Criteria and Ecological Consequences of Importing Natural Enemies*; Kauffman, W. C.; Nechols, J. R., Eds.; Thomas Say Publications in Entomology: Proceedings; Entomological Society of America, Lanham, MD, 1992; pp 53-70.

Gould, J. R.; Bellows, T. S. Jr.; Paine, T. D.; "Population dynamics of *Siphoninus phillyreae* in California in the presence and absence of a parasitoid *Encarsia partenopea*." Ecol. Ent., **1992**, *17*, 127-134.

Greathead, D. J. In *Insect Parasitoids*; Waage, J. K.; Greathead, D. J., Eds.; 13th Symp. Roy. Entomol. Soc. London; Academic Press: San Diego, California, 1986; pp 289-318.

Hall, R. W.; Ehler, L. E. "Rate of establishment of natural enemies in classical biological control." Bull. ent. Soc. Amer. **1979**, *27*, 280-282.

Hall, R. W.; Ehler, L. E.; Bisabri-Ershadi, B. "Rate of success in classical biological control of arthropods." Bull. ent. Soc. Amer. **1980**, *26*, 111-114.

Herren, H. R.; Neuenschwander, P. "Biological control of cassava pests in Africa." Annu. Rev. Entomol. **1991**, *36*, 257-284.

Howarth, F. G. "Environmental impacts of classical biological control". Annu. Rev. Entomol. **1991**, *36*, 485-510.

Ives, W. G. H. "The dynamics of larch sawfly (Hymenoptera: Tenthredinidae) populations in southeastern Manitoba." Can. Ent. **1976**, *108*, 701-730.

Luck, R. F., Shepard, B. M., Kenmore, P. E. "Experimental methods for evaluating arthropod natural enemies." Annu. Rev. Entomol. **1988**, *33*, 367-391.

Mackauer, M. *Biological Control Programs Against Insect Pests and Weeks in Canada, 1959-1968*; Tech. Comm. No. 4.; Commonwealth Institute of Biological Control, Trinidad, 1971, pp 1-10.

Smith, H. D.; Maltby, H. L.; Jimenez-Jimenez, E. "Biological control of the citrus blackfly in Mexico." USDA Technical bulletin no. 1311; 1964. U.S. Dept. of Agriculture, Washington. 30 pp.

Summy, K. R.; Gilstrap, F. E.; Hart, W. G.; Caballero, J. M.; Saenz, I. "Biological control of citrus blackfly (Homoptera: Aleyrodidae) in Texas." Environ. Entomol. **1983**, *12*, 782-786. et al. 1983

van den Bosch, R.; Frazer, B. D.; Davis, C. S.; Messenger, P. S.; Hom, R. "*Trioxys pallidus*...an effective new walnut aphid parasite from Iran." California Agric. November 1970, *24*, 8-10.

Van Driesche, R. G.; Ferro, D. N.; "Will the benefits of classical biological control be lost in the 'biotechnology stampede'". Am. J. Alter. Agric. **1987**, *2*, 50,96.

Varley, G. C.; Gradwell, G. R. "Population models for the winter moth." In *Insect Abundance*; Southwood, T. R. E., Eds.; Symp. Roy. ent. Soc. Lon., No. 4; Royal Entomological Society of London, London, 1968, pp 132-142.

Augmentation of Parasites and Predators for Suppression of Arthropod Pests

Edgar G. King, Subtropical Agricultural Research Laboratory, Agricultural Research Service, U. S. Department of Agriculture, 2301 South International Blvd., Weslaco, TX 78596

Number and diversity of arthropod predators and parasites (natural enemies) inherently are limited in ephemeral agricultural systems. Augmentation of natural enemies by propagation and periodic releases is a rational solution to many arthropod pest problems. Seven case studies of augmentation are examined; 37 species of predators and parasites are cited. Ability to mass propagate, store, transport, and distribute quality-assured natural enemies constrains economic feasibility of the augmentative approach; failure of synthetic chemicals to provide lasting control coupled with environmental safety concerns opens the path for expanded commercialization of the approach.

Augmentation often has been "defined as periodically increasing either the number of [arthropod] parasites or predators or the supply of their food resources..." (Ridgway and Vinson, 1977). Consequently, environmental manipulations, e.g., supplemental foods, supplemental hosts, behavioral chemicals, and provision of shelters are activities that augment populations of predators and parasites (DeBach, 1964). However, this review is restricted to augmentation by periodic release of insectary-reared parasites and predators (natural enemies).

Researchers often classify augmentative releases with terms such as inoculative releases, seasonal colonization, supplemental releases, strategic releases, programmed releases, inundative releases, and compensatory releases. The goal of this paper is to document advances in biological control of arthropod pests by augmentative releases of parasites and predators, identify impediments limiting utilization of this suppressive strategy, and provide guidance for researching, developing, and transferring this technology.

Ecological Basis for Augmentative Releases

The ecological basis for augmentation of predators and parasites is reviewed in Ridgway and Vinson (1977). Knipling (1992) focused further on the augmentation of parasites for the control of several pest species, emphasizing the concept that the rate of parasitism is primarily a function of the ratio of parasite numbers to host numbers--with host density being of little consequence, i.e., parasitism is often independent of host density. Recent findings support this hypothesis (Strong, 1984; Gross, 1990; Hopper *et al.*, 1991).

Price (1992) articulated the need for manipulating natural enemies to achieve biological control in ephemeral agricultural systems. Ephemeral habitats, encompassing most agricultural systems but not forestry and some orchards, limit the efficacy of natural enemies in number of species colonizing hosts in these habitats and the abundance of individuals per species. Pest potential is compounded by the availability of vast food supplies, yet low genetic diversity of agricultural systems. Augmentation of natural enemy populations, by propagation and release, offsets the often low number of natural enemies inherent to ephemeral systems thereby providing an environmentally-rational solution to many arthropod pest problems.

Economic Basis for Augmentative Releases

The diversity of predator and parasite species now being reared as compared to the number commercially produced as recently as 1977 (Ridgway and Vinson) is evidence of the increasing demand for them by growers, ranchers, homeowners, and pest management professionals. There has been a rapid expansion in the number of companies producing predators and parasites, including companies traditionally perceived as producers of synthetic chemicals. For example, Ciba-Geigy Ltd. (Press Release, January 1993) announced a partnership with Bunting Group Ltd. (a natural enemy producer) to "pursue further opportunities in biological crop protection with beneficial insects and mites on a world-wide basis."

Lisansky (1990), Thompson (1992), and Anonymous (1992) have compiled recent information on producers and distributors of biological control agents. Upwards of 85 firms are producing and/or distributing over 100 predator and parasite species for control of over 40 arthropod pest species.

Listings in these catologues are restricted to firms in the United States, Canada, and Europe. In fact, the augmentation approach is more widely used in the former USSR and People's Republic of China for pest control than all other countries, combined. On the other hand, propagation and release of phytoseiids for control of the twospotted spider mite, *Tetranychus urticae*, and *Encarsia formosa* for control of the greenhouse whitefly, *Trialeurodes vaporariorum*, in the greenhouse industry is widely used in Europe as well as the former USSR (van Lenteren, 1989). Also, the use of *Trichogramma* spp. for control of the European corn borer, *Ostrinia nubilalis*, in corn is prevalent in Europe, as well as in the former USSR and China. There is growing interest for this usage in the United States and Canada.

I believe that early widespread use of augmentation in the Soviet republics and China was based on (1) lack of hard currency exacerbated by high cost of synthetic chemicals, (2) release of natural enemies was better than no control, (3) insectaries do not necessarily require sophisticated and expensive equipment and could be located on the communes or cooperative farms where labor-intensive procedures were more acceptable, and (4) less emphasis was placed on cosmetic quality of commodities. More recently, other countries are being driven to greater use of biological-based methods of control by genetic resistance to synthetic chemicals, high cost to develop and register new chemicals, concern over food and water safety, and adverse effects of pesticides on nontarget organisms, particularly humans and endangered species.

Implementation of the augmentative release strategy in greenhouses in Europe initially was driven by the rapid development of acaricide resistance in the twospotted spider mite. Consequently, biological control of this pest necessitated biological control of other pests in the greenhouse. On the other hand, use of *Trichogramma* spp. to control the European corn borer in Europe apparently is being driven by cost effectiveness to the end-user, but costs for the product may be subsidized in justifiable attempts to sustain emerging businesses and technologies,particularlyenvironmentally-rational technology. For example, in Germany, one local state subsidizes the differential (as much as 60%) between high cost *Trichogramma* products and cheaper synthetic pesticide equivalents (Newton, 1992).

Case Studies

Expansion in the number of parasite and predator species reared for augmentative releases as well as producers for these natural enemies is expected to continue. Environmental, social, and economic drivers of this expansion are expected to become even more acute in the future.

Scientific and popular literature documenting successes in controlling pests by augmentative release of predators and parasites is expanding exponentially. The following are some case studies based on natural enemy (e.g., *Trichogramma*), pest (e.g., filth-breeding flies), or situation (e.g., greenhouse). This review, in part, constitutes an update of a previous analysis (King et al., 1985b) on augmentation, primarily in the United States. No comprehensive worldwide review of the subject area has been conducted since 1977 (Ridgway and Vinson), but there have been numerous reviews of selective parts.

Greenhouse This term incorporates glasshouse and plastic tunnels for holding plants in sheltered and controlled environments. Plants in greenhouses receiving augmentative release(s) of predators and parasites include ornamentals (cut flowers and house or interior decorative plants) and vegetable and fruit (cucumber, strawberry, legumes, grape, pepper, eggplant, and peach).

Augmentation of predators and parasites is practiced in about 10 thousand hectares, primarily in western Europe, Japan, and former USSR, of the world's approximately 150 thousand hectares of greenhouse space (van Lenteren and Woets, 1988). Filippov (1992) reports up to 6.2 thousand hectares of greenhouses being treated with *Phytoseiulus persimilis* alone in the former USSR; these numbers are probably inflated because the same surface area is counted each time a release is made (S.H. Greenberg, Pers. Comm.).

Reviews on augmentation in greenhouses include van Lenteren and Woets (1988), Sunderland et al. (1992), and Hussey and Scopes (1985). The two species most often released are *P. persimilis* and *E. formosa* followed by *Dacnusa sibirica* for control of the tomato leafminer, *Liromyza bryoniae, Diglyphus isaea* for control of the serpentine leafminer, *Liriomyza trifoli*, the predatory midge, *Aphidoletes aphidimyza*, for control of aphids, and the predatory mite, *Amblyseius cucumeris*, for control of the onion thrip, *Thrips tabaci*. Lisansky (1990) provides a brief discussion on the "biology" and "symptoms and damage" for each of the pests listed above as well as a brief discussion on the "biology" and "application" for each natural enemy.

The effectiveness of augmentative releases of predators and parasites for control of pests in the greenhouse is unambiguous. However, there are apparent limitations. Multiple pests occur; so, biological control must be generally used for control of all pests, if it is used for control of one pest. Some chemicals may be used selectively.

Guidance for using predators and parasites in greenhouses typically recommends release either before the pest or pest symptoms are observed or when they first appear. Thompson (1992) and Lisansky (1990) recommend hanging cards, with parasitized (by *E. formosa*) whitefly "scales" attached, on plants as soon as whitefly adults are first observed. Release rates ranging from 7500 to 100,000 parasites/hectare are recommended depending upon the whitefly density and crop infested. Ledieu *et al.* (1989) reported that most growers introduce *P. persimilis* when twospotted spider mite damage is first observed. The predator is then periodically released at a rate of one per plant or higher in areas of high pest density.

Hussey and Scopes (1985) report increases in cucumber and tomato yields of 10 to 15% where biological control replaced chemical pest control, though some increase was related to reduced chemical phytotoxicity. Other benefits included no personnel reentry and pre-harvest interval restrictions. van Lenteren (1989) reported that the cost for biological control in greenhouses can be far lower than costs of chemical pest control. Ramakers (1982) estimated that chemical control of the twospotted spider mite was 2.5 times more expensive than control by release of predatory mites.

Filth-Breeding Flies Many firms in North America and Europe produce or market several parasites for control of muscoid flies around homes, dairy installations, cattle feedlots, and poultry housing (Anonymous, 1992; Thompson, 1992). Certification of parasite species, and species composition of marketed product is a common problem. Other problems in achieving effective biological control relate to the number released, time of release, and the specific environmental conditions at the time of release. Augmentation of the pteromalid, *Spalangia endius*, at a ratio of one female parasite to five host pupae has resulted in high rates of parasitization and a reduction of house fly, *Musca domestica*, and stable fly, *Stomoxys calcitrans*, populations. Data from one poultry house demonstrated increased parasitism in conjunction with reduced pupal populations (Morgan and Patterson, 1990). This subject area is reviewed by Rutz and Patterson (1990) and J. J. Peterson (these Proceedings) also addresses it.

Trichogramma spp. are the most widely studied and augmented natural enemies in the world. Hassan *et al.* (1988) reported 15 million hectares, worldwide, of crop and forest areas being treated annually by augmentative releases of *Trichogramma* spp. for control of caterpillar pests. However, Greenberg and Nikonov (1988) reported that *Trichogramma* were released annually over 16.5 million hectares in the former Soviet republics, alone (cf. Filippov, 1992), but these numbers are inflated two to three-fold by the accounting method (S.H. Greenberg, ARS-USDA, Weslaco, TX, Personal Communication).

Twenty-two producers/suppliers are referred to by Thompson (1992) as providing six different species of *Trichogramma*. One producer, BASF Aktiengesellschaft, is located in Germany; others are in the United States and Canada. Olkowski and Zhang (1990) reported nine species of *Trichogramma* being reared for commercial use in augmentation programs. According to Newton (1992), most applications of *Trichogramma* are relatively small scale and largely restricted, except for about 40,000 hectares, to developing regions of the world. Three main producers of *Trichogramma* in France, Germany, and Switzerland supply *T. brassicae* for control of the European corn borer in about 20,000 hectares of corn.

The best evidence on the efficacy and "commercialization" of *Trichogramma* spp. augmentative releases is for control of *Ostrinia* spp. in corn. Li (1984) reported 50 to 92% suppression of *O. furnacalis* in China following releases of *T. dendrolimi* in 606.7 thousand hectares of corn. Greenberg (1992) reported that release of *Trichogramma* in the former Soviet republics at a ratio of 1-5 parasite females:6-10 European corn borer egg masses/100 plants resulted in 60 to 80% egg parasitism and increased yields by 180 to 230 kg/hectare. Filippov (1992) reported that *Trichogramma* were released for control of the European corn borer over 1.95 million hectares of corn in the former Soviet republics during 1989. Reports of success from releasing *Trichogramma* spp. to control the European corn borer in Germany (Hassan *et al.*, 1986), Switzerland (Bigler and Brunetti, 1986), and France (Voegele, 1981) opened the pathway for large-scale commercialization of the practice in Europe.

Prokrym *et al.* (1992) reported that three releases of *T. nubilale* (totaling 4.4 million parasites/hectare) in sweet corn, reduced the number of larvae per fruit (ear), number of tunnels per stalk, and number of larvae per stalk; consequently, ear damage met acceptable standards for use in cut-corn commercial processing. Development of a large-scale rearing procedure for *T. nubilale* (generally more selective for European corn borer eggs but does not develop well in Angoumois grain moth, *Sitotroga cerealella*, eggs) in eggs of the tobacco hornworm, *Manduca sexta*, could open the path for making augmentative releases of *T. nubilale* for control of the European corn borer economically feasible in the United States (Nagarakatti and Keeley, 1987).

The *Heliothis/Helicoverpa* complex causes billions of dollars loss each year in crops such as cotton, corn, sorghum, tomato, and pulses. The state-of-the-art for biological control, including augmentation technology, of this pest complex is reviewed in King and Jackson (1989). Filippov (1992) reports that nearly five million hectares [total surface area is probably two to three million] of crops in the former Soviet republics infested by *Helicoverpa armigera* are treated with natural enemies. Apparently, the primary natural enemy used is *Trichogramma*, though *Bracon hebetor* is also released for control of *H. armigera* in cotton. Greenberg (1992) reports 60 to 75 thousand female *Trichogramma* released per hectare for control of noctuids, primarily *H. armigera*, in cotton, tomato, and corn. Li (1984) reported 60 to 91% egg parasitism and 70 to 98% suppression of *H. armigera* larvae following release of 300 to 600 thousand *T. dendrolimi* or 215 to 645 thousand *T. confusum* per hectare of cotton in China. In King *et al.* (1985a) the ability to consistently mass produce, transport, and aerially distribute large numbers of *T. pretiosum* in cotton is reported. Increased egg parasitism was achieved in the test, and lint yield was greater in parasite-release fields than in nonrelease fields, but the practice was not economically feasible when compared to the insecticide check. Results from releases in tomato have consistently shown that it is feasible to control *H. armigera* and the bollworm, *H. zea*, by augmentative releases of *Trichogramma* (Oatman and Platner, 1978; Patel, 1975); economic feasibility was indicated.

There are many other reports on the effectiveness of *Trichogramma* spp., worldwide, in controlling numerous lepidopteran species (e.g., Wajnberg and Vinson, 1991). Ridgway *et al.* (1981) extensively reviewed the potential for use of *Trichogramma* spp. in the United States. Moreover, *Trichogramma* spp. are released in millions of hectares of sugarcane for control of stem borers, particularly *Chilo* spp. in the Eastern

Hemisphere and *Diatraea* spp. in the Western Hemisphere (David and Easwaramoorthy, 1990; Metcalfe and Breniere, 1969). However, Metcalfe and Breniere (1969) opined that the effectiveness of *Trichogramma* for stemborer control in sugarcane was variable and, where practiced should be reexamined.

King *et al.* (1986) reviewed impediments to the use of *Trichogramma* for control of the bollworm and tobacco budworm in cotton. Impediments identified are germane for other pests and crops, but continued progress has been made in advancing technologies for using this parasite. The collaborative project established between Ciba Canada, the Government of Ontario, and researchers at the University of Guelph, University of Toronto, and Forest Pest Management Institute to develop the use of *T. minutum* serves as an example of the type of developments required to bring *Trichogramma* augmentative release technology to fruition.

Carrow *et al.* (1990) state "it is clear that the release of *Trichogramma* against forest pests is biologically and technologically feasible." Smith *et al.* (1990) after review of a five-year study using *T. minutum* in augmentative releases for control of the spruce budworm, *Choristoneura fumiferana*, demonstrated a curvilinear relationship between parasite releases and parasitism of egg masses. Parasitism rates ranged from 60% to 80% after double releases totaling four to 10 million parasites released per hectare.

Orchards This setting provides a more stable environment to practice biological control. Indeed, there are numerous examples of success where exotic natural enemies have been imported, released, established, and controlled a pest.

On the other hand, some pests occur annually requiring remedial control to optimize yield and quality. *Macrocentrus ancylivorus* is marketed by several firms for control of the oriental fruit moth, *Grapholita molesta*, in stone fruits, particularly peach; suggested release rate for the parasite is 2500/hectare (Anonymous, 1992; Thompson, 1992). Several phytoseiid species, particularly *Metaseiulus* (=*Typhlodromus*) *occidentalis*, are propagated and marketed for control of phytophagous mites on apple, avocado, and almond. Hoy *et al.* (1982) released a carbaryl-resistant strain of *M. occidentalis* into almond orchards at a cost of about $49/ha and controlled two species of tetranychids. By releasing this strain, carbaryl could be used to control other pests without eliminating the predaceous mite. These releases were not necessary every year; savings ranged from $59 to $109/ha (Croft, 1990). The California red scale,

Aonidiella aurantii, is a pest in much of the arid and semiarid citrus production areas of the world. The exotic parasite, *Aphytis melinus*, proved to be more successful than *Aphytis lingnanensis* in controlling the scale by augmentative releases. (Rosen and DeBach, 1979). Hundreds of millions of *A. melinus* are released annually in California, Arizona, and Mexico for control of the California red scale. Moreno and Luck (1992) experimentally demonstrated that optimally-timed releases of 49 thousand to 198 thousand *A. melinus*/ha reduced the percentage of infested lemon fruit at season's end. They concluded that this practice was cost competitive with the use of synthetic chemicals.

Dietrick (1989) reviewed the development of the Fillmore Citrus Protective District in the late 1930s, including augmentative releases of the lady beetle, *Cryptolaemus montrouzieri*, for control of mealybugs in citrus. Other natural enemies used include *Metaphycus helvolus* for control of black scale, *Saissetia oleae*, and *A. melinus* for control of California red scale. Continued emphasis by this 3660-ha farmer association on augmentative releases is the basis of its successful pest management program.

In China, phytoseiids such as *Amblyseius newsami* and *Amblyseius nicholsi* and predatory thrips, viz., *Aleurodothrips fasciapennis*, may be augmented for control of mite and diaspid scale pests (cf Du *et al.*, 1992). Phytoseiid populations may be conserved by interplanting cover crops, particularly *Ageratum conyzoides*, to reduce the temperature and increase humidity; their pollen provide an alternative food source for the mites.

Strawberry *Phytoseiulus persimilis* is widely used by California strawberry growers for control of the twospotted spider mite. Trumble and Morse (1993) report three to four releases of *P. persimilis* at the rate of 4,050 predators/ha per release provided a net benefit per hectare of $2,170 to $4,315/ha over the untreated control, but chemical control was three times higher. Regardless, *P. persimilis* releases were complimentary and additive with two applications of abamectin each at 0.011 kg(AI)/ha resulting in a net benefit of $6,890 to $19,705/ha; this was substantially greater than abamectin alone.

Chrysoperla This genus includes three predators, *Chrysoperla carnea* and *C. rufilabris* in North America, *C. carnea* in Europe, and *C. sinica* in China, that are often propagated and augmentatively released. The IPM Practitioner (Anonymous, 1992) lists 20 different companies in four countries that produce and market *Chrysoperla* spp. One producer, Better Yield Insects in Canada, distributes *Chrysoperla* to 36 countries (Lisansky, 1990).

Numerous scientific reports attest to the value of *Chrysoperla* spp. used in augmentative releases (King and Nordlund, 1992). Ridgway and co-workers (1968 to 1976), reported 33 to 99% reduction in bollworm and tobacco budworm larval populations in cotton with releases of 10 thousand to 420 thousand *C. carnea* larvae per acre. Anonymous (1979) reported 67 to 83% reduction in *Helicoverpa armigera* larval populations in cotton after the application of 150 thousand to 450 thousand *C. sinica* eggs per hectare. Applications of one *C. sinica* egg to 600 to 700 citrus mite, *Panonychus citri*, life stages suppressed the mite population and reduced annual pesticide applications by 62 to 84% (Peng, 1985).

Chrysoperla spp. are not used over large surface areas, primarily because of rearing expense. Thompson (1992) suggests the application of 10 thousand to 50 thousand predator eggs per acre, one predator egg per 10 aphids, or 1000 predator eggs per 20m². Prices vary widely, with the least expensive being $2.75/1000 eggs (Lisansky, 1990). Consequently, the release of non-predaceous adults is often suggested, but data supporting this practice are lacking.

Sugarcane Stem Borers *Trichogramma* spp., tachinids, and a braconid, *Cotesia* (=*Apanteles*) *flavipes*, are mass produced and augmentatively released in several major sugarcane-producing countries. (See preceding section on *Trichogramma*) Bennett reported on the rearing and release of several species of tachnids but believed that reports of increased parasitism of borer larvae should be regarded circumspectly. Li (1984) reported release of 45 to 150 thousand *T. confusum*/ha reduced damage from three borer species in 20 thousand hectares of sugarcane by 85 to 95%. King *et al.* (1984) reported up to 70% parasitism of sugarcane borer, *Diatraea saccharalis*, larvae over a four-month period following augmentative releases of the tachinid, *Lixophaga diatraeae*, at the rate of 250 flies/ha for a 10-week period; parasitism by two previously introduced and established parasites ranged from 7 to 24% during the test. IAA/PLANALSUCAR implemented and conducted a mass propagation and release program using the braconid *C. flavipes*, and the tachinids, *Metagonistylum minense* and *Paratheresia claripalpis*, for control of the sugarcane borer in south-central Brazil (Botelho, 1992). From 1976 to 1989, more than 7 billion *C. flavipes* were released (rate of 6,000 per hectare per year) in a cumulative total of 1.2 million hectares of sugarcane. Larval parasitism by *C. flavipes* ranged from 8 to 38 percent, with an average of 23 percent observed for the last 11 years of the program. Parasitism by all larval parasites averaged 36 percent. This program is viewed as being of unquestionable importance in the sugarcane belt of Brazil where many growers lack other effective means of control of the borer (Botelho, 1992).

Other Opportunities Of particular interest is the use of exotic parasites, which may not become established, but are effective when seasonally introduced and augmented. There are three striking examples: (1) The eulophid, *Pediobius foveolatus*, from India, released early season caused 100% parasitism of the Mexican bean beetle, *Epilachna varivestis*, and a four-fold reduction in insecticide usage (Stevens *et al.*, 1975); (2) High numbers of Colorado potato beetle eggs were parasitized following inoculative/augmentative releases of the eulophid, *Edovum puttleri*, from Colombia, in potato and eggplant, but limited commercial usage is restricted to eggplant (Lashomb, 1989); and (3) King *et al.* (1993) report for the first time on the technical feasibility of controlling the boll weevil, *Anthonomus grandis grandis*, a key pest of cotton, by inoculative/ augmentative releases of the exotic pteromalid, *Catolaccus grandis*. Releases of 2500 *C. grandis* females/ha caused 98% mortality of boll weevil third instar and pupae. Consequently, boll weevil numbers and damage were reduced.

Losses in revenue to stored grain pests such as *Sitophilus* spp., lesser grain borer, *Rhyzopertha dominica*, and the Angoumois grain moth, *Sitotroga cerealella*, are estimated at up to three billion dollars annually in the United States--in spite of synthetic chemical usage. Experimental tests demonstrated that augmentative releases of the parasite *Anisopteromalus calandrae* could reduce rice weevil and lesser grain borer populations by 95%. Release of the predator, *Xylocoris flavipes*, at the rate of 10 pairs/week into a 500-bushel bin resulted in strong suppression of several pest species. *Bracon hebetor* effectively controlled the Indian meal moth, *Plodia interpunctella*, and almond moth, *Cadra cautella*. Heretofore, use of predators and parasites in stored grain violated the U. S. Food, Drug, and Cosmetic Act, which classified insect parts, pest or natural enemy, as filth. Now, as of April 1992, parasites and predators are officially classified as pesticides, and the U. S. Environmental Protection Agency has exempted them from pesticide registration requirements and the need for establishment of a tolerance. An ARS, USDA Pilot Test to assess the economic feasibility of the augmentative approach to control stored grain pests is now in progress (J. Brower, ARS-USDA, Savannah, GA, Personal Communication).

Other pests are proving amenable to biological control by augmentation. Examples include parasites for control of cockroaches in dwellings, predators for control of mosquitoes, and the use of parasites and predators for control of pests in plantscapings. Gross (1990) reported substantial increases in rates of parasitism of corn earworm larvae in whorl-stage corn after release of 340 or 170 *Archytas marmoratus* females/ha and suggested use of the parasite for areawide suppression of the corn earworm and fall armyworm, *Spodoptera frugiperda*.

PROPAGATION AND QUALITY CONTROL

Development of rational propagation/ augmentation programs proceed from initial identification of established predator and parasite species as well as exotic species which may be introduced and augmented, development of small-scale *in vivo* propagation procedures to facilitate the study of natural enemy attributes in the laboratory and field, and then into mass propagation of the "best" species (King and Nordlund, 1992). Understanding the chemical ecology of habitat selection and host/prey finding, acceptance, and suitability for development of the natural enemy is critical to selection of the "best" species for propagation and augmentation. Other key attributes include rate of search, host/prey range, synchronization with host/prey life cycle, sensitivity to pesticides or other chemicals used in the production system, and effects of commonly used cultural or husbandry practices. Mass production of a quality-assured product at an economical price is perceived as a critical event in transfer and commercialization of the augmentative approach.

Mass production capability is exemplified by the Angoumois grain moth: *Trichogramma* production system in the former USSR. Each parasite production unit (line) is capable of producing sufficient *Trichogramma* to treat 35 to 45 thousand hectares per season. Each line is divided into host production (grain storage, grain sterilization and infestation, larval development, moth collection, moth holding and egg collection) and parasite production (exposure of host eggs to the parasite, holding and distribution of the parasitized eggs) (Personal Observations; S.H. Greenberg, ARS-USDA, Weslaco, TX, Personal Communication). These technologies have evolved through interactions between key institutes conducting biological studies and evaluations (formerly All-Union Plant Protection

Institute, St. Petersburg, Russia), automation of the system (former All-Union Scientific Engineering Institute, Odessa, Ukraine), and implementation of the technology (former All-Union Biological Means of Plant Protection Institute, Kishinev, Moldova).

King and Leppla (1984) review aspects of insect rearing dealing comprehensively with genetics of reared insects, diets and containers, engineering of insect rearing facilities and systems, control of microbial contaminants and pathogens, and actual rearing systems for several insect and mite species. Within the context of this book, systems for production of eight species of parasites and predators are described. In fact, van Lenteren (1991) reports that mass production programs are available for about 15 predator and parasite species. He opined that "simple, representative and reliable quality control programs for natural enemies are not yet available."

King et al. (1984) stated that systems for mass producing parasites and predators typically do not measure traits or attributes such as genetic variation (adaptability), diurnal rhythmicity, flight propensity, or flight ability (motility), sexual activity or host/prey selection--yet these traits relate directly to performance by the natural enemies (products). Most measurements are production oriented and include parasitization or predation rate, hatch or emergence, sex ratio, adult size (head capsule), fecundity, and survival.

The IOBC Global Working Group "Quality Control of Mass Reared Arthropods" (Bigler, 1991) is attempting to develop criteria for product quality control criteria for natural enemies, emphasizing initially those used for controlling pests in greenhouse crops and Trichogramma spp. Hoy et al. (1991) outline issues concerning the need to certify identification of commercially produced and marketed natural enemies and assure the quality of the product. Explicit guidelines are needed for holding and release of the natural enemies.

King et al. (1985b) suggest that artificial diets and in vitro systems could be developed for mass rearing predators and parasites and media costs would be minimal compared to rearing them on host or prey. Potentially, mass production would be less complex and expensive and a standardized quality organism could be produced. Twenty-five species of parasites and predators have been reared on different kinds of artificial media (Bratti, 1990). Trichogramma dendrolimi has been reared for up to 35 generations on artificial media containing insect hemolymph (Gao et al., 1982), and Geocoris punctipes has been reared for over 90 consecutive generations on artificial diet

(Cohen, 1985). Concern has been expressed that continuous rearing on artificial diet may result in a narrow-genetic base, with possible reduced effectiveness.

RELEASE/DELIVERY SYSTEMS

Delivery of predators and parasites for effecting control of targeted pests is conducted in a variety of fashions, ranging from labor-intensive "hand" releases to highly automated aerial releases. The complex geometric shapes and delicate body parts and extremities of release "formulations" of entomophagous arthropods (i.e., adults, pharate adults, pupae, or other immatures) present unique problems as well. Delivery of "naked" entomophages as well as the addition of carriers (liquids or granulars) have been researched to optimize delivery of predators and parasites. Manual release allows precise placement of the natural enemy, yet is generally time-consumptive and labor-intensive; thus it is practical only on a small- scale. On the other hand, aerial releases, though rapidly covering large plantings of a commodity, may result in "drift" from the target (either delivered outside of the area or into an extreme environment such as hot, dry soil). So, tradeoffs exist and the development of release systems are dictated by parameters inherent in the entomophage, pest, and situation where control is required.

Manual Release In most augmentative research studies as well as for certain types of commercial releases (i.e., glasshouses, orchards, stored products, field-grown vegetables and fruits, and animal confinements), entomophages may be distributed manually with little or no motorized equipment. Insectary host eggs containing pharate adults of Trichogramma, glued to a substrate (generally paper or cardboard), may be attached to or hung from plants by workers who uniformly select release points in the protected commodity. Container release devices may include paper capsules or other minute "cages" (TrichoCaps, D. Orr, Personal Communication) which are dropped or placed upon plant surfaces. Predatory mites may be placed with a granular carrier and the mixture dispensed by a shaker device onto the plant crown of field-grown strawberries. Soybean leaves with predatory mites collected from field propagation sites are manually placed in the crotches of orchard trees for control of spider mites.

Aerial Release Much of the aerial release technology is a result of modifying and improving devices and techniques used in autocidal sterile-release programs. Airborne delivery systems may be mounted on or in fixed-winged or rotary-winged aircraft (piloted or remote-controlled) as well as ultralight aircraft.

Bouse and Morrison (1985) described a mechanized, refrigerated delivery system housed in a fixed-winged aircraft for release of *Trichogramma pretiosum* in cotton. This same system was also used to release *P. persimilis* for control of spider mites infesting corn (Pickett *et al.*, 1987). An airborne automatic release system was developed to rapidly deliver *Epidinocarsis lopezi* and predatory mites to control mealybug and green mite in the cassava belt of Africa (Herren *et al.*, 1987).

CONCLUSIONS

Over 100 parasites and predators are cited by the IPM Practitioner for use in augmentative releases (Anonymous, 1992), affirming Parrella *et al.'s* (1992) title "Biological Control through Augmentative Releases of Natural Enemies: A Strategy Whose Time Has Come." On the other hand serious concerns over the "Issues and Ethics in Commercial Releases of Arthropod Natural Enemies" (Hoy *et al.*, 1991) must be dealt with. Predators and parasites must be quality assured and explicit guidance must be given as part of the "label" on how to store, transport, and release the organisms. The development of standardized tests for measuring and maintaining predator and parasite (product) quality and regulations (self-imposed by the industry or externally imposed) ensuring uniform labeling of the commercial product are required for development of a credible, professional industry.

The ARS, USDA-conducted "Workshop on Research Priorities "in 1987 (King *et al.*, 1988) cited 10 priority critical research needs for biological control of insect, mite, and tick pests. Six of these recommendation relate directly to control by propagation and augmentative release(s) of predators and parasites, as follows:

- "Develop technology for mass propagation (*in vivo* and *in vitro*), harvesting, packaging, storage, and distribution of quality-assured biological agents for control of insect, mite, and tick pests by augmentative releases;
- Develop ... [crop production] systems that enhance survival and effectiveness of biological control agents ...;
- Develop the technology for application/release of biological agents and for maintaining their effectiveness in tests to assess the technical and economic feasibility of augmenting biological agent populations. Transfer the augmentation technology ...;
- Develop technology for species/biotype selection and genetic improvement ... for

improving the effectiveness of augmented biological agents [see Hoy, these Proceedings];
- Develop cost-benefit analyses ... to accelerate the transfer of biological control [augmentation] technology; and
- Develop computer-based decision-making technology that makes explicit use of biological agent populations ..."

These research needs continue to be critical for development of technology to mass propagate and inoculate/augment natural enemy populations. Conduct of natural enemy releases on an areawide basis enhances, and may be necessary, to offset ingression by the pest into the release area and egression by the natural enemy from the release area.

References

Anonymous. "A preliminary report on the use of *Chrysopa sinica* Tjeder to control agricultural and forest pests." *Natural Enemies of Insect* **1979**, 3, 11-18.

Anonymous. *EPA's pesticide programs.* United States Environmental Protection Agency. 21T-1005. Washington, DC. **1991**, 25 pp.

Anonymous. "Directory of producers of natural enemies of common pests." *IPM Practitioner* **1992**, 15, 8-18.

Bennett, F. D. 1969. "Tachinid flies as biological control agents for sugar cane moth borers," pp. 117-148. In Williams, J. R.; Metcalfe, J. R.; Mungomery, R. W.; Mathes, R., Eds.; *Pests of Sugar Cane*: Elsevier Publishing Company: Amsterdam, London, New York.

Bigler, F., Ed.; *Fifth workshop of the IOBC global working group "Quality control of mass reared arthropods"*; Swiss Federal Research Station for Agronomy: Wageningen, NL, 1991.

Bigler, F.; Brunetti, R. "Biological control of *Ostrinia nubilalis* Hbn. by *Trichogramma maidis* Pint. et Voeg on corn for seed production in southern Switzerland." *J. Appl. Entomol.* **1986**, 102, 303-308.

Botelho, P. S. M. "Quinze anos de controle biologico da *Diatraea saccharalis* utilizando parasitoides." *Pesq. Agropec. Bras.* **1992**, 27, 255-262.

Bouse, L. F.; Morrison, R. K. "Transport, storage, and release of *Trichogramma pretiosum*." *Southwest. Entomol. Suppl.* **1985**, 8, 36-48.

Bratti, A. "Teniche di allevamento *in vitro* per gli stadi larvali di insetti entomofagi parassitoidi." *Boll. Ist. Ent. Univ. Bologna* **1990**, 44, 169-220.

Carrow, J. R.; Smith, S. M.; Laing, J. E. 1990. "*Summary and prospects for the future*," pp. 82-87. In Smith, S. M.; Carrow, J. R.; Laing, J.

E., Eds.; *Inundative release of the egg parasitoid, Trichogramma minutum (Hymenoptera: Trichogrammatidae), against forest insect pests such as the spruce budworm, Choristoneura fumiferana (Lepidoptera: Tortricidae)*: The Ontatio Project 1982-1986.

Croft, B. A. *Arthropod biological control agents and pesticides*; Wiley: New York, 1990.

David, H.; Easwaramoorthy, S. "Biological control of Chilo spp. in sugar-cane." *Insect Science and Its Application* **1990**, 11, 733-748.

Debach, P., Ed.; *Biological control of insect pests and weeds*; Chapman & Hall: London, 1964.

Dietrick, E. J. 1989. "Commercialization of biological control in the United States," pp. 71-87. In *International symposium on biological control implementation: Proceedings and abstracts*.

Du, T.; Liang, W.; Li, M.; Lu, Yuanding. 1992. "Using natural enemies to control citrus pest in China (abstract)," pp. 344. In *XIX International Congress of Entomology Beijing, China*. Beijing, China.

Filippov, N. A. 1992. "Place of biocontrol in integrated pest management in the former Union of Soviet Socialist Republics," pp. 1-16. In Soper, R. S.; Filippov, N. A.; Alimukhamedov, S. N., Eds.; *Cotton-integrated pest management: proceedings of a symposium.* USDA, ARS, Beltsville, MD. Tashkent, Uzbekistan.

Finney, G. L.; Fisher, T. W. 1964. "Culture of entomphagous insects and their hosts," pp. 329-355. In DeBach, P., Ed.; *Biological control of insect pests and weeds.*

Gao, Y. G.; Dai, K. J.; Shong, L. S. 1982. "Studies on the artificial host egg for *Trichogramma*," pp. 181. In *Les trichogrammes. Ier Symposium International*, April 20-23, Antibes, France.

Greenberg, Sh. M. 1992. "Production and application of Trichogramma in the former Union of Soviet Socialist Republics," pp. 38-46. In Soper, R. S.; Filippov, N. A.; Alimukhamedov, S. N., Eds.; *Cotton-integrated pest management: proceedings of a symposium.* USDA, ARS, Beltsville, MD. Tashkent, Uzbekistan.

Greenberg, Sh. M.; Nikonov, P. V. "Trichogramma: vozmozhnosti, perspektivi (In Russian)." *Zaschita rastenii*, **1988**, 7, 23-27.

Gross, H. R. "Field release and evaluation of *Archytas marmoratus* (Diptera: Tachinidae) against larvae of *Heliothis zea* (Lepidoptera: Noctuidae) in whorl stage corn." *Environ. Entomol.*, **1990**, 19, 1122-1129.

Hassan, S. A.; Kohler, E.; Rost, W. M. "Mass production and utilization of Trichogramma: 10. Control of the codling moth *Cydia pomonella* and the summer fruit tortrix moth *Adoxophyes orana* (Lep.: Tortricidae)." *Entomophaga* **1988**, 33, 413-420.

Hassan, S. A.; Stein, E.; Dannemann, K.; Reichel, W. "Massenproduktion und anwendung von *Trichogramma*: 8. Optimierung des Einsatzes zur Bekampfung des Maiszunslers *Ostrinia nubilalis* Hbn." *Z. Angew. Entomol.*, **1986**, 101, 508-515.

Herren, H. R.; Neuenschwander, P.; Hennessey, R. D.; Hammond, W. N. O. "Introduction and dispersal of *Epdinocarsis lopezi* (Hym., Encyrtidae), an exotic parasitoid of the Cassava mealybug, *Phenacoccus manihoti* (Hom., Pseudococcidae), in Africa." *Agric. Ecosystems Environ.* **1987**, 19, 131-144.

Hopper, K. R.; Powell, J. E.; King, E. G. "Spatial density dependence in parasitism of *Heliothis virescens* (Lepidoptera: Noctuidae) by *Microplitis croceipes* (Hymenoptera: Braconidae) in the field." *Environ. Entomol.* **1991**, 20, 292-302.

Hoy, M. A.; Barnett, W. W.; Reil, W. O.; Castro, D.; Cahn, D.; Hendricks, L. C.; Coviello, R.; Bentley, W. J. "Large-scale releases of pesticide resistant spider mite predators." *Calif. Agric.* **1982**, 36, 8-10.

Hoy, M. A.; Nowierski, R. M.; Johnson, M. W.; Flexner, J. L. "Issues and ethics in commercial releases of arthropod natural enemies." *Am. Entomol.* **1991**, 37, 74-75.

Hussey, N. W.; Scopes, N. *Biological Pest Control: The Glasshouse Experience*; Cornell University Press: NY., 1985.

King, E. G.; Bouse, L. F.; Bull, D. L.; Coleman, R. J.; Dickerson, W. A.; Lewis, W. J.; Liapis, P.; Lopez, J. D.; Morrison, R. K.; Phillips, J. R. "Management of *Heliothis* spp. in cotton by augmentative releases of *Trichogramma pretiosum*." *Zeitschrift fur Angewandte Entomologie* **1986**, 101, 1-10.

King, E. G.; Bull, D. L.; Bouse, L. F.; Phillips, J. R., Eds. "Biological Control of Bollworm and Tobacco Budworm in Cotton by Augmentative Releases of *Trichogramma*." *Southwest. Entomol. Suppl.* **1985a**, 8.

King, E. G.; Coulson, J. R.; Coleman, R. J. *ARS National Biological Control Program, Proceedings of Workshop on Research Priorities*; USDA, ARS: Washington, D.C., 1988.

King, E. G.; Hopper, K. R.; Powell, J. E. 1985b. "Analysis of systems for biological control of crop arthropod pests in the U.S. by augmentation of predators and parasites," pp. 201-227. In Hoy, M. A.; Herzog, D. C., Eds.; *Biological control in agricultural IPM systems.*; Academic Press, Inc.: Orlando, FL.

King, E. G.; Jackson, R. D., Eds.; *Proceedings of*

the workshop on biological control of Heliothis: increasing the effectiveness of natural enemies. 11-15 Nov. 1985, New Delhi, India; Far Eastern Regional Research Office, U.S. Department of Agriculture: New Delhi, India, 1989.

King, E. G.; Leppla, N. C., Eds.; *Advances and Challenges in Insect Rearing.*; Agricultural Research Service (Southern Region), U. S. Department of Agriculture: New Orleans, LA, 1984.

King, E. G.; Nordlund, D. A. "Propagation and augmentative releases of predators and parasitoids for control of arthropod pests." *Pesq. Agropec. Bras.* **1992**, 27, 239-254.

King, E. G.; Ridgway, R. L.; Hartstack, A. L. 1984. "Propagation and release of entomophagous arthropods for control by augmentation," pp. 99-121. In Adkisson, P. L.; Shijun, M., Eds.; *Proceedings of the Chinese Academy of Sciences-United States National Academy of Sciences Joint Symposium on Biological Control of Insects*, September 25-28, 1982, Science Press: Beijing, China.

King, E. G.; Summy, K. R.; Morales-Ramos, J. A.; Coleman, R. J. 1993. Integration of boll weevil biological control by inoculative/ augmentative releases of the parasite *Catolaccus grandis* in short-season cotton. Proc. Beltwide Cotton Conferences, Memphis, TN.

Knipling, E. F. *Principles of insect parasitism analyzed from new perspectives practical implications for regulating insect populations by biological means*; USDA, ARS: Washington, DC, 1992.

Lashomb, J. H. 1989. "Suppression of Colorado potato beetle Leptinotarsa decemlineata populations in egg plant by an egg parasite Edovum putleri," pp. 106. In *International symposium on biological control implementation: proceedings and abstracts.*

Ledieu, M. S.; Helyer, N. L.; Derbyshire, D. M. 1989. "Pests and diseases of protected crops," pp. 405-511. In *Pest and disease control handbook*, 3rd edition.

Li, L. 1984. "Research and utilization of Trichogramma in China," pp. 204-223. In Adkisson, P. L.; Shijun, M., Eds.; *Proceedings of the Chineses Academy of Sciences-United States National Academy of Sciences Joint Symposium on Biological Control of insects,* September 25-28, 1982, Science Press: Beijing, China.

Lisansky, S. G., Ed.; *The worldwide directory of agrobiologicals: Green growers guide*; CPL Press: United Kingdom, 1990.

Menn, J. J. 1992. "Prospects and status for development of novel chemicals for integrated pest management in cotton," pp. 105-117. In Soper, R. S.; Filippov, N. A.; Alimukhamedov, S. N., Eds.; *Cotton-integrated pest management: proceedings of a symposium.* USDA, ARS, Beltsville, MD. Tashkent, Uzbekistan.

Metcalfe, J. R.; Breniere, J. 1969. "Egg parasites (Trichogramma spp.) for control of sugar-cane moth borers," pp. 81-116. In Williams, J. R.; Metcalfe, R. W.; Mungomery, R. W.; Mathes, R., Eds.; *Pests of Sugar-cane*; Elsevier Publishing Company: Amsterdam.

Moreno, D. S.; Luck, R. F. "Augmentative releases of *Aphytis melinus* (Hymenoptera: Aphelinidae) to suppress California red scale (Homoptera: Diaspididae) in Southern California lemon orchards." *J. Econ. Entomol.* **1992**, 85, 1112-1119.

Morgan, P. B.; Patterson, R. S. 1990. "Efficiency of target formulations of pesticides plus augmentative releases of Spalangia endius Walker (Hymenoptera: Pteromalidae) to suppress populations of Musca domestica L. (Diptera: Muscidae) at poultry installations in the Southeastern United States," pp. 69-78. In Rutz, D. A.; Patterson, R. S., Eds.; *Biocontrol of arthropods affecting livestock and poultry.*

Nagarkatti, S.; Keeley, T. 1987. "Rearing Trichogramma nubilale (Hymenoptera: Trichogrammatidae) and field release against Ostrinia nubilalie (Lepidoptera: Pyralidae)," pp. 13-14. In *1987 Report of Activities, North Carolina Department of Agriculture.*

Newton, P. J. 1992. "Increasing the use of trichogrammatids in insect pest management - a case study from the forests of Canada," In Society of Chemical Industry meeting: *Biological control: use of living organisms in the management of invertebrate pests, pathogens and weeds.*

Oatman, E. R.; Platner, G. R. "Effect of mass releases of *Trichogramma pretiosum* against lepidopterous pests on processing tomatoes in southern California, with notes on host egg population trends." *J. Econ. Entomol.* **1978**, 71, 896-900.

Olkowski, W.; Zhang, A. "*Trichogramma*-A modern day frontier in biological control." *The IPM Practitioner* **1990**, 12, 1-15.

Parrella, M. P.; Heinz, K. M.; Nunney, L. "Biological control through augmentative releases of natural enemies: A strategy whose time has come." *Am. Entomol.* **1992**, 38, 172-179.

Peng, Y. "Field release of *Chrysopa sinica* as a strategy in the integrated control of *Panonychus citri*." *Chin. J. Biol. Control* **1985**, 1, 2-7.

Pickett, C. H.; Gilstrap, F. E.; Morrison, R. K.;

Bouse, L. F. "Release of predatory mites (Acari: Phytoseiidae) by aircraft for the biological control of spider mites (Acari: Tetranychidae) infesting corn." *J. Econ. Entomol.* **1987**, 80, 906-910.

Price, P. W. "Three-trophic-level interactions affecting the success of biological control projects." *Pesq. Agropec. Bras.* **1992**, 27, 15-29.

Prokrym, D. R.; Adnow, D. A.; Ciborowski, J. A.; Sreenivasam, D. D. "Suppression of *Ostrinia nubilalis* by *Trichogramma nubilale* in Sweet Corn." *Entomol. Exp. Appl.* **1992**, 64, 73-85.

Ramakers, P. M. J. 1982. "Biological control in Dutch glasshouses: practical applications and progress in research," pp. 265-270. In *Proc. Symp. Integrated Crop Protection*, CEC. Valence, France.

Ridgway, R. L.; Ables, J. R.; Goodpasture, C.; Hartstack, A. W. 1981. "*Trichogramma* and its utilization for crop protection in the United States," In *Proc. Soviet-American Conference on use of beneficial organisms in control of crop pests.* Entomol. Soc. Am. Publ.

Ridgway, R. L.; Vinson, S. B., Eds.; *Biological Control by Augmentation of Natural Enemies*; Plenum Press: NY, 1977.

Rosen, D.; DeBach, P. *Species of Aphytis of the world (Hymenoptera: Aphelinidae)*; Junk: The Hague, 1979.

Rutz, D. A.; Patterson, R. S., Eds.; *Biocontrol of arthropods affecting livestock and poultry*; Westview Press: Boulder, San Francisco, & Oxford, 1990.

Smith, S. M.; Carrow, J. R.; Laing, J. E., Eds.; *Inundative release of the egg parasitoid, Trichogramma minutum (Hymenoptera: Trichogrammatidae), against forest insect pests such as the spruce budworm, Choristoneura fumiferana (Lepidoptera: Tortricidae)*: The Ontario Project 1982-1986; Entomological Society of Canada: Ottawa, Canada, 1990.

Stevens, L. M.; Steinhauer, A. L.; Coulson, J. R. "Suppression of Mexican bean beetle on soybeans with annual inoculative releases of *Prediobius foveolatus*." *Environ. Entomol.*, **1975**, 4, 947-952.

Strong, D. R. 1984. "Density-vague ecology and liberal populations regulation in insects," pp. 313-327. In Price, P. W.; Slobodchikoff, C. N.; Gaud, W. S., Eds.; *A new ecology: novel approaches to interactive systems.*

Sunderland, K. D.; Chambers, R. J.; Helyer, N. L.; Sopper, P. I. "Integrated pest management of greenhouse crops in Northern Europe." *Horticultural Reviews* **1992**, 13, 1-68.

Thompson, W. T. *A worldwide guide to beneficial animals (insects/mites/nematodes) used for pest control purposes*; Thompson Publications: Fresno, CA, 1992.

Trumble, J. T.; Morse, J. P. "Economics of integrating the predaceous mite, *Phyoseiulus persimilis* (Acari: Phytoseiidae), with pesticides in strawberries" *J. Econ. Entomol.* **1993**.

van Lenteren, J. C. 1989. "Implementation and commercialization of biological control in west Europe," pp. 50-70. In *International Symposium on Biological Control Implementation.* McAllen, TX.

van Lenteren, J. C.; Woets, J. "Biological and Integrated Pest Control in Greenhouses." *Annu. Rev. Entomol.* **1988**, 33, 239-269.

Voegele, J. "La lutte biologique contre *Ostrinia nubilalis* a l'aide des trichogrammes." *Bull. OEPP* **1981**, 11, 91-95.

Wajnberg, E.; Vinson, S. B., Eds.; *Trichogramma* and other egg parasitoids. 3rd International Symposium, San Antonio, TX, September 23-27, 1990; Les Colloques de l'I.N.R.A: 1991; Vol. 9.

Recent Examples of Conservation of Arthropod Natural Enemies in Agriculture

James D. Dutcher, Coastal Plain Experiment Station, Department of Entomology, University of Georgia, Tifton, GA 31973-0748

Outbreaks of phytophagous insects and mites often occur in agricultural systems when indigenous natural predators and parasitoids are reduced or eliminated by pesticides or climatic factors. Outbreaks of certain pests can be avoided when natural enemies are conserved. Conservation methods are increasing in importance with reductions in new pesticide registration and increased insecticide resistance. This paper reviews some recent attempts to conserve natural enemies in agriculture and suggests methods for integration of conservation into modern commercial production systems.

Natural controls are important regulators of insect and mite populations in arable crops, orchards and forests. Crop management practices alter natural controls and can change the population behavior of insects and mites. Among these natural controls, predators, parasitoids and pathogens (natural enemies) are often manipulated in biological control strategies as alternatives to chemical controls of pests (DeBach and Rosen, 1991). Among biological control tactics, conservation of natural enemies is a set of premediated actions to protect and maintain natural enemies (Rabb et al, 1976) by avoiding measures that impact negatively on natural enemies (Gross, 1987) and manipulating the environment to enhance the activity of natural enemies against the pest (DeBach and Rosen, 1991). Biological control, including conservation of natural enemies, has recently been reviewed in crops and orchards (DeBach and Rosen, 1991; Mackauer et al, 1990), forests (Speight and Wainhouse, 1989), and greenhouses (Hussey and Scopes, 1985).

The goals of conservation of predators and parasitoids are to reduce pest populations to low equilibrium levels and prevent pest epidemics. Conservation methods include: selective pesticides (Croft, 1990); establishing refugia (Pickett et al, 1990); intercropping (Vandemeer, 1989), cover cropping (Bugg et al, 1990a) and weedy culture (Altieri and Whitcomb, 1979); natural (Leius, 1967) and artificial (Hagan et al, 1970) food supplements; and, reduction of broad spectrum insecticide applications, using resistant cultivars, scouting,

economic thresholds and prediction models (Dent, 1991). Pathogens are often conserved to hasten the decline of an outbreak population. They are conserved by increasing survival of inoculum through selective pesticide use (Pickering et al, 1990), and by encouragement of inoculum dispersal, e.g. conservation of birds that spread disease by feeding on diseased prey and excreting viable inoculum away from the initial pest infestation (Entwistle et al, 1983). Ignoffo (1985) encourages research on methods to conserve the large amounts of inoculum produced by insect pathogens during an epizootic.

Phytophagous insects and mites interact with their natural enemies in crop systems against the backdrop of low plant diversity and broad spectrum pesticides. Countering these two main areas of adversity to natural enemies has been the focus of natural enemy conservation. This paper reviews some recent examples of conservation methods for natural enemies and suggests methods for integrating conservation into crop systems.

Low Plant Diversity

Plant monocultures in agriculture often create favorable conditions for populations of a few phytophagous insects and mites by concentrating host plants and reducing natural enemies. Planting alternate crops adjacent to the primary crop provides: food for adult parasitoids and predators; protected areas for epigeal predators; shelter during the winter for predators (van

Emden, 1990); and nesting sites for predatory birds (Spreight and Wainhouse, 1990). Conservation methods for increasing plant diversity include plant polyculture (Altieri et al, 1991; Mizell and Shiffhauer, 1987; Tedders, 1983; Bugg and Dutcher, 1989; 1993; Bugg et al, 1990a; 1990b); attraction of predators to low prey populations with supplemental prey (Bugg and Dutcher, 1989); and, providing nectar for adult parasitoids (Zandstra and Motooka, 1978; Leius, 1967). Reduced weed control and conversion of large fields to pasture to prevent overproduction of grain crops have contributed to natural enemy conservation by increasing plant diversity near agricultural areas (van Emden, 1990).

Broad Spectrum Pesticides

Chemical controls are important suppression tactics for insect pests when an injurious epidemic occurs or when the endemic population level is above the economic injury level. Many crops have key pests that can only be reduced below economic injury levels with pesticides. The development of pesticide resistance and subsequent pest resurgence (Toner, 1992) have increased the importance of conservation methods in many agricultural systems. Judicious use of broad spectrum insecticides will often conserve natural enemies and lower the risk of the resurgence of secondary pests. In specific examples, the negative effects of insecticides have been reduced by: timing insecticide applications to coincide with presence of key pests and absence of natural enemies (Johnson et al, 1976; Heyerdahl and Dutcher, 1985); reducing application rates (Poehling, 1990) and frequency (Boethel and Ezell, 1978). The toxicity of insecticides is often higher for natural enemies than for the pest. Certain insecticide controls of the same pest vary in toxicity to the natural enemies (Hassan et al, 1991; Mizell and Schiffhauer, 1990; Niemczyk et al, 1990, Powell and Scott, 1991). Certain insecticides have a favorable selective toxicity to the natural enemies from the pest (Croft, 1990). Expert systems are often used to organize knowledge on pesticide toxicity so that crop managers will be able to select a tactic that will reduce pests

and conserve natural enemies (Messing and Croft, 1990)

Integration of Conservation and Pest Control

Polyculture methods to conserve natural enemies have had varying degrees of success. A faster decline in green peach aphid populations occurred in polyculture of tomato and zucchini over populations in tomato alone. An increase in convergent lady beetle density in the polyculture was responsible (Altieri et al, 1991). Weeds were an important reservoir for certain parasitoids of leafminers, *Lyriomyza* spp., in tomato fields (Schuster et al, 1991). Intercropping maize with rice, legumes or weeds often reduced populations of oriental maize borer. These reductions were not associated with an increase in predator abundance in the intercrop over the monocrop, but were due to interactions between the plants and the oriental maize borer (Litsinger et al, 1991). Intercropping beans with maize reduced bean fly infestation. The reduction may have been due to parasitoids and restricted movement of the bean fly in the intercrop (Karel, 1991). Wireworm damage to potato increased after early-planting of summer cover crops and decreased after late-planting of the cover crop (Jansson and Lecorne, 1991).

Changes in cultural practices can conserve natural enemies. Predators of pea aphid and *Lygus* spp. are conserved on alfalfa by replacing a pre-harvest insecticide treatment with harvesting by swathing or clear-cutting (Harper et al, 1990), or strip-cutting (Summers, 1976). Harvesting date was timed to selectively kill more alfalfa weevils than parasitoids (Casagrande and Stehr, 1973). Cultivar selection often influences the efficacy of natural enemies. Nectariless cotton has 16 % lower bollworm egg parasitism by *Trichogramma* spp. than nectaried cotton. Nectar and honey increase adult parasitoids longevity (Treacy et al, 1987).

Insecticide materials and/or time of application can be changed to conserve natural enemies. Rice stink bug control with methyl parathion reduced emergence of a scelionid parasitoid of eggs more than carbaryl (Sudarsono et al, 1992). Carbaryl sprays for filbert leafroller reduced predators of the filbert aphid. Early season sprays with carbaryl against filbert leafroller could be replaced with formulations of the selective microbial insecticide, *Bacillus thuringiensis* (Bt), to conserve parasitoids and predators (AliNiazee, 1983). The mummified shell of green peach aphids protects immature parasitoids from insecticide sprays. Conservation is possible in the greenhouse by spraying when mummy density is high (Shean and Cranshaw, 1991).

Recently, a predator was combined with a microbial insecticide to control a primary pest. The twospotted stink bug, a native predator, controlled Colorado potato beetle at low and not at high population densities. The combination of twospotted stink bug and a treatment with Bt extended the period of control by two weeks. This prevented an outbreak that occurred after treatment with Bt alone (Hough-Goldstein and Keil, 1991).

The combination of generalist predators that are abundant during pest outbreaks and specialist parasitoids and/or predators in endemic pest populations is particularly effective in reducing pest populations. Two separate successful biological control introductions of the aphidiid parasitoid, *Trioxys pallidus*, are reported against filbert and walnut aphids (Messing and Aliniazee, 1988; van den Bosch et al, 1979). Initial surveys of the importance of various natural enemies (Messing and Aliniazee, 1985; Sluss, 1967) indicated a lack of important aphid parasitoids in the two systems. In both systems generalist predators were abundant but did not regulate aphid populations.

Generalist predators and parasitoids often are not synchronized with their prey populations and have variable control ability. *Aphis pomi* is not consistently controlled on apple (Carroll and Hoyt, 1984) by generalists in the tree and on the ground. High populations of the *A. pomi* occur when generalist aphidiid parasitoids and coccinellids fail to attack fundatrix colonies in the Spring. In a second example, indigenous generalist Hemipteran predators of *Lygus* spp. in apple (Arnoldi et al, 1991) were associated with high populations of damaging populations of *Lygus* spp. during the early season and not during low populations of *Lygus* spp. The predators moved to sheltered habitats surrounding the apple orchards when *Lygus* spp. populations were low in apples.

Generalist predators and parasitoids can regulate pest populations effectively in some cases. Predators were important natural controls of cabbage looper, a secondary pest of cotton (Ehler, 1973). Ants and other generalist predators increased in importance as regulators of boll weevil as the season progressed. Planting late cotton exposed weevils to higher natural mortality, primarily predation by red imported fire ants (Sturm et al, 1990). Ant predation also reduced damage by fall armyworm to irrigated maize even though ant foraging was reduced by broad spectrum insecticide treatments (Perfecto, 1991). Conservation of *Bracon mellitor*, an important generalist parasitoid of the boll weevil, may be possible by sowing or conserving host plants of a large number of alternate host insects of the parasitoid during the spring and fall in Texas when boll weevil larvae are not available as hosts (Tillman and Cate, 1989).

Negative effects of predators preying on natural enemies (e.g., ants, Tedders et al, 1990) can be reduced by partitioning their foraging behavior with barriers (Samways et al, 1981; Phillips and Sherk, 1991), or by culturing alternate prey and primary predators on plants that repel ants (Bugg and Dutcher, 1989; Kaakeh and Dutcher, 1992). Secondary parasitoids can reduce the efficacy of natural controls in systems where a few primary parasitoids regulate a single host (Oliver, 1964). The negative effects of secondary parasitoids appear to be diluted in more complex systems with an array of host species and many oligophagous parasitoids (Dutcher and Heyerdahl, 1988; Nickels et al, 1950).

Several conservation methods are combined to conserve predators of pecan aphids. Aphid populations can double every 2-3 days and thus outbreaks are common (Kaakeh and Dutcher, 1992a). Natural enemies of these aphids occur in high numbers after the outbreak has peaked (Edelson and Estes, 1987). Carbaryl sprays for pecan weevil reduce predator populations and multiple applications appear to be more detrimental than single applications (Dutcher and Payne, 1983). Scouting can reduce the number of carbaryl sprays for effective pecan weevil control (Eikenbary et al, 1978). Propagation of alternate prey aphids on cover crops to sustain high populations of predators when pecan aphids populations are low has been reported (Tedders, 1983; Bugg et al, 1990a; Bugg and Dutcher 1989). Only cool season crops have been associated with reductions in pecan aphid populations. Warm season crops can increase coccinellids by as much as 125 times (Bugg and Dutcher, 1993) but a technique is needed to attract them into the tree crown. One problem is secondary predation by red imported fire ants on primary predators in the cover crops and in the trees (Tedders et al, 1990, Perfecto and Sediles, 1992). Red imported fire ants are also important predators of pecan weevil larvae on the soil surface (Dutcher and Sheppard, 1981). Foraging by red imported fire ants can easily be partitioned by excluding them from the tree crown with insecticide trunk sprays (after Samways et al, 1981) and still benefit from their foraging on the ground. The ants are naturally excluded from a cover crop of hemp sesbania (Bugg and Dutcher, 1989) and predator populations

increase without interference from red imported fire ants. Chemical ant repellents and ant toxins have been demonstrated in this plant (Kaakeh and Dutcher, 1992b). Another problem is that cover crops are also hosts, and often preferred hosts, for tarnished plant bugs (Fleischer and Gaylor, 1987), leaffooted bugs (Bugg and Dutcher, 1989) and velvetbean caterpillars (Waters and Barfield, 1992).

Crops, like many large-scale systems (Ryaciotaki-Boussalis, 1980), have unstable biological components with stability preserving interconnections. Conservation of arthropod natural enemies preserves stability between the crop and arthropod pests by reducing the pest population level and preventing pest population outbreaks. The goal of the crop manager is to stabilize phytophagous insects and mites below their economic injury levels. Natural enemies will often control pests at a steady state population level that often is above the economic injury level. Crop managers often apply chemical controls in these situations. In my view of the pecan system (Fig. 1), the biological components are naturally unstable

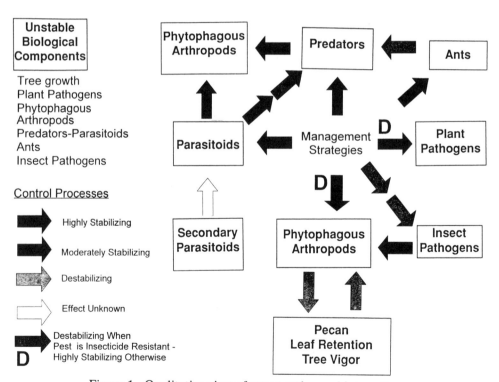

Figure 1. Qualitative view of pecan arthropod interactions.

104

without the negative feedback mechanisms of natural enemies and management strategies. Interactions between components are viewed as stabilizing or destabilizing based on the effect on pest population level. Natural controls contribute to the stability of a crop by reducing pest populations. These reductions are not usually sufficient when the phytophagous insect is pernicious or attacks marketable plant parts. Natural enemies are a subset of natural controls that can be manipulated by various methods. The diverse array of conservation methods defies generalization. Conservation methods have to be devised for each crop and tailored for each growing region. However, the more successful programs incorporate a broad base of knowledge on natural pest controls.

References

Aliniazee, M. T. "Pest status of filbert (hazelnut) insects: a 10-year study." *Can, Entomol.* **1983**, *115*, 1155-1162.

Altieri, M. A.; Whitcomb, W. H. "The potential use of weeds in manipulation of beneficial insects. *HortSci.*, **1979**, *14*, 12-18.

Altieri, M. A.; Trujillo, J. A.; Astier, M. A.; Gersper, P. L.; Bak, W. A. "Low input technology proves viable for limited-resource farmers in Salinas Valley." *Calif. Agric.*, **1991**, *45*,20-23.

Arnoldi, D.; Stewart, R. K.; Boivin, G. "Field survey and laboratory evaluation of the predator complex of *Lygus lineolaris* and *Lygus communis* (Hemiptera: Miridae) in apple orchards." *J. Econ. Entomol.*, **1991**, *84*, 830-836.

Boethel, D. J.; Ezell, J. E. "Influence of spray programs on seasonal abundance of pecan leaf scorch mite and phytoseiids in Louisiana." *J. Econ.Entomol.*, **1978**, *71*, 508-517.

Bugg, R. L.; Dutcher, J. D. "Warm-season cover crops for pecan orchards: horticultural and entomological implications." *Biolog. Agric. Hortic.* **1989**, *6*, 123-148.

Bugg, R. L.; Dutcher, J. D. "*Sesbania exaltata* (Rafinesque-Schmaltz) Cory (Fabaceae) as a warm-season cover crop in pecan orchards: effects on aphidophagous Coccinellidae and pecan aphids." *Biolog. Agric. Hortic.*, **1993**, *9*, 215-229.

Bugg, R. L.; Dutcher, J. D.; McNeill, P. J. "Cool-season cover crops in the pecan orchard understory: effects on Coccinellidae (Coleoptera) and pecan aphids (Homoptera: Aphididae)." *Biol. Control.* **1990a**, *1*, 8-15.

Bugg, R. L.; S. C. Phatak, S. C.; Dutcher, J. D. "Insects associated with cool-season cover crops in southern Georgia: implications for pest control in truck-farm and pecan agroecosystems." *Biol. Agric. Hortic,* **1990b**, *7*, 17-45.

Carroll, D. P.; Hoyt, S. C. "Natural enemies and their effects on apple aphid, *Aphis pomi* DeGeer (Homoptera: Aphididae), colonies on young apple trees in central Washington." *Environ. Entomol.*, **1984**, *13*, 469-481.

Croft, B. A. *Arthropod Biological Control Agents and Pesticides.* Wiley & Sons; New York, 1990.

DeBach, P.; Rosen. D. *Biological Control by Natural Enemies.* Cambridge University Press; Cambridge, UK, 1991.

Dent, D. *Insect Pest Management;* C. A. B.International; Oxon, UK, 1991.

Dutcher, J. D.; Heyerdahl, R. In *Advances in Parasitic Hymenoptera Research*; Gupta, V. K., Ed.; E. J. Brill: New York, NY, 1988; pp. 445-458.

Dutcher, J. D.; Payne, J. A. "Impact assessment of carbaryl, dimethoate, and dialifor on foliar and nut pests of pecan orchards." *J. Georgia Entomol. Soc.*, **1983**, *18*, 495-507.

Dutcher, J. D.; Sheppard, D. C. "Predation of pecan weevil larvae by red imported fire ants." *J. Georgia Entomol. Soc.*, **1981**, *16*, 211-213.

Edelson, J. V.; Estes, P. M. "Seasonal distribution of predators and parasites associated with *Monelliopsis pecanis* Bissell and *Monellia caryella* (Fitch) (Homoptera: Aphidae [sic])." *J. Entomol. Sci.*, **1987**, *22*, 336-347.

Eikenbary, R. D.; Morrison, R. D.; Hedger, G.H.; Grovenburg, W. G. "Development and validation of prediction equations for estimation and control of pecan weevil populations." *Environ. Entomol.*, **1978**, 7, 113-120.

Ehler, L. E.; Eveleens, K. G.; van den Bosch, R. "An evaluation of some natural enemies of cabbage looper on cotton in California." *Environ. Entomol.*, **1973**, *2*, 1009-1015.

Entwistle, P. F.; Adams, P. H. W.; Evans, H. F.; Rivers, C. F. "Epizootiology of a nuclear polyhedrosis virus (Baculoviridae) in European spruce sawfly (*Gilpinia hercyniae*)." *J. Appl. Ecol.*, **1983**, *20,* 473-487.

Fleischer, S. J.; M. J. Gaylor. "Seasonal abundance of *Lygus lineolaris* (Heteroptera: Miridae) and selected predators in early season uncultivated hosts: implications for managing movement into cotton." *Environ. Entomol.*, **1987**, *17*, 246-253.

Gross, H. R. "Conservation and enhancement of entomophagous insects." *J. Entomol. Sci.* **1987**, *22*, 97-105.

Hagan, K. S.; Sawell, Jr., E. F.; Tassan, R. L. "The use of food sprays to increase effectiveness of entomophagous insects." *Proc. Tall Timbers Conf. Ecol. Anim. Control Habitat Manage.* **1970**, *2,* 59-82.

Harper, A. M.; Schaber, B. D.; Story, T.P.; Entz, T. "Effect of swathing and clear-cutting alfalfa on insect populations in southern Alberta." *J. Econ. Entomol.*, **1991**, *84*, 2050-2057.

Hassan, S. A.; Bigler, F.; Bogenschutz, H.; Boller, E.; Brun, J.; Calis, J. N. M.; Chiverton, P.; Coremans-Pelseneer, J.; Duso, C.; Lewis, G.B.; Mansour, F.; Moreth, L.; Oomen, P. A.; Overmeer, W. P. G.; Polgar, L.; Rieckmann, W.; Samsoe-Petersen, L.; Staubli, A.; Sterk, G.; Tavares, K.; Tuset, J. J.; Viggiana, G. "Results of the fifth joint pesticide testing programme carried out by the IOBC/WPRS-working group 'Pesticides and Beneficial Organisms'." *Entomophaga*, **1991**, *36*, 55-67.

Heyerdahl, R.; Dutcher, J. D. "Management of the pecan serpentine leafminer (Lepidoptera: Nepticulidae)." *J. Econ. Entomol.* **1985**, *78*, 1121-1124.

Hough-Goldstein, J.; Keil, C. B. "Prospects for integrated control of the Colorado potato beetle (Coleoptera: Chrysomelidae) using *Perillus bioculatus* (Hemiptera: Pentatomidae) and various pesticides." *J. Econ. Entomol.*, **1991**, *84*, 1645-1651.

Hussey, N. W.; Scopes, N. E. A. *Biological Pest Control: The Glasshouse Experience.* Poole, Dorset, Blanford, 1985.

Ignoffo, C. M. In *Biological Control in Agricultural IPM Systems.* Hoy, M. A.; Herzog, D. C. Ed.; Academic Press, New York, 1985.

Jansson, R. K.; Lecorne, S. H. "Effects of summer cover crop management on wireworm (Coleoptera: Elateridae) abundance and damage to potato." *J. Econ. Entomol.*, **1991**, *84*, 581-586.

Johnson, E. F.; Laing, J. E.; Trottier, R. "The seasonal occurrence of *Lithocolletis blancardella* (Gracillariidae), and its major natural enemies in Ontario apple orchards." *Proc. Entomol. Soc. Canada*, **1976**, *107*, 31-45.

Kaakeh, W.; Dutcher, J. D. "Estimation of life parameters of *Monelliopsis pecanis, Monellia caryella,* and *Melanocallis caryaefoliae* (Homoptera: Aphididae) on single pecan leaflets." *Environ. Entomol.*, **1992a**, *21*, 632-639

Kaakeh, W.; Dutcher, J. D. "Foraging preference of red imported fire ants (Hymenoptera: Formicidae) among three species of summer cover crops and their extracts." *J. Econ. Entomol.*, **1992b**, *85*, 389-394.

Karel, A. K. "Effects of plant populations and intercroppping on the population patterns of bean flies on common bean." *Environ. Entomol.*, **1991**, *20*, 354-357.

Leius, K. "Influence of wild flowers on parasitism of the tent caterpillar and codling moth." *Can. Entomol.*, **1967**, *99*, 865-871.

Litsinger, J. A.; Hasse, V.; Barrion, A. T.; Schmutterer, H. "Response of *Ostrinia furnacalis* (Guenee) (Lepidoptera: Pyralidae) to intercropping." *Environ. Entomol.*, **1991**, *20*, 988-1004.

Mackauer, M.; Ehler, L. E.; Roland, J. eds. *Critical Issues in Biological Control,* Intercept, Andover, Hants 1990, 330 p.

Messing, R. H.; Aliniazee, M. T. "Natural enemies of *Myzocallis coryli* (Homoptera: Aphididae) in Oregon hazelnut orchards." *J. Entomol. Soc. Brit. Columbia,* **1985**, *82*, 14-18.

Messing, R. H.; Aliniazee, M. T. 1988. "Hybridization and host susceptibility of two biotypes of *Trioxys pallidus* (Hymenoptera: Aphidiidae)." *Ann. Entomol. Soc. Amer.*, **1988**, *81*, 6-9.

Messing, R. H.; Croft, B. A. "NERISK: An expert system to enhance the integration of pesticides with arthropod biological control." *Acta Hortic.*, **1990**, *276*, 15-20.

Mizell, R. F.; Schiffhauer, D. E. "Seasonal abundance of the crapemyrtle aphid, *Sarucallis kahawaluokalani*, in relation to the pecan aphids, *Monellia caryella* and *Monelliopsis pecanis* and their common predators." *Entomophaga,* **1987**, *32*, 511-520.

Mizell, R. F.; Schiffhauer, D. E. "Effects of pesticides on pecan aphid predators, *Chrysoperla rufilabris* (Neuroptera: Chrysopidae), *Hippodamia convergens, Cycloneda sanguinea* (L.), *Olla v-nigrum* (Coleoptera: Coccinellidae), and *Aphelinus perpallidus* (Hymenoptera: Encyrtidae)." *J. Econ. Entomol.* **1990**, *83*, 1806-1812.

Nickels, C. B.; Pierce, W. C.; Pinkney, C. C. "Parasites of the pecan nutcasebearer in Texas." USDA Tech. Bull. No. 1011, **1950**, 21p.

Niemczyk, E.; Koslinska, M.; Maciesiak, A.; Nowakowski, Z.; Olsak, R.; Szufa, A. "Some growth regulators: their effectiveness against orchard pests and selectivity to predatory and parasitic arthropods." *Acta Hortic.* **1990**, *285,* 157-164.

Oliver, A. D. "Studies on biological control of the fall webworm, *Hyphantria cunea* in Louisiana." *J. Econ. Entomol.*, **1964**, *57*, 314-318.

Perfecto, I. "Ants (Hymenoptera: Formicidae) as natural control agents of pests in irrigated maize in Nicaragua." *J. Econ. Entomol.*, **1991**, *84*, 65-70.

Perfecto, I.; Sediles, A. "Vegetational diversity, ants (Hymenoptera: Formicidae), and herbivorous pests in a Neotropical agroecosystem." *Environ. Entomol.* **1992**, *21*, 1-67.

Pickering, J.; Dutcher, J. D.; Ekbom, B. S. "The effect of a fungicide on fungal-induced mortality of pecan aphids (Homoptera: Aphididae) in the field." *J. Econ. Entomol.*, **1990**, *83*, 1801-1805.

Pickett, C. H.; Wilson, L. T.; Flaherty, D. L. In *Monitoring Integrated Management of Arthropod Pests of Small Fruit Crops.* Bostanian, N. J.; Wilson, L. T.; Dennehy, T. J. eds. Intercept, Andover, Hampshire 1990, 301 pp.

Phillips, P. A.; Sherk, C. J. "To control mealy-bugs, stop honeydew-seeking ants." *Calif. Agric.*, **1991**, *45*, 26-28.

Poehling, H. M. "Use of reduced rates of pesticides for aphid control: economic and ecological aspects." *Monog. Br. Crop Prot. Counc.*, **1990**, *45*, 77-86.

Powell, J. E.; Scott, W. P. "Survival of *Microplitis croceipes* (Hymenoptera: Braconidae) in contact with residues of insecticides on cotton." *Environ. Entomol.*, **1991**, *20*, 346-348.

Rabb, R. L.; Stinner, R. E.; van den Bosch, R.; In *Theory and Practice of Biological Control.* Huffaker, C. B.; Messenger, P. S., Ed.; Plenum Press, New York, 1976, pp. 294-311.

Ryaciotaki-Boussalis, H. A. "Stability of large-scale systems." Ph.D. Dissertation, **1980**, N. Mex. State Univ., Las Cruces. 149p.

Samways, M. J.; Weaving, A. J. S.; Nel, M. "Efficacy of chemical and sticky banding in preventing ants entering guava trees." *Subtropica*, **1981**, *2,* 3pp.

Schuster, D. J.; Gilreath, J. P.; Wharton, R. A.; Seymour, P. J. "Agromyzidae (Diptera) leafminers and their parasitoids in weeds associated with tomato in Florida." *Environ. Entomol.*, **1991**, *20*, 720-723.

Shean, B.; Cranshaw, W. S. "Differential susceptibilities of green peach aphid (Homoptera: Aphididae) and two endo-parasitoids (Hymenoptera: Encyrtidae and Braconidae) to pesticides." *J. Econ. Entomol.*, **1991**, *84*, 844-850.

Sluss, R. R. "Population dynamics of the walnut aphid, *Chromaphis juglandicola* (Kalt.) in northern California." *Ecology*, **1967**, *48*, 41-58.

Spreight and Wainhouse, 1989. *Ecology and Management of Forest Insects.* Clarendon Press, Oxford, UK, 395 pp.

Sturm, M. M.; Sterling, W. L.; Hartstack, A. W. "Role of natural mortality in boll weevil (Coleoptera: Curculionidae) management programs." *J. Econ. Entomol.*, **1990**, *83*, 1-7.

Sudarsono, H.; Bernhardt, J. L.; Tugwell, N. P. "Survival of immature *Telonomus podisi*

(Hymenoptera: Scelionidae) and rice stink bug (Hemiptera: Pentatomidae) embryos after field applications of methyl parathion and carbaryl." *J. Econ. Entomol.*, **1992**, *85*, 375-378.

Tedders, W. L. "Insect management in deciduous orchard ecosystems: habitat manipulation." *Environ. Manage.* **1983, 7**, 29-34.

Tedders, W. L.; Reilly, C. C.; Wood, B. W.; Morrison, R. K.; Lofgren, C. S. "Behavior of *Solenopsis invicta* (Hymenoptera: Formicidae) in pecan orchards." *Environ. Entomol.*, **1990**, *19*, 44-53.

Tillman, P. G.; Cate, J. R. "Six new hosts of *Bracon mellitor* (Hymenoptera: Braconidae) with a review of recorded hosts." *Environ. Entomol.*, **1989**, *18*, 328-333.

Toner, M. "When bugs fight back." *The Atlanta Journal-Constitution*, **1992**, *42(25)*, A1, A8.

Treacy, M. F.; Benedict, J. H.; Walmsley, M. H.; Lopez, J. D.; Morrison, R. K. "Parasitism of bollworm (Lepidoptera: Noctuidae) eggs on nectaried and nectariless cotton." *Environ. Entomol.*, **1987**, *16*, 420-423.

van den Bosch, R.; Hom, R.; Matteson, P.; Frazer, B. D.; Messenger, P. S.; Davis, C. S. "Biological control of walnut aphid in California: impact of the parasite, *Trioxys pallidus*." *Hilgardia*, **1979**, *47*, 1-13.

van Emden, H. F. In *Critical Issues in Biological Control*. Mackauer, M.; Ehler, L. E.; Roland, J. eds. Intercept, Andover, Hants, 1990.

Vandermeer, J. *The Ecology of Intercropping*. Cambridge Univ. Press, Cambridge, UK, 1989.

Waters, D. J.; Barfield, C. S. "Larval development and consumption by *Anticarsia gemmatalis* (Lepidoptera: Noctuidae) fed various legume species." *Environ. Entomol.*, **1989**, *18*, 1006-1010.

Zandstra, B. H.; Motooka, P. S. "Beneficial effects of weeds in pest management - a review." *PANS*, **1978**, *24*, 333-338.

Aspects of Tritrophic Interactions of Russian Wheat Aphid

David K. Reed, John D. Burd, and Norman C. Elliott, USDA-Agricultural Research Service, 1301 N. Western, Stillwater, OK 74075
Russell K. Campbell, Oklahoma State University, Dept. of Entomology, Present address Guam Dept. of Agriculture, PPQ Unit, P.O. Box 2950, Agana, Guam 96910

Classical biological control strategies and resistant germplasm development are primary components of an IPM program against Russian wheat aphid (RWA). Studies of interactions among three trophic levels were initiated using esistant and susceptible wheats, barleys, triticales, and slender wheatgrasses along with the RWA and a parasitoid. Host plants resistant by antibiosis conferred detrimental effects to parasitoids including decreased adult emergence, reduced adult size and mummy weights and increased preoviposition periods. Drought stress caused reduced parasitization on resistant wheats. A tolerant wheat with reduced leaf rolling is a good candidate for IPM. Greater coordination is needed between biocontrol and resistance programs.

Within any management system of plant pests, a knowledge of interactions between the major trophic levels is important. Tritrophic relationships in cereal ecosystems have been reported (Starks *et al.*,1972). These studies were limited to relationships between greenbug, *Schizaphis graminum*, its hosts, and a parasitoid, *Lysiphlebus testaceipes*. Since the Russian wheat aphid, *(RWA) Diuraphis noxia,* is a recently (1986) introduced pest to the U.S., we initiated experiments to determine beneficial or detrimental effects of resistant and susceptible hosts on natural enemies.

Materials and Methods

Interactions among three trophic levels of resistant wheat (PI 372129), triticale (PI 386148), slender wheatgrasses (PI 440100 and PI 440102), and susceptible wheat (TAM W-101), triticale (Beagle 82), and wheatgrasses (PI 387888 and Pryor) were studied in three experiments. RWA and a parasitoid, *Diaeretiella rapae (Hymenoptera:Aphidiidae)* were reared in the greenhouse at 24 ± 5^0 C and L16:D8 photoperiod provided by metal halide lamps. The plant entries were grown under the same conditions in a fritted clay medium in cone-tainers (Supercell Cone-tainer, Ray Leach Cone-Tainer Nursery, Canby, OR). Treatments in the first two experiments consisted of host entries subjected to (A) control (no aphids or parasitoids); (B) aphids alone: and (C) aphids plus parasitoids (1 female/cage). Mated female parasitoids were confined in the (C) cages with ca 50 aphids for 24 h and removed for measurements. Aphids and parasitoids were confined with clear vented plastic tubing (3.5 x 50 cm). After seven days, aphids and mummies were counted and weighed (Mettler UM3 Microbalance, Hightstown, NJ), and parasitoids allowed to emerge. After emergence, parasitoids were sexed and measurements of head width and femur length were taken. Analysis of variance and frequency analysis (PROC GLM and PROC FREQ, SAS Institute, 1985) were used to test treatment differences ($P < 0.05$). In the third experiment, methods were identical, but treatments included combinations of drought, RWA infestation, and *D. rapae*. Drought stress treatments received only minimal maintenance water.

Results and Discussion

These experiments led to the conclusion that (1) the trophic level represented by plants can influence the third level represented by parasitoids, and (2) parasitoids can decrease damage to plants by lowering aphid levels.

Growth rate of aphids weighed daily (Fig. 1) was influenced by plant entries. Aphids reared on wheat were significantly heavier than those reared on resistant grasses. Aphids reared on the most resistant entries had an almost identical growth rate.

Table 1 shows the number of aphids on plants, with and without parasitoids, 7-10 days after parasitoid introduction. Aphid populations were influenced by both the plant entry and by parasitoids. Generally, plants resistant by antibiosis had lower aphid populations than susceptible and tolerant plant entries, and drought stressed plants contained greater populations of aphids. Overall, significantly fewer aphids were found on plants with parasitoid activity, and it should be remembered that this reduction occurred with a 24 h activity period of one *D. rapae* female.

The mean weights of parasitoid mummies from the three tests are presented in Table 2. Mummy weights were lighter from entries resistant by antibiosis. PI 387888 mummy weights were erratic, indicating that such data might not be valid criteria for evaluating immature parasitoid development; further research has borne this out (Reed *et al.,* in press). Mummy weights from drought stressed plants were greater than those from well watered ones. Emergence time of parasitoids were generally lengthened by exposure to resistant entries, and particularily to those resistant by antibiosis. The parasitoids also took longer to emerge from drought stressed resistant plants than from well watered susceptible wheat. There were no significant differences between emergence times of parasitoids reared on wheatgrasses, whether resistant or susceptible, although all times were greater than those from wheat. Possibly, the

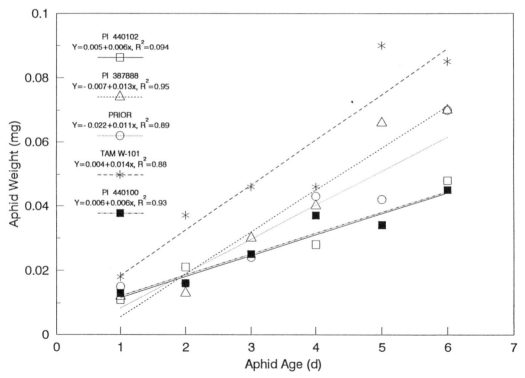

Fig. 1. Relationship between age and weight of Russian wheat aphids confined on five plant entries varying in resistance expression. (PI 440102 and PI 440100 are resistant wheatgrasses, PI 387888 and Prior are susceptible wheatgrasses, and TAM 'W-101' is a susceptible wheat.

(Reproduced with permission from Reed et al., 1992). Copyright 1992 Kluwer Academic Publishers.

Table 1. Mean number (SEM) of Russian wheat aphids on resistant and susceptible entries with aphids only and aphids plus *Diaeretiella rapae*.

Plant entry	Host status	Aphid only	Aphid + parasitoids	Percent Reduction
Experiment 1				
'Beagle 82'	Triticale–S	260 (120) b	77 (61) b	70.4
PI 386148	Triticale–R	111 (64) c	16 (8) c	85.6
'TAM W–101'	Wheat–S	478 (67) a	110 (67) a	77.0
PI 372129	Wheat–R	437 (81) a	97 (35) a	77.8
Experiment 2				
PI 440100	Wheatgrass–R	45 (6) d	27 (4) c	40.0
PI 440102	Wheatgrass–R	52 (7) d	31 (7) c	20.4
'Pryor'	Wheatgrass–S	303 (52) b	99 (36) b	67.6
PI 387888	Wheatgrass–S	110 (29) c	38 (13) c	65.5
'TAM W–101'	Wheat–S	681 (64) a	206 (91) a	69.8
Experiment 3				
PI 372129				
Watered	Wheat–R	206 (41) b	184 (30) a	10.7
Drought		342 (61) a	161 (19) a	53.0
'TAM W–101'				
Watered	Wheat–S	170 (42) b	76 (28) b	55.3
Drought		335 (38) a	173 (30) a	48.4

SOURCE: Adapted from Campbell et al., 1992, Reed et al., 1991, and Reed et al., 1992. For each experiment, means followed by a different letter are significantly different ($P < 0.05$).

cause of lengthened life cycle is inherent to the wheatgrasses and not present in wheat.

Adult parasitoid size (Table 3) was greater for the wheats and the susceptible triticale than for the resistant triticale, and greater in wheat than for the wheatgrasses, irrespective of the resistance status. Drought stress did not influence adult parasitoid size, however, we demonstrated that smaller aphids on resistant hosts produced smaller parasitoids.

Our research indicates that the mechanisms that operate in a detrimental fashion to the second trophic level can be carried over to the third. Thus, detailed observations of such effects must be made prior to release of germplasm, especially when antibiosis is the mechanism of resistance. Moreover, we found that entries that were capable of maintaining flat (unrolled) leaves during RWA infestation have lower aphid populations with greater parasitoid activity. Such germplasm should be actively developed since a positive impact upon natural enemy interactions will result, whereas development of antibiotic lines may actually interfere with such relationships.

Table 2. Mean weight and days to emergence of *Diaeretiella rapae* mummies collected from Russian wheat aphid on resistant and susceptible entries.

Plant entry	Host status	Mean weight (mg)	Days to emergence
Experiment 1			
'Beagle 82'	Triticale−S	0.24 a	10.5 b
PI 386148	Triticale−R	0.14 b	11.2 a
'TAM W−101'	Wheat−S	0.26 a	10.3 bc
PI 372129	Wheat−R	0.26 a	10.1 c
Experiment 2			
PI 440100	Wheatgrass−R	0.17 c	14.1 a
PI 440102	Wheatgrass−R	0.19 b	14.3 a
'Pryor'	Wheatgrass−S	0.19 b	14.0 ab
PI 387888	Wheatgrass−S	0.16 c	13.7 b
'TAM W−101'	Wheat−S	0.22 a	13.2 c
Experiment 3			
PI 372129			
Watered	Wheat−R	0.16 a	15.2 b
Drought		0.17 a	17.0 a
'TAM W−101'			
Watered	Wheat−S	0.16 a	15.2 b
Drought		0.17 a	15.5 b

SOURCE: Adapted from Campbell et al., 1992, Reed et al., 1991, and Reed et al., 1992. For each experiment, means followed by a different letter are significantly different ($P < 0.05$).

Table 3. Mean (SEM) head capsule width and femur length of *Diaeretiella rapae* F1 generations collected from Russian wheat aphids on resistant and susceptible entries.

Plant entry	Host status	Mean head capsule width (um)	Mean femur length (um)
Experiment 1			
'Beagle 82'	Triticale−S	372 (24) a	332 (28) a
PI 386148	Triticale−R	351 (27) b	300 (26) b
'TAM W−101'	Wheat−S	372 (22) a	342 (31) a
PI 372129	Wheat−R	373 (14) a	338 (22) a
Experiment 2			
PI 440100	Wheatgrass−R	331 (5) b	277 (5) b
PI 440102	Wheatgrass−R	345 (7) b	277 (6) b
'Pryor'	Wheatgrass−S	347 (8) b	278 (9) b
PI 387888	Wheatgrass−S	345 (8) b	284 (8) b
'TAM W−101'	Wheat−S	372 (10) a	306 (10) a
Experiment 3			
PI 372129			
Watered	Wheat−R	349 (9) a	285 (8) a
Drought		337 (25) a	297 (25) a
'TAM W−101'			
Watered	Wheat−S	346 (13) a	289 (12) a
Drought		346 (8) a	289 (8) a

SOURCE: Adapted from Campbell et al., 1992, Reed et al., 1991, and Reed et al., 1992. For each experiment, means followed by a different letter are significantly different ($P < 0.05$).

References

Campbell, R.K.; Reed, D.K.; Burd, J.D.; Eikenbary, R.D. "Russian wheat aphid and drought stresses in wheat: tritrophic interactions with plant resistance and a parasitoid." Proc. 5th Annual Russian Wheat Aphid Conf. Fort Worth, TX. 1992 .

Reed, D.K.; Webster, J.A.; Jones, B.J.; Burd, J.D. "Tritrophic relationships of Russian wheat aphid (Homoptera:Aphididae), a hymenopterous parasitoid (*Diaeretiella rapae* McIntosh), and resistant and susceptible small grains." *Biol. Control.* **1991**, 1:35-41.

Reed, D.K.; Kindler, S.D.; Springer, T.L. "Interactions of Russian wheat aphid, a hymenopterous parasitoid and resistant and susceptible slender wheatgrasses." *Entomol. exp. appl.* **1992**, 64:239-246.

SAS Institute; SAS User's Guide; Statistics Version 6.03. **1985.** Cary, NC.

Starks, K.J.; Muniappan, R.; Eikenbary, R.D.; "Interaction between plant resistance and parasitism against the greenbug on barley and sorghum." Ann. Entomol. Soc. Am. **1972,** 65:650-655.

Sterilization of the Gypsy Moth by Disruption of Sperm Release from Testes

J. M. Giebultowicz, University of Maryland, Dept. Zoology, College Park, MD 20742
M. B. Blackburn, P. A. Thomas-Laemont and A. K. Raina, USDA, ARS, Insect Neurobiology and Hormone Lab., Beltsville, MD 20705

In gypsy moth, *Lymantria dispar (L.),* release of sperm bundles from the testis into the upper vas deferens (UVD) occurs in daily cycles, initiated several days before adult eclosion. We investigated two aspects of the regulation of sperm release. First, RH 5992, an ecdysteroid agonist, was injected into pupae and inhibited release of sperm from the testis in a dose dependent fashion, rendering males sterile. Second, muscle activity of the UVD was recorded during the cycle of sperm release. In males kept in a 16h light: 8h dark, transfer of sperm from the UVD to the seminal vesicles, 2-4 h after lights-on, was accompanied by a characteristic pattern of contractions by the UVD muscles. In males kept in constant light, in which sperm fail to leave the UVD, this pattern of the UVD contractions was absent.

Knowledge of the reproductive physiology of pest insects should provide new targets for designing biologically based pest control methods. In our effort toward this goal, we have studied the process of sperm release from testis which, when disrupted, causes sterility in male moths. In moths, spermatozoa differentiate in pre-adult stages and the resulting sperm bundles are contained in the testes. A few days before males are ready for mating, sperm bundles are released from the testes, stored in the duplex and subsequently transferred to the female. The process of sperm release from the testes is one of the critical steps in achieving male fertility and, yet, it has received very little attention from reproductive physiologists. One exception was a study of the flour moth, *Ephestia kuehniella,* which demonstrated that release of sperm from the testes occured only at a restricted time of day controlled by light-dark cycles (Riemann et al.,1974) and that sperm release was disrupted in constant light (Rieman and Ruud, 1974). Since then, similar periodic release of sperm has been shown in other moth species (Giebultowicz et al., 1992); the process has been most extensively studied in the gypsy moth, *L. dispar.*

In gypsy moth males kept under 16h light and 8h darkness, release of sperm is rhythmic and occurs in 2 steps (Giebultowicz et al.,1988). First, sperm bundles are released from testis into the upper vas deferens (UVD) during the 4h period before lights-off and remain in the UVD overnight. Second, sperm bundles are transferred from the UVD to the seminal vesicles (SV) 2-4h after lights-on. We have studied the mechanism of sperm release from testis and sperm transfer from the UVD to the SV. Disruption of either process would prevent sperm from reaching the duplex, rendering males sterile.

Inhibition of Sperm Release by Ecdysone Agonist, RH -5992

In the gypsy moth, release of sperm from the testis is initiated in developing adults 5 days before adult eclosion. Injections of insect molting hormone, 20-hydroxyecdysone (20-HE), delays initiation of sperm release, suggesting that the release is controlled by the decreasing titer of 20-HE (Giebultowicz et al., 1990). Injected 20-HE is rapidly metabolized so that sperm release resumes in a normal, rhythmic fashion. Recently, a new nonsteroidal agonist of 20-HE, RH-5992, has been discovered by Rohm and Haas Co (Philadelphia, PA). This compound is more resistant to metabolic clearance then 20-HE and we have studied its effects on sperm release in developing adults. RH-5992 (a gift from T. Dhadialla, Rhom and Haas Co.) was dissolved in

DMSO and injected into male pupae in the morning of the day when sperm release was first expected to occur. Sperm bundles present in the UVD were counted 1 and 2 days after injection. RH-5992 inhibited the release of sperm in a dose dependent fashion (Fig. 1) With a dose of 50 ng the release of sperm was prevented in the first cycle after injection, but partially recovered during the second cycle. Nearly complete inhibition of sperm release, lasting for at least 3 cycles after injection, was achieved with 100 ng or higher doses. Preliminary studies suggest that RH-5992 acts by binding to ecdysteroid receptors (T. Dhadialla, personal communication). Since RH-5992 is less prone to metabolic clearance than 20-HE, its prolonged presence in the male pupae might prevent molecular events involved in initiation of sperm release. RH-5992 causes mortality when fed to lepidopterous larvae. We plan to study whether sublethal doses of this compound fed to gypsy moth larvae will prevent sperm release in the ensuing adults.

UVD Movement during Sperm Release Cycle

Each batch of sperm released from testis remains in the UVD overnight and then is transferred to the SV 2-4h after lights-on. The UVD wall consists of secretory epithelium overlaid by muscles which, although not innervated, cause considerable motility of the UVD. To determine the pattern of muscle activity during different phases of sperm release cycle, testis-UVD-SV complexes were dissected and attached to a displacement transducer connected to a chart recorder. During most of the 24 h sperm release cycle, UVD muscles showed alternating slow waves of contractions and relaxations with a frequency of 40-50/h. Superimposed on these slow waves were more rapid contractions of varying amplitude and frequency. A dramatic change in this basic pattern of muscle contractions was consistently observed prior to the transfer of sperm from the UVD to the SV. At this time, the UVD displayed a prolonged period of rapid contractions with high and even amplitude and frequency of 5-7/min. This characteristic period of contractions lasted 0.5-2 h in different preparations and was often accompanied by the actual transfer of sperm *in vitro* (Fig.2).

Gypsy moth males kept in constant light (LL) show several abberations in the release of sperm

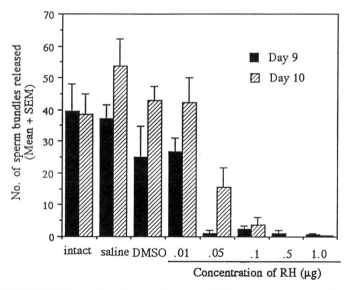

Fig.1 Effects of RH 5992 (RH) on the release of sperm bundles from the testis. RH was injected into day 8 pupae; sperm bundles released into the UVD were counted on day 9 and 10. Each bar represents average count from 8 pupae.

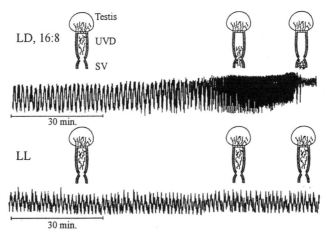

Fig.2 Recording of the UVD muscle activity from male kept in light:dark cycles (LD, 16:8) or in constant light (LL). Drawings above each trace indicate location of sperm bundles at the moment.

(Giebultowicz, et al, 1990). In some LL males sperm is released from the testis but fails to transfer from the UVD to the SV. Recordings of the UVD muscle activity from such males demonstrated that the pattern of muscle contractions characteristic for transfer of sperm to the SV, was absent (Fig. 2). Thus, we concluded that the enhanced activity of the UVD muscles, which occurs spontaneously at a certain phase of sperm release cycle, is necessary for the transfer of sperm from the UVD to the SV. We intend to study the mechanism controlling this enhanced activity which, when absent, renders males sterile.

Acknowledgements We thank Dr.T. Dhadialla from Rohm and Haas Co. for his advice and a gift of RH-5992. This study was partially suported by grants to JMG: DCB 9105932 and 91-37302-6211.

References

Giebultowicz, J.M.; Bell, R.A.; Imberski, R.B.; " Circadian rhythm of sperm movement in the male reproductive tract of the gypsy moth, *Lymantria dispar.*" *J. Insect Physiol.* 1988,vol. 34, pp. 527-532.

Giebultowicz,J.M.;Feldlaufer,M.;Gelman,D.B "Role of ecdysteroids in the regulation of sperm release from testis of the gypsy moth, *Lymantria dispar.* " J. Insect Physiol. 1990, vol.36, pp. 567-571.

Giebultowicz, J.M.; Joy, J.E.; Riemann, J.G. In *Advances in Regulation of Insect Reproduction ;* Bennettova, B.; Gelbic, I.; Soldan, T., Eds.; Institute of Entomology, Czech Acad. Sci.: 1992; pp. 91-94.

Giebultowicz, J.M.; Ridgway, R.L.; Imberski, R.B. " Physiological basis for sterilizing effects of constant light in *Lymantria dispar.*" *Physiol. Entomol.* 1990, vol.15, pp. 149-156.

Riemann, J.G.; Ruud, R.L. " Mediterranean flour moth: effects of continuous light on the reproductive capacity." *Ann. Entomol. Soc. Amer.* 1974, vol.67, pp. 857-860.

Riemann, J.G.; Thorson, B.J.; Ruud, R.L. " Daily cycle of release of sperm from the testes of the Mediterranean flour moth." *J. Insect Physiol.* 1974, vol. 20, pp. 195-207.

Control of Reproductive Behavior of Female Moths by Factors in Male Seminal Fluids

Timothy G. Kingan, Ashok K. Raina, Patricia Thomas-Laemont,
USDA ARS PSI INHL, BARC-East Bldg. 306 Rm. 322, Beltsville, MD 20705

In many species of moths mating causes a cessation of the production and release of sex pheromone by females. By surgical removal of portions of the reproductive tract of males of the corn earworm *Helicoverpa zea,* we have shown that the accessory glands and duplex contain factors that are necessary for initiating the normal post-mating depletion of sex pheromone. The cessation of pheromone release (calling behavior) apparently does not require the transfer of seminal fluid from these structures. Instead, a soluble component of, or transferred with, the spermatophore is required for shutting off calling behavior. Additionally, fractionated extracts of the lower portion of the reproductive tract (ejaculatory duct) are active in shutting off calling behavior while having no effect on the reduction of sex pheromone. Thus, the alteration of specific components of the female's post-mating behavior may be regulated by distinct chemical messengers.

The females of many species of insects are transiently or permanently unreceptive to new sexual encounters following mating. This change in behavior is thought to be adaptive, because the female receives sufficient spermatozoa after a single mating to fertilize most or all the eggs she is capable of producing. Among the moths, the female corn earworm, *Helicoverpa zea,* is sexually unreceptive for 24 hr following mating. In addition, she becomes depleted of sex pheromone, and the calling behavior that results in release of pheromone ceases. These changes in behavior ensure that the female will not attract new potential mates at a time when oviposition is beginning.

The changes from 'virgin' to 'mated' behavior in insects are of great interest to reproductive physiologists, as well as to those studying sexual selection and mating systems. Thus, sexual selection has led to the evolution of a diversity of mechanisms for communicating "availability" of the female; similar forces have apparently led to the evolution of mechanisms whereby males are not attracted to mated females, thus maximizing the male's reproductive potential. The interest of our laboratory is in the proximate mechanisms which insure maximum mating efficiency; we have specifically sought to determine the chemical basis for the male's ability to evoke the switch in female behavior after mating.

Materials and Methods

Corn earworms were reared in long days (16:8, light:dark) under a reversed photoperiod as previously described (Raina, et al., 1986).

In the first or second photophase males were anesthetized with CO_2 and pinned under saline solution in a dissecting dish. The fused testes, accessory glands and duplex (storage organ for spermatozoa and seminal fluids) were removed through a dorsolateral incision in the 3rd abdominal segment. The wound was blotted dry and sealed with wax. These "radically gonadectomized" (RG) males were mated to 2nd scotophase virgin females the following day. Two hr later the ovipositors from some of the females were removed and extracted with heptane; the level of sex pheromone (Z11-hexadecenal) was quantified by gas chromatography as described earlier (Raina and Kempe, 1992). The remaining females mated to RG males were retained in observation cages for the remainder of the scotophase and checked at 30 min intervals for calling behavior. At the end of this time all females were checked for the presence of a spermatophore in their bursae.

Extracts of the caudal portion of the male's primary simplex (ejaculatory duct) were prepared by sonication in *Heliothis virescens* saline (Bindokas and Adams, 1988) diluted 1:1 with water. The latter extract was brought to 33% acetone (ice-cold), with stirring; the resulting precipitate was separated by centrifugation and the supernatant was then brought to 66% acetone. This second precipitate was also removed by centrifugation. The supernatant was reduced in volume by 75% under a stream of N_2, diluted 3-fold with 0.1% trifluoroacetic acid, and desalted in a C18 Sep-Pac cartridge (Millipore Corp.; see Kingan et al., 1993). Residual acetone was removed from the 33% and 66% acetone precipitates under a stream of N_2. Samples were dissolved in saline and tested for anti-calling activity in virgin females in their second scotophase. A portion of the 66% acetone precipitate was held in boiling water for 5 min. Control females were injected with saline. Calling was recorded at 15 min intervals for 2 hr after injection. At this time the ovipositors were extracted for quantitation of pheromone as described above. A calling score was calculated as the percent of # individuals x observations in which calling behavior was observed.

Results and Discussion

Female *H. zea* become depleted of sex pheromone within 2 hr after termination of mating; this response to mating was shown to occur even in females mated to males castrated as larvae (Raina, 1989). Thus, we wanted to determine if this depletion (pheromonostasis) was associated with the transfer of seminal fluids. As can be seen in Fig. 1, females mated to RG males do not become depleted of pheromone to the extent found in females mated to intact males (Kingan, et al., 1993). Moreover, 85% of these females receive apparently normal, though empty, spermatophores (those females not receiving spermatophores were excluded from the

calculation of pheromone levels). Thus, the normal course of pheromonostasis after mating depends on the transfer of seminal fluids from the accessory glands and/or duplexes. In addition, we have recently shown (Kingan et al., 1993) that peptidic factors in partially purified extracts of these tissues are potent in evoking the depletion of pheromone from headless females that have been injected with the pheromonotropic peptide PBAN (pheromone biosynthesis activating neuropeptide; Raina, et al., 1989).

When RG-mated females were retained for the remainder of the scotophase and observed, it was found that very few females resumed calling (Fig. 1). In this experiment, the few females that did call were found not to have a spermatophore in their bursae. This observation appears to implicate the spermatophore itself, rather than the seminal fluids contained therein, in the cessation of calling after mating. However, in this experiment we obtained two mated females that did not resume calling and that did not contain spermatophores; they did, however, have in their bursae a small amount of a milky fluid, possibly derived from the male's ejaculatory duct (see below). Moreover, we have recently found that when males are "vasectomized" by surgical ligation of the simplex, we can produce females with a similar appearing fluid in their bursae. These

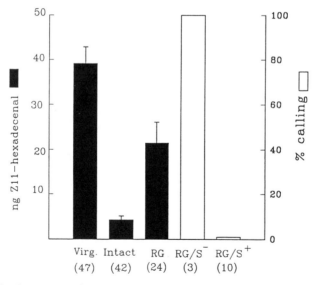

Fig. 1 Levels of pheromone (2 hr after separation of the mating pair) and incidence of calling (in remaining scotophase) in females mated to RG males. Number of females in each group shown in parentheses. The smallest bar (RG/S+) represents 0% calling. Abbreviations: Virg., age-matched virgins; Intact, females

mated to intact males; RG, females mated to RG males; S, presence or absence of spermatophore in the females' bursae after mating; "N" for each group is shown in parentheses. The data are reformated by permission from tables contained in Kingan et al (1993).

118

mated females either do not call or call only briefly in the minutes after separation (T. G. Kingan, unpublished observations). Thus, it appears that it is not the spermatophore per se that evokes the cessation of calling; rather, our findings suggest that either the inflation provided to the bursa by the "milky fluid" or a chemical component of this fluid is the requisite signal that leads to cessation of calling.

The transparent cephalad segment of the simplex (second secretory area of primary simplex, Callahan and Cascio, 1963) gives rise to the walls of the spermatophore. On the other hand, the expanded caudal segment of the primary simplex (first secretory area, Callahan and Cascio, 1963; also known as the ejaculatory duct) contains a globular secretory material which is the first to pass into the female during copulation. If soluble components transferred during mating participate in evoking the cessation of calling, then the secretion of the ejaculatory duct must be considered a candidate for the source of the putative chemical messenger. Accordingly, we fractionated an extract of the ejaculatory duct and tested the action of these fractions in calling females. In three experiments the 66% acetone precipitate was very active in evoking the cessation of calling, while it does not deplete pheromone (Fig. 2). The anti-calling activity is at least partially labile to boiling (data not shown). The 66% acetone supernatant contains little anti-calling activity (1 experiment, score=20.0).

The detailed mechanism(s) by which sex pheromone is depleted and calling is shut off in moths after mating have not been determined. In principle, mating could lead to a block in the release of the pheromonotropic neuropeptide PBAN and to an active clearing of pheromone from the glandular tissue. Calling, which is regulated in moths by descending input to the terminal abdominal ganglion (TAG; Itagaki and Conner, 1986), is likely to be controlled at the level of these regulatory elements or the motor circuitry for the behavior itself in the TAG. Any effect of mating on PBAN release and calling could be mediated by inhibition via sensory mechanisms or by humoral factor(s) transferred with the male seminal fluid that ultimately act centrally to mediate an inhibition. Our findings show that the seminal fluid does participate in the depletion of pheromone; in addition, we have shown that soluble components from the ejaculatory duct, transferred prior to the spermatophore, are sufficient to shut off calling after mating. An understanding of the switch from 'virgin' to 'mated' behavior will require the identification of the active components in these fluids and the characterization of their target tissues in the female.

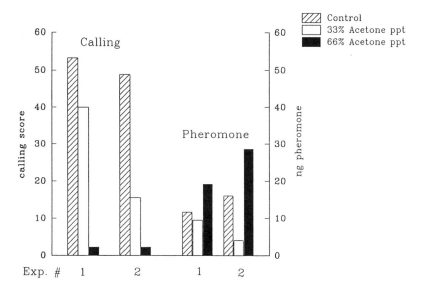

Fig. 2 Two experiments showing incidence of calling and levels of pheromone in virgin females injected with fractionated extracts of ejaculatory ducts. Females were injected with: 0.25 tissue equivalent in 10 μl saline of 33% acetone precipitate or 66% acetone precipitate (see Materials and Methods); controls received 10 μl saline. N=5 for each treatment in each experiment.

References

Bindokas, V.P.; Adams, M.E. Hemolymph composition of the tobacco budworm, *Heliothis virescens* F.(Lepidoptera:Noctuidae). *Comp. Biochem. Physiol.*, **1988**, *90A*, 151-155.

Callahan, P.S.; Cascio, T. Histology of the reproductive tracts and transmission of sperm in the corn earworm, *Heliothis zea. Ann. Ent. Soc. Amer.,* **1963**, *56*, 535-556.

Itagaki, H.; Conner, W.E. Physiological control of pheromone release behaviour in *Manduca sexta* (L.). *J. Insect Physiol.*, **1986**, *32*, 657-664.

Kingan, T.G.; Thomas-Laemont, P.; Raina, A.K. Male accessory gland factors elicit change from 'virgin' to 'mated' behaviour in female corn earworm moth, *Helicoverpa zea. J. Exp. Biology*, **1993**, (in press).

Raina, A.K. Male-induced termination of sex pheromone production and receptivity in mated females of *Heliothis zea. J. Insect Physiol.*, **1989**, *35*, 821-826.

Raina, A.K.; Jaffe, H.; Kempe, T.G.; Keim, P.; Blacher, R.W.; Fales, H.M.; Riley, C.T.; Klun, J.A.; Ridgway, R.L.; Hayes, D.K. Identification of a neuropeptide hormone that regulates sex pheromone production in female moths. *Science*, **1989**, *244*, 796-798.

Raina, A.K.; Kempe, T.G. Structure activity studies of PBAN of *Helicoverpa zea* (Lepidoptera:Noctuidae). *Insect Biochem. Molec. Biol.*, **1992**, *22*, 221-225.

Raina, A.K.; Klun, J.A.; Stadelbacher, E.A. Diel periodicity and effect of age and mating on female sex pheromone titer in *Heliothis zea* (Lepidoptera:Noctuidae). *Ann. Entomol. Soc. Amer.*, **1986**, *79*, 128-131.

Dairy Manure Control of Western Corn Rootworm Damage

L. L. Allee, P. M. Davis, Entomology Department, Cornell University, Ithaca, NY 14853
N. V. Bushamuka, Soil, Crop & Atmospheric Sciences Department, Cornell University, Ithaca, NY 14853
R. W. Zobel, Rhizobotany Project, USDA-ARS, Ithaca, NY 14853

Research was conducted to evaluate the effectiveness of using manure to reduce losses from western corn rootworm. Manure accelerated corn root regeneration and reduced lodging and yield losses in rootworm infested plots. Possible mechanisms for reduced losses involving changes in soil CO_2 and predacious arthropods were investigated.

Although the western corn rootworm (WCRW), <u>Diabrotica virgifera virgifera</u>, is the major pest of corn throughout most of the U.S., the only currently reliable control methods are soil insecticides and crop rotation. In many farming situations, a two-year rotation is not feasible. Chiang (1970) found manure applied at 50 tons/acre reduced rootworm larval and adult populations approximately 50% compared with plots receiving no manure. WCRW larvae exhibit positive chemotaxis towards CO_2 and may use a CO_2 gradient to orient toward corn roots in the soil (Strnad et al. 1986). Previous research on other crops indicates that elevated CO_2 can alter root growth, orientation and subsequent plant growth (reviewed by Zobel, 1992). This paper presents the results from the first year of a three year study to evaluate the effectiveness of manure to reduce losses from WCRW.

Materials and Methods

In 1992, treatments were arranged in a split/split plot design with four replications. Corn hybrids Pioneer 3733 and Cornell 281 were planted in plots treated with 0, 20, 40, and 60 tons per acre of cow manure. Each hybrid/manure combination contained single rows infested with 0, 500 and 700 WCRW eggs per foot. All plots received nitrogen (supplied by manure and/or inorganic fertilizer) in excess of soil test recommendations.

On July 6 and 20, WCRW larvae were sampled from one 7-inch cube of soil per plot. WCRW adult emergence was monitored from July 20 to September 3 using two emergence cages per rootworm plot. Soil arthropods and nematodes were sampled nine times throughout the season by placing soil cores in Berlese funnels.

On July 29 and August 12, three roots per rootworm plot were dug, washed, and rated for both damage and regrowth using a 1-6 and 0-4 scale (Hills and Peters, 1971), respectively. Soil CO_2 was measured each month by drawing 1cc samples with a syringe from rubber capped PVC tubes (1cm in diameter) placed at depths of 10, 20, 30, 40, and 50cm within each manure plot. Samples were analyzed using a thermal conductivity gas chromatograph.

After recording lodging, a 6 meter section of each rootworm plot was hand harvested, chopped and weighed for silage yield. Five ears per rootworm row were shelled and weighed for grain yield.

Results and Discussion

On July 6, larval counts averaged 10.8 larvae/plant and did not differ between manure rates (F=1.03; df=3, 18; p=0.40). On July 20, larval counts significantly differed across manure rates (F=5.21; df=3, 18; p=0.0091). The 20 and 40 t/a manure plots averaged 7.9 larvae/plant, and were significantly less than 11.5 larvae/plant in the 0 t/a manure plots. However, the 60 t/a plots did not differ from the 0 t/a manure plots. In contrast, adult WCRW emergence declined linearly with increasing manure rate (F=9.73; df=1, 18; p<0.01) averaging 14.3, 9.2, 9.1 and 5.3 in 0, 20, 40, and 60 t/a manure plots, respectively.

Predation may have contributed to lower adult emergence in manured plots. During the WCRW larval and early pupal stages, the numbers of potentially predatory beetle larvae and large soil mites in manured plots were variable but generally higher than those in unmanured plots. However, predatory soil arthropod populations in manured plots crashed to unmanured plot population levels at the end of July, the time of pupation. Precipitation during July was twice the thirty year average for the area and may also have contributed to WCRW mortality in manured plots.

On both sample dates, root damage ratings averaged 5.7 and did not differ significantly across manure rates (F=2.94; df=3, 18; p=0.06 and F=.71; df=3, 18; p=0.56). However, on July 29 root regrowth increased linearly with increasing manure rate (F=7.55; df=1, 18; p<0.025). By August 12, root regrowth in the 40 and 60 t/a manure plots was about two weeks ahead of root regrowth in the 0 and 20 t/a manure plots (F=18.52; df=3, 18; p=0.0001). Hybrids did not differ in root damage or regrowth ratings. On all sample dates, soil CO_2 content significantly differed with manure rate (F>5.72; df=3, 59; p<0.006) (Fig. 1). In general, on each date,

soil CO_2 increased with manure rate at depths down to 40cm. Along with CO_2, differences in nutrient availability, soil permeability, temperature and water holding capacity between manured versus unmanured soils also may have contributed to accelerated root regrowth in manured plots.

Lodging was significantly reduced in manured plots of 40 and 60 t/a of both hybrids (Fig. 2). Unmanured plots suffered significant silage yield losses due to WCRW (F=9.82; df=3, 18; p=0.005). These losses were reduced or eliminated in manured plots of both hybrids (Fig. 3). Grain yield of Pioneer 3733 was not reduced by WCRW. In contrast, significant grain yield losses due to WCRW in unmanured plots (F=9.40; df=3, 18; p=0.0006) were reduced or eliminated in manured plots of Cornell 281.

In conclusion, manure did not reduce larval numbers or root damage ratings. Although predacious arthropod numbers were higher in manured plots, numbers of WCRW larvae were not consistently reduced. Disorientation of WCRW larvae was not evident despite higher CO_2 levels in manured plots.

Adult emergence was reduced in manured plots. Predation and/or effects of precipitation are possible mortality factors and are under study.

Manure did increase the corn plant's ability to recover from injury by WCRW larvae. Root

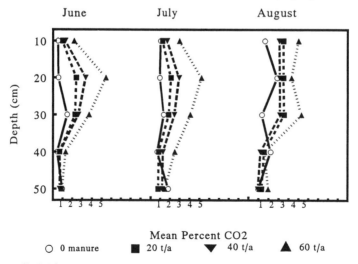

Fig. 1. Percent soil CO_2 by manure rate and soil depth on June 30, July 9 and August 13, 1992.

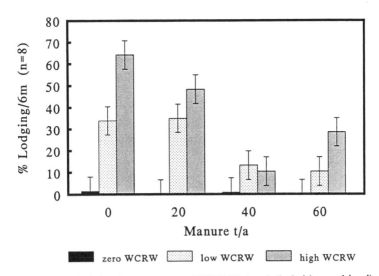

Fig. 2. Percent lodging by manure and WCRW level (hybrids combined).

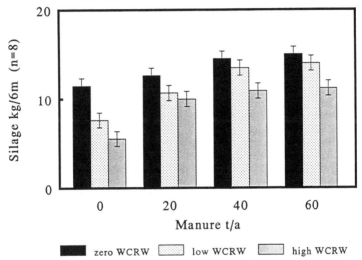

Fig. 3. Silage yield by manure and WCRW level (hybrids combined).

regrowth ratings were higher and lodging and yield losses were reduced in manured plots.

Future research is needed to further evaluate the mechanism(s) of accelerated plant growth in manured plots.

References

Chiang, H. C. "Effects of manure applications and mite predation on corn rootworm populations in Minnesota." *J. Econ. Entomol.* **1970**, *63*, 934-936.

Hills, T. M.; Peters D. C. "A method of evaluating postplanting insecticide treatments for control of western corn rootworm larvae." *J. Econ. Entomol.* **1971**, *64*, 764-765.

Strnad, S. P.; Bergman, M. K.; Fulton, W. C. "First-instar western corn rootworm (Coleoptera: Chrysomelidae) response to carbon dioxide." *Environ. Entomol.* **1986**, *15*, 839-842.

Zobel, R. W. In *Limitations to Plant Root Growth*; Hatfield, J. L.; Stewart, B. A., Eds.; Advances in Soil Science; Springer-Verlag New York Inc.: New York, NY, **1992**, Vol.19; 27-51.

Biocontrol for Insect Pests of Livestock and Poultry

James J. Petersen, Midwest Livestock Insects Research Unit, ARS, USDA, Department of Entomology, University of Nebraska, Lincoln, NE 68553

Biocontrol of insect pests of livestock and poultry is of great interest because insecticide resistance, increasing restrictions on insecticide use, and concern for the environment, and because livestock and poultry production compose a large portion of farm income. This discussion is limited to house flies, stable flies, horn flies and face flies because they cause the greatest economic loss to the livestock and poultry industry. Although few scientists are working on alternative control methods for these flies, progress has been made in the past 10 years. This review discusses the status of biocontrol research on flies associated with poultry, pastured livestock, and confinement dairy and beef cattle.

The damage to livestock and poultry caused by insects in the U.S. is difficult to assess because it is rarely direct with few overt symptoms or death. Although stable flies, *Stomoxys calcitrans,* and house flies, *Musca domestica,* have been incriminated in the transmission of a number of diseases in livestock, disease transmission is of little economic importance. Of more importance are animal stress and nuisance problems associated with these flies. Stress is manifested in reduced weight gains and feed efficiency in feeder cattle (Campbell et al., 1987) and reduced milk production in dairy cattle (Bruce and Decker, 1958). Estimates of these losses are in excess of 4 billion dollars in the U.S. (Steelman, 1976), and 10 billion dollars worldwide (Patterson, 1990). Another area of increasing concern is the encroachment of the urban environment into the rural environment. This movement has resulted in increased pressure to reduce pest problems associated with confined livestock and poultry, and increased litigation caused by the pest problems (Thomas and Skoda, 1993).

The single most important group of insects affecting livestock and poultry are the muscoid flies. Because of space limitations, discussions will be limited to the face fly, *Musca autumnalis,* horn fly, *Haematobia irritans,* stable fly and house fly.

The state of the art of biological control of livestock pest flies lags well behind that of plant pests. Of the 210 scientists in the U.S. government working on biological control only 16 are involved with livestock pests and many of these only marginally (Patterson, 1990). However, substantial research progress is being made and will be highlighted here.

Biocontrol Associated with Poultry

Meyer (1990) stated that indirect losses caused by arthropods were probably more important than direct losses. These indirect losses include fines and other fees associated with regulatory inspection, control costs, and court and legal fees incurred through criminal and civil activities against poultry facilities that support excessive fly populations.

The poultry environment, of all animal production systems, is best suited to the development of the biological control component of arthropod pest management systems because of the relative stable environmental conditions. Because of the increased pressure to control flies and the controlled environments of poultry facilities, substantial effort has been directed toward implementing biocontrol into the management of pest fly populations. The two general approaches for biocontrol of filth flies in accumulations of poultry manure are enhancement of natural enemies through manure management and habitat stability, and augmentation or inundative releases of pathogens, predators or parasites (Legner and Dietrick, 1974).

Geden (1990) reported that the predatory fauna in poultry manure is similar across broad geographic areas; but in spite of a great species diversity, the majority of individual predators belong to a small number of key species. The two most important predatory species are the histerid beetles in the genus *Carcinops* and the macrochelid mite *Macrocheles muscaedomesticae*. Both have high attack rates on fly immatures, are present when fly pressure on the manure habitat is great, occupy microhabitats that assure contact with house flies, and have mechanisms for dispersal (Geden, 1990).

A number of factors affect the success of these predators. *Carcinops pumilio* adults are present in manure having a 10-70% moisture and larvae are most abundant in a 50-70% moisture range and generally prefer drier manure (Geden and Stoffolano, 1988). *M. muscaedomesticae* is most common in manure with 50-70% moisture (Stafford and Bay, 1987). Beetles show a strong preference for old manure with fresh manure of less than 2 days old having a repellent effect. *M. muscaedomesticae* prefers the fresher manure and will rapidly move from old manure to fresh droppings. This behavior has led to the practice of leaving a portion of the manure behind when poultry facilities are cleaned thus permitting a more rapid buildup of predator populations.

M. muscaedomesticae attack *C. pumilio* in the laboratory. This predatory behavior and competition for prey undoubtedly limit populations of both species in the field (Geden, 1990). Thus, the prospects for use of these predators in fly management are good but require an understanding of their biology and behavior. The mites generally invade fresh manure rapidly and multiply quickly. The beetles are slow to colonize fresh manure accumulations and build up populations more slowly than mites. Predator conservation is more practical than augmentation at present. Good manure management practices including ventilation, moisture control, discreet use of insecticides (Meyer et al., 1984), and leaving a manure residue at cleanout can often provide adaquate control.

Where predator augmentation is planned, Geden (1990) suggests that predators not be released into old manure accumulations; that proper manure management practices be in place; and efforts should be aimed at repopulating houses after cleanout where sanitation concerns prohibit leaving residuals of old manure. He also suggested that releases should occur in two phases, *M. muscaedomesticae* should be released within a few days after cleanout followed by *C. pumilio*. Although no data are available on application rates, Geden (1990) recommends about 20 mites and 10 beetles per cage should be sufficient to establish these predators.

To date these agents are not commercially available and transportation from one poultry unit to another poses biosecurity problems.

However, Ho et al. (1990) reported on a successful method for the mass-production of *M. muscaedomesticae*. They were able to produce 2500 mites in 8 days from an average of 34.5 females at 30°C.

Pteromalid wasps also have received considerable attention as biocontrol agents of filth flies associated with poultry. Pteromalids occur commonly in poultry manure. Four species, *Muscidifurax raptor, Spalangia cameroni, S. endius* and *S. nigroaenea*, account for most of the parasitism; but relative abundance of a given species varies with location and type of poultry facility. In North Carolina *M. raptor* is the predominat species in caged-layer houses whereas *S. cameroni* is the predominat species in broiler-breeder houses. In Florida, *M. raptor* and *S. endius* are the predominat species (Morgan and Patterson, 1975). Legner and Olton (1971) reported *M. raptor, S. cameroni, S. endius,* and *S. nigroaenea* were most abundant in poultry manure in southern California. Generally, parasitism ranges between 20 and 30% (Legner and Dietrick, 1974, Rutz and Axtell, 1979).

A number of studies have been made to evaluate the effectiveness of parasite releases for the control of flies associated with poultry facilities. Legner and Dietrick (1974) released *S. endius, M. raptor* and *Tachinaephagus zealandicus* at six poultry houses in California which resulted in a lowering of fly densities, and increased parasitism from 12.9 to 22.5%. Olton and Legner (1975) released the same three species as Legner and Dietrick (1974) from December through April in a poultry house resulting in 46% parasitism of house flies. Morgan et al. (1975) released 44,500 *S. endius* weekly for 10 weeks in a small poultry house resulting in 100% parasitism after 4 weeks and complete adult fly suppression after 35 days. Releases of the Florida strain of *S. endius* did not result in increased parasitism by that species following releases in North Carolina (Rutz and Axtell, 1980). Release attempts with an indigenous strain of *M. raptor* were successful in increasing the rates of parasitism by that species (Rutz and Axtell, 1979, 1981). The *Muscidifurax* used in these studies were indigenous to the area while the *S. endius* were imported leading the authors to speculate that perhaps the *S. endius* were not adapted to the North Carolina environment (Axtell and Rutz, 1986). Field tests integrating diflubenzuron applied to resting surfaces and the release of a commercial strain of *M. raptor* in South Carolina resulted in a significant reduction in fly populations (Shepard and Kissam, 1981). However, a majority of the parasitism was due to indigenous species. Shepard and Kissam (1981) concluded that the program could be improved by including releases of more abundant and possibly better adapted indigenous species.

Meyer (1990) stated that for biocontrol to become a larger and more reliable component of poultry IPM, much more research is necessary. He saw little progress in this area in the near future because there is little emphasis on the part of state or federal granting agencies to provide the necessary funding.

Biocontrol Associated with Pastured Livestock.

The major pests of pastured livestock in the U.S. are the horn fly and face fly, both introduced pests that breed in cattle dung. The horn fly is an obligate blood-sucking insect causing an estimated loss in cattle production of $730 million (Drummond et al., 1981). The face fly feeds on mucous secretions of cattle, especially around the eyes and nose causing various infections and considerable irritation to the animal. The annual losses caused by this fly are estimated to be in excess of $53 million (Drummond et al., 1981).

Although some 43 species of parasitic Hymenoptera are known to be associated with cattle dung (Blume, 1985), and 26 species have been recovered from horn flies and face flies (Fincher, 1990), parasitism is usually low and only occasionally is parasitism sufficiently high to reduce fly populations (Thomas and Wingo, 1968, Thomas and Morgan, 1972). No successful attempts have been made to control these flies through parasite releases.

Predators belonging to the families Staphylinidae, Histeridae, and Hydrophilidae are important biological control agents of pasture breeding horn flies and face flies (Thomas and Morgan, 1972, Thomas et al., 1983, Roth, 1989). Legner (1978, 1986) listed seven exotic species of Histeridae and one Staphylinidae that were released in California for the control of horn flies and face flies, and Fincher (1990) released two species of Staphylinidae in Texas for horn fly control. It is unknown if any of these species have become established.

Competitors have received the most attention for pasture environments because many of them, in addition to their fly control potential, remove dung accumulations; thus increasing pasture productivity. The principle competitors are dung-burying beetles which

have the potential to remove cattle dung and eliminate fly breeding (Fincher, 1981, 1986). Many dung-burying scarab species occur in the U.S., but few can bury sufficient amounts of dung in short enough time to eliminate fly breeding. Therefore, efforts have been made to introduce exotic dung-burying beetles. Fifteen species have been released in several states and at least five of the 15 are known to be established (Fincher, 1990) (Table 1). The effects of these releases on populations of pest flies have not been fully evaluated. However, decreased horn fly populations have been recorded on cattle in several states when dung beetle populations are sufficient to bury most dung pats within 24 hours after deposition. Roth et al. (1983) and Roth (1989) reported that scarabs were a significant mortality factor of horn flies in east-central Texas.

Biocontrol Associated with Confined Livestock

Biocontrol of filth flies associated with confinement (dairy and beef) cattle is a difficult proposition because of the open nature of the environment compared with poultry facilities, the often temporary and continually changing environment, and the highly variable conditions from facility to facility. Predators are acknowledged to be important elements of the natural control component of flies associated with confinement cattle. Hall et al. (1989) recovered 45 species of arthropods from feedlots with macrochelid mites and staphylinid beetles being the most numerous predators. These researchers determined that egg-to-adult mortality for stable flies was between 95 and 97%, and that predation accounted for about 27%. Predators and competitors are difficult to manipulate as biocontrol agents. Also, little is known about their population dynamics on dairy and beef cattle confinements or the impact that livestock management practices have on these beneficial arthropods.

Pathogens as biocontrol agents of flies have received only minimal attention. Jepersen and Keiding (1990) reported that *Bacillus thuringiensis* may be useful for control of larvae of house flies, but until now long lasting effective control has not been demonstrated in animal units. Studies with the fungal pathogen *Entomophthora muscae* suggest that this fungus dramatically reduces fecundity in house flies and that it is a primary factor in cool-season regulation of fly populations in some locations (Mullens et al., 1987). However, because of behavioral fever

Table 1. Exotic species of dung beetles released in Texas to aid in the control of horn flies[a]

Species	Year released	Country of origin
*Onthophagus gazella**	1972	South Africa
*Euoniticellus intermedius**	1979	South Africa
Onthophagus bonasus	1980	Pakistan
*Onitis alexis**	1980	South Africa
Liatongus militaris	1984	South Africa
*Onthophagus taurus***	1985	Europe
Onthophagus sagittarius	1985	Sri Lanka
Gromphas lacordairei	1985	Argentina
Onthophagus binodis	1986	South Africa
*Onthophagus depressus***	1987	South Africa
Onthophagus nigriventris	1987	Kenya
Ontherus sulcator	1987	Argentina
Copris incertus	1987	Mexico
Sisyphus rubrus	1987	South Africa
Onitis vanderkelleni	1987	Kenya

[a]Adapted after Fincher 1990.
*Released and established in one or more states.
**Accidental introductions.

response of house flies to infections by *E. muscae*, house flies apparently have the ability to cure themselves of the disease during the warmer portions of the fly breeding season (Watson et al., 1993). This greatly reduces the potential usefulness of the fungus as a biocontrol agent.

Pupal parasites of the family Pteromalidae are the primary parasites associated with confined cattle. Assuming that fly pupae represent 1-5% of the total house fly and stable fly eggs deposited (Hall et al., 1989), then these pupal parasites must exert a greater natural control effect per individual than an agent destroying any previous stage of the host (Legner and Brydon, 1966). The species guild is essentially the same for confined cattle as it is for poultry. Legner et al. (1967) found that four species of parasites, *M. raptor, S. cameroni, S. endius* and *S. nigroaenea* were commonly associated with confined livestock installation in the western hemisphere. However, the relative abundance of a given species varies widely even within a region (Table 2). In the midwestern U.S., seven species were reported from dairies and beef cattle confinements with mean seasonal parasitism of 14.2% in house fly pupae and 8.2% in stable fly pupae. Parasitism varied seasonally and generally increased throughout the fly season (Petersen and Meyer, 1983) (Table 3). Much higher levels of natural parasitism (monthly means up to 59%) have been reported (Petersen et al., 1992a, Greene, 1990).

Successful releases of pteromalids on cattle facilities were reported in Florida during the 1970's. Releases of 6000 *S. endius* three time each week for 5 weeks in a small calf barn (13x7x3 m) resulted in an eventual 93% reduction in adult fly populations (Morgan et al., 1976); releases of 266,000 *S. endius* weekly for 8 weeks followed by 509,000 *S. endius* per week for an additional 7 weeks at a pasture feeding station for dairy cattle resulted in a 84-100% and 40-100% parasitism of house fly pupae and stable fly pupae, respectively, for weekly samples (Morgan and Patterson, 1977). These early successes were accomplished by releasing large numbers of an endemic species to very small fly problem areas. They served to show that augmentative releases could play an important roll in the integrated management of fly populations in open environments. These results served to stimulate a rapid increase in commercial sources of these parasites for fly control on feedlots (Bezark and Rey, 1980). As a result considerable effort was expended in the promotion of these agents for fly control on large midwestern beef cattle and dairy facilities before their efficacy was documented.

Studies at midwestern feedlots indicated that released *S. endius* were marginally effective or totally ineffective (Stage and Petersen,

Table 2. Relative abundance (%) of pteromalid wasps from stable fly pupae in bovine feedlots and dairies[a]

Species	Central Missouri	Southeastern Nebraska	Western Nebraska	Western Kansas
S. cameroni	0	14	10	12
S. nigra	68	2	4	2
S. nigroaenea	11	34	60	25
Muscidifurax sp.	17	49	9	42
Others	4	2	16	19[b]

[a] Adapted from Petersen (1989).
[b] Unidentified or aborted.

1981, Petersen et al., 1983). Mass releases of *S. endius* in 1982 and *M. raptor* in 1985 did not reduce stable fly populations in western Kansas (G. L. Greene, Kansas State University, personal communication, in Petersen, 1986). These latter studies stressed the need for basic data on such parameters as parasite species best adapted to the particular pest host and environment, application rates, dispersal rates, environmental effects, release procedures, timing of releases, and better methods of evaluating the impact of released parasites on host populations.

Recently a number of these factors have been studied in an effort to improve the effectiveness of these agents. Releases of parasite species endemic to the release area have improved results. Greene (1985) released five endemic *S. cameroni* per animal per week on open feedlots and reduced fly eclosion from 95 to 20%. This release was the first documented release to have a significant effect on stable fly populations. Petersen et al. (1992a) released an endemic *M. zaraptor* on two feedlots resulting in mean fly mortality in sentinel hosts of 37 and 26% compared with 4% for untreated confine-

ments. In an additional study, Petersen et al. (1992b) released the same endemic strain of *M. zaraptor* and an introduced strain of *Pachycrepoideus vindemiae*. Mean mortality of sentinel pupae for four release feedlots ranged from 28 to 42% compared with 5% for two untreated feedlots. *M. zaraptor* comprised 87% of the parasites recovered from host puparia. Also, Greene and Cilek (1993) reduced stable fly populations in western Kansas 47% with releases of native *S. nigroaenea*.

Guzman and Petersen (1986) showed that parasite survival was low on midwestern confinements, and that because of their protracted rate of development compared with that of their hosts, their populations will have to be augmented if they are to be effective during peak fly periods. Furthermore, studies have shown that mean weekly temperatures above 27 - 28°C greatly reduce parasite activity (Petersen et al., 1992a, 1992b). This temperature threshold is often reached during the summer on midwestern feedlots and may account for some of the results that were lower than expected during parasite releases.

Table 3. Percentage parasitism of house fly and stable fly pupae by the principal species of pteromalid wasps at feed-lots in eastern Nebraska[a]

Species	May	June	July	Aug	Sept	Oct
M. zaraptor	0	2.7	2.9	4.5	5.3	9.7
S. cameroni	0	1.0	0.4	1.8	1.8	1.2
S. nigroaenea	0	1.1	1.2	3.7	4.2	4.6
Total	0	4.8	4.5	10.0	11.3	15.5

[a] Adapted from Petersen and Meyer (1983).

Dosage responses are based on the numbers of animals or the size of the facility often without regard to animal density or waste management practices. Development of recommendations for application rates will depend on these factors as will behavioral traits of the parasite species being released. This information is difficult to assess because studies vary in methods of parasite application, size of facilities and methods of measuring parasite activity. Few studies have been made to measure dosage response. In a study where three confinements received 2,400, three received 11,900 and three received 20,100 *M. zaraptor* per week, mortality in sentinel hosts averaged 26, 31 and 46% over an 11 week period (Fig. 1) (Unpublished data). However, in a follow up study in which four parasite concentrations of 11,000, 21,000, 34,000 and 42,000 were released four times on individual feedlots resulted in mean host mortalities of 35, 43, 38 and 44%, respectively (Unpublished data). Application rate recommendations remain guess work until a better understanding of factors influencing parasitism are understood.

Recent studies also have shown the need for improved parasite application methods. It is likely that simply broadcasting parasitized host pupae as a method of releasing parasites is ineffective and often results in high mortality of the emerging parasites. Petersen et al. (1992b) reported that when protection from the elements and predators, and when a humidity source was provided, parasite emergence was substantially improved, especially during the warmer and dryer part of the summer.

Recent studies also have shown that released pteromalids disperse rapidly from the release point and within 1 to 2 weeks have dispersed over the confinement surface depending on temperature. In a study with a recently isolated strain of gregarious *Muscidifurax*, parasitism peaked 2 weeks after a release in late May, and parasitism peaked 1 week after two releases later in the season (Fig. 2) (Unpublished data).

Considerable field data have been accumulated in the past few years on the use of pteromalids for biocontrol of flies associated with cattle confinements. Endemic species have given the best results; however, few efforts have been made to introduce exotic species into these environments.

Biocontrol of insect pests of livestock and poultry have been slow in developing, in part because of heavy reliance on insecticides (i.e., ear tags, area sprays, feed throughs). The greatest emphasis has been directed toward filth flies associated with confined livestock and poultry because they offer the greatest potential for success. At present there are no effective pathogens for controlling fly populations, but this area of research remains largely unexplored. Predators and competitors also have received little attention because they generally fail to provide satisfactory control and are difficult to augment to increase their effectiveness. However, the dung-burying beetles and macrochelid mites show considerable potential but need more work before they can be effectively employed in fly control. The parasitic wasps have received the greatest amount of study because they attack the pupal stage of the

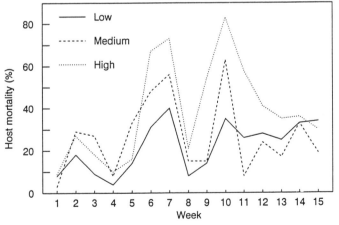

Fig. 1. Percentage mortality of sentinel house fly pupae resulting from mean weekly releases of 2,400 (low), 11,900 (medium), and 20,100 (high) *Muscidifurax zaraptor* in beef cattle confinements over a 15 week period.

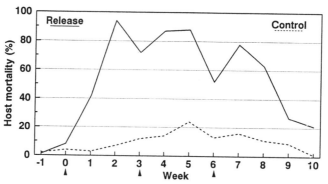

Fig. 2. Mean percentage host mortality in sentinel house fly pupae resulting from three releases (Arrows) of an endemic, gregarious strain of *Muscidifurax* at three beef cattle confinements.

host, can be easily and inexpensively mass produced, and releases can produce high levels of parasitism in host populations. However, their effectiveness remains questionable. Considerable research is needed on the effects of environmental factors on released parasites, most suitable parasite species to be released, parasite quality, timing of releases, parasite application rates, parasite dispersal and persistence, and improved methods of parasite releases. Biocontrol approaches will only be affective as part of an integrated control program. Fly management practices will need to include proper sanitation, management practices that encourage and enhance natural populations of fly predators and parasites and augmented with releases of agents as they are developed (Geden et al. 1992).

References

Axtell, R.C.; Rutz, D. A. Factors affecting the use of an IPM scheme for poultry installations in a semitropical climate. *Misc. Publ. Entomol. Soc. Am.*, **1986**, *61*, 99-100.

Bezark, L.G.; Rey, E.J. Suppliers of beneficial organisms in North America. *Calif. Dept. Food Agri.*, **1980**, 4 p.

Blume, R.R. A checklist, distributional record, and annotated bibliography on the insects associated with bovine droppings on pastures in America north of Mexico. *Southwest. Entomol.*, **1985**, *Suppl. 9*, 55 pp.

Bruce, W.N.; Decker, G. C. The relationship of stable fly abundance to milk production in dairy cattle. *J. Econ. Entomol.*, **1958**, *51*, 269-274.

Campbell, J.B.; Berry, I.L.; Boxler, D.J.; Davis, R.L.; Clanton, D. C.; Deutscher, G.H. Effects of stable flies (Diptera: Muscidae) on weight gains and feed efficiency of feedlot cattle. *J. Econ. Entomol.*, **1987**, *70*, 592-594.

Drummond, R.O.; Lambert, G.; Smalley, H.E., Jr.; Terrill, C.E. Estimated losses of livestock to pests. In *CRC Handbook of Pest Management in Agriculture*; Pimentel, D. Ed.; CRC Press, Boca Raton, FL, **1981**, pp 111-127.

Fincher, G.T. The potential value of dung beetles in pasture ecosystems. *J. Ga. Entomol. Soc.*, **1981**, *16(Suppl.)*, 316-333.

Fincher, G.T. Importation, colonization, and release of dung-burying scarabs. *Misc. Publ. Entomol. Soc. Am.*, **1986**, *61*, 69-76.

Fincher, G.T. Biological control of dung-breeding flies: pests of pastured cattle in the United States. In *Biocontrol of Arthropods Affecting Livestock and Poultry*; Rutz D.A., Patterson, R.S., Eds.; Westview Press, Boulder, CO, **1990**, pp 137-151.

Geden, C.J. Coleopteran and acarine predators of house fly immatures on poultry production systems. In *Biocontrol of Arthropods Affecting Livestock and Poultry*; Rutz D.A., Patterson, R.S., Eds.; Westview Press, Boulder, CO, **1990**, pp 177-200.

Geden, C.J.; Rutz, D.A.; Miller, R.W.; Steinkraus, C.D. Suppression of house flies (Diptera: Muscidae) on New York and Maryland dairies using releases of *Muscidifurax raptor* (Hymenoptera: Pteromalidae) in an integrated management program. *Environ. Entomol.*, **1992**, *21*, 1419-1426.

Geden, C.J.; Stoffolano, J. G., Jr. Dispersion patterns of arthropods associated with poultry manure in enclosed houses in

Massachusetts: spatial distribution and effects of manure moisture and accumulation time. *J. Entomol. Sci.*, **1988**, *23*, 136-148.

Greene, G.L. Naturally occurring fly parasites in western Kansas feedlots during 1983. *Kansas State Univ. Agri. Exper. Stat. Rpt.*, **1985**, *474*, 45-48.

Greene, G.L. Biological control of filth flies in confined cattle feedlots using pteromalid parasites. In *Biocontrol of Arthropods Affecting Livestock and Poultry*; Rutz D.A., Patterson, R.S., Eds.; Westview Press, Boulder, CO, **1990**, pp 29-42.

Greene, G.L.; Cilek, J.J. Management of stable flies in cattle feedlots with releases of parasitic wasps. *Kansas State Univ. Agri. Exper. Stat. Rpt.*, **1993**, *678*, 148-150.

Guzman, D.R.; Petersen, J.J. Overwintering of filth fly parasites (Hymenoptera: Pteromalidae) in open silage in eastern Nebraska. *Environ. Entomol.*, **1986**, *15*, 1296-1300.

Ho, C.C.; Cromroy, H.L.; Patterson, R.S. Mass production of the predaceous mite *Macrocheles muscaedomesticae* (Scopoli) (Acarina: Macrochelidae), a predator of the house fly. In *Biocontrol of Arthropods Affecting Livestock and Poultry*; Rutz D.A., Patterson, R.S., Eds.; Westview Press, Boulder, CO, **1990**, pp 201-213.

Hall, R.D.; Smith, J.P.; Thomas, G.D. Effect of predatory arthropods on the survival of immature stable flies (Diptera: Muscidae). *Misc. Publ. Entomol. Soc. Am.*, **1989**, *74*, 33-40.

Jespersen, J.B.; Keiding, J. The effect of *Bacillus thuringiensis var thuringiensis* on *Musca domestica* L. larvae resistant to insecticides. In *Biocontrol of Arthropods Affecting Livestock and Poultry*; Rutz, D.A., Patterson, R.S., Eds.; Westview Press, Boulder, CO. **1990**, pp 215-229.

Legner, E.F. Parasites and predators introduced against arthropod pests. Diptera-Muscidae. In *Introduced Parasites and Predators of Arthropod Pests and Weeds: A World Review;* Clausen, C.P. Ed.; Agri. Handb. No. 480, USDA, U.S. Govt. Printing Office, Wash., DC, **1978**, pp 346-355.

Legner, F.E. The requirement for reassessment of interactions among dung beetles, symbovine flies, and natural enemies. *Misc. Publ. Entomol. Soc. Am.*, **1986**, *61*, 88-100.

Legner, E.F.; Bay, E.C.; White, E.B. Activity of parasites from Diptera: *Musca domestica, Stomoxys calcitrans, Fannia canicularis,* and *F. femoralis,* at sites in the western hemisphere. *Ann. Entomol. Soc. Am.*, **1967**, *60*, 462-468.

Legner, E.F.; Brydon, H.W. Suppression of dung-inhabiting fly populations by pupal parasites. *Ann. Entomol. Soc. Am.*, **1966**, *59*, 638-651.

Legner, E.F.; Dietrick, E.I. Effectiveness of supervised control practices in lowering population densities of synanthropic flies on poultry ranches. *Entomophaga*, **1974**, *19*, 467-478.

Legner, E.F.; Olton, G.S. Distribution and relative abundance of dipterous pupae and their parasitoids in accumulations of domestic animal manure in the southwestern United States. *Hilgardia*, **1971**, *40*, 505-535.

Meyer, J.A. Biological control as a component of poultry integrated pest management. In *Biocontrol of Arthropods Affecting Livestock and Poultry*; Rutz D.A., Patterson, R.S., Eds.; Westview Press, Boulder, CO, **1990**, pp 45-57.

Meyer, J.A.; Rooney, W.F.; Mullens, B.A. Effect of Larvadex feed-through on cool-season development of filth flies and beneficial Coleoptera in poultry manure in southern California. *Southwest. Entomol.*, **1984**, *9*, 52-55.

Morgan, P.B.; Patterson, R.S. Field parasitization of house flies by natural populations of *Pachycrepoideus vindemiae* (Rondani), *Muscidifurax raptor* Girault and Sanders, and *Spalangia nigroaenea* Curtis. *Fl. Entomol.*, **1975**, *58*, 202.

Morgan, P.B.; Patterson, R.S. Sustained releases of *Spalangia endius* to parasitize field populations of three species of filth breeding flies. *J. Econ. Entomol.*, **1977**, *70*, 450-452.

Morgan, P.B.; Patterson, R.S.; LaBrecque, G.C. Controlling house flies at a dairy installation by releasing a protelean parasitoid, *Spalangia endius* (Hymenoptera: Pteromalidae). *J. Ga. Entomol. Soc.*, **1976**, *11*, 39-43.

Morgan, P.B.; Patterson, R.S.; LaBrecque, G.C.; Weidhass, D.E.; Benton, A. Suppression of a field population of house flies with *Spalangia endius. Science*, **1975**, *198*, 388-389.

Mullens, B.A.; Rodriguez, J.L.; Meyer, J.A. An epizootiological study of *Entomophthora muscae* in muscoid fly populations on southern California poultry facilities, with emphasis on *Musca domestica. Hilgardia*, **1987**, *55*, 1-41.

Olton, G.S.; Legner, E.F. Winter inoculative releases of parasitoids to reduce house flies in poultry manure. *J. Econ. Entomol.*, **1975**, *68*, 35-38.

Patterson, R. S. Status of biological control for livestock pests. In *Biocontrol of Arthropods Affecting Livestock and Poultry*; Rutz D.A., Patterson, R.S., Eds.; Westview Press, Boulder, CO, **1990**, pp 1-10.

Petersen, J.J. Potential for the biological control of stable flies associated with confined livestock. *Misc. Publ. Entomol. Soc. Am.*, **1989**, *74*, 41-45.

Petersen, J.J.; Meyer, J.A. Host preference and seasonal distribution of pteromalid parasites (Hymenoptera: Pteromalidae) for control of house flies and stable flies (Diptera: Muscidae) associated with confined livestock in eastern Nebraska. *Environ. Entomol.*, **1983**, *12*, 567-571.

Petersen, J.J.; Meyer, J.A.; Stage, D.A.; Morgan, P.B. Evaluation of sequential releases of *Spalangia endius* (Hymenoptera: Pteromalidae) for control of house flies and stable flies (Diptera: Muscidae) associated with confined livestock in eastern Nebraska. *J. Econ. Entomol.*, **1983**, *76*, 283-286.

Petersen, J.J.; Watson, D. W.; Pawson, B.M. Evaluation of field propagation of *Muscidifurax zaraptor* (Hymenoptera: Pteromalidae) for control of flies associated with confined beef cattle. *J. Econ. Entomol.*, **1992a**, *85*, 451-455.

Petersen, J.J.; Watson, D.W.; Pawson, B.M. Evaluation of *Muscidifurax zaraptor* and *Pachycrepoideus vindemiae* (Hymenoptera: Pteromalidae) for controlling flies associated with confined beef cattle. *Biol. Cont.*, **1992b**, *2*, 44-50.

Roth, J.P. Field mortality of the horn fly on unimproved central Texas pasture. *Environ. Entomol.*, **1989**, *18*, 98-102.

Roth, J.P.; Fincher, G.T.; Summerlin, J.W. Competition and predation as mortality factors of the horn fly, *Haematobia irritans* (L.) (Diptera: Muscidae) in a central Texas pasture habitat. *Environ. Entomol.*, **1983**, *12*, 106-109.

Rutz, D.A.; Axtell, R.C. Sustained releases of *Muscidifurax raptor* (Hymenoptera: Pteromalidae) for house fly (*Musca domestica*) control in two types of caged-layer poultry houses. *Environ. Entomol.*, **1979**, *8*, 1105-1110.

Rutz, D.A.; Axtell, R.C. House fly parasites (Hymenoptera: Pteromalidae) associated with poultry manure in North Carolina. *Environ. Entomol.*, **1980**, *9*, 175-180.

Rutz, D.A.; Axtell, R.C. House fly (*Musca domestica*) control in broiler-breeder poultry houses by pupal parasites (Hymenoptera: Pteromalidae): Indigenous parasite species and releases of *Muscidifurax raptor*. *Environ. Entomol.*, **1981**, *10*, 343-345.

Shepard, M.; Kissam, J. B. Integrated control of house flies on poultry farms: Treatment of house fly resting surfaces with diflubenzuron plus releases of the parasitoid, *Muscidifurax raptor*. *J. Ga. Entomol. Soc.*, **1981**, *16*, 222-227.

Stafford, K.C., III; Bay, D.E. Diversion patterns and association of house fly, *Musca domestica* (Diptera: Muscidae), larvae and both sexes of *Macrocheles muscaedomesticae* (Acari: Macrochelidae) in response to poultry manure moisture, temperature, and accumulation. *Environ. Entomol.*, **1987**, *16*, 159-164.

Stage, D.A.; Petersen, J.J. Mass release of pupal parasites for control of stable flies and house flies in confined feedlots in Nebraska; In *Status of Biological Control of Filth Flies* Patterson R.S., Ed.; U.S. Dept. Agri. Sci. Educ. Adm. Publ. A106.1:F64, **1981**; pp 52-58.

Steelman, C. D. Effects of external and internal arthropod parasites in domestic livestock production. *Ann. Rev. Entomol.* **1976**, *21*, 155-178.

Thomas, G.D.; Hall, R.D.; Wingo, C.W.; Smith, D.B.; Morgan, C.E. Field mortality studies of the immature stages of the face fly (Diptera: Muscidae) in Missouri. *Environ. Entomol.*, **1983**, *12*, 823-830.

Thomas, G.D.; Skoda, S.R. (Eds.). *Rural Flies in Urban Environments?* Univ. Nebra. Agri. Exper. Stat. Res. Bull. **1993**, *317*, 99 pp.

Thomas, G.D.; Morgan, C.E. Parasites of the horn fly in Missouri. *J. Econ. Entomol.*, **1972**, *65*, 169-174.

Thomas, G.D.; Wingo, C.W. 1968. Parasites of the face fly and two other species of dung-inhabiting flies in Missouri. *J. Econ. Entomol.*, **1968**, *61*, 147-152.

Watson, D.W.; Mullens, B.A.; Petersen, J.J. Behavioral fever of *Musca domestica* (Diptera: Muscidae) to infection by *Entomophthora muscae* (Zygomycetes: Entomophthorales). *J. Invert. Pathol.* **1993**, *61*, 10-16.

BIOCONTROL AGENTS FOR SUPPRESSION
OF PLANT PATHOGENS

Viral Satellites, Molecular Parasites for Plant Protection

J.M. Kaper, Molecular Plant Pathology Laboratory, Plant Sciences Institute, Agricultural Research Service, U.S. Dept. of Agriculture, Beltsville, MD 20705

Viral satellites are small nucleic acid molecules naturally associated with certain plant viruses to which they are sequence-unrelated but which they need for replication support. Satellites are also known for their capability to modify viral symptoms. They can be considered molecular parasites of the viruses with which they are associated. Satellites introduced into plants via preinoculation have proven to be effective in large-scale protection of field crops from virus infection. Satellite-transgenic plants with improved resistance to virus disease are under intensive study. This paper shows how the use of satellites in biologically based control of virus infections has placed cucumber mosaic virus (CMV) satellites at the leading edge of these novel crop protection technologies.

"Great fleas have little fleas, upon their backs to bite 'em, and little fleas have lesser fleas, and so *ad infinitum*" (Scott, 1986).

Exactly 18 years ago, in the first of this series of Beltsville symposia, I.R. Schneider[*] (1977) reviewed the work on a viral satellite he had discovered several years before. The satellite of tobacco ringspot virus was only the second RNA molecule of this type reported (Schneider, 1969), the first dating back to the discovery of the satellite of tobacco necrosis virus in Britain in the early sixties (Kassanis, 1962). Viral satellites commonly occur in association with different plant viruses (Collmer; Howell, 1992; Roossinck *et al.*, 1992) and, as far as presently known, less frequently with animal viruses (Kaper; Tousignant, 1984). From a molecular and functional standpoint they can best be defined as nucleic acids which, for their replication in appropriate host cells, depend upon the replication of a specific "helper virus" nucleic acid with which they do not share any substantial sequence homology.

In his symposium paper Schneider (1977) made reference to ongoing Beltsville research on a third satellite associated with cucumber mosaic virus (CMV) (Lot; Kaper, 1976; Kaper *et al.*, 1976), an economically important plant virus causing severe damage to fruit and vegetable crops worldwide (Tomlinson, 1987). Within one year of the first report on this CMV satellite (Kaper *et al.*, 1976) a second publication linked it to a lethal crop disease, tomato necrosis (Kaper; Waterworth, 1977). Five years before an epidemic of tomato necrosis was reported to have annihilated the field tomato crop in the Alsace region of France in 1972 (Putz *et al.*, 1974; Marrou; Duteil, 1974).

----
[*]deceased

Satellite-Mediated Viral Symptom Modulation and Plant Protection

Until 1977 the biological effects of viral satellites had received relatively little attention, since only induction of smaller local lesions on certain host plants had been noted with the two earlier satellites (Kassanis, 1962; Schneider, 1969). The experimental linkage of the CMV satellite CARNA 5 (CMV Associated RNA 5) to the tomato necrosis disease (Kaper; Waterworth, 1977) marked the first time that helper virus-supported replication of a satellite was shown to result in a highly noticeable change in systemic viral symptoms. Surprisingly, only two years later tomato necrosis-inducing CARNA 5 was shown to be also capable of attenuating systemic CMV symptoms in several other plant species (Waterworth *et al.*, 1979; Waterworth; Kaper, 1980). Thus it became clear that these simple satellite RNA molecules possessed fascinating disease regulatory properties, and could be valuable for plant protection (Kaper; Tousignant, 1984).

For CMV and the wide spectrum of crop diseases this virus causes, the above expectations have been borne out. On the one hand CMV satellites have been responsible for several major tomato necrosis epidemics, particularly in the Mediterranean region [(Kaper *et al.*, 1990b; Jorda *et al.*, 1992); C. Varveri, Greece, personal communication]. On the other hand, in China CMV satellites are presently used on a practical scale as biological control agents to protect more than 10,000 ha of crop plants against the yearly onslaughts of CMV (Tien; Wu, 1991). In several other countries practical application of CMV satellites is imminent.

In the first sections of this paper a brief introduction to the molecular structure of naturally occurring CMV satellites and their competition with the helper virus CMV in the host cell will be given. This will be used to explain how viral satellites should be regarded as "molecular parasites" of their own helper viruses. Then, in later sections it will be shown how at the present time this form of molecular parasitism is rapidly being exploited, not only to gain a better understanding of the epidemiology of CMV, but also in the development of biologically based technologies for control of the devastating losses caused by this crop virus.

Satellite Structure and Viral Symptom Modulation

Almost simultaneously with the discoveries of the dualistic (Dr. Jekyll/Mr. Hyde-like) biological behavior of CARNA 5 in different host plants (Waterworth *et al.,* 1979; Waterworth; Kaper, 1980), from Australia another CMV satellite was reported which attenuated the viral symptoms in tomato (Mossop; Francki, 1979). This finding suggested the existence of CMV satellite variants or "strains" that were capable of eliciting different responses from the same CMV-infected host species. This notion was confirmed when the first nucleotide sequences of variants necrogenic (Richards *et al.,* 1978) and nonnecrogenic in tomato (Collmer *et al.,* 1983; Gordon; Symons, 1983) were determined. At the present time the sequences of some 40 CARNA 5 variants with a size-range of 334-405 nt are known. Many of these variants induce the lethal tomato necrosis disease upon coinfection with CMV. Some also induce symptoms of a different kind, mostly distinct forms of chlorosis in tomato or tobacco. However, most variants ameliorate symptoms to the point that host plants appear entirely healthy (Collmer; Howell, 1992).

The dualistic nature of the symptom modulating effects of CARNA 5 sequence variants in CMV infections of tomato are illustrated in Fig. 1. Plants infected by CMV strain 1 in the absence of CARNA 5 are stunted, its leaves show chlorosis, and as the plant matures they will become deformed to assume a so-called "shoestring" shape (center). While coinfection with a necrogenic CARNA 5 variant will kill plants (right), another CARNA 5 variant attenuates the CMV symptoms till they virtually disappear (left). The two variants in Fig. 1 were only recently characterized. They are nearly identical in size (387 nt and 391 nt, respectively) and have some 51 nt differences in their sequences (Tousignant; Kaper, 1993). Important from a structural standpoint is the fact that the necrogenic variant has a ca. 15-nucleotide stretch in its 3' half, which is conserved in all CMV satellites known to induce lethal tomato necrosis (Collmer; Howell, 1992). Although

apparently an important determinant for lethal tomato necrosis, the presence of sequence elements in the 5' half and other parts of the CARNA 5 molecule also seems to be required. Apparently these elements together constitute the structural determinants for tomato necrosis induction (Sleat; Palukaitis, 1990; Wu; Kaper, 1992).

Satellite Replication and Viral Symptom Modulation

CMV satellites drastically influence viral RNA multiplication (Fig. 2), when CMV + CARNA 5 are coinoculated in tobacco. In the course of a 100-hour experiment the rate of replication of CMV-RNA and CARNA 5 was monitored by ^{32}P pulse labeling of semisynchronously infected leaf strips (Piazzolla *et al.,* 1982). Viral replication was clearly overtaken by that of satellite (Fig. 2). If this type of infection is monitored long-term in a systemically infected plant, the CMV-RNA titer in the plant tissue can be shown to decrease with time. It is followed by a slower but inevitable decrease in the CARNA 5 titer, although replication of both RNAs continues at very low levels for the life of the plant (Kaper; Tousignant, 1984; White; Kaper, 1989; Montasser *et al.,* 1991). Such a course of events can basically be explained by the replicative competition of CMV-RNA and CARNA 5 for the virus-encoded RNA-dependent RNA polymerase or replicase enzyme (Kaper, 1982). This enzyme, although designed to specifically replicate the viral RNA by which it is encoded, has always been suspected and has recently been shown capable of replicating CMV satellite RNAs *in vitro* with equal specificity and efficiency as the viral RNA (Wu *et al.,* 1991; Hayes *et al.,* 1992). This dependence of the satellite upon an enzyme encoded by the viral RNA, which then is outcompeted because of the enzyme's greater affinity for the satellite, generates a relationship where the satellite in essence can be regarded a molecular parasite of the virus. In such a parasitic relationship, with the satellite depending upon the virus and not vice versa, a proliferation ceiling is automatically built-in. Thus, satellite automatically limits virus replication, which in turn limits satellite, after which virus can re-emerge, but then is immediately outcompeted again by satellite, and so on. In practice this results in a low-level, and symptomless, persistent infection during the plant's lifetime. It would explain the fact that CMV satellites in most cases exert an attenuating effect on viral symptoms in the plant. However, with lethal tomato necrosis the satellite sequence not only imposes a CARNA 5 structure that is recognized by the viral repli-

135

Fig. 1. Dualistic nature of symptom modulation caused by two satellite variants in tomato (*Lycopersicon esculentum* cv. UC82B) infected by CMV strain 1. Left: Attenuation of CMV-1 symptoms by nonnecrogenic D27-CARNA 5 variant. Middle: Symptoms caused by CMV-1 in the absence of satellite. Right: Aggravation of CMV-1 symptoms by necrogenic J876-CARNA 5 variant resulting in lethal necrosis. Plants photographed 21 days after inoculation.

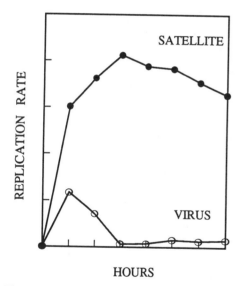

Fig. 2. Satellite overtaking CMV RNA replication. Data were adapted from an experiment in which leaf strips from tobacco plants infected with CMV and its satellite were pulse labeled with ^{32}P in the course of a 100-hour experiment (Piazzolla *et al.*, 1982). Plotted replication rates of virus and satellite represent relative rates of ^{32}P incorporation into CMV-RNA 3 and (+) stranded CARNA 5.

case, it also encodes and expresses its own symptoms. Thus a pathogenic response is elicited via direct interaction of satellite with the host plant, and these symptoms may mix with the viral symptoms. Alternatively the host plant's response to the virus infection is modified after the satellite first interacts with the viral symptom inducing mechanism.

CMV Satellites as Molecular Parasites in the Etiology of Crop Epidemics

The first recorded example of "molecular parasitism" on a scale large enough to destroy an entire field crop probably dates back to the tomato necrosis epidemic of 1972 in the Alsace region of France (Vuittenez; Putz, 1972). French investigators, who also provided the first systematic description of the tomato necrosis disease and its epidemiology (Marrou *et al.*, 1973), related this epidemic to the spread of what they thought was a new CMV strain (Putz *et al.*, 1974; Marrou; Duteil, 1974). The possibility of satellite involvement was not recognized until four years later when the first CMV satellite was discovered (Kaper *et al.*, 1976), and its linkage to lethal tomato necrosis proven (Kaper; Waterworth, 1977). While strongly suggesting the involvement of a necrogenic CARNA 5 in the 1972 epidemic, and establishing the cause-effect relationship conceptually, that work could not deliver direct proof because of the wide separation in time and place of the two events.

The second recorded tomato necrosis epidemic occurred in the Basilicata region of southern Italy in 1988 and subsequent years (Gallitelli *et al.*, 1988b). This time the nucleic acid technology developed over a period of 11 years of research on CMV satellites, was able to show a direct correlation of the disease spread with the presence of a low molecular weight RNA in plants from the tomato necrosis affected areas (Gallitelli *et al.*, 1988c). It allowed for the direct experimental characterization of this RNA from the necrosis-affected fields as a CMV satellite or CARNA 5 (Kaper *et al.*, 1990a), and for the first time demonstrated a 334-nucleotide viral satellite to be the principal etiological agent in a major outbreak of crop disease.

Due to the occurrence of several massive CMV outbreaks among vegetable crops in southern Italy, eastern Spain, and other Mediterranean countries, this region in the past five years has become a focal point for field studies of the fascinating role of satellites as molecular parasites in CMV epidemics (Gallitelli *et al.*, 1988a; Gallitelli *et al.*, 1988b; Gallitelli *et al.*, 1988c; Kaper *et al.*, 1990a; Jorda *et al.*, 1992; Crescenzi *et al.*, 1993b). Since in the open field natural selection determines their parasitic role, sometimes both mild and severe variants and their symptoms are displayed and may predominate depending on factors as yet not understood. Because of the enormous economic importance of the tomato processing industry to southern Italy, and of fresh market tomato production in eastern Spain, several research groups have focused their efforts on CMV epidemiology as well as on attempts to control such epidemics (see below).

Several types of CMV disease in tomato have been observed in different surveys. In addition to the presently well-recognized lethal necrosis syndrome of satellite etiology (Gallitelli *et al.*, 1988c; Kaper *et al.*, 1990a), a deceptive and economically perhaps much more devastating "tomato fruit necrosis" disease also seems to be quite prevalent (Jorda *et al.*, 1992; Crescenzi *et al.*, 1993b; Crescenzi *et al.*, 1993a). In the Campania region of Italy this disease is characterized by the presence of variously extended internal browning of fruits which makes them unmarketable. The deceptiveness of the disease lies in the fact that these fruits are borne by vigorous plants that are virtually symptomless, or show only a mild mottling of the leaves. This type of disease pattern correlated invariably with the presence of a 390 nt satellite named Tfn-CARNA 5, which, however, in experimen-

tal infections of tomato proved to be attenuative rather than necrogenic (Crescenzi *et al.*, 1993b; Crescenzi *et al.*, 1993a). Thus, one could hypothesize that Tfn-CARNA 5, while attenuating the CMV disease to the point that the plants themselves look almost healthy, is unable to prevent necrosis from occurring inside the fruits. This theory perhaps has been borne out by the fact that in Spain there was no satellite present in plants bearing necrotic fruits, suggesting that this disease may be caused by the CMV strain itself (Jorda *et al.*, 1992).

Several other vegetable crops affected by the CMV epidemic in Campania were also shown to have significant satellite presence. However, the satellite role, if any, in the symptomatology of the diseased crops has not been established thus far. Using very sensitive molecular probes for detection, overwintering weeds surrounding the affected fields, and probably the principal reservoir for CMV infections in the spring, were only occasionally found to have CARNA 5. This relative paucity of satellite, as opposed to the abundance of its helper virus CMV in weeds, rather suggested that cultivated crops were more likely to function as the main natural inoculum source for CARNA 5 variants and thus play a role in its epidemiology (Crescenzi *et al.*, 1993b).

At this point the dualistic nature of the symptomatology observed for CMV + CARNA 5 infections should perhaps be emphasized once more. Because of their inherently parasitic nature, these two agents elsewhere have been aptly described as "..nested parasitic nucleic acids competing for genetic expression.." (Kaper, 1992). Thus satellites should be regarded as (molecular) parasites of their helper viruses, while both in turn parasitize the host plant. An ameliorative or Dr. Jekyll-type effect should always occur when a satellite outcompetes the virus without expressing its own symptoms. However, this scenario may have an entirely different outcome if the virus is not outcompeted and displays it own symptoms, or if severe satellite symptoms or a Mr. Hyde-type effect are expressed. To make matters more complicated, in a natural setting, where several helper virus and/or satellite strains may occur simultaneously, there will be also be competition among satellites. Recent tests mimicking such competition have shown that this may determine which symptoms ultimately prevail (Smith *et al.*, 1992).

Satellites as Agents in Biologically Based Control of CMV Disease

The disease regulating capability of CMV satellites has followed directly from the earlier publications which clearly demonstrated the ameliorative effects of their presence in infections of CMV in a variety of host plants (Waterworth *et al.,* 1979; Waterworth; Kaper, 1980; Mossop; Francki, 1979; Jacquemond; Leroux, 1982). Soon after that work the first report appeared of successful protection of tomato plants preinfected with a CMV strain containing a nonnecrogenic CARNA 5 against challenge infection with another strain of CMV (Jacquemond, 1982). An example of a similar test taken from recent work in Beltsville (Montasser et al., unpublished) is shown in Fig. 3. Pepper plants were successfully protected against the stunting effects of challenge infection by CMV strain 16 by preinoculating seedlings with the combination of CMV strain S + S-CARNA 5. Rationalization for this type of satellite-induced resistance follows from S-CARNA 5 spreading systemically throughout the plant in a low-level symptomless infection initiated by preinoculation. The challenge virus CMV-16, upon invasion will support the replication of the satellite which lies in wait, and immediately be outcompeted. Experiments such as these have set the stage for the emergence of a novel satellite-based CMV control strategy. At this time this technology is only beginning to establish itself as one of the more powerful, yet deceivingly simple, weapons in the quest for finding resistance against viral crop disease.

Control Strategy for Preventing CMV disease Using Satellite Preinoculation

Reduction in CMV disease symptoms mediated by satellites and the introduction of the protective CARNA 5 via preinoculation or "vaccination" of seedlings in the presence of a CMV helper [(Jacquemond, 1982) and Fig. 3] has become the basis for practical field strategies for the satellite-based control of CMV infections. Due to a superficial resemblance with conventional cross-protection (which is mediated in part by viral coat protein), several direct comparisons of these two approaches to control CMV infections have been reported (Yoshida *et al.,* 1985; Wu *et al.,* 1989; Montasser *et al.,* 1991). These show consistently that the protective effects of satellite-free CMV (cross protection) are considerably enhanced when the preinoculation mixture also contains CARNA 5. Furthermore, conventional cross-protection is totally ineffective against a CARNA 5-induced disease such as tomato necrosis (M. Montasser, unpublished).
Field tests demonstrating the effectiveness of satellite-mediated control of CMV in tomato have been carried out under conditions of

simulated (Montasser *et al.,* 1991) as well as natural disease pressure (Gallitelli *et al.,* 1991). Testing is still continuing in Italy (D. Gallitelli, personal communication). Field tests were recently also completed in Japan (Sayama *et al.,* 1993), where satellite-based preinoculation technology to control CMV will be commercialized in 1993 (H. Sayama, personal communication). In China where CMV infestations are endemic, satellite-mediated biocontrol strategies have been applied on a practical scale since 1983 (Tien; Chang, 1983). At the present time in China the technology is state-licensed for the protection of tomato, different pepper varieties, cucurbits, cabbage and tobacco plants in over 10,000 hectares in at least 16 localities, apparently with outstanding results (Tien; Wu, 1991).

Control Strategy for the Prevention of CMV disease Using Satellite Transgenic Plants

Tomato, the same crop plant for which CMV satellite control strategies with preinoculation have yielded such favorable results, was also one of the first plant species targeted for CMV satellite insertion using *Agrobacterium tumefaciens* transformation techniques. Such insertions were initially carried out successfully in tobacco (Baulcombe *et al.,* 1986). Insertions of a necrogenic CARNA 5 in tomato yielded perfectly healthy-looking plants. Only upon challenge infection with (satellite-free) CMV, did they develop lethal tomato necrosis (McGarvey *et al.,* 1990; Tousch *et al.,* 1990). This means that for necrogenic CARNA 5 to reach sufficiently high titers to express tomato necrosis the satellite sequences first have to be amplified by the replicase of the challenge virus.
The logical next step has been the insertion of constructs containing an ameliorative CARNA 5 variant into *Agrobacterium tumefaciens* Ti plasmids, which then were used to transform certain targeted plant species. In China a variant 1-CARNA 5, originally isolated and characterized in Beltsville (Kaper *et al.,* 1981; Collmer *et al.,* 1983), has been used to transform tobacco as well as tomato. The resulting transgenic plants when tested for CMV resistance under greenhouse conditions, and also in preliminary field assays, showed significant decreases in disease index for both tobacco and tomato, and in the latter case a 50% improvement in fruit yield (Tien; Wu, 1991). Fig. 4 (center plant) shows an example of this type of CMV resistance in a tomato line similarly transformed with the prototype ameliorative variant S-CARNA 5 (Collmer; Kaper, 1986) and subsequently challenge infected with CMV . It should be compared with the transgenic plant on the left,

138

Fig. 3. Control of CMV-16 infection in pepper (*Capsicum frutescens* cv. Cal. Wonder) by preinoculating seedlings with CMV strain S + S-CARNA 5. From left to right. Row 1: Mock inoculated control plants. Row 2: Plants preinoculated with CMV-S + S-CARNA 5 at seedling stage. Rows 3-5: Plants preinoculated with CMV-S + S-CARNA 5 and challenge inoculated with CMV-16 1, 2, and 3 weeks later, respectively. Row 6: Non-preinoculated control plants, challenge inoculated at the same time as plants in row 5. Note that CMV resistance sets in about 2 weeks after preinoculation.

Fig. 4. Test of S-CARNA 5 transgenic tomato (*Lycopersocon esculentum* cv. UC82B) resistance against CMV strain 1. Left: Mock inoculated transgenic plant functioning as control. Middle: Transgenic plant after challenge inoculation with CMV-1. Right: Nontransgenic plant after inoculation with CMV-1. Plants were photographed 24 days after inoculation.

which was not challenged, and with the non-transgenic control plant on the right which had no resistance to CMV challenge (McGarvey *et al.*, submitted for publication).

Safety Considerations in Satellite Based Strategies for Control of CMV Disease

There have been justifiable concerns about the safety of releasing viral satellites in large-scale agricultural practice, either via preinoculation of seedlings or via transgenic plants. This is because of the danger of a potentially pathogenic virus/satellite combination replicating at low levels and spreading in the protected plant and being transmitted to other nonprotected plants (McGarvey; Kaper, 1993). However, different CMV/CARNA 5 combinations bioassayed by different investigators in numerous native species and crop plants have thus far not shown harmful effects (Waterworth *et al.*, 1979; Waterworth; Kaper, 1980; Tien; Wu, 1991; Sayama *et al.*, 1993), except for select CARNA 5 variants in tobacco and tomato (Collmer; Howell, 1992). It is also possible that an ameliorative CARNA 5 variant could undergo mutations converting it into a harmful variant, especially because of the higher mutation rate of RNA genomes as compared to DNA (Domingo; Holland, 1988). This potential danger would be more pronounced with the satellite preinoculation technology because of the presence of CMV and CARNA 5 (albeit at low levels) in all protected plants. Here the experience of eight years of practical-scale satellite-mediated control of CMV in China, using primarily preinoculation technology, has been invaluable (Tien; Wu, 1991). It has taught us that first, none of the specific CMV strain/satellite combinations used (all of natural origin), have thus far caused damage to the target crop or other surrounding plant species. Second, since no new types of disease nor lethal tomato necrosis were observed in extensive field surveys of protected areas, and since sensitive molecular probes had detected no obvious increase in satellite RNA titers, there has been no uncontrolled spread and replication of new harmful satellite variants (generated by mutation) (Tien; Wu, 1991). Third, any such surprises could be effectively prevented by periodic and detailed quality control of the stock virus/satellite cultures.

Comparison of Satellite Transgenic Plant and Preinoculation Technologies

Although there is not much experience with the transgenic plant approach, its obvious advantage is that virus resistance is encoded in the plant's genome, presumably as a stable trait, and much less subject to mutation than the replicating satellite RNA in the preinoculation approach. Moreover, transgenic plants only attain significant satellite titers following a virus infection. Thus in areas without high natural disease pressure there is very little or no unnecessary release of a potentially harmful satellite. However, from a protection standpoint this may be disadvantageous because resistance will probably be weaker in the early stages of virus attack. In cases that have been tested, suppression of virus symptoms was only clearly seen at later stages after the challenge infection (Tien; Wu, 1991). This disadvantage may be inconsequential for many crops, particularly those that derive economic value from their physiological state at maturity. It has recently also been overcome in tobacco via the use of chimeric coat protein/satellite constructs providing early as well as late resistance against CMV infection (Yie *et al.*, 1992). A more important disadvantage is the complexity and time-consuming nature inherent in the design and engineering of protective satellite constructs to transform different crop plants. Thus, there would be less flexibility in combating disease induced by different virulent CMV strains that prevail in different localities. Hence the preinoculation approach, which is easy to manipulate and adapt to local conditions, will probably remain a first line of defense. Also, this approach produces rapid antiviral effects and promotes yields. Only with large-scale production of a uniform target crop and a well defined virus menace, would the transgenic plant approach offer definitive advantages over preinoculation.

Epilogue

In the course of 30-plus years that their existence has been recognized, viral satellite RNAs have evolved conceptually from their status of being mere [in the words of their discoverer Kassanis (1981)] "..objects of academic interest.." to that of more respectable "lesser fleas." Indeed, in retrospect it seems that some fifteen years ago the bite of one such lesser flea, the CMV satellite (Vuittenez; Putz, 1972; Kaper; Waterworth, 1977), may have opened a window on a growing new world of molecular size parasites with fascinating biological properties (Collmer; Howell, 1992). The satellites of CMV, which in addition to their virulent Mr. Hyde-like demeanor clearly also play an ameliorative Dr. Jekyll role in the epidemiology of crop diseases caused by CMV (Crescenzi *et al.*, 1993b), seem to occupy an ecological niche that has guaranteed their indefinite survival. Thus, in such a world of molecular parasites each member may have its natural role of functioning as the lesser flea of its own helper virus. If this notion could be extended to each newly discovered satellite/virus combination, and put to agricultural practice, as with the CMV satellites, perhaps we will witness the emergence of a new biologically based plant protection technology applicable to many other viral crop diseases.

References

Baulcombe, D.C.; Saunders, G.R.; Bevan, M.W.; Mayo, M.A.; Harrison, B.D. "Expression of biologically active viral satellite RNA from the nuclear genome of transformed plants." *Nature*, **1986**, *321*, 446-449.

Collmer, C.W.; Tousignant, M.E.; Kaper, J.M. "Cucumber mosaic virus-associated RNA 5. X. The complete nucleotide sequence of a CARNA 5 incapable of inducing tomato necrosis." *Virology*, **1983**, *127*, 230-234.

Collmer, C.W.; Howell, S.H. "Role of satellite RNA in the expression of symptoms caused by plant viruses." *Ann. Rev. Phytopathol.*, **1992**, *30*, 419-442.

Collmer, C.W.; Kaper, J.M. "Infectious RNA transcripts from cloned cDNAs of cucumber mosaic viral satellites." *Biochem. Biophys. Res. Commun.*, **1986**, *135*, 290-296.

Crescenzi, A.; Barbarossa, L.; Cillo, F.; DiFranco, A.; Volvas, N.; Gallitelli, D. "Role of cucumber mosaic virus and its satellite RNA in the etiology of tomato fruit necrosis in Italy." *Archives of Virology*, **1993a**, (In Press).

Crescenzi, A.; Barbarossa, L.; Gallitelli, D.; Martelli, G.P. "Cucumber mosaic cucumovirus populations in Italy under natural epidemic conditions and after a satellite-mediated protection test." *Plant Dis.*, **1993b**, *77*, 28-33.

Domingo, E.; Holland, J. J. High error rates, population equilibrium, and evolution of RNA replication systems. In *RNA genetics. Vol. III, Variability of RNA genomes*; Domingo, E.; Holland, J. J.; Ahlquist, P. , Eds.; CRC Press: Boca Raton, 1988, pp 3-36.

Gallitelli, D.; DiFranco, A.; Vovlas, C. "Epidemie del virus del mosaico del cetriolo e di potyvirus in Italia meridionale repertorio dei virus e aspetti ecologici." *L'Inform. Agr.*, **1988a**, *44*, 40-45.

Gallitelli, D.; DiFranco, A.; Vovlas, C.; Crescenzi, A.; Ragozzino, A. "Una grave virosi del pomodoro in Italia meridionale." *L'Inform. Agr.*, **1988b**, *44*, 67-70.

Gallitelli, D.; DiFranco, A.; Vovlas, C.; Kaper, J.M. "Infezioni miste del virus del mosaico del cetriolo (CMV) e di potyvirus in colture ortive di Puglia e Basilicata." *Inform. Fitopatol.*, **1988c**, *38*, 57-64.

Gallitelli, D.; Vovlas, C.; Martelli, G.P.; Montasser, M.S.; Tousignant, M.E.; Kaper, J.M. "Satellite-mediated protection of tomato against cucumber mosaic virus. II. Field test under natural epidemic conditions in Southern Italy." *Plant Dis.*, **1991**, *75*, 93-95.

Gordon, K.H.J.; Symons, R.H. "Satellite RNA of cucumber mosaic virus forms a secondary structure with partial 3'-terminal homology to genomal RNAs." *Nucleic Acids Research*, **1983**, *11*, 947-959.

Hayes, R.J.; Tousch, D.; Jacquemond, M.; Pereira, V.C.; Buck, K.W.; Tepfer, M. "Complete replication of a satellite RNA in vitro by a purified RNA-dependent RNA polymerase." *J. Gen. Virol.*, **1992**, *73*, 1597-1600.

Jacquemond, M. "Phénomenes d'interferences entre les deux types d'ARN satellite du virus de la mosaïque du concombre. Protection des tomates vis a vis de la nécrose létale." *C. R. Acad. Sci. Paris*, **1982**, *290*, 991-994.

Jacquemond, M.; Leroux, J. "L'ARN satellite du virus de la mosaïque du concombre II. Étude de la relation virus-ARN satellite chez divers hôtes." *Agronomie*, **1982**, *2*, 55-62.

Jorda, C.; Alfaro, A.; Aranda, M.A.; Moriones, E.; Garcia-Arenal, F. "An epidemic of cucumber mosaic virus plus satellite RNA in tomatoes in eastern Spain." *Plant Dis.*, **1992**, *76*, 363-366.

Kaper, J.M.; Tousignant, M.E.; Lot, H. "A low molecular weight replicating RNA associated with a divided genome plant virus: Defective or satellite RNA?" *Biochem. Biophys. Res. Commun.*, **1976**, *72*, 1237-1243.

Kaper, J.M.; Tousignant, M.E.; Thompson, S.M. "Cucumber mosaic virus-associated RNA 5. VIII. Identification and partial characterization of a CARNA 5 incapable of inducing tomato necrosis." *Virology*, **1981**, *114*, 526-533.

Kaper, J.M. "Rapid synthesis of double-stranded cucumber mosaic virus-associated RNA 5: Mechanism controlling viral pathogenesis?" *Biochem. Biophys. Res. Commun.*, **1982**, *105*, 1014-1022.

Kaper, J.M.; Gallitelli, D.; Tousignant, M.E. "Identification of a 334-ribonucleotide viral satellite as principal aetiological agent in a tomato necrosis epidemic." *Res. Virol.*, **1990a**, *141*, 81-95.

Kaper, J.M.; Tousignant, M.E.; Geletka, L.M. "Cucumber mosaic virus-associated RNA 5. XII. Symptom modulating effect is codetermined by satellite replication support function of helper virus." *Res. Virol.*, **1990b**, *141*, 487-503.

Kaper, J.M. "Satellite-induced viral symptom modulation in plants: a case of nested parasitic nucleic acids competing for genetic expression." *Res. Virol.*, **1992**, *143*, 5-10.

Kaper, J.M.; Tousignant, M.E. "Viral satellites: parasitic nucleic acids capable of modulating disease expression." *Endeavour New Series*, **1984**, *8*, 194-200.

Kaper, J.M.; Waterworth, H.E. "Cucumber mosaic virus-associated RNA 5: Causal agent for tomato necrosis." *Science*, **1977**, *196*, 429-431.

Kassanis, B. "Properties and behavior of a virus depending for its multiplication upon another." *J. Gen. Microbiol.*, **1962**, *27*, 477-488.

Kassanis, B. "Portraits of viruses: tobacco necrosis virus and its satellite virus." *Intervirology*, **1981**, *15*, 57-70.

Lot, H.; Kaper, J.M. "Physical and chemical differentiation of three strains of cucumber mosaic virus and peanut stunt virus." *Virology*, **1976**, *74*, 209-222.

Marrou, J.; Duteil, H.; Lot, H.; Clerjeau, M. "La nécrose de la tomate: Une grave virose des tomates cultivées en plein champ." *Pepin. Hortic. Maraich.*, **1973**, *137*, 37.

Marrou, J.; Duteil, H. "La nécrose de la tomate." *Ann. Phytopathol.*, **1974**, *6*, 155-171.

McGarvey, P.B.; Kaper, J.M.; Avila-Rincon, M.J.; Pena, L.; Diaz-Ruiz, J.R. "Transformed tomato plants express a satellite RNA of cucumber mosaic virus and produce lethal necrosis upon infection with viral RNA." *Biochem. Biophys. Res. Commun.*, **1990**, *170*, 548-555.

McGarvey, P. B.; Kaper, J. M. Transgenic plants for conferring virus tolerance: Satellite approach. In *Transgenic Plants, Vol. 1*; Academic Press: New York, 1993, pp 277-296.

Montasser, M.S.; Tousignant, M.E.; Kaper, J.M. "Satellite-mediated protection of tomato against cucumber mosaic virus. I. Greenhouse experiments and simulated epidemic conditions in the field." *Plant Dis.*, **1991**, *75*, 86-92.

Mossop, D.W.; Francki, R.I.B. "Comparative studies on two satellite RNAs of cucumber mosaic virus." *Virology*, **1979**, *95*, 395-404.

Piazzolla, P.; Tousignant, M.E.; Kaper, J.M. "Cucumber mosaic virus-associated RNA 5. IX. The overtaking of viral RNA synthesis by CARNA 5 and dsCARNA 5 in tobacco." *Virology*, **1982**, *122*, 147-157.

Putz, C.; Kuszala, J.; Kuszala, M.; Spindler, C. "Variation du pouvoir pathogène des isolats du virus de la mosaïque du concombre associée à la nécrose de la tomate." *Ann. Phytopathol.*, **1974**, *6*, 139-154.

Richards, K.E.; Jonard, G.; Jacquemond, M.; Lot, H. "Nucleotide sequence of cucumber mosaic virus-associated RNA5." *Virology*, **1978**, *89*, 395-408.

Roossinck, M.J.; Sleat, D.; Palukaitis, P. "Satellite RNAs of plant viruses: structures and biological effects." *Microbiological Reviews*, **1992**, *56*, 265-279.

Sayama, H.; Sato, T.; Kominato, M.; Natsuaki, T.; Kaper, J.M. "Field testing of a satellite-containing attenuated strain of cucumber mosaic virus for tomato protection in Japan." *Phytopathology*, **1993**, (In Press)

Schneider, I.R. "Satellite-like particle of Tobacco Ringspot Virus that resembles Tobacco Ringspot Virus." *Science*, **1969**, *166*, 1627-1629.

Schneider, I. R. Defective Plant Viruses. In *Beltsville Symposia in Agricultural Research [1] Virology in Agriculture*; Romberger, J. A., Ed.; Allanheld, Osmun & Co.: Montclair, N.J., 1977, pp 201-219.

Scott, A. "At the limits of infection." *New. Scient.*, **1986**, *1508*, 41-44.

Sleat, D.E.; Palukaitis, P. "Site-directed mutagenesis of a plant viral satellite RNA changes its phenotype from ameliorative to necrogenic." *Proc. Natl. Acad. Sci. USA*, **1990**, *87*, 2946-2950.

Smith, C.R.; Tousignant, M.E.; Geletka, L.M.; Kaper, J.M. "Competition between cucumber mosaic virus satellite RNAs in tomato seedlings and protoplasts: A model for satellite-mediated control of tomato necrosis." *Plant Dis.*, **1992**, *76*, 1270-1274.

Tien, P.; Chang, X.H. "Control of two seed-borne virus diseases in China by the use of protective inoculation." *Seed Sci. & Technol.*, **1983**, *11*, 969-972.

Tien, P.; Wu, G-S. Satellite RNA for the biological control of plant disease. In *Advances in Virus Research, Vol. 39*; Maramorosch, K.; Murphy, F.; Shatkin, A. , Eds.; Academic Press: New York, 1991, pp 321-339.

Tomlinson, J.A. "Epidemiology and control of virus diseases of vegetables." *Ann. Appl. Biol.*, **1987**, *110*, 661-681.

Tousch, D.; Jacquemond, M.; Tepfer, M. "Transgenic tomato plants expressing a cucumber mosaic virus (CMV) satellite RNA gene: inoculation with CMV induces lethal necrosis." *C. R. Acad. Sci. Paris, Serie III*, **1990**, *311*, 377-384.

Tousignant, M.E.; Kaper, J.M. "Cucumber mosaic virus associated RNA 5. XIII. Opposite necrogenicities in tomato of variants with large 5' half insertion/deletion regions." *Res. Virol.*, **1993**, (In Press)

Vuittenez, A.; Putz, C. "Catastrophe pour les producteurs Alsaciens de tomates." *Alsace*, **1972**, *August 20 and 22*,

Waterworth, H.E.; Kaper, J.M.; Tousignant, M.E. "CARNA 5, the small cucumber mosaic virus-dependent replicating RNA, regulates disease expression." *Science*, **1979**, *204*, 845-847.

Waterworth, H.E.; Kaper, J.M. "Cucumber mosaic virus from *Prunus domestica*: Some diseases incited in herbaceous species in the presence and absence of a small replicating RNA." *Acta Phytopathol. Acad. Sci. Hung.*, **1980**, *15*, 123-127.

White, J.L.; Kaper, J.M. "A simple method for detection of viral satellite RNAs in small plant tissue samples." *J. Virol. Meth.*, **1989**, *23*, 83-94.

Wu, G-S.; Kang, L-Y.; Tien, P. "The effect of satellite RNA on cross-protection among cucumber mosaic virus strains." *Ann. Appl. Biol.*, **1989**, *114*, 489-496.

Wu, G-S.; Kaper, J.M.; Jaspars, E.M.J. "Replication of cucumber mosaic virus satellite RNA *in vitro* by an RNA-dependent RNA polymerase from virus infected tobacco." *FEBS Letters*, **1991**, *292*, 213-216.

Wu, G-S.; Kaper, J.M. "Widely separated sequence elements within cucumber mosaic virus satellites contribute to their ability to induce lethal tomato necrosis." *J. Gen. Virol.*, **1992**, *73*, 2805-2812.

Yie, Y.; Zhao, F.; Zhao, S.Z.; Liu, Y.Z.; Liu, Y.L.; Tein, P. "High resistance to cucumber mosaic virus conferred by satellite RNA and coat protein in transgenic commercial tobacco cultivar G-140." *Mol. Plant-Microbe Interact.*, **1992**, *5*, 460-465.

Yoshida, K.; Goto, T.; Iizuka, M. "Attenuated isolates of cucumber mosaic virus produced by satellite RNA and cross-protection between attenuated isolates and virulent ones." *Ann. Phytopath. Soc. Japan*, **1985**, *51*, 238-242.

Roles of Competition and Antibiosis in Suppression of Plant Diseases by Bacterial Biological Control Agents

Joyce E. Loper, USDA-ARS Horticultural Crops Research Laboratory, and Department of Botany and Plant Pathology, Oregon State University, Corvallis, OR 97331
Steven E. Lindow, Department of Plant Pathology, University of California, Berkeley, CA 94720

Bacterial antagonists commonly are selected as potential agents for biological control of plant disease based on their production of antibiotics on one or more culture media. Nevertheless, the significance of antibiotic production by a bacterial antagonist in culture to its biological control activity on plant surfaces is variable. Many successful biological control agents exhibit no antibiosis against a target pathogen in culture and suppress plant disease primarily through competition for limiting resources, such as carbon compounds or iron. For example, biological control of Pythium damping-off disease of cotton by certain strains of *Pseudomonas fluorescens* is attributed to siderophore-mediated iron competition. Similarly, biological control of frost injury is attributed to preemptive competitive exclusion of ice-nucleation-active strains of *Pseudomonas syringae* through prior acquisition of limiting resources by antagonistic bacteria on leaf surfaces. Strains of *P. syringae* compete for the same limiting resources in the phylloplane but differ in the efficiency of acquiring these resources and thus in competitive fitness.

Biological control offers promise for the suppression of soilborne and bacterial diseases of plants, many of which are not managed effectively or economically by conventional chemical pesticides. Many genera of bacteria have been selected from the phyllosphere or rhizosphere and identified as antagonists useful in biological control of foliar diseases caused by bacterial phytopathogens or of soilborne diseases, respectively. Of these, a handful are currently used in commercial agriculture. *Agrobacterium radiobacter* strain K84 (Moore, 1988) has been used commercially in various parts of the world for more than a decade for the biological control of crown gall disease, caused by *Agrobacterium tumefaciens*. The remarkable success of K84 has done much to stimulate enthusiasm in the potential of bacterial antagonists for biological control of plant diseases. *Streptomyces griseoviridis* strain K-61 (Tahvonen and Avikainen, 1987) is the active ingredient of "Mycostop," developed by Kemira Oy (Espoo, Finland). Mycostop is marketed in Europe primarily for suppression of diseases of

ornamentals and vegetable crops caused by *Fusarium* spp. A strain of *Bacillus subtilis* was registered by the U.S. Environmental Protection Agency (EPA) in 1992 as a product called "Kodiak" (Gustafson, Inc., Plano, Texas) for use as a seed treatment or hopper box formulation for suppression of diseases of several crops caused by *Rhizoctonia solani* or *Fusarium solani*. A strain of *Pseudomonas cepacia* also was registered by EPA in 1992 under the trade name "Blue Circle" (Stine Microbial Products, Madison, Wisconsin) for suppression of damping-off diseases of a variety of vegetable and field crops caused by *Rhizoctonia* sp. and *Fusarium* sp. and of lesion, lance and sting nematodes. A mixture of three strains of *Pseudomonas* spp. was registered by the EPA in 1992 as a product called "Frostguard" (Frost Technologies, Oakland, California) for the control of frost damage to several deciduous tree and vegetable crops and of fire blight disease of pear and apple. A strain of *Pseudomonas fluorescens* (Hagedorn *et al.*, 1993) was sold as "Dagger-G" (Ecogen,

Langhorne, Pennsylvania) for suppression of damping-off diseases of cotton, but is no longer available commercially. Companies presently developing biological agents for management of plant disease are confronting a number of obstacles, including difficulties in evaluating efficacy of products in fields having variable disease pressure, in developing effective delivery systems that are compatible with current grower practices, and in acceptance of this new technology by a cautious agricultural industry. Biological control proponents are optimistic that obstacles will be successfully surmounted and these pioneering biological control agents will be widely adopted by growers.

Bacteria commonly are selected for evaluation as biological control agents because they produce antibiotics toxic to a target pathogen in culture (Andrews, 1985; Fravel, 1988; see Weller and Thomashow, this volume). This selection strategy is based on the assumption that antibiotics produced in culture also are produced on plant surfaces. Nevertheless, the significance of antibiotic production by a bacterial antagonist to its biological control activity on plant surfaces is often variable. Strains that produce antibiotics in culture are not always effective biological control agents whereas non-antibiotic producers may be quite effective (Andrews, 1985). Over the past decade, the application of molecular approaches to questions regarding the *in situ* production of antibiotics by bacterial antagonists inhabiting plant surfaces and the role of these compounds in biological control has done much to clarify our views of microbial interactions in the phyllosphere and rhizosphere (Défago and Haas, 1990; Gutterson, 1990).

In contrast to antibiotic production, a phenotype that is relatively conducive to genetic analysis, competitive fitness is an extremely complex phenotype that is difficult to characterize physiologically or genetically. Competition is considered to be an important mechanism determining microbial interactions in natural habitats, although it is also a mechanism of default; ie. competition is commonly proposed to explain interactions that cannot be explained by known antibiosis or parasitism. Difficulties encountered in studies evaluating the role of competition in biological control are

reviewed in Handelsman and Parke (1989). To a large extent, studies evaluating the role of iron competition in biological control have avoided these difficulties because the production of siderophores, which generally mediate microbial competition for iron, is amenable to genetic analysis. The role of iron competition in biological control of plant disease has been reviewed extensively (Loper and Buyer, 1991; Loper and Ishimaru, 1991; Schippers *et al.*, 1987); the role in biological control of soilborne diseases is summarized briefly here. The recent application of ecological models to interactions between isogenic bacterial strains now provides compelling evidence for the importance of competition in microbial interactions on leaf surfaces (Wilson and Lindow, 1991).

In this chapter, we offer our perspectives on the relative contributions of competition and antibiosis to the biological control of selected bacterial epiphytes or soilborne pathogens. No attempt was made to exhaustively review the literature on this subject. We refer the reader to some of the excellent reviews that have been published recently (Andrews, 1992; Baker and Dunn, 1990; Handelsman and Parke, 1989; O'Sullivan and O'Gara, 1992; Weller, 1988).

Biological Control of Foliar Diseases caused by Bacterial Phytopathogens

Many plant pathogenic bacteria reside epiphytically on leaf surfaces and initiate disease only when environmental conditions become favorable for infection (Hirano and Upper, 1983). Disease is likely only if epiphytic populations of bacterial pathogens are large (greater than 10^4 to 10^6 cells per leaf or flower) (Rouse *et al.*, 1985). Similarly, ice nucleation active (Ice$^+$) strains of bacterial species such as *Pseudomonas syringae*, *Pseudomonas fluorescens*, *Erwinia herbicola*, and *Xanthomonas campestris* occur commonly as epiphytes on plant surfaces (Gross *et al.*, 1983; Kim *et al.*, 1987; Lindow *et al.*, 1978a). Ice$^+$ bacteria incite frost damage to the plants on which they reside by catalyzing ice formation at temperatures as warm as -2 to -5 C (Lindow, 1983a; 1983b; Lindow *et al.*, 1978c). The incidence of frost damage to plants exposed to a given subfreezing temperature increases proportionally with the logarithm of the population size of Ice$^+$ bacteria on plants at the time of freezing (Lindow *et al.*, 1978c; 1982). Thus, the reduction of the epiphytic population size of bacteria on plants is an effective strategy for the control of plant bacterial diseases and frost damage.

Several examples of the successful biological control of foliar diseases and frost damage have been reported. Strains of *E. herbicola* and *P. fluorescens* inhibit the growth of *Erwinia amylovora* in pear and apple flowers, thereby reducing the incidence of fire blight disease (Beer *et al.*, 1984; Johnson *et al.*, 1993; Thomson *et al.*, 1976; Wilson and Lindow, 1993). The incidence of bacterial brown spot disease of bean and bacterial speck disease of tomato is reduced by the use of non-pathogenic strains of *P. syringae* (Cooksey, 1988; Lindow *et al.*, 1987). The incidence of frost damage to several plant species under field conditions is reduced by the application of non-ice nucleation active bacterial strains (Lindow, 1982; 1983b; 1985b; Lindow *et al.*, 1978c; 1983b). Reductions in incidence of frost damage are proportional to the logarithm of the population size of the non-ice nucleation active bacterial antagonist on plants (Lindow, 1985a). Although not always demonstrated, it is presumed that the application of an antagonistic bacterium reduces the population size of phytopathogenic or Ice$^+$ bacterial strains, thereby reducing the probability of infection or ice nucleation associated with these populations.

Relative Importance of Competition and Antibiosis. Genetic approaches have been very useful in examining the importance of antibiosis in antagonism observed between bacterial strains on plant surfaces. For example, Lindow (1985a, 1988) determined that less than half of the bacterial strains that inhibited the growth of an Ice$^+$ strain of *P. syringae* on leaf surfaces produced antibiotics inhibitory towards the strain in culture. Antibiosis-negative mutants of certain antibiotic-producing strains were selected; antagonism of *P. syringae* on leaf surfaces by parental and antibiosis-negative derivative strains was equivalent (Lindow, 1988). Thus, antibiotics that were produced by bacterial strains in culture contributed little to the antagonism of *P. syringae* on leaf surfaces.

In contrast, antibiotic production is a determinant of the biological control of *Erwinia amylovora* by *Erwinia herbicola* (Ishimaru *et al.*, 1988; Vanneste *et al.*, 1992). *E. herbicola* strain C9-1 produces two antibiotics and suppresses disease caused by a strain of *E. amylovora* that is sensitive to the antibiotics; it is less effective in suppression of disease caused by antibiotic-resistant mutants of *E. amylovora* (Ishimaru *et al.*, 1988). Similarly, *E. herbicola* strain Eh252 inhibits the multiplication of *E. amylovora* in pear fruit tissue whereas mutants of Eh252 that do not produce antibiotics were less inhibitory (Vanneste *et al.*, 1992). These studies indicate that antibiosis plays a role in certain interactions among bacteria on plant surfaces. In no case, however, has antibiosis accounted for all of the biological control activity exhibited by a bacterial antagonist.

Many bacterial strains that inhibit the growth of a target bacterial strain on leaf surfaces exhibit no antibiosis against the target strain in culture. For example, *E. herbicola* strain M232A inhibits the growth of Ice$^+$ strains of *E. herbicola* and *P. syringae* on corn leaves (Lindow *et al.*, 1983a). This antagonist exhibits no antibiosis towards these strains on any bacteriological culture media (Lindow *et al.*, 1983a). Similarly, *P. fluorescens* strain A506 exhibits no antibiosis against either *P. syringae* or *E. amylovora* in culture, yet strain A506 reduces the population size of *E. amylovora* on pear blossoms compared to that on flowers not pre-inoculated with an antagonist (Wilson and Lindow, 1993). In more directed studies, isogenic bacterial strains exhibited substantial antagonism towards each other on leaf surfaces. Non-ice nucleation active (Ice$^-$) mutants inoculated onto plants prior to inoculation with parental Ice$^+$ strains of *P. syringae* largely prevented growth of the Ice$^+$ strains (Lindemann and Suslow, 1987; Lindow, 1987). Because the Ice$^+$ and Ice$^-$ strains were isogenic, antibiosis could be dismissed as a possible explanation for the inhibition of Ice$^+$ strains. In contrast, competition appears sufficient to account for the observed interactions between Ice$^+$ and Ice$^-$ strains on leaf surfaces.

Pre-emptive Competitive Exclusion. Several studies have demonstrated that target bacterial populations are affected by pre-emptive competitive exclusion. That is, population sizes of target epiphytes are maximally reduced if the antagonist establishes a large population size on plants before the arrival of the target strain. For example, *E. herbicola* strain M232A and *P. fluorescens* strain A506 have little effect on the final population size of *E. herbicola* or *E. amylovora,* respectively, if they are co-inoculated onto plants with the target strains (Lindow *et al.,* 1983a; Wilson and Lindow, 1993). In contrast, population sizes of both target strains are reduced over 100 fold if the antagonist is applied from 12 to 24 hours prior to inoculation with the target strain. In such experiments, the antagonist has ample opportunity to increase in population size before the arrival of the target bacterial strain. An inverse relationship has been observed between the logarithm of the population size of antagonistic bacterial strains on plants and the logarithm of the population size of Ice$^+$ *P. syringae* strains that are supported on the same plants (Lindow, 1985b). Thus, if bacteria are competing for the same limiting resources on leaves, then antagonists that establish a large population size before the arrival of a target strain have more opportunity to usurp resources otherwise accessible to the target. Preliminary data suggests that carbon compounds are the most limiting nutrient resource on leaf surfaces. Addition of carbon compounds to leaf surfaces increases the maximum population size attained by epiphytic bacterial strains, whereas addition of nitrogenous compounds has little effect (Wilson and Lindow, unpublished).

Bacterial strains appear to compete unequally for the same limiting nutrient resources on leaf surfaces. Significant differences in the abilities of non-ice nucleation active strains of *P. syringae* to exclude a given Ice$^+$ *P. syringae* strain on leaf surfaces have been observed (Lindow, 1986a; Kinkel and Lindow 1989a; 1989b). Detailed examinations of the competitive interactions between bacterial strains using DeWit replacement designs also reveal significant differences in competitive ability among *P. syringae* strains (Wilson and Lindow, 1991).

Because the competitive abilities of *P. syringae* strains differ, a given Ice$^-$ *P. syringae* strain might not be effective in preemptive competitive exclusion of all Ice$^+$ *P. syringae* strains. Strains of *P. syringae* are diverse genetically (Cooksey and Graham 1989; Denny *et al.,* 1988; Hirano and Upper, 1986); whether or not this genetic diversity mediates differential competitive abilities on leaf surfaces is largely unknown. Potential differences in the ability of *P. syringae* strains to coexist with a given Ice$^-$ strain on plant surfaces were tested in field experiments. Ice$^-$ mutants were constructed by introducing genomic deletions into the *iceC* gene of several Ice$^+$ strains of *P. syringae* (Orser *et al.,* 1985; Lindow, 1985c). The recombinant Ice$^-$ mutants of *P. syringae* were inoculated onto young potato plants to achieve a large population size of this single genotype on leaf surfaces of field-grown plants (Lindow and Panopoulos, 1988; Lindow, 1986b, 1991). Wheat, barley, and pea plants that surrounded the experimental site harbored large numbers of indigenous Ice$^+$ *P. syringae* strains that immigrated via dry aerosol particles to inoculated potato plants. The size of the heterogenous population of Ice$^+$ *P. syringae* on potato plants inoculated with an Ice$^-$ *P. syringae* strain was approximately 50-fold less than that on non-inoculated potato plants during the first four weeks after inoculation with the antagonist (Lindow and Panopoulos, 1988). In contrast, the population size of an isogenic Ice$^+$ *P. syringae* strain was 1,000-fold less on plants harboring large numbers of an isogenic Ice$^-$ mutant than on non-inoculated plants (Lindow, unpublished data). These results suggest that indigenous Ice$^+$ *P. syringae* strains generally were less affected by competition with a previously established Ice$^-$ strain of *P. syringae* than was an isogenic Ice$^+$ strain, which had substrate requirements and ecological adaptations that were identical to those of the Ice$^-$ antagonist. Nevertheless, reductions in the population size of heterogenous Ice$^+$ *P. syringae* strains were sufficient to cause reductions in the incidence of frost damage (Lindow and Panopoulos, 1988). The nature of *P. syringae* strains that escape competition with Ice$^-$ mutants and the significance of this escape to the control of frost injury need further study.

Biological Control of Soilborne Plant Pathogens

Propagules of soilborne plant pathogens exist

in the soil environment in a quiescent state and require exogenous nutrients to germinate and grow saprophytically prior to infection of a susceptible host. Because soil is generally deplete of carbon and nitrogenous compounds, nutrients required for growth and development are provided by plants, in the form of root exudates, sloughed-off root cells, mucigel, senescent plant parts, and crop residues. Thus, like epiphytic bacteria, many soilborne fungal pathogens have a saprophytic phase in association with a plant prior to infection. A pathogen probably is most susceptible to antagonism by a bacterial biological control agent during this period of saprophytic growth. Bacterial antagonists are thought to protect plant surfaces from infection by competition or antibiosis of target plant pathogens in the spermosphere or rhizosphere. Of the many potential bacterial antagonists associated with plant roots, the fluorescent pseudomonads have received prominent attention due to their abundance in the plant rhizosphere (Suslow, 1982; Weller, 1988), their striking capacity to utilize a variety of organic substrates found in root exudates (Loper and Schroth, 1986), and their prolific production of secondary metabolites that are toxic against phytopathogenic fungi and bacteria (Défago and Haas, 1990; Fravel, 1988; Gutterson, 1990; O'Sullivan and O'Gara, 1992).

Antibiosis. During the past decade, many research groups have studied mechanisms by which fluorescent pseudomonads suppress soilborne diseases. These efforts have focussed on the roles of antifungal compounds produced by fluorescent pseudomonads in suppression of target fungal phytopathogens. Commonly, the biological control activity of one or more mutants, which are deficient in anti-fungal metabolite production, are compared to that of a parental strain, which produces the metabolite. For example, derivatives of *P. fluorescens* strain Hv37a that are deficient in oomycin A production exhibit only 50% of the biocontrol activity of strain Hv37a against Pythium damping-off of cotton (Howie and Suslow, 1991). Similarly, hydrogen cyanide (Voisard *et al.*, 1989), 2,4-diacetylphloroglucinol (Fenton *et al.*, 1992; Keel *et al.*, 1992; Vincent *et al.*, 1991), phenazine (Thomashow and Weller,

1988; Pierson and Thomashow, 1992), and pyrrolnitrin (Homma and Suzui, 1989) production contribute to biocontrol activities of fluorescent pseudomonads (see Weller and Thomashow, this volume). Antifungal metabolites generally contribute to, rather than account for all of, the biocontrol activity of parental strains; mutants commonly exhibit some disease suppression.

In contrast to the antifungal metabolites listed above, pyoluteorin production contributes little to the biocontrol activity of strain Pf-5 against Pythium damping-off of cucumber (Kraus and Loper, 1992). Nevertheless, pyoluteorin is toxic to *Pythium* spp. on seed surfaces and in culture (Howell and Stipanovic, 1980), accounts for the inhibition of *P. ultimum* by Pf-5 on certain culture media (Kraus and Loper, 1992), and contributes to the biocontrol activity of strain Pf-5 against Pythium damping-off of cotton (C. Howell, personal communication). Strain Pf-5 may not produce concentrations of pyoluteorin on cucumber seed surfaces that are adequate for suppression of *P. ultimum*. Because Pf-5 produces a number of antifungal metabolites, the influence of pyoluteorin may be masked by suppressive effects of other compounds to which *P. ultimum* is sensitive. Alternatively, the mechanism(s) by which Pf-5 suppresses Pythium damping-off of cucumber may be physical exclusion or competition for nutrients required for germination of sporangia or subsequent mycelial growth. The lack of correlation between pyoluteorin production and biological control activity of strain Pf-5 on cucumber provides an important example of a case in which antagonism of a fungus in culture is not related to the biological control of a soilborne plant disease.

Competition. Competition for iron is another mechanism by which fluorescent pseudomonads may inhibit the growth of plant pathogens. The fluorescent pseudomonads are characterized by their production of yellow-green pigments, termed pyoverdines or pseudobactins, that fluoresce under UV irradiation and function as siderophores (Abdallah, 1991). Pyoverdines produced by fluorescent pseudomonads inhibit the growth of a variety of microorganisms in iron-deplete culture media (Buyer *et al.*, 1989; Kloepper *et al.*, 1980; Meyer *et al.*, 1987;

Misaghi *et al.*, 1982). Pyoverdines produced in the rhizosphere by fluorescent pseudomonads are thought to sequester iron in a form that is unavailable to target pathogens (Loper and Buyer, 1991; Schippers *et al.*, 1987). By comparing the biological control activities of pyoverdine-biosynthesis mutants to that of pyoverdine-producing parental strains, convincing evidence for a role of pyoverdines in biological control of certain plant diseases has been obtained. For example, pyoverdine production contributes up to 95% of the biocontrol activity of certain strains (Table 1). Nevertheless, pyoverdines do not have a universal role in biological control activity of *Pseudomonas* spp. (Hamdan *et al.*, 1991; Keel and Défago, 1991; Kraus and Loper, 1992; Paulitz and Loper, 1991).

In a recent review, Paulitz (1990) summarized studies demonstrating the importance of competition for nitrogen and carbon in biological control of certain soilborne plant pathogens. In addition to these, Paulitz (1991) and Nelson and Maloney (1992) proposed that bacterial antagonists may catabolize specific components of seed exudates that function as chemical "signals" in triggering sporangial germination of *Pythium ultimum*. These observations have exciting implications for the potential specificity of interactions between bacterial antagonists and target soilborne fungi that is reminiscent of the specific interactions

Table 1. Role of Pyoverdines in Disease Suppression by Rhizosphere Pseudomonads: Selected Examples

Strain	Pathogen	Host	% Suppression Attributed to Pyoverdine[a]	References
JL3551	*Pythium ultimum*	Cotton	95	Loper, 1988
B324	*Pythium* spp.	Wheat	82	Becker and Cook, 1988
Pf-5	*P. ultimum*	Cucumber	5	Kraus and Loper, 1992
N1R	*P. ultimum*	Cucumber	27	Paulitz and Loper, 1992
WCS358	Uncharacterized	Potato	84	Bakker *et al.*, 1986a, 1986b
CHA0	*Gaeumannomyces graminis* var. *tritici*	Wheat	8	Keel and Défago, 1991
2-79	*G. graminis* var. *tritici*	Wheat	30	Hamdan *et al.*, 1991

[a] "% Suppression attributed to pyoverdine" was calculated as follows: Disease incidence or severity was assessed for plants that were untreated (D^u), plants that were treated with pyoverdine-producing parental strains in column 1 (D^p), and plants that were treated with mutants of the strains listed in column 1 that no longer produced a pyoverdine (D^m). Disease suppression by the pyoverdine-producing parental strain was $D^p - D^u$. Disease suppression by the pyoverdine-non-producing mutant strain was $D^m - D^u$. % Suppression attributed to pyoverdine production was $[(D^p - D^u) - (D^m - D^u)]/(D^p - D^u)$. For experiments in which more than one mutant was compared to the parental strain, the mean of disease assessment values for all mutants with identical phenotypes is presented.

between epiphytic bacterial populations discussed above.

Conclusions

Bacterial antagonists can suppress populations of plant pathogens in the phyllosphere, rhizosphere, and spermosphere by *in situ* production of antibiotics and by competition. In this chapter, we presented examples of antagonistic interactions between beneficial and pathogenic microorganisms on both aerial and root surfaces of plants that are mediated by antibiosis or competition. Nevertheless, current studies focus on the process of pre-emptive exclusion of foliar phytopathogenic bacteria and on antibiosis of soilborne pathogens. This focus simply may reflect the dominant interests of researchers studying biological control of aerial or soilborne diseases. On the other hand, based on our present understanding of the microbiology of plant surfaces, we can speculate on why these mechanisms may take on differential importance in the very different environments of seed, root or leaf surfaces.

At present, we have little understanding of the chemical and physical nature of microbial habitats on plant surfaces. Nevertheless, there has been tremendous speculation on the relative physical stresses and nutrients available to microbes inhabiting the spermosphere, rhizosphere, and phyllosphere. Because media composition profoundly affects antibiotic production by bacteria in culture, chemical composition of microhabitats on plant surfaces must influence the *in situ* production of antibiotics and thus the importance of antibiosis as a mechanism in biological control. The *in situ* production of phenazine-1-carboxylate (Thomashow *et al.*, 1990), 2,4-diacetylphloroglucinol (Keel *et al.*, 1992), and pseudobactin (Buyer *et al.*, 1993) by *Pseudomonas* spp. inhabiting the plant rhizosphere has been demonstrated. These data indicate that the chemical composition of microhabitats occupied by *Pseudomonas* spp. in the plant rhizosphere are conducive to antibiotic and siderophore production. *In situ* antibiotic production by bacteria occupying microhabitats in the phyllosphere, however, is poorly understood. It is tempting to speculate that such

habitats may not be conducive to the production of antibiotics in concentrations needed to antagonize target phytopathogens. This speculation is consistent with the observation that antibiosis has at best a minor role in biological control of foliar plant pathogens.

Leaf and floral surfaces often are devoid of epiphytic bacteria shortly after emergence or flowering, respectively (Gross *et al.*, 1983; Lindow, 1982; Lindow *et al.*, 1978a). Antagonists can be readily established on these surfaces due to the relative absence of competing microbes. Once established, the population size of introduced antagonists generally comprises ca. 90% of that of aerobic heterotrophic bacteria that can be isolated from leaf surfaces in greenhouse experiments. Bacterial antagonists also can be readily established on seed surfaces. Radicles emerging from seed are relatively sterile but they immediately contact an abundance of soil microbes. Although inoculation of seed with large numbers of bacterial cells results in the establishment of populations of the inoculated strain in the rhizosphere (Suslow, 1982), the size of these populations is generally only a fraction of that of total heterotrophic bacteria that can be cultured from the rhizosphere (Loper *et al.*, 1984). Thus, the impact of the introduced population on precluding establishment of a competing microbe through competition alone may be far less in the rhizosphere, in which the introduced bacterium is a small component of the total population of culturable bacteria, than in the phyllosphere, in which the introduced bacterium is a relatively large component of the culturable bacterial population.

The fluorescent pseudomonads have received prominent attention during the past decade as antagonists of plant disease. Advances in the molecular genetics of this group have provided powerful tools for analysis of mechanisms by which fluorescent pseudomonads interact with related and unrelated microbes on plant surfaces. The success of directed studies identifying ecological roles of specific bacterial phenotypes and testing ecological models of competition is due largely to our current capability to derive marked, well-characterized, isogenic bacterial strains that differ only in a characteristic of interest. Fluorescent

150

pseudomonads may not be the optimal bacterial antagonists of many plant diseases. Currently, many successful programs are directed at antagonistic strains of other genera of gram-negative and gram-positive bacteria as potential biological control agents. Nevertheless, fluorescent pseudomonads have served biological control admirably in model systems through which our understanding of microbial interactions has been advanced.

References

Abdallah, M.A. "Pyoverdins and pseudobactins." In *Handbook of Microbial Iron Chelates;* Winkelmann, G., Ed; CRC Press: Boca Raton, FL, **1991**; pp 139-153.

Andrews, J.H. "Strategies for selecting antagonistic microorganisms from the phylloplane." In *Biological Control On The Phylloplane*; Windels, C.E.; Lindow, S.E., Eds.; American Phytopathological Society Press: St. Paul, MN, **1985**; pp 31-44.

Andrews, J.H. "Biological control in the phyllosphere." *Annu. Rev. Phytopathol.* **1992**, *30*, 603-635.

Baker, R.R.; Dunn, P.E. Eds.; *New Directions in Biological Control*, UCLA Symposia on Molecular and Cellular Biology New Series, Vol. 12; Alan R. Liss, Inc.: New York, NY, **1990**.

Bakker, P.A.H.M.; Lamers, J.G.; Bakker, A.W.; Marugg, J.D.; Weisbeek, P.J.; Schippers, B. "The role of siderophores in potato tuber yield increase by *Pseudomonas putida* in short rotation of potato." *Neth. J. Plant Path.* **1986a**, *92*, 249-256.

Bakker, P.A.H.M.; Weisbeek, P.J.; Schippers, B. "The role of siderophores in plant growth stimulation by fluorescent *Pseudomonas* spp." *Med. Fac. Landbouww. Rijksuniv. Gent* **1986b**, *51*, 1357-1362.

Becker, J.O.; Cook, R.J. "Role of siderophores in suppression of *Pythium* species and production of increased-growth response of wheat by fluorescent pseudomonads." *Phytopathology* **1988**, *78*, 778-782.

Beer, S.V.; Rundle, J.R.; Norelli, J.L. "Recent progress in the development of biological control of fire blight - a review." *Acta. Hortic.* **1984**, *151*, 195-201.

Buyer, J.S.; Sikora, L.J.; Chaney, R.L. "A new growth medium for the study of siderophore-mediated interactions." *Biol. Fert. Soils* **1989**, *8*, 98-101.

Buyer, J.S.; Kratzke, M.G.; Sikora, L.J. "A method for detection of pseudobactin, the siderophore produced by a plant-growth-promoting *Pseudomonas* strain in the barley rhizosphere." *Appl. Environ. Microbiol.* **1993**, *59*, 677-681.

Cooksey, D.A. "Reduction of infection by *Pseudomonas syringae* pv. *tomato* using a nonpathogenic, copper-resistant strain combined with a copper bactericide." *Phytopathology* **1988**, *78*, 601-603.

Cooksey, D.A.; Graham, J.H. "Genomic fingerprinting of two pathovars of phytopathogenic bacteria by rare-cutting restriction enzymes and field inversion gel electrophoresis." *Phytopathology* **1989**, *79*, 745-750.

Défago, G.; Haas, D. "Pseudomonads as antagonists of soilborne plant pathogens: Modes of Action and genetic analysis." In *Soil Biochemistry*; Bollag, J.-M.; Stotzky, G., Eds.; Marcel Dekker, Inc.: New York, NY, **1990**, Vol. 6; pp 249-291.

Denny, T.P.; Gilmour, M.N.; Selander, R.K. "Genetic diversity and relationships of two pathovars of *Pseudomonas syringae.*" *J. Gen. Microbiol.* **1988**, *134*, 1949-1960.

Fenton, A.M.; Stephens, P.M.; Crowley, J.; O'Callaghan; O'Gara, F. "Exploitation of gene(s) involved in 2,4-diacetylphloroglucinol biosynthesis to confer a new biocontrol capability to a *Pseudomonas* strain." *Appl. Environ. Microbiol.* **1992**, *58*, 3873-3878.

Fravel, D.R. "Role of antibiosis in the biocontrol of plant diseases." *Annu. Rev. Phytopathol.* **1988**, *26*, 75-91.

Gross, D.C.; Cody, Y.S.; Proebsting, E.L.; Radamaker, G.K.; Spotts, R.A. "Distribution, population dynamics, and characteristics of ice nucleation active bacteria in deciduous fruit tree orchards." *Appl. Environ. Microbiol.* **1983**, *46*, 1370-1379.

Gutterson, N. "Microbial fungicides: Recent approaches to elucidating mechanisms." *Crit. Rev. Biotechnol.* **1990**, *10*, 69-90.

Hagedorn, C.; Gould, W.D.; Bardinelli, T.R. "Field evaluations of bacterial inoculants to control seedling disease pathogens on cotton." *Plant Dis.* **1993**, *77*, 278-282.

Hamdan, H.; Weller, D.M.; Thomashow, L.S. "Relative importance of fluorescent siderophores and other factors in biological control of *Gaeumannomyces graminis* var. *tritici* by *Pseudomonas fluorescens* 2-79 and M4-80R." *Appl. Environ. Microbiol.* **1991**, *57*, 3270-3277.

Handelsman, J.; Parke, J.L. "Mechanisms in biocontrol of soilborne plant pathogens." In *Plant Microbe Interactions, Molecular and Genetic Perspectives*; Kosuge, T.; Nester, E., Eds.; McGraw Hill: New York, NY, **1989**, vol 3; pp 27-61.

Hirano, S.S.; Upper, C.D. "Ecology and epidemiology of foliar plant pathogens." *Annu. Rev. Phytopathol.* **1983**, *21*, 243-269.

Hirano, S.S.; Upper, C.D. "Temporal, spatial and genetic variability of leaf-associated bacterial populations." In *Microbiology of the Phyllosphere*; Fokkema, N.J.; Van Den Heuvel, J., Eds.; Cambridge University Press: London, **1986**, pp 235-251.

Homma, Y.; Suzui, T. "Role of antibiotic production in suppression of radish damping-off by seed bacterization with *Pseudomonas cepacia*." *Ann. Phytopath. Soc. Japan* **1989**, *55*, 643-652.

Howell, C.R.; Stipanovic, R.D. "Suppression of *Pythium ultimum*-induced damping-off of cotton seedlings by *Pseudomonas fluorescens* and its antibiotic, pyoluteorin. *Phytopathology* **1980**, *70*, 712-715.

Howie, W.J.; Suslow, T.V. "Role of antibiotic biosynthesis in the inhibition of *Pythium ultimum* in the cotton spermosphere and rhizosphere by *Pseudomonas fluorescens*." *Mol. Plant-Microbe Interact.* **1991**, *4*, 393-399.

Ishimaru, C. A.; Klos, E. J.; Brubaker, R. R. "Multiple antibiotic production by *Erwinia herbicola*." *Phytopathology* **1988**, *78*, 746-750.

Johnson, K. B.; Stockwell, V. O.; McLaughlin, R. J.; Sugar, D. Loper, J.E.; Roberts, R. G. "Effect of antagonistic bacteria on establishment of honey-bee dispersed *Erwinia amylovora* in pear blossoms and on fire blight control." *Phytopathology* **1993**, *83*, (in press).

Keel, C.; Défago, G. "The fluorescent siderophore of *Pseudomonas fluorescens* strain CHAO has no effect on the suppression of root diseases of wheat." In *Plant Growth-Promoting Rhizobacteria-Progress and Prospects*; Keel, C.; Koller, B.; Défago, G. Eds.; Union Internationale des Sciences Biologiques: Zurich, Switzerland, **1991**. pp. 136-142.

Keel, C.; Schnider, U.; Maurhofer, M.; Voisard, C.; Laville, J.; Burger, U.; Wirthner, P.; Haas, D.; Défago, G. "Suppression of root diseases by *Pseudomonas fluorescens* CHAO: importance of the bacterial secondary metabolite 2,4-diacetylphloroglucinol." *Mol. Plant-Microbe Interact.* **1992**, *5*, 4-13.

Kim, H.K.; Orser, C.; Lindow, S.E.; Sands, D.C. "*Xanthomonas campestris* pv. *translucens* strains active in ice nucleation." *Plant Disease* **1987**, *71*, 994-997.

Kinkel, L.L.; Lindow, S.E. "The role of competitive interactions in bacterial survival and establishment on the leaf surface." In *Recent Advances in Microbial Ecology*; Hattori, T.; Ishida, Y.; Maruyama, Y.; Morika, R.Y.; Uchida, A., Eds., Japan Sci. Soc. Press: Tokyo, **1989a**, pp 634-638.

Kinkel, L.L.; Lindow, S.E. "Dynamics of *Pseudomonas syringae* strains coexisting on leaf surfaces." *Phytopathology* **1989b**, *79*, 1162.

Kloepper, J. W.; Leong, J.; Teintze, M.; Schroth, M.N. "Enhanced plant growth by siderophores produced by plant growth-promoting rhizobacteria. *Nature (London)* **1980**, *286*, 885-886.

Kraus, J.; Loper, J.E. "Lack of evidence for a role of antifungal metabolite production by *Pseudomonas fluorescens* Pf-5 in biological control of Pythium damping-off of cucumber." *Phytopathology* **1992**, 264-271.

Lindemann, J.; Suslow, T.V. "Competition between ice nucleation active wild-type and ice nucleation deficient deletion mutant strains of *Pseudomonas syringae* and *P. fluorescens* biovar I and biological control of frost injury on strawberry blossoms." *Phytopathology* **1987**, *77*, 882-886.

Lindow, S. E. "Population dynamics of epiphytic ice nucleation active bacteria on frost sensitive plants and frost control by means of antagonistic bacteria." In *Plant Cold Hardiness*; Li, P.H.; Sakai, A., Eds.;

152

Academic Press: New York, NY, **1982**, pp 395-416.

Lindow, S.E. "The role of bacterial ice nucleation in frost injury to plants." *Annu. Rev. Phytopathol.* **1983a,** *21,* 363-384.

Lindow, S.E. "Methods of preventing frost injury caused by epiphytic ice nucleation active bacteria." *Plant Disease* **1983b,** *67,* 327-333.

Lindow, S.E. "Integrated control and role of antibiosis in biological control of fire blight and frost injury." In *Biological Control on the Phylloplane*; Windels, C.; Lindow, S.E., Eds.; American Phytopathological Society Press: St. Paul, MN, **1985a,** pp 83-115.

Lindow, S.E. "Strategies and practice of biological control of ice nucleation active bacteria on plants." In *Microbiology of the Phyllosphere*; Fokkema, N., Ed.; Cambridge University Press: London, **1985b,** pp 293-311.

Lindow, S.E. "Ecology of *Pseudomonas syringae* relevant to the field use of Ice⁻ deletion mutants constructed *in vitro* for plant frost control." In *Engineering Organisms in the Environment: Scientific issues*; Pramer, D. Ed.; Am. Soc. Microbiol.: Washington, D.C., **1985c,** pp 23-35.

Lindow, S.E. "Specificity of epiphytic interactions of *Pseudomonas syringae* strains on leaves." *Phytopathology* **1986a,** *76,* 1136.

Lindow, S.E. "Construction of isogenic Ice⁻ strains of *Pseudomonas syringae* for evaluation of specificity of competition on leaf surfaces." In *Microbial Ecology*; Megusar, F.; Gantar, M., Eds.; Slovene Society for Microbiology: Ljubljana, **1986b,** pp 509-515.

Lindow, S. E. "Competitive exclusion of epiphytic bacteria by Ice⁻ mutants of *Pseudomonas syringae*." *Appl. Environ. Microbiol.* **1987,** *53,* 2520-2527.

Lindow, S.E. "Lack of correlation of antibiosis in antagonism of ice nucleation active bacteria on leaf surfaces by non-ice nucleation active bacteria." *Phytopathology* **1988,** *78,* 444-450.

Lindow, S.E. "Tests of specificity of competition among *Pseudomonas syringae* strains on plants using recombinant Ice⁻ strains and use of ice nucleation genes as probes of *in situ* transcriptional activity." In *Advances in Molecular Genetics of Plant-Microbe Interactions*; Heinrich, M., Ed.; Kluwer

Academic Publishers: London, **1991**, Vol. 1; pp 457-465.

Lindow, S.E.; Panopoulos, N.J. "Field tests of recombinant Ice⁻ *Pseudomonas syringae* for biological frost control in potato." In *Proc. First International Conference on Release of Genetically Engineered Microorganisms*; Sussman, M.; Collins, C.H.; Skinner, F.A., Eds.; Academic Press: London, **1988**, pp 121-138.

Lindow, S.E.; Arny, D.C.; Upper, C.D. "Distribution of ice nucleation active bacteria on plants in nature." *Appl. Environ. Microbiol.* **1978a,** *36,* 831-838.

Lindow, S.E.; Arny, D.C.; Upper, C.D. *"Erwinia herbicola*: a bacterial ice nucleus active in increasing frost injury to corn." *Phytopathology* **1978b,** *68,* 523-527.

Lindow, S.E.; Arny, D.C.; Barchet, W.R.; Upper, C.D. "The role of bacterial ice nuclei in frost injury to sensitive plants." In *Plant Cold Hardiness and Freezing Stress*; Li, P., Ed.; Academic Press: New York, NY, **1978c,** pp 249-263.

Lindow, S.E.; Arny, D.C.; Upper, C.D. "Bacterial ice nucleation: a factor in frost injury to plants." *Plant Physiol.* **1982,** *70,* 1084-1089.

Lindow, S.E.; Arny, D.C.; Upper, C.D. "Biological control of frost injury I: An isolate of *Erwinia herbicola* antagonistic to ice nucleation-active bacteria." *Phytopathology* **1983a,** *73,* 1097-1102.

Lindow, S.E.; Arny, D.C.; Upper, C.D. "Biological control of frost injury II: Establishment and effects of an antagonistic *Erwinia herbicola* isolate on corn in the field." *Phytopathology* **1983b,** *73,* 1102-1106.

Lindow, S.E.; Willis, D.K.; Panopoulos, N.J. "Biological control of bacterial brown spot disease of bean with Tn*5*-induced avirulent mutants of the pathogen." *Phytopathology* **1987,** *77,* 1768.

Loper, J.E. "Role of fluorescent siderophore production in biological control of *Pythium ultimum* by a *Pseudomonas fluorescens* strain." *Phytopathology* **1988,** *78,* 166-172.

Loper, J.E.; Buyer, J.S. "Siderophores in microbial interactions on plant surfaces." *Molec. Plant-Microbe Interact.* **1991,** *4,* 5-13.

Loper, J.E.; Ishimaru, C.A. "Factors influencing siderophore-mediated biocontrol activity of rhizosphere *Pseudomonas* spp." In *The Rhizosphere and Plant Growth*; Keister, D.L.; Cregan, P.B., Eds. Kluwer Academic Publishers: Dordrecht, The Netherlands, **1991**; pp 253-261.

Loper, J.E.; Schroth, M.N. "Importance of siderophores in microbial interactions in the rhizosphere." In *Iron, Siderophores, and Plant Diseases*; Swinburne, T.R., Ed.; Plenum Press: New York, NY, **1986**; pp 85-98.

Loper, J.E.; Suslow, T.V.; Schroth, M.N. "Lognormal distribution of bacterial populations in the rhizosphere." *Phytopathology* **1984**, *74*, 1454-1460.

Meyer, J.-M.; Hallé, F.; Hohnadel, D.; Lemanceau, P.; Ratefiarivelo, H. "Siderophores of *Pseudomonas* - Biological Properties." In *Iron Transport in Microbes, Plants and Animals*; Winkelmann, G.; van der Helm, H.; Neilands, J.B., Eds.; VCH Publishers: Weinheim, FRG, **1987**; pp 189-205.

Misaghi, I.J.; Stowell, L.J.; Grogan, R.G.; Spearman, L.C. "Fungistatic activity of water-soluble fluorescent pigments of fluorescent pseudomonads." *Phytopathology* **1982**, *72*, 33-36.

Moore, L.W. "Use of *Agrobacterium radiobacter* in agricultural ecosystems." *Microbiol. Sci.* **1988**, *5*, 92-95.

Nelson, E.B.; Maloney, A.P. "Molecular approaches for understanding biological control mechanisms in bacteria: Studies of the interaction of *Enterobacter cloacae* with *Pythium ultimum. Can J. Plant Pathol.* **1992**, *14*, 106-114.

Orser, C.S.; Staskawicz, B.J.; Panopoulos, N.J.; Dahlbeck, D.; Lindow, S.E. "Cloning and expression of bacterial ice nucleation genes in *Escherichia coli." J. Bacteriol.* **1985**, *164*, 359-366.

O'Sullivan, D.J.; O'Gara, F. "Traits of fluorescent *Pseudomonas* spp. involved in suppression of plant root pathogens." *Microbiol. Rev.* **1992**, *56*, 662-676.

Paulitz, T.C. "Biochemical and ecological aspects of competition in biological control." In *New Directions in Biological Control*, UCLA Symposia on Molecular and Cellular Biology New Series, Vol. 12; Baker, R.R.; Dunn, P.E. Eds.; Alan R. Liss, Inc.: New York, NY, **1990**; pp 713-724.

Paulitz, T.C. "Effect of *Pseudomonas putida* on the stimulation of *Pythium ultimum* by seed volatiles of pea and soybean." *Phytopathology* **1991**, *81*, 1282-1287.

Paulitz, T.C.; Loper, J.E. "Lack of a role for fluorescent siderophore production in the biological control of Pythium damping-off of cucumber by a strain of *Pseudomonas putida." Phytopathology* **1991**, *81*, 930-935.

Pierson, L.S. III; Thomashow, L. S. "Cloning and heterologous expression of the phenazine biosynthesis locus from *Pseudomonas aureofaciens* 30-84." *Molec. Plant-Microbe Interact.* **1992**, *5*, 330-339.

Rouse, D.I.; Nordheim, E.V.; Hirano, S.S.; Upper, C.D. "A model relating the probability of foliar disease incidence to the population frequencies of bacterial plant pathogens." *Phytopathology* **1985**, *75*, 505-509.

Schippers, B.; Bakker, A.W.; Bakker, P.A.H.M. "Interactions of deleterious and beneficial rhizosphere microorganisms and the effect of cropping practices." *Annu. Rev. Phytopathol.* **1987**, *25*, 339-358.

Suslow, T.V. "Role of root colonizing bacteria in plant growth." In *Phytopathogenic Prokaryotes*, Mount, M.S.; Lacy, G.H., Eds; Academic Press: New York, NY, **1982**, Vol. 1; pp 187-223.

Tahvonen, R.; Avikainen, H. "The biological control of seed-borne *Alternaria brassicola* of cruciferous plants with a powdery preparation of *Streptomyces* sp." *J. Agric. Sci. Finland* **1987**, *59*, 199-208.

Thomashow, L.S.; Weller, D.M. "Role of a phenazine antibiotic from *Pseudomonas fluorescens* in biological control of *Gaeumannomyces graminis* var. *tritici." J. Bacteriol.* **1988**, *170*, 3499-3508.

Thomashow, L.S.; Weller, D.M.; Bonsall, R.F.; Pierson, L.S.III "Production of the antibiotic phenazine-1-carboxylic acid by fluorescent *Pseudomonas* species in the rhizosphere of wheat." *Appl. Environ. Microbiol.* **1990**, *56*, 908-912.

Thomson, S.V.; Schroth, M.N.; Moller, W.J.; Reil, W.O. "Efficacy of bactericides and saprophytic bacteria in reducing colonization and infection of pear flowers by *Erwinia amylovora." Phytopathology* **1976**, *66*, 1457-1459.

Vanneste, J.L.; Yu, J.; Beer, S.V. "Role of

antibiotic production by *Erwinia herbicola* Eh252 in biological control of *Erwinia amylovora.*" *J. Bacteriol.* **1992,** *174,* 2785-2796.

Vincent, M.N.; Harrison, L.A.; Brackin, J.M.; Kovacevich, P.A.; Mukerji, P.; Weller, D.M.; Pierson, E.A. "Genetic analysis of the antifungal activity of a soilborne *Pseudomonas aureofaciens* strain." *Appl. Environ. Microbiol.* **1991,** *57,* 2928-2934.

Voisard, C.; Keel, C.; Haas, D.; and Défago, G. "Cyanide production by *Pseudomonas fluorescens* helps suppress black root rot of tobacco under gnotobiotic conditions." *EMBO J.* **1989,** *8,* 351-358.

Weller, D.M. "Biological control of soilborne plant pathogens in the rhizosphere with bacteria." *Annu. Rev. Phytopathol.* **1988,** *26,* 379-407.

Wilson, M.; Lindow, S.E. "Resource partitioning among bacterial epiphytes in the phyllosphere." *Phytopathology* **1991,** *81,* 1170.

Wilson, M.; Lindow, S.E. "Interactions between the biological control agent *Pseudomonas fluorescens* A506 and *Erwinia amylovora* in pear blossoms." *Phytopathology* **1993,** *83,* 117-123.

Plant Growth-Promoting Rhizobacteria as Inducers of Systemic Disease Resistance

Joseph W. Kloepper, Sadik Tuzun, Li Liu, and **Gang Wei.** Department of Plant Pathology, Biological Control Institute, Auburn University, Auburn, AL 36849

Plant growth-promoting rhizobacteria (PGPR) may increase crop productivity by growth promotion or biological disease control. Mode of action studies with documented PGPR strains have revealed that biological control by PGPR involves production of bacterial metabolites which reduce pathogen numbers or activities. Effects of PGPR on host plant physiology indicate that select PGPR strains may enhance production of host defense-related compounds, raising the potential that some PGPR strains may be used as agents to induce disease resistance. In 1991, 3 laboratory groups published results that inoculation of plants by specific PGPR strains was associated with increased resistance to distally inoculated pathogens. Collectively, these results support the conclusion that some PGPR strains may induce systemic disease resistance. Implications of this conclusion for biological control research and development efforts are discussed.

Various groups of plant-associated microorganisms have been evaluated for potential use in enhancing crop production systems. Rhizobacteria are the subset of rhizosphere bacteria which exhibit root colonization (Schroth and Hancock 1982). Root colonization is an active process which may involves multiplication in or on roots in the presence of a competitive indigenous microflora (Kloepper and Beauchamp 1992). Rhizobacteria which exert a beneficial effect on crop development are termed "plant growth-promoting rhizobacteria (PGPR)" (Kloepper and Schroth 1978). PGPR encompass specific strains within a broad group of bacterial taxa. The beneficial effects of PGPR on crops may be evident through biological disease control or plant growth-promotion. Several review articles published in the 1990s discuss aspects of growth promotion (Jagnow *et al.* 1991, Kapulnik 1991, Kloepper *et al.* 1991, Lugtenberg *et al.* 1991, Schippers *et al.* 1991, Schroth and Becker 1990) and biological control (Bakker *et al.* 1991, Défago *et al.* 1990, Jagnow *et al.* 1991, Kloepper 1991, Kloepper 1993, Lugtenberg *et al.* 1991, Schippers *et al.* 1991, Schroth and Becker 1990, Thomashow and Weller 1990) by PGPR. As demonstrated in these reviews, PGPR-related research has progressed along one of three principal areas: screening and determination of efficacy for new strains, characterizing the microbial ecology of introduced PGPR, and mechanisms of action.

Collectively, the published information on mechanisms of action for PGPR, discussed in the reviews cited above, suggests the following broad conclusions. (1) Individual PGPR strains frequently demonstrate both growth promotion and biological control in experiments involving nonsterile soils. (2) Most PGPR strains do not have a single mechanisms which completely accounts for the observed beneficial effects on plants. (3) Most underlying mechanisms for biological control by PGPR involve production of bacterial metabolites which reduce populations or activities of pathogens or deleterious rhizosphere microflora; these metabolites include siderophores, HCN, antibiotics, lytic enzymes, and uncharacterized antifungal factors.

Over the past 5 years, evidence has been mounting that some PGPR strains may affect biological control through stimulation of the host plant's natural defenses, i.e. via induced disease resistance. This manuscript summarizes evidence from several laboratory groups that induced resistance may be achieved through applications of PGPR.

Terminology Related to Induced Resistance

Before discussing experimental evidence related to induced resistance activity of PGPR, it is useful to review the terminology and concepts related to induced resistance. Several different terms have been used basically as synonyms to describe resistance triggered by infection of a necrosis-causing pathogen. Chester (1933) referred to "acquired physiological immunity" and "acquired immunity" when describing this resistance. Ross used the terms "localized acquired resistance" (Ross 1961a) and "systemic acquired resistance" (Ross 1961b) to describe plant resistance which resulted from prior inoculation of *Nicotiana tabacum* with tobacco mosaic virus. "Systemic acquired resistance (SAR)" continues to be used by many researchers as the general term for such resistance (Malamy *et al.* 1990, Métraux *et al.* 1990, Uknes *et al.* 1992). "Induced resistance" and "induced systemic resistance" are also widely used to describe pathogen-induced resistance (Falkhof *et al.* 1988, Heller and Gessler 1986, Métraux *et al.* 1991, Smith and Hammerschmidt 1988, Steiner *et al.* 1988, Tuzun *et al.* 1988, Wieringa-Brants and Dekker 1987, Ye *et al.* 1992). "Immunization" is another synonym which has been used to describe induced resistance (Kuć 1987, Tuzun and Kuć 1991).

Participants at a discussion session in a NATO Advanced Research Workshop in 1991 suggested definitions for several terms related to induced resistance. The purpose of the definitions is to facilitate scientific communication among physiological plant pathologists, who have worked for many years on induced resistance, and biological control researchers, who are beginning induced resistance investigations. For this review, the term *induced disease resistance* will be defined according to the proposed definition as "the process of active resistance dependent on the host plant's physical or chemical barriers, activated by biotic or abiotic agents (inducing agents)" (Kloepper *et al.* 1992).

Suggestions of Bacterial-Mediated Induced Systemic Resistance

Suggestions that some beneficial bacteria may induce disease resistance came from results of several different research groups. Scheffer (1983) found elm trees inoculated with four strains of fluorescent pseudomonads prior to inoculation with the Dutch elm pathogen, *Ophiostoma ulmi*, had significantly less symptoms than trees not treated with bacteria. Although all 4 bacterial strains exhibited some antibiosis *in vitro* to the pathogen, the levels were weak with some strains, and Scheffer suggested that induced or enhanced host resistance may be operable. Voisard *et al.* (1989) reported that cyanide-producing wild-type *Pseudomonas fluorescens* strain CHA0, but not the Hcn⁻ mutant, induced physiological changes in roots in gnotobiotic assays. These changes were evidenced by increased root hair formation, and the authors suggested that HCN might also induce plant defense mechanisms. Altered plant physiology evidenced by various root growth parameters has also been reported under gnotobiotic conditions with other PGPR strains of *Pseudomonas* and *Azospirillum* (Bashan and Levanony 1989, de Freitas and Germida 1990, Frommel *et al.* 1991, and Lifshitz *et al.* 1987).

Biochemical Evidence for Induced Resistance

Complementary evidence that specific bacteria may induce physiological changes in plants is provided from studies with specific biochemical compounds associated with host defense. Hynes and Lazarovits (1989) found increased foliar accumulation of a pathogenesis-related protein in bean following seed treatment with PGPR. Increased lignification of tissues distal from bacterial treatments was reported in bean (Anderson and Guerra 1985) and wheat (Frommel *et al.* 1991). Zdor and Anderson (1992) used Northern hybridization to reveal enhancement of mRNA encoding PR1a protein in bean leaves following root inoculation with a *P. putida* strain.

Pathological Evidence for Induced Resistance

While these studies clearly demonstrate that specific bacteria may elicit systemic physiological changes in plants, they do not provide direct evidence for systemic disease protection. Three laboratory groups published such direct evidence in 1991.

van Peer *et al.* (1991) reported that root colonization of carnation by *Pseudomonas* sp. strain WCS417 resulted in a significant reduction in Fusarium wilt incidence compared to nonbacterized controls. Spatial separation of strain WCS417 and the pathogen was provided by applying WCS417 to rockwool cubes one week prior to stem injection of the pathogen, and separation was confirmed by failure to isolate WCS417 from stem tissues. Phytoalexins accumulated faster in stems of plants treated with WCS417 and inoculated with the pathogen, compared to stems of nonbacterized, pathogen-inoculated plants.

Alström (1991) reported that numbers of foliar lesions caused by *Pseudomonas syringae* pv. *phaseolicola* were significantly reduced on plants grown from seed treated with a PGPR strain of *P. fluorescens* compared to nonbacterized, pathogen-inoculated controls. The conclusion was reached in the reports of van Peer *et al.* and Alström that the observed disease protection resulted from induction of systemic resistance by bacterial treatments.

Another group has been investigating known PGPR strains as inducing agents against *Colletotrichum orbiculare* on cucumber. In the first report from this group (Wei *et al.* 1991), 6 of 94 PGPR strains applied as seed treatments significantly reduced foliar symptoms compared to nonbacterized controls. PGPR treated plants exhibited reductions in lesion number and total lesion diameter (Fig. 1.) on the second leaf which was challenge inoculated with *C. orbiculare*. In the cucumber system, the effect of inducing PGPR strains on lesion development was compared to that from "classical induced resistance" which was achieved by prior inoculation of the first leaf with the pathogen. No root necrosis was observed on plants induced by PGPR. In a subsequent study with the same PGPR strains (Kloepper *et al.* 1992), none of the strains was recovered from petioles of leaves which expressed reduced disease severity, suggesting that protection resulted from induced systemic resistance and not from competition or antibiotic production. Some PGPR strains which induced resistance were recovered from inside surface-disinfested roots, demonstrating that they have limited internal root colonization. Such internal colonists may have improved inducing activity resulting from the close association with the host plant. Plants treated with the inducing PGPR strains appear to have enhanced activity of defense-related compounds. Increased peroxidase activity in leaves of PGPR-treated plants was associated with induced resistance activity of some strains (Wei *et al.* 1992).

Because it has been used for many years as one of the major model systems for classical induced resistance research (Dean and Kuć 1987, Hammerschmidt *et al.* 1982, Kuć and Richmond 1977, Kuć *et al.* 1975, Métraux and Boller 1986), cucumber can be used for comparisons between PGPR-mediated and "classical" induced resistance. Work at Auburn University is currently involved with such comparisons in an effort to determine if multiple underlying mechanisms for induced resistance may be operating.

Preliminary evidence with two PGPR strains which induce systemic resistance against *C. orbiculare* indicates that protection is also achieved in cucumber against cucumber mosaic virus (CMV) (Liu *et al.* 1992), *Fusarium oxysporum* f. sp. *cucumerinum*, *Erwinia tracheiphila*, and *Pseudomonas syringae* pv. *lachrymans* (Liu, unpublished). With angular leaf spot, caused by *P. s.* pv. *lachrymans*, seed treatment of cucumber with both PGPR strains led to significant reductions in lesion number and lesion size (Fig. 2.) compared to noninduced, pathogen-inoculated controls. The level of protection was equivalent to that obtained by the classical induced control (induced by a prior inoculation with *P. s.* pv. *lachrymans)*. With CMV, both PGPR strains protected plants when the virus was inoculated onto cotyledons, but only one strain provided significant protection when viral inoculation was conducted on the first, second, or third leaves. One of the PGPR strains provided significant reductions in the number of plants killed by Fusarium wilt in a series of repeated experiments using split root systems to separate PGPR from the pathogen. In composite, these results suggest similarities and differences with classical induced resistance. Both classical and PGPR-mediated

induced systemic resistance may lead to protection against a wide spectrum of pathogens. Based on preliminary evidence, the plant response to PGPR-mediated induced resistance may be strain-specific, unlike with classical resistance where a common plant response appears to occur.

Another area where comparisons of classical and PGPR-mediated induced systemic resistance would be useful is determining the biochemical response of plants. With classical induction, cucurbits exhibit increased activity of several defense-related compounds, including peroxidases (Hammerschmidt 1982), chitinases (Métraux and Boller 1986), and beta-1,3-glucanase (Rabenantoandro *et al.* 1976). In an initial investigation of plant biochemical changes associated with inducing PGPR, Wei *et al.* (1992) reported that induced plants had elevated peroxidase activity compared to noninduced control plants. A full comparison of the biochemical responses of plants to classical and PGPR-mediated induced systemic resistance will require substantially more information.

Field Trials

Unlike organismal biological control, where individual agents are frequently tested in experimental field trials early in the investigatory process, classical induced systemic resistance has predominately been examined as a laboratory and greenhouse phenomenon (Tuzun and Kuć 1991). The most comprehensive field work with classical induced resistance has been with tobacco induced by stem-injection of *Peronospora tabacina* for protection against blue mold, caused by *P. tabacina* (Tuzun and Kuć 1987, Tuzun *et al.* 1986, Tuzun *et al.* 1990).

The history of PGPR research contrasts to that of classical induced resistance by the emphasis on field evaluations. It is predictable, therefore, that as more PGPR researchers investigate induced resistance, more field reports will be published. Three PGPR strains, shown to induce systemic resistance in the greenhouse with the cucumber/anthracnose systems, were evaluated in 1992 field trials (Table 1. and 2.). For both trials, PGPR were applied as seed treatments, and induced resistance controls consisted of plants induced by inoculating the first leaf with *C. orbiculare*. In the first trial (Table 1.), plants were transplanted from the greenhouse to the field at the second-leaf stage, and challenge inoculation was done on fully expanded new leaves with *P. s.* pv. *lachrymans* two weeks after transplanting. In the second trial (Table 2.), seeds were directly planted in the field. Symptoms of bacterial wilt were evident at the planned time for challenge inoculation, and

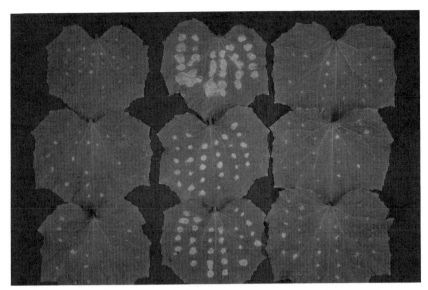

Fig. 1. PGPR-mediated induced systemic resistance in cucumber against anthracnose. Center leaves were nonbacterized, noninduced controls; leaves on the left were nonbacterized and induced by inoculation of the first leaf with *Colletotrichum orbiculare*; leaves on the right were induced by seed treatment with one PGPR strain. All leaves shown were challenge inoculated with *C. orbiculare* (Wei et al., 1991).

Fig. 2. PGPR-mediated induced systemic resistance in cucumber against angular leaf spot. Center leaves were nonbacterized, noninduced controls; leaves on the left were nonbacterized and induced by inoculation of the first leaf with *Pseudomonas syringae* pv. *lachrymans*; leaves on the right were induced by seed treatment with PGPR strains. The second leaf of all plants (shown) was challenge inoculated with *P. s.* pv. *lachrymans*.

Table 1. Field trial of induced systemic resistance mediated by PGPR (plant growth-promoting rhizobacteria): growth promotion and control of angular leafspot[a]

Treatments	Length (cm) of Main Runner/Plant	Leaf Number/ Plant	Total Lesion Diameter (mm)[b]	Yield (kg fruit)
89B-61	96.0	64.9*	110.3*	20.8*
90-166	101.5*	63.6*	107.2*	20.1*
INR-5	102.1*	59.8*	100.8*	18.4
IR control[c]	94.8	51.0	235.8	14.9
Disease control[d]	91.1	49.3	150.3	16.8
LSD$_{0.05}$	6.9	7.2	13.1	2.1

[a]Mean of six replications per treatment; each replication had 6 plants.

[b]On challenge-inoculated leaf.

[c]Induced resistance control where induction was by inoculation of cotyledons with *C. orbiculare*.

[d]Nonbacterized, noninduced control challenged with *Pseudomonas syringae* pv. *lachrymans*.

*Significantly different from disease control (P=0.05).

160

Table 2. Field trial of induced systemic resistance mediated by PGPR (plant growth-promoting rhizobacteria): growth promotion and control of bacterial wilt[a]

Treatments	Length (cm) of Main Runner/Plant	Leaf Number/Plant[a]	Bacterial Wilt Incidence		Yield (kg fruit)
			Percent Wilt	Arcsin √% Transformed Mean	
89B-61	96.1*	73.3	9.2	0.275*	13.2*
90-166	107.6*	77.4*	10.1	0.317*	15.4*
INR-5	103.4*	74.7*	14.9	0.350*	12.1
IR control[c]	100.4*	66.3	25.0	0.513	13.3
Disease control[d]	90.8	66.1	31.3	0.595	10.0
LSD$_{0.05}$	3.9	7.7		0.165	2.8

[a]Mean of six replications per treatment; each replication had 12 plants.

[b]Plants were naturally infested and not challenge inoculated. Statistical analysis was conducted using arcsin √% transformation of each replication.

[c]Induced resistance control where induction was by inoculation of cotyledons with *C. orbiculare*.

[d]Nonbacterized, noninduced control.

*Significantly different from disease control (*P*=0.05).

Erwinia tracheiphila was isolated from symptomatic tissues. Therefore, the naturally occurring disease was allowed to progress, and no challenge inoculation was conducted. All three PGPR strains resulted in growth promotion as evidenced by increased runner length (Table 2.) or leaf number per plant (Table 1.). Significant protection against angular leaf spot, measured by total lesion diameter, occurred with all PGPR treatments (Table 1., Fig. 3.). Similarly, all PGPR strains significantly reduced the incidence of bacterial wilt (Table 2.). In these trials, classical induced resistance did not provide significant disease protection, and in the case of angular leaf spot, it resulted in significant disease enhancement. Two of the PGPR strains resulted in significant yield increases in both trials. These results indicate that PGPR-mediated induced resistance may occur under field conditions, and that plant growth promotion may occur in addition to systemic disease control.

Future Research and Development Priorities

A limitation to practical implementation of classical induced systemic resistance (induction through prior inoculation of pathogens) is obvious: wide-scale inoculation of crops with a pathogen is contrary to basic principals of crop protection. Research is currently underway in several corporate R & D groups to discover chemical inducers which could be sprayed on crops. This approach will likely lead to successful wide-scale induced systemic resistance; however, the requirement for necrosis to initiate induced resistance may restrict grower acceptability of such chemicals. PGPR-mediated induced resistance should be investigated as a strategic alternative or additional method for achieving the goal of agricultural applications. Seed treatments of inducing PGPR may lead to induced systemic resistance with the lowest input costs for labor and energy.

Research on bacterial-mediated induced systemic resistance discussed here is clearly only in the preliminary stages. Many questions remain to be answered by future studies, including: What is the length of protection afforded by PGPR? What is the nature of the bacterial trigger for induced resistance? Is the translocatable signal with PGPR-mediated induced resistance the same as classical induced resistance? Does internal colonization of plant tissues increase the level of protection? To assess fully the potential value of PGPR as inducing agents in agriculture, it will be necessary to answer these questions.

Fig. 3. Induced systemic resistance against angular leaf spot in a field trial. Fig. A shows typical disease symptoms on nonbacterized, noninduced controls. Fig. B shows a plant from PGPR-treated seed. Both plants were challenge inoculated with *P. s.* pv. *lachrymans.*

162

B

Figure 3. *Continued.*

References

Alström, S. "Induction of disease resistance in common bean susceptible to halo blight bacterial pathogen after seed bacterization with rhizosphere pseudomonads." J. Gen. Appl. Microbiol. 1991, *37*, 495-501.

Anderson, A.J.; Guerra, D. "Responses of bean to root colonization with *Pseudomonas putida* in a hydroponic system." Phytopath. 1985, *75*, 992-995.

Bakker, P.A.H.M.; van Peer, R.; Schippers, B. In *Biotic Interactions and Soil-Borne Diseases*; Beemster, A.B.R.; Bollen, G.J.; Gerlagh, M.; Ruissen, M.A.; Schippers, B.; Tempel, A., Eds.; Elsevier: Amsterdam, The Netherlands, 1991, pp. 217-223.

Bashan, Y.; Levanony, H. "Factors affecting adsorption of *Azospirillum brasilense* Cd to root hairs as compared with root surface of wheat." Can. J. Microbiol. 1989, *35*, 936-944.

Chester, K.S. "The problem of acquired physiological immunity in plants." Quar. Rev. Biol. 1933, *VIII*, 129-154.

de Freitas, J.R.; Germida, J.J. "A root tissue culture system to study winter wheat-rhizobacteria interactions." Appl. Microbiol. Biotechnol. 1990, *33*, 589-595.

Dean, R.A.; Kuć, J.A. "Immunization against disease: the plant fights back." In *Fungal Infection of Plants;* Pegg, G.F.; Ayres, P.G., Eds.; Cambridge University Press: Cambridge, U.K., 1987; pp. 383-410.

Défago, G.; Berling, C.H.; Burger, U.; Haas, D.; Kahr, G.; Keel, C.; Voisard, C.; Wirthner, P.; and Wüthrich, B. In *Biological Control of Soil-Borne Plant Pathogens*; Hornby, D., Eds.; CAB International: Wallingford, Oxon, U.K., 1990; pp. 93-108.

Falkhof, A.G.; Dehne, H.W.; Schonbeck, F. "Dependence of the effectiveness of induced resistance on environmental factors." J. Phytopathol. 1988, *123*, 311-321.

Frommel, M.I.; Nowak, J.; Lazarovits, G. "Growth enhancement and developmental modifications of *in vitro* grown potato (*Solanum tuberosum* ssp. *tuberosum*) as affected by a nonfluorescent *Pseudomonas* sp." Plant Physiol. 1991, *96*, 928-936.

Hammerschmidt, R.; Nuckles, E.; Kuć, J. "Association of enhanced peroxidase activity with induced systemic resistance of cucumber to *Colletotrichum lagenarium*." Physiol. Plant Pathol. 1982, *20*, 73-82.

Heller, W.E.; Gessler, C. "Induced systemic

resistance in tomato plants against *Phytophthora infestans*." J. Phytopathol. 1986, *116*, 323-328.

Hynes, R.K.; Lazarovits, G. "Effect of seed treatment with plant growth promoting rhizobacteria on the protein profiles of intercellular fluids from bean and tomato leaves. Can. J. Plant Pathol. 1989, *11*, 191.

Jagnow, G.; Höflich, G.; Hoffman, K.-H. "Inoculation of non-symbiotic rhizosphere bacteria: possibilities of increasing and stabilizing yields." Angew. Botanik 1991, *65*, 97-126.

Kapulnik, Y. In *Plant Roots: The Hidden Half*; Waisel, Y.; Eshel, A.; Kafkafi, U., Eds.; Marcel Dekker, Inc.: New York, NY, 1991; pp. 717-729.

Kloepper, J.W.; Beauchamp, C.J. "A review of issues related to measuring colonization of palnt roots by bacteria." Can. J. Microbiol. 1993, *38*, 1219-1232.

Kloepper, J.W. In *Soil Microbial Ecology*; Metting, F.B., Jr., Ed.; Marcel-Dekker, Inc.: New York, NY, 1993; pp. 255-274.

Kloepper, J.W. In *The Biological Control of Plant Diseases*; Bay-Petersen, J., Ed.; FFTC Book Series No. 42; Food and Fertilizer Technology Center: Taipai, Taiwan, 1991, pp. 142-152.

Kloepper, J.W.; Schroth, M.N. "In promoting rhizobacteria on radishes". Proc. 4th Int. Conf. Plant Path. Bact.: Angers, 1978, pp. 879-882.

Kloepper, J.W.; Tuzun, S.; Kuć, J.A. "Proposed definitions related to induced disease resistance." Biocon. Sci. Technol. 1992, *2*, 347-349.

Kloepper, J.W.; Wei, G.; Tuzun, S. In *Biological Control of Plant Diseases*; Plenum Press: New York, NY, 1992, pp. 185-191.

Kloepper, J.W.; Zablotowicz, R.M.; Tipping, E.M.; Lifshitz, R. In *The Rhizosphere and Plant Growth*; Keister, D.L.; Cregan, P.B., Eds.; Kluwer Academic Publishers: The Netherlands, 1991; pp. 315-326.

Kuć, J. In *Innovative Approaches to Plant Disease Control*; Chet, I., Ed.; John Wiley: New York, NY, 1987, pp. 255-274.

Kuć, J.; Richmond, S. "Aspects of the protection of cucumber against *Colletotrichum lagenarium* by Colletotrichum lagenarium. Phytopathol. 1977, *67*, 533-536.

Kuć, J.; Shockley, G.; Kearney, K. "Protection of cucumber against *Colletotrichum lagenarium* by *Colletotrichum lagenarium*. Physiol. Plant Pathol. 1975, *7*, 195-199.

Lifshitz, R.; Kloepper, J.W.; Kozlowski, M.; Simonson, C.; Carlson, J.; Tipping, E.M.; Zaleska, I. "Growth promotion of canola (rapeseed) seedlings by a strain of *Pseudomonas putida* under gnotobiotic conditions." Can. J. Microbiol. 1987, *33*, 390-395.

Liu, L.; Kloepper, J.W.; Tuzun, S. "Induction of systemic resistance against cucumber mosaic virus by seed inoculation with select rhizobacteria strains." Phytopath. 1992, *82*, 1108.

Lugtenberg, B.J.J.; deweger, L.A.; Bennett, J.W. "Microbial stimulation of plant growth and protection from disease." Curr. Opin. Biotech. 1991, *2*, 457-464.

Malamy, J.; Carr, J.P.; Klessig, D.F.; Raskin, I. "Salicylic acid: a likely endogenous signal in the resistance response of tobacco to viral infection." Science 1990, *250*, 1002-1004.

Métraux, J.P.; Boller, T.; "Local and systemic induction of chitinase in cucumber plants in response to viral, bacterial and fungal infections." Physiol. Mol. Plant Pathol. 1986, *28*, 161-169.

Métraux, J.P.; Ahl Goy, P.; Staub, Th.; Speich, J.; Steinemann, A.; Ryals, J.; Ward, E. In *Advances in Molecular Genetics of Plant-Microbe Interactions*; Hennecke, H.; Verma, D.P.S., Eds.; Kluwer Academic Publishers: The Netherlands, 1991, Vol. 1; pp. 432-439.

Métraux, J.P.; Signer, H.; Ryals, J.; Ward, E.; Wyss-Benz, M.; Gaudin, J.; Raschdorf, K.; Schmid, E.; Blum, W.; Inverardi, B. "Increase in salicylic acid at the onset of systemic acquired resistance in cucumber." Science 1990, *250*, 1004-1006.

Rabenantoandro, Y.; Auriol, P.; Tozue, A. "Implication of β-1,3-glucanase in melon anthracnose." Physiol. Plant Pathol. 1976, *8*, 313-324.

Ross, A.F. "Localized acquired resistance to plant virus infection in hypersensitive hosts." Virology 1961, *14*, 329-339.

Ross, A.F. "Systemic acquired resistance induced by localized virus infections in plants." Virology 1961, *14*, 340-358.

Scheffer, R.J. "Biological control of Dutch elm disease by *Pseudomonas* species." Ann. Appl. Biol. 1983, *103*, 21-30.

Schippers, B.; Bakker, A.W.; Bakker, P.A.H.M.; van Peer, R. In *The Rhizosphere and Plant Growth*; Keister, D.L.; Cregan, P.B., Eds.; Kluwer Academic Publishers: The Netherlands, 1991; pp. 211-219.

Schroth, M.N.; Becker, J.O. In *Biological Control of*

Soil-Borne Plant Pathogens; Hornby, D., Ed.; CAB International: Wallingford, Oxon, U.K., 1990; pp. 389-414.

Schroth, M.N.; Hancock, J.G. "Disease-suppressive soil and root-colonizing bacteria." Science 1982, *216*, 1376-1381.

Smith, J.; Hammerschmidt, R. "Comparative study of acidic peroxidases associated with induced resistance in cucumber, muskmelon and watermelon." Physiol. Mol. Plant Pathol. 1988, *33*, 255-261.

Steiner, J.; Oerke, E.C.; Schonbeck, F. "The efficiency of induced resistance under practical culture conditions. IV. Powdery mildew grain yield of winter barley cultivars with induced resistance and fungicide treatment." J. Plant Dis. Protec. 1988, *95*, 506-517.

Thomashow, L.S.; Weller, D.M. In *Biological Control of Soil-Borne Plant Pathogens*; Hornby, D., Eds.; CAB International, Wallingford, Oxon, U.K., 1990; pp. 109-124.

Tuzun, S.; Juarez, J.; Nesmith, W.C.; Kuć, J. In *Blue Mold Disease of Tobacco*; Main, C.E.; Spurr, H.W., Jr., Eds.; Tobacco Literature Service, N.C. State Univ.: Raleigh, NC, 1990; pp. 157-159.

Tuzun, S.; Kuć, J. In *The Biological Control of Plant Diseases*; Bay-Petersen, J., Ed.; FFTC Book Series No. 42; Food and Fertilizer Technology Center: Taipai, Taiwan, 1991, pp. 1-11.

Tuzun, S.; Kuć, J. "Persistence of induced systemic resistance to blue mold in tobacco plants derived via tissue culture." Phytopathol. 1987, *77*, 1032-1035.

Tuzun, S.; Nesmith, W.; Ferriss, R.S.; Kuć, J. "Effects of stem injections with *Peronospora tabacina* on growth of tobacco and protection against blue mold in the field." Phytopathol. 1986, *76*, 938-941.

Tuzun, S.; Rao, M.N.; Vogeli, U.; Schardl, C.L.; Kuć, J. "Induced systemic resistance to blue mold: early induction and accumulation of β-

1,3-glucanases, chitinases, and other pathogenesis related proteins (b-proteins) in immunized tobacco." Phytopathol. 1988, *79*, 979-983.

Uknes, S.; Mauch-Mani, B.; Moyer, M.; Potter, S.; Williams, S.; Dincher, S.; Chandler, D.; Slusarenko, A.; Ward, E.; Ryals, J. "Acquired resistance in Arabidopsis." The Plant Cell 1992, *4*, 645-656.

van Peer, R.; Niemann, G.J.; Schippers, B. "Induced resistance and phytoalexin accumulation in biological control of ´Fusarium wilt of carnation by *Pseudomonas* sp. strain WCS417r." Phytopathol. 1991, *81*, 728-734.

Voisard, C.; Keel, C.; Haas, D.; Défago, G. "Cyanide production by *Pseudomonas fluorescens* helps suppress black root rot of tobacco under gnotobiotic conditions." The EMBO J. 1989, *8*, 351-358.

Wei, G.; Tuzun, S.; Kloepper, J.W. "Comparison of biochemical responses in cucumber systemically protected against *Colletotrichum orbiculare* by prior leaf inoculation with the pathogen or seed treatment with rhizobacteria." Phytopathol. 1992, *82*, 1109.

Wei, G.; Kloepper, J.W.; Tuzun, S. "Induction of systemic resistance of cucumber to *Colletotrichum orbiculare* by select strains of plant growth-promoting rhizobacteria." Phytopathol. 1991, *81*, 1508-1512.

Wieringa-Brants, D.H.; Dekker, W.C. "Induced resistance in hypersensitive tobacco against tobacco mosaic virus by injection of intercellular fluid from tobacco plants with systemic acquired resistance." J. Phytopathol. 1987, *118*, 165-170.

Ye, X.S.; Pan, S.Q.; Kuć, J. "Specificity of induced systemic resistance as elicited by ethephon and tobacco mosaic virus in tobacco." Plant Sci. 1992, *84*, 1-9.

Zdor, R.E.; Anderson, A.J. "Influence of root colonizing bacteria on the defense responses of bean." Plant and Soil 1992, *140*, 99-107.

Biological Control of Plant Parasitic Nematodes with Plant-Health-Promoting Rhizobacteria

Richard A. Sikora and **Sabine Hoffmann-Hergarten**, Institut für Pflanzenkrankheiten, Universität Bonn, Nussallee 9, 5300 Bonn 1, F.R.Germany.

One strategy for the biological control of nematodes is based on the introduction of bacteria colonizing the rhizosphere of the host plant. These plant-health-promoting rhizobacteria (PHPR) adversely influence the intimate relationship between the plant-parasitic nematode and its host by regulation of nematode behavior during the early root penetration phase of parasitism. Significant levels of control have been obtained with the sugar beet and potato cyst nematodes. Two mechanisms of action are thought to be responsible for reduction in nematode infection 1) production of metabolites which reduce hatch and attraction and/or 2) degradation of specific root exudates which control nematode behavior.

Attempts to control plant parasitic nematodes biologically have been based mainly on the inundative approach (Kerry, 1990). The application of bacterial or fungal antagonists by these traditional delivery systems has proven inadequate. Factors limiting practical development include: 1) high amounts of inoculum required for adequate control 2) low levels of spread in the competitive soil ecosystem and 3) adverse influence of abiotic factors on efficacy. Therefore, although significant research has been conducted on fungal and bacterial antagonists of plant parasitic nematodes, effective systems are not yet available to the grower.

Alternatively, it has been suggested that crop production systems should be designed to manage endemic populations of fungal or bacterial antagonists for nematode control purposes (Sikora, 1992). Although this approach is still in its infancy, progress has been made in management of the obligate bacterial parasite *Pasteuria* spp. for control of root-knot nematodes (Oostendorp et al., 1991).

The use of rhizosphere bacteria for the biological control of plant parasitic nematodes is a dramatically different control approach. Plant parasitic nematodes are obligate parasites dependent on a host plant for development and reproduction. This obligate parasitism is based on an intricate host parasite interrelationship controlled in many cases by specific chemical components of the root exudates or root tissue.

Compounds emitted from the root, acting singly or in combination, may influence hatch, emergence, attraction, recognition, penetration, development and sexual differentiation (Table 1). The rhizobacteria based system utilizes microorganisms that directly or indirectly regulate or alter activity of root exudate, thereby influencing specific behavioral aspects of the parasite.

Rhizosphere bacteria have been studied for their plant growth and health promoting effects and successful control of several fungal and bacterial plant pathogens has been attained. The mechanisms of action responsible for control include: siderophore activity, suppression of deleterious rhizobacteria, antibiotic production, plant hormone production, toxin development and induced resistance (Weller, 1988). Attempts to use rhizobacteria for nematode control have been scarce (Table 2). Investigations have been conducted on direct activity through toxin production and indirect activity based on disruption of certain phases of the host parasite interrelationship.

Development of effective nematode control with rhizobacteria should be based on 1) development of efficient and sensitive isolation techniques favoring rhizosphere competence, 2) formulation of screening techniques based on mechanisms of action that lead to detection of isolates with biological activity toward specific nematode behavior patterns, 3) regulation of

Table 1: Typical life-cycle of a cyst nematode

Developmental stage	Behaviour	Host stimulus	Location of interrelationship
J_2^* in egg	Hatch	Root exudate	Soil
Free-living J_2	Attraction to root	Root exudate	Soil
	Host recognition/ Penetration	Root exudate + Lectin-carbohydrate-recognition	Rhizosphere
Sedentary $J_2/J_3/J_4$	Giant cell formation/ Juvenile development	Host suitability	Root tissue
Adult female	Pheromone release/ Embryogenesis	Giant cell formation	Root tissue
Adult male	Copulation		Soil/ Rhizosphere

* J_2, J_3, J_4 = second, third, fourth stage juvenile

biotic and abiotic factors in favor of the control system, and 4) development of acceptable formulations for seed and planting material. Recent commercialization of rhizobacteria has raised prospects of successful control of diseases and nematodes with rhizobacteria (McIntyre & Press, 1991).

Mechanism of action

Besides adding to our general knowledge of the interrelationship, determination of the mechanism of action increases chances of isolating effective strains and improves screening procedures thereby enhancing attempts to develop control systems. Progress has been made in the elucidation of the mechanisms involved in biological control of cyst nematodes and to a lesser extent root-knot nematodes with plant health promoting rhizobacteria (PHPR) (Table 3).

Metabolite production: Some bacteria are known to produce metabolites that have nematicidal activity. Avermectin from *Streptomyces avermitilis* is a highly effective anthelminthic (Campbell et al., 1983). The *Bacillus thuringiensis* toxin has also been shown to be somewhat active toward species of *Meloidogyne* (Devidas & Rehberger, 1992).

In vitro bioassays have been used to detect nematistatic or nematicidal metabolites produced by microorganisms. However, metabolite production in nutrient media is different in both quality and quantity than that occurring in the rhizosphere. Testing, therefore, can lead to results which are not reproducible in field trials. The production of nematicidal compounds by an isolate of *B. subtilis* toward *Meloidogyne incognita*, for example, only occurred on one specific medium (Keuken & Sikora, 1993 unpublished). The pH of the medium also has been shown to affect nematicidal activity (Spiegel et al., 1991).

Mechanisms of action can be determined using bioassays which simulate more closely natural conditions existing around the host parasite interaction. For example a number of bioassays have been used for this purpose: 1)

Table 2: Studies concerned with the mode-of-action of rhizobacteria exhibiting biological control activity toward plant parasitic nematodes

Bacteria	Nematode[*]	Host plant	Mode-of-action	Reference
Different species	*M. incognita*	Tomato Cucumber		Zavaleta-Meija & Van Gundy 1982
B. subtilis	*M. incognita* *M. arenaria* *R. reniformis*	Cotton		Sikora 1988
Different species	*M. incognita*	Tomato Cucumber White clover	Nematicidal components	Becker et al. 1988
P. fluorescens	*H. schachtii*	Sugar beet	Alteration of root exudates	Oostendorp & Sikora 1990
P. chitinolytica	*M. javanica*	Tomato	Toxic metabolites	Spiegel et al. 1991
A. radiobacter *B. sphaericus*	*G. pallida*	Potato	Alteration of root exudates	Racke & Sikora 1992
Different species	*H. glycines* *M. incognita*	Soybean		Kloepper et al. 1992

[*] *M. = Meloidogyne; R. = Rotylenchulus; H. = Heterodera; G. = Globodera*

hatching tests using root exudates from greenhouse cultures to detect stimulatory or inhibitory substances 2) sand-block, agar or sephadex substrate chambers with host plants to examine factors influencing nematode attraction (Oostendorp & Sikora, 1990). There is a need for improvement of bioassay testing techniques.

Reduction of hatching factor activity: Nematode behavior is greatly dependent on specific components in the root exudates of the host. Bacterial metabolism of these components can breakdown the chain of command coding for recognition of specific behavioral pathways between parasite and host pertinent to survival.

Bacteria have been isolated that reduce hatching of two species of cyst nematodes (Oostendorp & Sikora, 1989; Racke & Sikora, 1992), either by bacterial pre-treatment of the host plant before collection of hatching factor or by bacterial treatment of root exudate solutions containing hatching factor (Oostendorp & Sikora, 1990). A complication in both experimental designs was difficulty in differentiating between hatch reduction caused by direct action of bacterial inhibitors/toxins on the nematode versus indirect activity via bacterial metabolism of the hatching factor. Although the chemical composition of substances directing nematode hatching behavior are in most cases not known, hatching factors for potato cyst nematode (*Globodera rostochiensis*) and soybean cyst nematode (*Heterodera glycines*) have been purified and identified. Utilization of purified hatching factor would facilitate selective isolation of bacteria metabolizing these substances and aid in separating out mechanisms of action.

Table 3: Possible modes of action of rhizobacteria towards sedentary endoparasitic nematodes

Bacterial influence / Mode of action	Nematode stage affected	Site of action
Toxic/inhibitory metabolites	Direct activity on: J_2^* in egg/ free-living J_2/ males	Soil/ Rhizosphere
Repellent metabolites	Direct activity on: Free-living J_2/ males	
Metabolism of factors influencing life-cycle	Egg: Disruption of hatch J_2: Interferrence with attraction to root J_2: Masking host recognition	Soil/ Rhizosphere
Blockage of lectin-carbohydrate recognition sites	J_2: Suppression of host recognition leads to reduced penetration rate	Soil/ Rhizosphere
Alteration of root exudate pattern	J_2: Effect on hatch/ attraction/ host recognition through altered exudate pattern	Soil/ Rhizosphere
Change in host suitability	J_2-J_4/ Adult: Influence on development through induced resistance mechanism - Retared or suppressed juvenile development - Transformation of sex ratio - Hypersensitive reaction in giant cell	Root tissue

* J_2, J_3, J_4 = second, third, fourth stage juvenile

Alteration in attraction or recognition: Rhizobacteria can alter root exudates or produce repellents affecting nematode attraction to or recognition of the host. Although Oostendorp and Sikora (1990) were unable to show any influence of *Pseudomonas* spp. on attraction of *Heterodera schachtii* to sugar beet, Franken et al. (1990) demonstrated that attraction of *Globodera pallida* was reduced following tuber treatment with *Agrobacterium radiobacter*.

Recognition is thought to be controlled by the interaction between root surface lectins and nematode cuticular carbohydrates (Zuckerman, 1983). Rhizobacterial influence on the host recognition process has not been studied in depth. PHPR may induce biological control of nematodes by binding lectins required for host recognition (Sikora, 1992).

Host suitability: PHPR have been shown to induce resistance in plants to fungal pathogens (van Peer et al., 1991). Similar alterations in host plant susceptibility to nematodes have been proposed by Sikora and Hoffmann-Hergarten (1992). The delicate nature of the nematode host parasite interrelationship, as indicated by a complex cellular syncytium, hypersensitivity mechanisms in resistant cultivars, and sex reversal to adverse environmental conditions favors the possible use of rhizobacteria induced resistance as a mechanism for nematode control. Microscopic and video based technology exists that permit detection of alterations in syncytial development or sex ratios within the root tissue (Wyss, 1987). Tests to determine stimulation of stress proteins are also available and could be used to test for PHPR induced resistance.

Isolation and screening methodology

Once mechanisms of action are known, efficient techniques for isolation of active strains and screening for efficacy can be developed. Isolation methodology could be designed to detect rhizobacteria with specific activity such as hatch inhibition or production of nematode repellents. Isolates demonstrating high levels of biological activity are often rhizosphere competent. Therefore, isolation from the root surface of the host plant and from specific root areas known to be important for nematode hatch, attraction and infection can increase detection of active isolates.

Although suppressiveness in soils to nematodes has not been correlated with the presence of specific PHPR, there are indications that specific soil amendments can increase nematode-controlling bacteria in and near the rhizosphere (Spiegel et al., 1991). Screening of bacteria from bacteriological collections or from non-rhizosphere soil has not proven as successful as when bacterial strains are obtained from the host plant. Bacteria from such sources may exhibit activity in vitro, but they are usually poor rhizosphere colonizers and seldom active in field soil. Isolation from nematode antagonistic plants, e.g. castor bean (*Rhizinus communis*), may lead to a high percentage of plants producing nematicidal activity (Kloepper et al., 1992).

Screening systems used to test for PHPR activity toward nematodes are of two types, each having advantages and disadvantages. In vivo bioassays are usually based on greenhouse tests where the interactions between the rhizobacterium, parasite and host occurs under semi-field conditions. In vitro bioassay laboratory tests are targeted at detection of toxins, inhibitors or metabolization of specific behavior regulating components of root exudates.

In vivo bioassay techniques are labor intensive and time consuming and substantially reduce the number of isolates screened with time. Limited replication and high variation in data ultimately lead to an inadvertent loss of active isolates. However, the test is rigorous; isolates are exposed to complex biotic and abiotic factors existing in the field which influence activity. In addition, selection is extended beyond the single mechanism of action as characteristic of laboratory tests.

In vitro techniques are designed to select for a single mechanism such as toxin production or hatch inhibition, and they drastically reduce the number of isolates requiring expensive in vivo testing. Views differ on their efficiency as it relates to detection of active isolates. Becker et al. (1988) demonstrated, for example, that although only one percent of the in vitro screened bacteria exhibited nematicidal activity, 20 percent of these isolates were effective when tested further in vivo. Without pre-selection, detection of active isolates was below 5 percent. Conversely, activity detected in in vitro tests is often non-reproducible in in vivo trials (Sikora et al., 1990). The nematode species for testing should also be selected with care, as it can lead to dubious results with no practical application. Metabolites toxic toward microphagous nematodes, for example, are not necessarily active against plant parasites and vice versa.

Abiotic and biotic factors

Environmental factors (temperature, moisture, soil type etc.) can influence level of control in an isolate-specific manner as in studies examining the influence of soil moisture on *Pseudomonas fluorescens* control of *Heterodera schachtii* (Oostendorp & Sikora 1989). However, Hackenberg and Sikora (1992) detected only minor temperature and soil moisture influences on the levels of biological control of *Globodera pallida* with *Agrobacterium radiobacter*. In situations where control is only required for short intervals after planting, optimum conditions for plant growth are also usually favorable for host, parasite and PHPR.

Biotic factors, such as cultivar selection and microbial competition within the rhizosphere, may influence efficacy. Cultivar influenced the level of biological control of the potato cyst nematode caused by *Agrobacterium radiobacter* (Hackenberg & Sikora, 1992). Variation in cultivar-bacterial interactions are probably caused by differences in the composition of the hatching factor, the amount produced and the time of release.

Rhizosphere competence and density

Rhizosphere colonization and multiplication after germination of the seed influence efficacy. The root zone colonized by the rhizobacterium and the density of the bacterium in this zone probably influence control attained (Racke, 1988). The importance of rhizosphere competence for nematode control is still poorly understood, however, nematodes often are attracted to and penetrate specific sites along the root. Therefore, if rhizosphere competence is important then colonization and multiplication of PHPR in these critical zones should lead to higher and more consistent control.

Spectrum of activity

A broad spectrum of activity against simultaneously occurring pathogens and parasites is considered advantageous and a commercial necessity for marketing. PHPR have been shown to have combined activity against *Heterodera schachtii* and damping-off fungi on sugar beet (Sikora et al., 1990). Similarity in the infection process, in that both organisms penetrate early after germination, is probably the reason for the effectiveness of a single PHPR. Furthermore, *Bacillus subtilis* reduced penetration of three species of nematodes in two different genera on a range of host plants in the greenhouse (Sikora, 1988). Other rhizobacteria may affect fewer pathogens. For example, Rhizobacteria isolated from potato showed different activity toward two species of potato cyst nematode. Whereas *Bacillus sphaericus* and *Agrobacterium radiobacter* reduced infection of the potato cyst nematode *Globodera pallida*, they did not effectively control a closely related species *Globodera rostochiensis* (Franken et al., 1990).

Conclusions

Plant parasitic nematodes should be considered excellent targets for biological control programs based on the application of plant health promoting rhizosphere bacteria. The intricate host-parasite interrelationship is highly sensitive to rhizosphere competent microorganisms. The fact that plant damage caused by some species of nematodes can be avoided with short term disruption of early root penetration favors biological control with rhizobacterial systems. Many of the behavioral aspects of a nematode's life-cycle can be studied in bioassay chambers that can be used for screening purposes. The use of seed dressing or transplant drench formulation of PHPR permits targeted application and reduces inoculum levels. Furthermore, introducing one endemic component of the antagonistic potential in soils back into the agro-ecosystem is an environmentally sound approach to crop protection.

References

Becker, J.O.; Zavaleta-Mejia, E.; Colbert, S.F.; Schroth, M.N.; Weinhold, A.R.; Hancock, J.G.; van Gundy, S.D. "Effects of rhizobacteria on root-knot nematodes and gall formation." *Phytopathology*, **1988**, 78, 1466-1469.

Campbell, W.C.; Fisher, M.H.; Stapley, E.O.; Albers-Schönberg, G.; Jacob, T.A. "Ivermectin: a potent new antiparasitic agent." *Science*, **1983**, 221, 823-828.

Devidas, P.; Rehberger, L.A. "The effects of exotoxin (thuringiensin) from *Bacillus thuringiensis* on *Meloidogyne incognita* and *Caenorhabditis elegans*." *Plant and Soil*, **1992**, 145, 115-120.

Franken, S.; Hackenberg, C.; Sikora, R.A. "Untersuchungen an Rhizosphärebakterien mit antagonistischer Wirkung gegen *Globodera* spp." *Phytomedizin*, **1990**, 20/4, p.23.

Hackenberg, C.; Sikora, R.A. "Influence of cultivar, temperature and soil moisture on the antagonistic potential of *Agrobacterium radiobacter* against *Globodera pallida*." *J. Nematol.*, **1992**, 24, p.594.

Kerry, B.R. "An assesment of progress toward microbial control of plant-parasitic nematodes." *J. Nematol. Suppl.*, **1990**, 22, 621-631.

Kloepper, J.W.; Rodriguez-Kabana, R.; McInroy, J.A.; Young, R.W. "Rhizosphere bacteria antagonistic to soybean cyst (*Hetero-*

dera glycines) and root-knot (*Meloidogyne incognita*) nematodes: Identification by fatty acid analysis and frequency of biological control activity." *Plant and Soil*, **1992**, 139, 75-84.

McIntyre, J.L.; Press, L.S. "Formulation, delivery systems and marketing of biocontrol agents and plant growth promoting rhizobacteria (PGPR)." In *The rhizosphere and plant growth*; Keister, D.L.; Cregan, P.B., Eds.; Beltsville Symposia in Agricultural Research, 14; Kluwer Academic Publishers: Dordrecht, The Netherlands, 1991, 289-295.

Oostendorp, M.; Sikora, R.A. "Seed-treatment with antagonistic rhizobacteria for the suppression of *Heterodera schachtii* early root infection of sugar beet." *Revue Nématol.*, **1989**, 12, 77-83.

Oostendorp, M.; Sikora, R.A. "In-vitro interrelationships between rhizosphere bacteria and *Heterodera schachtii*." *Revue Nématol.*, **1990**, 13, 269-274.

Oostendorp, M.; Dickson, D.W.; Mitchell, D.J. "Population development of *Pasteuria penetrans* on *Meloidogyne arenaria*." *J. Nematol.*, **1991**, 23, 58-64.

Peer, R. van; Niemann, G.J.; Schippers, B. "Induced resistance and phytoalexin accumulation in biological control of Fusarium wilt of carnation by *Pseudomonas* sp. strain WCS417r." *Phytopathology*, **1991**, 81, 728-734.

Racke, J. Untersuchungen zur biologischen Bekämpfung von *Globodera pallida* (Stone) an Kartoffeln durch Pflanzgutbehandlung mit antagonistisch wirkenden Rhizobakterien. 1988, Dissertation Universität Bonn

Racke, J.; Sikora, R.A. "Isolation, formulation and antagonistic activity of rhizobacteria toward the potato cyst nematode *Globodera pallida*." *Soil Biol. Biochem.*, **1992**, 24, 521-526.

Sikora, R.A.: "Interrelationship between plant health promoting rhizobacteria, plant parasitic nematodes and soil microorganisms." *Med. Fac. Landbouww. Rijksuniv. Gent*, **1988**, 53/2b, 867-878.

Sikora, R.A.: "Management of the antagonistic potential in agricultural ecosystems for the biological control of plant parasitic nematodes." *Annu. Rev. Phytopathol.*, **1992**, 30, 245-270.

Sikora, R.A.; Bodenstein, F.; Nicolay, R.: "Einfluß der Behandlung von Rübensaatgut mit Rhizosphärebakterien auf den Befall durch Pilze der Gattung *Pythium*. I. Antagonistische Wirkung verschiedener Bakterienisolate gegenüber *Pythium* spp." *J. Phytopathology*, **1990**, 129, 111-120.

Sikora, R.A.; Hoffmann-Hergarten, S. "Importance of plant health-promoting rhizobacteria for the control of soil-borne fungal diseases and plant parasitic nematodes." *Arab. J. Pl. Prot.*, **1992**, 10, 53-48.

Spiegel, Y.; Chon, E.; Galper, S.; Sharon, E.; Chet, I. "Evaluation of a newly isolated bacterium, *Pseudomonas chitinolytica* sp. nov., for controlling the root-knot nematode *Meloidogyne javanica*." *Biocontrol Sci. Technol.*, **1991**, 1, 115-125.

Weller, D.M. "Biological control of soilborne plant pathogens in the rhizosphere with bacteria." *Annu. Rev. Phytopathol.*, **1988**, 26, 379-407.

Wyss, U. "Video assessment of root cell responses to dorylaimid and tylenchid nematodes." In *Vistas on Nematology*; Veech, J.A.; Dickson, D.W., Eds.; Society of Nematologists, Inc.: Hyattsville, Maryland, 1987, 211-220.

Zavaleta-Mejia, E.; van Gundy, S.D. "Effects of rhizobacteria on *Meloidogyne* infection." *J. Nematol.*, **1982**, 14, 475A-475B.

Zuckerman, B.M. "Hypotheses and possibilities of intervention in nematode chemoresponses." *J. Nematol.*, **1983**, 15, 173-182.

Microbial Metabolites with Biological Activity Against Plant Pathogens

David M. Weller and Linda S. Thomashow, USDA-ARS, Root Disease and Biological Control Unit, Pullman, Washington 99164-6430

The use of recombinant DNA technology has provided unequivocal evidence that some metabolites produced by biocontrol agents contribute to disease suppression. Determining the importance of a metabolite in biocontrol activity involves generation of mutants deficient in production, complementation of mutants with wild-type DNA and comparison of the activity of the parent, mutant and complemented mutant. Antibiotics have been isolated from sites where biocontrol occurs. Biosynthetic loci for phenazine-1-carboxylic acid and 2,4-diacetylphloroglucinol have been cloned from *Pseudomonas* spp. and expressed in other strains. Some of the recombinant strains have improved biocontrol activity.

Biocontrol agents produce a wide variety of metabolites that may have a role in biological control. Some of the compounds that have been studied include antibiotics (secondary metabolites) such as phenazines, 2,4-diacetylphloroglucinol, pyoluteorin, pyrrolnitrin, oomycin A, agrocin 84, herbicolin A, chaetomin, gliovirin, gliotoxin, viridin and iturin A; volatiles such as hydrogen cyanide, ammonia, and ethanol; enzymes such as chitinases, β-glucanases, cellulases, and proteases; and siderophores such as the pyoverdins. Metabolites contribute to the biocontrol activity of an agent by 1) inhibiting the growth or activity of the target pathogen (e.g. antibiotics or enzymes), 2) inhibiting nontarget organisms thus increasing the growth and competitiveness of the introduced agent (e.g. antibiotics or enzymes), 3) inducing resistance mechanisms in the host (e.g. hydrogen cyanide? or antibiotics?), and/or 4) mediating competition for essential nutrients (e.g. siderophores)

The role of metabolites, particularly antibiotics, in biocontrol has been studied for nearly 60 years. However, the availability of only equivocal evidence for antibiosis (Fravel, 1988; Williams and Vickers, 1986), during most of this time, fostered continual debate until recently about the importance of antibiotics in the biocontrol process. Weller and Thomashow (1990) summarized that evidence and it includes the following observations: 1) that many biocontrol agents produced antibiotics, 2) that for some agents there was a correlation between antibiotic production and biocontrol ability

(Cullen and Andrews, 1984; Di Pietro et al., 1992; Kerr, 1980), 3) that purified antibiotics or cell-free filtrates or extracts of filtrates of some agents duplicated the effect of the whole agent (Baker et al., 1985; Cullen and Andrews, 1984; DeCal et al., 1988; Gueldner et al., 1988; Howell and Stipanovic, 1979, 1980; Janisiewicz and Roitman, 1988), and 4) that antibiotic-deficient mutants generated by chemical mutagens or UV-light (which causes multiple mutations) were less suppressive than wild-type strains (Howell and Stipanovic, 1983; Weller et al., 1988). The following are selected examples of these types of studies. The level of suppression of apple scab, caused by *Venturia inaequalis* (Cullen and Andrews, 1984) or Pythium damping-off of sugar beet (Di Pietro et al., 1992) by strains of *Chaetomium globosum* was directly related to the amount of chaetomin produced in vitro. Further, *Pseudomonas fluorescens* Pf-5, which produces both pyrrolnitrin and pyoluteorin, protected cotton against damping-off caused by *Rhizoctonia solani* and *Pythium ultimum*. In soil infested with *R. solani* or *P. ultimum*, pyrrolnitrin and pyoluteorin, respectively, applied to the seed duplicated the protection of Pf-5 (Howell and Stipanovic, 1979,1980). Finally, Howell and Stipanovic (1983) reported that a gliovirin-deficient mutant of *Gliocladium virens* was less effective at protecting cotton against Pythium damping-off than the wild type.

Recombinant DNA technology has revolutionized research in biological control of plant pathogens by providing techniques to identify traits responsible for disease suppression and to clone, characterize and transfer these important "biocontrol genes." In the last several years, application of this technology along with the use of modern analytical techniques, has provided unequivocal evidence that production of antibiotics, siderophores, and/or hydrogen cyanide is the primary mechanism of pathogen suppression by many biocontrol agents (Lugtenberg et al., 1991; O'Sullivan and O'Gara, 1992; Weller and Thomashow, 1993). Our understanding of the molecular basis of biocontrol is more advanced with bacterial agents as compared to fungal agents because bacterial genetic systems currently are more assessable. Phenazines, 2,4-diacetylphloroglucinol and hydrogen cyanide are the metabolites under the most intensive study at the genetic level.

The genetic strategy now commonly used to determine the role of a specific metabolite in the biocontrol process involves: 1) mutagenesis of a biocontrol agent (e.g. transposon mutagenesis), 2) screening for loss of the trait, 3) preparation of a genomic library of wild-type DNA, 4) genetic complementation of mutants to restore the target trait, and 5) comparison of the biocontrol abilities of the parental strain, mutant and complemented mutant. The mutant will be impaired in suppressiveness and complementation will restore activity if a trait is important.

Phenazines

Studies of *Pseudomonas fluorescens* 2-79, which produces phenazine-1-carboxylic acid (PCA) and suppresses take-all of wheat, caused by *Gaeumannomyces graminis* var. *tritici*, provided the first conclusive evidence that production of an antibiotic in situ contributed to biocontrol activity (Thomashow and Weller, 1988). Phenazine-deficient (Phz⁻) Tn5 mutants of 2-79 failed to inhibit *G. g. tritici* in vitro and were significantly less suppressive of take-all than 2-79. Mutants that were complemented with homologous DNA from a 2-79 genomic library were restored for production of PCA, inhibition of *G. g. tritici* and suppression of take-all. Subsequently, Bull et al. (1991) demonstrated an inverse linear relationship between the population size of Phz⁺ 2-79 or 2-79-B46R (complemented mutant) on roots of wheat seedlings and the number of take-all lesions resulting from primary infections. In contrast, no relationship existed when the Phz⁻

mutant 2-79-B46 colonized the roots and no dose of this strain reduced lesion number. PCA accounts for 60-90% of the biocontrol activity of 2-79. A pyoverdin siderophore and anthranilic acid also produced by this strain make at best minor contributions to take-all suppression; most residual activity probably results from nutrient competition (Hamdan et al., 1991; Ownley et al., 1992), however, induced resistance also needs to be explored. Ownley et al. (1992) found that strain 2-79 suppressed take-all across a range of soil pH from 4.9-8.0 indicating that pH is not a limitation to the activity of PCA as had been suggested by Brisbane et al. (1987).

Pseudomonas aureofaciens 30-84 like 2-79 produces PCA but in addition produces 2-hydroxyphenazine-1-carboxylic acid (2-OH-PCA) and 2-hydroxyphenazine (2-OH-PZ). A genetic approach also was used to show that phenazine production was the primary mechanism of take-all suppression by 30-84. Genetic analysis so far has demonstrated that a cosmid containing a 9.2 kb *Eco*RI fragment from 30-84 contains a significant portion of the phenazine biosynthetic pathway (Pierson and Thomashow, 1992). Two genes, *phzB* and *phzC*, involved in the production of PCA and 2-OH-PCA, respectively, were localized to a 2.8 kb region of the 9.2 kb fragment. A putative activator, *phzA*, was identified upstream of *phzB* and *phzC* (Pierson and Keppene, 1992; L. S. Pierson, personal communication). A 12 kb *Hind*III-*Bam*HI fragment in pPHZ108A containing phenazine biosynthetic genes from 2-79 hybridized with the 9.2 kb *Eco*RI fragment from 30-84. Sequences required for PCA production were localized to two divergently transcribed units of approximately 5 kb and 0.75 kb that appear to correspond functionally to *phzB* and *phzA*, respectively, in 30-84 (Thomashow et al, 1993).

Production of the phenazine pyocyanine is involved in the suppression of *Septoria tritici* on wheat by *Pseudomonas aeruginosa* LEC1 (Flaishman et al., 1990).

2,4-Diacetylphloroglucinol and Hydrogen Cyanide

Production of 2,4-diacetylphloroglucinol (Phl) is an important mechanism of suppression of several *Pseudomonas* strains that control soilborne pathogens. Phl inhibits a variety of fungi and bacteria and also has herbicidal activity (Keel et al., 1992). *P. fluorescens* CHA0 is the most extensively studied of the Phl producers; it also produces pyoluteorin (Plt), hydrogen cyanide (HCN), indoleacetic acid and a pyoverdin. The role of these metabolites in suppression of black root rot of tobacco, caused by *Thielaviopsis basicola*, take-all, and damping-off of cucumber caused by *Pythium ultimum* was determined. Production of Phl is the primary mechanism of take-all suppression and both Phl and HCN contribute to the control of black root rot (Haas et al., 1991, Keel et al., 1992; Voisard et al., 1989). Plt is involved in suppression of damping-off (Maurhofer et al., 1992) but roles for the siderophore and indoleacetic acid were not found (Haas et al., 1991). A locus in CHA0 termed *gacA* (global antibiotic and cyanide control) was characterized, that appears to function as a global regulator of secondary metabolism. The *gacA* mutants were deficient in production of Phl, HCN, and Plt and suppression of black root rot. The *gacA* locus showed homology to the *E. coli uvrY* gene and the predicted GacA protein belongs to the FixJ/DegU family of two component regulatory systems (Laville et al., 1992).

Phl also contributes to the biocontrol activity of *P. aureofaciens* Q2-87 against take-all (Harrison et al., 1993; Vincent et al., 1991) and *Pseudomonas* sp. F113 against damping-off of sugar beet caused by *Pythium ultimum* (Fenton et al., 1992; Shanahan et al., 1992). Q2-87 also produces HCN. Putative biosynthetic genes in Q2-87 and F113 were localized to a 4.8 kb *Bam*HI fragment in plasmid pMON5123 and to a 6 kb fragment in pCU203, respectively. Interestingly, three of six fluorescent pseudomonads isolated from the same soil as Q2-87 and selected for their biocontrol ability also produced Phl and HCN.

Isolation of Antibiotics from Natural Habitats

Confirmation of the above described genetic evidence for the involvement of antibiotics in biocontrol has been provided by direct isolation of these compounds from sites of biocontrol activity. Thomashow et al. (1990) planted wheat seeds treated with strain 2-79, 2-79-B46 (Phz⁻), 2-79-B46R (Phz⁺ complemented mutant), 30-84 or 30-84-44-8 (Phz⁻), in raw and steamed soil. Extracts were prepared from the roots and rhizosphere soil and analyzed by high pressure liquid chromatography. PCA was recovered only from roots treated with the Phz⁺ strains. More PCA was produced on roots in steamed as compared to raw soil. In a growth chamber study with raw soil, amounts ranged from 28-133 ng PCA/g root. In a field study amounts ranged from 5-27 ng PCA/g root. Similarly, Keel et al. (1992) isolated Phl from the rhizosphere and roots of wheat grown in an artificial soil and colonized by CHA0 (0.94-1.36 μg/g root) but not from roots colonized by the Phl⁻ Tn5 mutant CHA625.

Erwinia herbicola B247 suppresses *Fusarium culmorum* and *Puccinia recondita* f. sp. *tritici* on wheat and produces herbicolin A, an antibiotic involved in the suppression of the later pathogen. Herbicolin A was extracted from washed crowns and roots (2.7 and 0.9 μg/g dry weight, respectively) of seedling grown in the field and growth chamber in nonamended raw soil. No antibiotic was detected when an antibiotic-deficient transposon mutant was used. This water-soluble antibiotic is apparently absorbed by the plant (Kempf et al., 1993).

Gliocladium virens suppresses several soilborne pathogens and produces a variety of metabolites including gliotoxin, gliovirin, gliocladic acid, heptelidic acid (avocetin), viridin, viridiol and valinotricin. *G. virens* strain GL-21 is registered for commercial use in greenhouse-grown vegetable and ornamental bedding plants and production of gliotoxin is thought to be an important mechanism of suppression (Lumsden et al., 1992). Lumsden et al. (1992) studied the spatial-temporal pattern of production of gliotoxin in soil and soilless mix by strain GL-21 introduced in a wheat bran alginate-prill formulation. Following introduction of 0.1% w/v of the prill formulation into a soilless medium, composted soil, silty clay loam and gravely loamy sand, gliotoxin was detected at 0.42, 0.36, 0.20 and 0.02 μg/cm³, respectively. Increasing the amount of prill increased the gliotoxin. Suppression of Pythium damping-off of zinnia was correlated with the presence of the antibiotic and as the concentration of the gliotoxin increased, so also did suppression. The antibiotic was detected 3-4 cm from an individual prill in soilless mix, which was the distance that *G. virens* had grown out.

Traditionally, research on the role of antibiotics and other secondary metabolites has focused on their direct effects on target pathogens. However, finding that herbicolin A is absorbed by the plant (Kempf et al.,1993) and that PGPR strains are able to induce resistance in cucumber to *Colletotrichum orbiculare* (Wei et al., 1991) and in carnation to *Fusarium oxysporum* f. sp. *dianthi*, and increase accumulation of phytoalexins in carnation stems (van Peer et al., 1991) raises the possibility that disease suppression by secondary metabolites may also be mediated through induction of host resistance. One metabolite of special interest in this regard in HCN. Wei et al. (1991) noted that four of the six PGPR strains that induced resistance produced HCN.

Role of Antibiotics in Ecological Competence

Mazzola et al. (1992) demonstrated that phenazine production contributes positively to the persistence of *P. fluorescens* 2-79 and *P. aureofaciens* 30-84 in soil habitats. Strains 2-79, 30-84, phenazine-deficient mutants (2-79-B46 and 30-84.44-8) and genetically complemented mutants (2-79-B46R and 30-84.44-8R) were introduced individually into a raw Thatuna silt loam with or without added *G. g. tritici*. Up to five cycles of wheat were sown, each lasting twenty days from planting to harvest. After each growth cycle, shoots were removed and the soil and roots were mixed, repotted and again sown to wheat. Populations of the Phz⁻ mutants 2-79-B46 and 30-84.44.8 declined more rapidly than the populations of 2-79, 30-84 and their respective complemented mutants. Differences between populations of Phz⁺ and Phz⁻ strains occurred in an earlier cycle and were of a greater magnitude for 30-84 than for 2-79 indicating strain specific differences in the relative importance of the contribution of phenazines to survival. Further, phenazine was less critical to survival of these strains when wheat roots were infected with *G. g. tritici* than when healthy. Interestingly, when populations trends were compared in steam-pasteurized soil populations of Phz⁺ and Phz⁻ strains were similar. Pasteurization reduces the level of microbial competition. Thus, it appears that phenazines contribute to the ability of 2-79 and 30-84 to compete with the resident microflora. The findings of this study of colonization contrast with previous studies lasting no longer than 14 days that demonstrated that loss of antibiotic production did not alter the ability of *Pseudomonas* spp. to colonize the rhizosphere (Howie et al., 1991; Thomashow and Weller, 1988). It is now obvious that studies designed to determine the role of a trait in ecological competence must be conducted for much longer periods of time than previously used.

Enzymes

Chitinases, β-glucanases, cellulases, and proteases are enzymes most frequently considered to have a role in biological control. Of particular interest are chitinolytic enzymes because chitin is a major structural component of cell walls of many fungi. Enzyme production is considered to be an important component in the process of mycoparasitism. *Gliocladium* and *Trichoderma* spp. as well as other fungi and bacteria, produce high concentrations of a variety of chitinolytic enzymes in vitro. For example, the biocontrol agents *G. virens* 41 and *T. harzianum* P1 produce endochitinases, glucosaminidases and chitobiosidases in media containing chitin (Di Pietro et al., 1993; Harman et al., 1993). Lorito et al. (1993) reported that the ED_{50} of the endochitinase and chitobiosidase from strain P1 against *Botrytis cinerea*, *Fusarium solani*, *F. graminearum*, *Ustilago avenae*, and *Uncinula necator* was 35-135 μg/ml. However, combining these two enzymes resulted in a synergistic increase of antifungal activity against these pathogens. These data along with our understanding of fungal wall structure suggest that maximum biocontrol activity requires the coordinate action of mixtures of enzymes including chitinases with complementary modes of action, β-glucanases and possibly polysaccharidases, lipases and proteases (Cherif and Benhamov, 1990). Interestingly, Di Pietro et al. (1993) also noted a synergistic inhibitory effect when gliotoxin and an endochitinase from *G. virens* 41 were tested together against *B. cinerea*. Cell wall degradation by a chitinase may facilitate more rapid diffusion of the antibiotic into the cell. Although considerable research on lytic enzymes produced by fungi is ongoing, conclusive evidence supporting a role for them in biocontrol is still lacking. Genetic studies using the aforementioned protocol are needed. Of interest, however, is the report of Jones et al. (1986) that inactivation of the chitinase gene *chiA* in *Serratia marcesens* QMB1466 reduced its chitinase production and suppressiveness of *Fusarium oxysporum* f. sp. *pisi* on pea. Further, a recombinant *Escherichia coli* containing *chiA* under the control of a strong promoter provided significant suppression of *Sclerotium rolfsii* in bean (Shapira et al., 1989).

Biotic and Abiotic Factors Affect Production

Performance of introduced biocontrol agents is often inconsistent. One reason may be that the occurrence of a threshold concentration of a critical metabolite is temporally separated from the occurrence of pathogen infection or spread. This may often occur because production of secondary metabolites is highly dependent on cultural conditions and in natural habitats the physical and chemical environment in a microsite can change rapidly. Thus, even if a biocontrol agent is positioned in an infection court, expression of an antibiotic biosynthetic gene may not occur at the critical time. For example, production of oomycin A, which is involved in the suppression of Pythium damping-off of cotton by *P. fluorescens* HV37a (Howie and Suslow, 1991), is induced by glucose but inhibited by amino acids, all of which are found in the rhizosphere. Reporter gene fusion systems are powerful "biological sensors" for evaluating the influence of environmental factors on expression of genes critical to biocontrol activity in natural habitats. A transcriptional fusion between of the key oomycin A biosynthetic gene, *afuE* and the *lux* operon of *Vibrio fisheri* was used to monitor indirectly antibiotic expression by HV37a in the cotton spermosphere. Bioluminescence occurred 10 hr after planting, however, *P. ultimum* infection was detected at six hr. Expression of *afuE* varied with the soil type, water potential, and temperature. For example, expression was greatest at 20 C and was significantly reduced at 16 and 24 C (Gutterson et al., 1990). In contrast, production of gliotoxin by *G. virens* GL-21 in a soilless mix increased as the temperature of incubation increased from 15 to 30 C (Lumsden et al., 1992). Ownley et al. (1991) reported that seven soil variables were directly related to the biocontrol activity of 2-79 against take-all and several, including zinc and sulfate sulfur, were thought to be related to phenazine production in the rhizosphere. In support of this was the finding that the addition of 50 μg zinc (as Zn EDTA/g soil) to a soil naturally low in zinc increased the biocontrol activity of 2-79 in that soil as compared to the nonamended soil (B.H. Ownley and D.M. Weller, unpublished). Further, the addition of $ZnSO_4$ to liquid culture of 2-79 increased PCA production without increasing cell growth (Slininger and Jackson, 1992).

Metabolites such as antibiotics may be rapidly inactivated due to chemical instability or degradation by other organisms. In soil, some antibiotics, particularly basic ones, may be adsorbed and inactivated by clay and humus colloids (Williams and Vickers, 1986). For example, when 100 μg of gliotoxin was added to a soilless mix only 27.7 μg were recovered.

Further, following the introduction of *G. virens* GL-21 into a soilless mix, the amount of gliotoxin in the mix peaked between 1 and 4 days and then declined to low levels between 9 and 18 days (Lumsden et al., 1992). Suppression of take-all by 2-79 was inversely related to the percentage of clay or organic matter in a soil, suggesting that PCA might be tied up (Ownley et al., 1991).

Genetic Engineering to Improve Biocontrol

Considerable progress has been made in the last several years in understanding the molecular basis of biocontrol and it is now clear that heterologous expression of important biocontrol genes can be achieved more easily than previously anticipated. For example, Fenton et al. (1992) demonstrated that transfer of pCU203 containing the putative Phl biosynthetic locus from *Pseudomonas* strain F113, which suppresses Pythium damping-off of sugar beet, into the nonproducer M114 resulted in a recombinant that produced Phl and was more suppressive than the parent. Similarly, transfer of biosynthetic genes from Q2-87 into *P. fluorescens* 2-79 and *Pseudomonas* sp. 5097 resulted in synthesis of Phl and increased inhibition of *G. g. tritici*, *P. ultimum* and *R. solani* in vitro (Vincent et al., 1991). Recently, the same locus was expressed in all of 11 *Pseudomonas* spp. into which it was introduced and at least one strain showed improved biocontrol of take-all (H. Hara, M. Bangera, L. S. Thomashow and D. M. Weller, unpublished). Similarly, all of 27 pseudomonads into which the phenazine biosynthetic locus from 2-79 was introduced produced PCA and some showed enhanced suppression of take-all (H. Hara, L. S. Thomashow, D. M. Weller and D. E. Essar, unpublished). Voisard et al. (1989) reported only marginal disease suppression of black root rot with *P. fluorescens* P3 but a recombinant with HCN genes from CHA0 had improved biocontrol activity. It should be noted that insertion of regulatory genes can activate "silent" antifungal genes in heterologous strains (Weller and Thomashow, 1993).

Another approach to improve biocontrol activity involves increasing the production of important biocontrol traits. For example, phenazine production by 2-79 and 30-84 was increased by introducing extra copies of biosynthetic (Thomashow and Pierson, 1991) or activator genes (Pierson and Keppenne, 1992).

Similarly, introduction of a 22-kb fragment of CHA0 DNA back into CHA0 increased Plt production and the recombinant protected cucumber against damping-off caused by *Pythium ultimum* better than the parental strain. Introduction of extra plasmid-borne copies of the Phl biosynthetic locus from *P. aureofaciens* Q2-87 back into Q2-87 resulted in increased Phl production and suppression of take-all (H. Hara, M. Bangera, L. S. Thomashow and D. M. Weller, unpublished). Production of oomycin A and biocontrol of Pythium damping-off of cotton by *P. fluorescens* HV37a was increased by placing the biosynthetic gene cluster under the control of the constitutive *tac* promoter from *E. coli* (Gutterson, 1990).

Increasing the production of an important metabolite or transferring a trait into another strain will not necessarily enhance biocontrol activity and recombinant strains will need to be screened in order to select the best.

References

Baker, C.J.; Stavely, J.R.; Mock, N. "Biocontrol of bean rust by *Bacillus subtilis* under field conditions." *Plant Dis.* **1985**, 69:770-772.

Brisbane, P.G.; Janik, L.J.; Tate, M.E.; Warren, R.F.O. "Revised structure for the phenazine antibiotic from *Pseudomonas fluorescens* 2-79 (NRRL B-15132)." *Antimicrob. Agents Chemother.* **1987**, 31:1967-1971.

Bull, C.T.; Weller, D.M.; Thomashow, L.S. "Relationship between root colonization and suppression of *Gaeumannomyces graminis* var. *tritici* by *Pseudomonas fluorescens* strain 2-79." *Phytopathology.* **1991**, 81:954-959.

Cherif, M.; Benhamou, N. "Cytochemical aspects of chitin breakdown during the parasitic action of a *Trichoderma* sp. on *Fusarium oxysporum* f. sp. *radicis-lycopersici*." *Phytopathology.* **1990**, 80:1406-1414.

Cullen, D.; Andrews, J.H. "Evidence for the role of antibiosis in the antagonism of *Chaetomium globosum* to the apple scab pathogen, *Venturia inaequalis*". Can. J. Bot. **1984**, 62:1819-1823.

DeCal, A.; M-Sagasta, E.; Melgarejo, P. "Antifungal substances produced by *Penicillium frequentans* and their relationship to the biocontrol of *Monilinia laxa*." *Phytopathology.* **1988**, 78:888-893.

Di Pietro, A.; Gut-Rella, M.; Pachlatko, J.P.; Schwinn, F.J. "Role of antibiotics produced by *Chaetomium globosum* in biocontrol of *Pythium ultimum*, a casual agent of damping-off." *Phytopathology.* **1992**, 82:131-135.

Di Pietro, A.; Lorito, M.; Hayes, C.K.; Broadway, R.M.; Harman, G.E. "Endochitinase from *Gliocladium virens*: Isolation, characterization, and synergistic antifungal activity in combination with gliotoxin." *Phytopathology.* **1993**, 83:308-313.

Fenton, A. M.; Stephens, P. M.; Crowley, J.; O'Callaghan, M.; O'Gara, F. "Exploitation of gene(s) involved in 2,4-diacetylphloroglucinol biosynthesis to confer a new biocontrol capability to a *Pseudomonas* strain." *Appl. Environ. Microbiol.* **1992**, 58:3873-3878.

Flaishman, M.; Eyal, Z.; Voisard, C.; Haas, D. "Suppression of *Septoria tritici* by phenazine- or siderophore-deficient mutants of *Pseudomonas*." *Current Microbiology.* **1990**, 20:121-124.

Fravel, D.R. "Role of antibiosis in the biocontrol of plant diseases." *Ann. Rev. Phytopathol.* **1988**, 26:75-91.

Gueldner, R.C.; Reilly, C.C.; Pusey, P.L; Costello, C.E.; Arrendale, R.F.; Cox, R.H.; Himmelsbach, D.S.; Crumley, F.G.; Culter, H.G. "Isolation and identification of iturins as antifungal peptides in biological control of peach brown rot with *Bacillus subtilis*." *J. Agric. Food Chem.* **1988**, 36:366-370.

Gutterson, N. "Microbial fungicides: recent approaches to elucidating mechanisms." *Crit. Rev. Biotechnol.* **1990**, 10:69-91.

Gutterson, N.; Howie, W.; Suslow, T. "Enhancing effects of biocontrol agents by use of biotechnology." In *New Directions in Biocontrol: Alternatives for Suppressing Agricultural Pests and Diseases;* Baker, R.R., Dunn, P.E., Eds.; Alan R. Liss Inc.: New York, NY, 1990; pp. 749-765.

Haas, D; Keel, C.; Laville, J.; Maurhofer, M.; Oberhansli, T.; Schnider, U.; Voisard, C.; Wuthrich, B.; Defago, G. "Secondary metabolites of *Pseudomonas fluorescens* strain CHA0 involved in the suppression of root diseases." In *Advances in Molecular Genetics of Plant-Microbe Interactions;* Hennecke, H., Verma, D.P.S., Eds.; Kluwer Academic Publishers, Dordrecht, 1991, Vol.1; pp. 450-456.

Hamdan, H.; Weller, D.M.; Thomashow, L.S. "Relative importance of fluorescent siderophores and other factors in biological control of *Gaeumannomyces graminis* var. *tritici* by *Pseudomonas fluorescens* 2-79 and M4-80R." *Appl. Environ. Microbiol.* **1991**, 57:3270-3277.

Harman, G.E.; Hayes, C.K.; Lorito, M.; Broadway, R.M.; Di Pietro, A.; Peterbauer,

C.; Tronsmo, A. "Chitinolytic enzymes of *Trichoderma harzianum*: purification of chitobiosidase and endochitinase." *Phytopathology.* **1993**, 83:313-318.

Harrison, L.A.; Letendre, L.; Kovacevich, P.; Pierson, E.A.; Weller, D.M. "Purification of an antibiotic effective against *Gaeumannomyces graminis* var. *tritici* produced by a biocontrol agent, *Pseudomonas aureofaciens.*" *Soil Biol. Biochem.* **1993**, 25:215-221.

Howell, C.R; Stipanovic, R.D. "Control of *Rhizoctonia solani* on cotton seedlings with *Pseudomonas fluorescens* and with an antibiotic produced by the bacterium." *Phytopathology.* **1979**, 69:480-482.

Howell, C.R.; Stipanovic, R.D. "Suppression of *Pythium ultimum*-induced damping-off of cotton seedlings by *Pseudomonas fluorescens* and its antibiotic, pyoluteorin." *Phytopathology.* **1980**, 70:712-715.

Howell, C.R.; Stipanovic, R.D. "Gliovirin, a new antibiotic from *Gliocladium virens*, and its role in the biological control of *Pythium ultimum.*" *Can. J. Microbiol.* **1983**, 29:321-324.

Howie, W.J.; Suslow, T.V. "Role of antibiotic biosynthesis in the inhibition of *Pythium ultimum* in the cotton spermosphere and rhizosphere by *Pseudomonas fluorescens.*" *Mol. Plant-Microbe Interact.* **1991**, 4:393-399.

Janisewicz, W.J.; Roitman, J. "Biological control of blue mold and gray mold on apple and pear with *Pseudomonas cepacia.*" *Phytopathology.* **1988**, 78:1697-1700.

Jones, J.D.G.; Grady, K.L.; Suslow, T.V.; Bedbrook, J. R. "Isolation and characterization of genes encoding two chitinase enzymes from *Serratia marcescens.*" *EMBO J.* **1986**, 5:467-473.

Keel, C.; Schnider, U.; Maurhofer, M.; Voisard, C.; Laville, J.; Burger, U.; Wirthner, P.; Hass, D.; Defago, G. "Suppression of root diseases by *Pseudomonas fluorescens* CHA0: importance of the bacterial secondary metabolite 2,4-diacetylphloroglucinol." *Mol. Plant-Microbe Interact.* **1992**, 5:4-13.

Kempf, H.-J.; Bauer, P.H.; Schroth, M.N. "Herbicolin A associated with crown and roots of wheat after seed treatment with *Erwinia herbicola* B247." *Phytopathology.* **1993**, 83:213-216.

Kerr, A. "Biological control of crown gall through production of agrocin 84." *Plant Dis.* **1980**, 64:25-30.

Laville, J.; Voisard, C.; Keel, C.; Maurhofer, M.; Defago, G.; Hass, D. "Global control in *Pseudomonas fluorescens* mediating antibiotic synthesis and suppression of black root rot of tobacco." *Proc. Natl. Acad. Sci.* **1992**, 89:1562-1566.

Lorito, M.; Harman, G.E.; Hayes, C.K.; Broadway, R.M.; Tronsmo, A.; Woo, S.L.; Di Pietro, A. "Chitinolytic enzymes produced by *Trichoderma harzianum*: antifungal activity of purified endochitinase and chitobiosidase." *Phytopathology.* **1993**, 83:302-307.

Lugtenberg, B.J.J.; de Weger, L.A.; Bennett, J.W. "Microbial stimulation of plant growth and protection from disease." *Curr. Opin. Biotechnol.* **1991**, 2:457-464.

Lumsden, R.D; Locke, J.C.; Adkins, S.T.; Walter, J.F.; Ridout, C.J. "Isolation and localization of the antibiotic gliotoxin produced by *Gliocladium virens* from alginate prill in soil and soilless media." *Phytopathology.* **1992**, 82:230-235.

Maurhofer, M.; Keel, C.; Schnider, U.; Voisard, C.; Haas, D.; Defago, G. "Influence of enhanced antibiotic production in *Pseudomonas fluorescens* strain CHA0 on its disease suppressive capacity." *Phytopathology.* **1992**, 82:190-195.

Mazzola, M.; Cook, R.J.; Thomashow, L.S.; Weller, D.M.; Pierson, L.S.,III. "Contribution of phenazine antibiotic biosynthesis to the ecological competence of fluorescent pseudomonads in soil habitats." *Appl. Environ. Microbiol.* **1992**, 58:2616-2624.

O'Sullivan, D.J.; O'Gara, F. "Traits of fluorescent *Pseudomonas* spp. involved in suppression of plant root pathogens." *Microbiol. Rev.* **1992**, 56:662-676.

Ownley, B.H.; Weller, D.M.; Alldredge J.R. "Relation of soil chemical and physical factors with suppression of take-all by *Pseudomonas fluorescens* 2-79." In *Plant Growth-Promoting Rhizobacteria- Progress and Prospects*; Keel, C., Koller, B., Defago, G., Eds.; WPRS Bulletin 1991/XIV/8: 1991; pp. 299-301.

Ownley, B.H.; Weller, D.M.; Thomashow, L.S. "Influence of in situ and in vitro pH on suppression of *Gaeumannomyces graminis* var. *tritici* by *Pseudomonas fluorescens* 2-79." *Phytopathology* **1992**, 82:178-184.

Pierson, L.S., III; Keppenne, V.D. "Identification of a locus that acts *in trans* to stimulate phenazine gene expression in *Pseudomonas aureofaciens* 30-84." **1992**, Abstract 197, 6th Internationas Symposium on Molecular Plant-Microbe Interactions, Seattle.

Pierson, L.S., III; Thomashow, L.S. "Cloning and heterologous expression of the phenazine biosynthetic locus from *Pseudomonas aureofaciens* 30-84. *Mol. Plant-Microbe Interact.* **1992**, 5:330-339.

Shanahan, P.; O'Sullivan, D.J.; Simpson, P.; Glennon, J.D.; O'Gara, F. "Isolation of 2,4-diacetylphloroglucinol from a fluorescent pseudomonad and investigation of physiological parameters influencing its production." *Appl. Environ. Microbiol.* **1992**, 58:352-358.

Shapira, R.; Ordentlich, A.; Chet, I.; Oppenheim, A.B. "Control of plant diseases by chitinase expressed from cloned DNA in *Escherichia coli*." *Phytopathology.* **1989**, 79:1246-1249.

Slininger, P.J.; Jackson, M.A. "Nutritional factors regulating growth and accumulation of phenazine 1-carboxylic acid by *Pseudomonas fluorescens* 2-79." *Appl. Microbiol. Biotechnol.* **1992**, 37:388-392.

Thomashow, L.S.; Essar, D.W.; Fujimoto, D.K.; Pierson, L.S., III; Thrane, C.; Weller, D.M. "Genetic and biochemical determinants of phenazine antibiotic production in fluorescent pseudomonads that suppress take-all disease of wheat." In *Advances in Molecular Genetics of Plant-Microbe Interactions*; Nester, E.W., Verma, P.S., Eds.; Kluwer Academic Publishers, Dordrecht, 1993, Vol.2; pp. 535-541.

Thomashow, L.S.; Pierson, L.S., III. "Genetic aspects of phenazine antibiotic production by fluorescent pseudomonads that suppress take-all disease of wheat." In *Advances in Molecular Genetics of Plant-Microbe Interactions;* Hennecke, H., Verma, D.P.S., Eds.; Kluwer Academic Publishers: Dordrecht, 1991, Vol.1, pp.443-449.

Thomashow, L.S.; Weller, D.M. "Role of a phenazine antibiotic from *Pseudomonas fluorescens* in biological control of *Gaeumannomyces graminis* var. *tritici*. *J. Bacteriol.* **1988**, 170:3499-3508.

Thomashow, L.S.; Weller, D.M.; Bonsall, R.F.; Pierson, L.S., III. "Production of the antibiotic phenazine-1-carboxylic acid by

fluorescent *Pseudomonas* species in the rhizosphere of wheat." *Appl. Environ. Microbiol.* **1990**, 56:908-912.

van Peer, R.; Niemann, G.J.; Schippers, B. "Induced resistance and phytoalexin accumulation in biological control of Fusarium wilt of carnation by *Pseudomonas* sp. strain WCS417r." *Phytopathology.* **1991**, 81:728-734.

Vincent, M.N.; Harrison, L.A.; Brackin, J.M.; Kovacevich, P.A.; Mukerji, P.; Weller, D.M.; Pierson, E.A. "Genetic analysis of the antifungal activity of a soilborne *Pseudomonas aureofaciens* strain." *Appl. Environ. Microbiol.* **1991**, 57:2928-2934.

Voisard, C.; Keel, C.; Haas, D.; Defago, G. "Cyanide production by *Pseudomonas fluorescens* helps suppress black root rot of tobacco under gnotobiotic conditions." *EMBO J.* **1989**, 8:351-358.

Wei, G.; Kloepper, J.W.; Tuzun, S. "Introduction of systemic resistance of cucumber to *Colletotrichum orbiculare* by select strains of plant growth-promoting rhizobacteria." *Phytopathology.* **1991**, 81:1508-1512.

Weller, D.M.; Howie, W.J.; Cook, R.J. "Relationship between in vitro inhibition of *Gaeumannomyces graminis* var. *tritici* and suppression of take-all of wheat by fluorescent pseudomonads." *Phytopathology.* **1988**, 78:1094-1100.

Weller, D.M.; Thomashow, L.S. "Antibiotics: Evidence for their production and sites where they are produced." In *New Directions in Biological Control: Alternatives for Suppressing Agricultural Pests and Diseases;* Baker, R.R., Dunn, P.E., Eds.; Alan R. Liss, Inc.: New York, NY, **1990**, pp. 703-711.

Weller, D.M.; Thomashow, L.S. "Use of rhizobacteria for biocontrol." *Curr. Opin. Biotechnol.* **1993**, 4:(in press).

Williams, S.T.; Vickers, J.C. "The ecology of antibiotic production." *Microb. Ecol.* **1986**, 12:43-52.

Multifaceted Biological Control of Postharvest Diseases of Fruits and Vegetables

Charles L. Wilson and Ahmed El Ghaouth, USDA, ARS, Appalachian Fruit Research Station, Kearneysville, WV 25530 and Centre de recherche en horticulture, Université Laval, Quebec, G1K7P4, Canada

Utilizing a broad definition of biological control, a variety of methods are available as alternatives to synthetic chemicals for the control of postharvest diseases of fruits and vegetables. Among these are the use of: (1) antagonistic microorganisms; (2) natural plant- and animal-derived fungicides; and (3) induced resistance. Often these methodologies are not as effective alone as synthetic fungicides. However, when utilized in combination additive and synergistic control can be realized. A multifaceted approach toward the control of postharvest diseases is presented. Arguments are presented to validate this thesis and suggestions are made for future research.

Public concern over food safety has placed the use of pesticides to control postharvest diseases of fruits and vegetables under scrutiny. A National Academy of Sciences report (National Research Council, 1987) indicates that fungicides pose 60% of the oncological risk from the use of pesticides on our food — a greater risk than insecticides and herbicides together. It is clear that effective alternatives to synthetic fungicides for the control of postharvest diseases of fruits and vegetables are needed.

Where are these alternatives going to come from? Biological control presents an attractive possibility. Because conditions are more controllable in the postharvest environment as opposed to the field, it has been argued that biological control may have a better chance of working (Wilson and Wisniewski, 1989). In fact, considerable success has been realized recently utilizing antagonistic microorganisms to control postharvest diseases (Wisniewski and Wilson, 1992).

By limiting our definition of biological control, we also limit our search for effective alternatives to synthetic pesticides. Baker (1987) has defined biological control as "...the decrease of inoculum or the disease-producing activity of a pathogen accomplished through one or more organisms, including the host plant but excluding man." With this broad definition of biological control in mind, we set out to find: (1) antagonistic microorganisms to control postharvest pathogens; (2) plant-derived fungicides; (3) and means of inducing resistance in harvested commodities. What started out as research in three divergent areas has evolved into a convergent biological control methodology.

This paper examines the various components of a multifaceted biological control strategy and its advantages.

Antagonistic microorganisms

Antagonistic bacteria, yeasts, and filamentous fungi have been found which are capable of protecting fruit against postharvest rots (Wisniewski and Wilson, 1992). Some of these antagonists have been patented and tested on a large scale under commercial conditions (Hofstein et al., 1990; Pusey et al., 1988). Attempts are now being made to develop and market antagonistic microorganism as postharvest treatments to control rots of fruits and vegetables.

The so called "silver bullet" approach in utilizing a single antagonists has its limitations when compared to synthetic fungicides (Spurr, 1991). Often the antagonist has a more limited spectrum of activity and the duration of its effectiveness is less than that provided by synthetic fungicides. Furthermore, antagonists are unlikely to affect the resumption of quiescent infections.

In order to more intelligently apply antagonists and enhance their activity we have studied the mode-of-action of a number of antagonists to postharvest pathogens. It was found that the antagonist *Bacillus subtilis* , which controls brown rot of peaches, produces an antibiotic, Iturin, that inhibits spore germination of the brown rot fungus *Monilinia fructicola* (Gueldner et al., 1988). Iturin alone is able to reduce infection by the brown rot fungus, therefore, it is thought to play an important role in the mode-of-action of *Bacillus subtilis*.

We have also studied a number of antagonistic yeasts which do not produce antibiotics. In some cases their biocontrol activity can be diminished by the addition of nutrients to the wound site where infection occurs (Chalutz and Wilson, 1989). Some of the same antagonistic yeasts are able to attach to the walls of the pathogen and produce fungal wall hydrolyzing enzymes such as ß-1,3 glucanase (Wisniewski et al., 1991). In addition, they have been shown to induce PAL activity in citrus fruit (Chalutz, personal communication) and stimulate chitinase activity in apple (El Ghaouth and Wilson, unpublished data).

The yeast antagonists we have studied have a complex mode-of-action involving nutrient competition, direct parasitism, and induced resistance. The development of resistance to such antagonists should be more difficult than to those having a simpler mode-of-action such as antibiotic production.

Plant- and Animal-derived Fungicides

Entomologists have developed a number of effective plant-derived insecticides for the control of insects. Because of past perceptions that fungicides were innocuous, plant pathologists have not been aggressive in exploring plant- or animal-derived fungicides. It is well know that plants contain a number of effective fungicides (Grainge and Ahmed, 1988), yet there are none available commercially. Also, El Ghaouth (1992) has recently demonstrated that chitosan, a derivative of chitin, exhibits fungicidal activity against plant pathogens.

One negative concern in the use of plant-derived fungicides has been the argument that they are potentially as carcinogenic as synthetic fungicides (Ames and Magan, 1987). Counter arguments for the greater safety of natural compounds have been presented based on the co-evolution of animals and plants (Basson, 1987). Also, it might be expected that natural plant fungicides would be recycled more readily in the environment and would be less likely to accumulate in the food chain.

Plant-derived fungicides have been effective in controlling postharvest diseases. Ark and Thompson (1959) were successful in controlling brown rot of peaches with a deodorized garlic extract. Sholberg and Shimizu (1992) also controlled peach brown rot with a cedar extract, hinokitiol. Wilson et al. (1986) found that a number of volatile compounds produced by ripening peaches were fungicidal. Three of these compounds (benzaldehyde, methyl salicylate, ethyl benzoate) completely inhibited growth of two major rot fungi, *Monilinia fructicola* and *Botrytis cinerea*. Essential oils produced by plants are often fungicidal and have been suggested for use to control postharvest diseases (Dikshit, 1983).

Induced Resistance in Postharvest Commodities

Using Baker's broad concept (Baker, 1987), host plant resistance can be categorized as biological control. Little attention has been given to understanding the fundamental nature of postharvest disease resistance. In fact, it has been suggested that in our breeding programs to reduce astringency and increase tenderness in fruits and vegetables we have robbed plants of phenolic compounds and thick cell walls which impart resistance (Wilson et al., 1991).

Most of our understanding of resistance in plants comes from studies of the vegetative plant body. We need to exercise caution in assuming that resistance in reproductive structures is the same. The vegetative plant body exists primarily to produce reproductive organs. Once produced, reproductive organs exist primarily to protect the seed and serve in its dispersal. Plants could conceivably have evolved different resistant mechanism for these different functions.

Dr. Clauzell Stevens and his colleagues at Tuskegee University discovered that resistance in onions and sweet potatoes could be induced by treating them with low-dose UV light (Lu et al., 1987; Stevens et al., 1990). Subsequently, in cooperative studies with our laboratory and Dr. Edo Chalutz's laboratory at the Volcani Center in Israel it was demonstrated that this phenomenon also occurs with peach, apple, citrus, and tomato fruit (Stevens et al., 1990; Chalutz et al., 1992). Mercier et al. (1993) at Laval University induced a phytoalexin and resistance to storage pathogens in carrots with low-dose UV-C irradiation.

It has been concluded that the primary protective effect of UV light is induced resistance and not germicidal. This is based on the fact that the effect is not dose related and that lesion development is inhibited when UV-treated commodities are subsequently inoculated with pathogens.

El Ghaouth et al. (1992) demonstrated that chitosan can induce resistance in a number of fruits and vegetables. Chitosan forms a film that can be used to coat fruit and vegetables and protect them against decay. The coating is fungicidal and also induces defensive enzymes in the harvested commodity such as chitinase and ß-1,3 glucanase along with structural barriers (El Ghaouth, 1991).

Combining Different Biological Control Methodologies

A variety of biological control approaches are available for developing alternatives to synthetic chemicals in the control of post harvest diseases. Each has certain limitations. Our research has focused recently on means of enhancing antagonists, natural fungicides, and induced resistance used to control postharvest diseases. We are also examining methods to combine these treatments for additive and synergistic effects.

The effectiveness of antagonists can be enhanced and their activity made more comparable to synthetic fungicides with various additives. One of the most effective has been CaCl2 (McLaughlin et al., 1990). The addition of this salt can enhance the effectiveness of certain antagonists and greatly reduce the populations of yeast required to give effective control. We have also found antagonists which are resistant to synthetic and natural fungicides. This presents the possibility of using reduced rates of fungicides (synthetic and plant- and animal-derived) in combination with antagonists. In Israeli citrus packinghouses it was possible to reduce the rate of TBZ fungicide 90% in combination with the yeast antagonist *Pichia guilliermondii* and obtain control better than that with the full amount of fungicide alone (Hofstein et al., 1991).

One limitation to the use of UV-light induced resistance is that the protective effect diminishes over time. We have found that combining UV-light treatments in apples with antagonistic yeasts increases and prolongs the protection of the fruit against postharvest rots (unpublished data). Intriguingly, Dr. Stevens at Tuskegee University (personal communication) has found that the optimum dosage of UV light for induction of resistance in peaches promotes high populations of a naturally occurring antagonistic yeast on the surface of the peach. UV-light treatments have also been found to induce fungicidal compounds in the citrus peel (Chalutz and Stevens, personal communications) and stimulate antifungal hydrolases in bell pepper and tomato fruit.

A Multifaceted Biological Control Strategy

It has become apparent that a multifaceted approach has advantages over the use of a solitary biological control method for postharvest diseases. Even when we have applied antagonists, natural fungicides, and induced resistance singly an inter relatedness in these methodologies has emerged. For example, UV-light treatments in some instances induce natural fungicidal compounds in the fruit and promote naturally occurring antagonistic yeast populations on the surface of the fruit.

Because of the inherent variability in living systems, biological control methodologies often do not give as consistent results as synthetic

chemicals in pest control. In order for biological control to gain broader acceptance, this variability must be minimized. The multifaceted biological control strategy that we are proposing can be utilized to provide greater consistency and efficacy to biological control for postharvest diseases.

Pathogens are known to develop resistance to new control methods. It can be argued that the more complex the biological control methodology is, the less likely the pathogen will develop resistance to it. This appears to be the case with certain yeast antagonists which compete with the pathogen for nutrients, induce resistance responses in the host, and directly parasitize the pathogen (Wisniewski et al., 1991).

Combining antagonists having multiple modes of action with natural fungicides and treatments which induce resistance should present pathogens with more formidable resistance barriers. Such multifaceted resistance strategies should also be expected to have greater stability and effectiveness than solitary methodologies.

References

Ark, P. A., Thompson, J. P., "Control of certain diseases of plants with antibiotics from garlic (*Allium sativum* L.)" , *Plant Dis. Reptr.* **1959**, 43, 276.

Ames, B. N., Magan, R., and Gold, L. S., "Ranking possible carcinogenic hazards", *Science*, **1987**, 236, 271.

Baker, K. F., "Evolving concepts of biological control of plant pathogens", *Ann. Rev. Phytopathol.* **1987**, 25, 67-85.

Basson, P. A., "Poisoning of wildlife in southern Africa", *Afr. J. South. Afr. Vet. Assoc.*, **1987**, 58, 219.

Chalutz, E., Wilson, C. L., "Postharvest biocontrol of green and blue mold and sour rot of citrus fruit by *Debaryomyces hansenii*", *Plant Disease* ,**1989**, 74, 134-137.

Chalutz, E., Droby, S., Wilson, C. L., Wisniewski, M. E., "UV-induced resistance to postharvest diseases of citrus fruit", *J. Photochem. Photobiol.*, **1992**, B, 15, 367-374.

Dikshit, A., Dubey, N., Tripathi, N., Dixit, S. N., "Cedrus oil–a promising storage fungitoxicant", *Stored Produ. Res.,* **1983**, 19, 159-162.

El Ghaouth, A., Arul, J., Asselin, A., "Potential

use of chitosan in postharvest preservation of fruits and vegetables", *Proc. 1991 Chitin and Chitosan Conf.* ,**1992,** (in press).

El Ghaouth, A., Arul, J., Grenier, J., Asselin, A., "Use of chitosan coating to reduce weight loss and maintain quality for cucumbers and bell peppers", *J. Food Process. and Preserv.,* **1991,** (in press).

Grainge, M., and Ahmed, S., *Handbook of Plants with Pest Control Properties*, Wiley, New York, NY, 1988.

Gueldner, R. C., Reilly, C. C., Pusey, P. L., Costello, C. E., Arrendale, R. F., "Isolation and identification of iturins as antifungal peptides in biological control of peach brown rot with *Bacillus subtilis.*" *J. Agric. Food Chem.* **1988,** 36, 366-370.

Hofstein, R. S., Droby, S., Chalutz, E., Wilson, C. L., Friedlender, T., "Scaling-up the production of an antagonist — From basic research to R and D.", Biological Control of Postharvest Diseases of Fruits and Vegetables Wrkshp. Proc., Shepherdstown, WV, Sept. *U. S. Dept. Agr., Agr. Res. Serv. Publ.,* **1990,** 92, 188-201.

Lu, J. Y., Stevens, C., Yakabu, P., Loretan, P. A., and Eskin, D., "Gamma, electron beam and ultraviolet radiation on control of storage rots and quality of Walla Walla onions", *J. Food Process. Preserv.* , **1987,** 12, 53.

McLaughlin, R. J., Wisniewski, M. E., Wilson, C. L., Chalutz, E. "Effect of inoculum concentration and salt solutions on biological control of postharvest diseases of apple with *Candida* sp.", *Phytopathology*, **1990,** 80, 451-456.

McLaughlin, R. J., Wilson, C. L., Droby, S., Ben-Arie, R., Chalutz, E., "Biological control of postharvest diseases of grape, peach, and apple with the yeasts *Kloechera apiculata* and *Candida guilliermondii*", *Plant Disease* , **1992,** 76,470-473.

Mercier, J., Arul, J., Ponnampalam, Boulet, M."Induction of σ-methoxymellein and resistance to storage pathogens in carrot slices by UV-C", *J. Phytopahol.* , **1993,** 137: 44-54.

National Research Council, Committee on Scientific and Regulatory Issues Underlying Pesticide Use Patterns and Agricultural Innovation, *"Regulating Pesticides in Food"*

The Delany Paradox, Nat. Acad. Sci., Washington, D.C., 1987.

Pusey, P. L., Wilson, C. L., "Postharvest biological control of stone fruit brown rot by *Bacillus subtilis*", *Plant Disease* , **1984,** 68,753-756.

Pusey, P. L., Hotchkiss, M. W., Dulmage, H. T., Baumgardner, R. A., Zehr, E. I., Reilly, C. C., Wilson, C. L., "Pilot test for commercial production and application of *Bacillus subtilis* (B-3) for postharvest control of peach brown rot", *Plant Disease* , **1988,** 72, 622-626.

Spurr, H. W., Jr., Elliott, V. J. and W. M. Thal "Managing epiphytic microflora for biocontrol",*Biological Control of Postharvest Diseases of Fruits and Vegetables Wrkshp. Proc., Shepherdstown, WV, Sept.* ,*U. S. Dept. Agr., Agr. Res. Serv. Publ.,* **1991,** 92, 3-13.

Stevens, C., Lu, J. Y., Kahn, V. A., Wilson, C. L., Chalutz, E., Droby, S. Kawbe, M. K., Haung, Z., Adeyeye, O., and Liu, J., "Ultraviolet light induced resistance against postharvest diseases in vegetables and fruits", *Biological Control of Postharvest Diseases of Fruits and Vegetables Wrkshp. Proc., Shepherdstown, WV, Sept.* , *U. S. Dept. Agr., Agr. Res. Serv. Publ.,* **1991,** 92, 160-176.ᐟ

Stevens, C., Khan, V. A., Tang, A. Y., and Lu, J., "The effect of ultraviolet radiation on mold rots and nutrients of stored sweet potatoes", J. *Food Prot.*, **1990,** 53, 223.

Wilson, C. L. and M. E. Wisniewski , "Biological control of postharvest diseases of fruits and vegetables: An emerging technology", *Ann. Rev. Phytopathology* , **1989,** 27, 425-441.

Wilson, C. L., Wisniewski, M., Biles, C., McLaughlin, R., Chalutz, E. and Droby, S., "Biological control of postharvest diseases of fruits and vegetables — Alternatives to synthetic fungicides", *Crop Protection* , **1991,** 10, 172-177.

Wilson, C. L. Chalutz, E., "Postharvest biological control of Penicillium rots of citrus with antagonistic yeasts and bacteria", *Scientia Horticulturae* ,**1989,** 40, 105-112.

Wilson, C. L., Otto, B. E., "Fruit volatiles inhibitory to *Monilinia fructicola* and *Botrytis cinerea*" , *Plant Disease*, **1986,** 71, 316-319.

Wisniewski, M., Biles, C. L., Droby, S., McLaughlin, R., Wilson, C. L., and Chalutz, E. "Mode of action of postharvest biocontrol yeast, Pichia guilliermondii. I. Characterization of attachment to Botrytis cinerea", *Physiological and Molecular Plant Pathology*, **1991**, 39, 245-258.

Wisniewski, M. E. and Wilson, C. L., "Biological control of postharvest diseases of fruits and vegetables: Recent advances", *Hortscience*, **1992**, 27, 94-98.

Toward a Concept of Genus and Species in Trichoderma

Gary J. Samuels, Stephen A. Rehner, United States Department of Agriculture, Agriculture Research Service; Systematic Botany and Mycology Lab.; Rm. 304, B-011A, BARC-W; Beltsville, MD 20705

The expansion of the generic concept of *Trichoderma* to include *Gliocladium virens* is explained and justified. A concept of species limits in *Trichoderma* is lacking. Current morphological species concepts do not account for the wide variety of activities found in the genus. Species should be regarded more narrowly than at present and there are far more species than are currently recognized. Taxonomy must combine classical with non-morphological characters. Applied mycologists must be prepared for new names.

A quick review of the biological control or biodeterioration literature reveals the importance of *Trichoderma*. Unfortunately taxonomy has not kept up with progress in applied fields with a consequence that the number of species of *Trichoderma* that we recognize does not adequately account for the diversity of activities attributed to the genus (see reviews in Papavizas, 1985; Taylor, 1986). A taxonomy is urgently needed that will ensure that the concept of the genus can encompass all the species that phylogenetically belong in it, and that species names actually refer to discrete species and not to complexes. Applied research depends upon strains that are accurately named if they are to be legally registered, and if results and activities of named species are to be repeated in different laboratories. Moreover, a useful taxonomy must be predictive; it must provide insight as to the relationships of the organisms in order to rationally direct the search for improved biological activity. While the genus *Trichoderma* is easily recognized, some species have not been included in it because the generic limits have been obscure, and individual species are poorly defined. Following is an overview of problems in *Trichoderma* taxonomy and recent efforts to attain a useful classification.

Concept of Genus

Rifai's (1969) concept of *Trichoderma*, followed by Bissett (1991a), is largely accepted today. Rifai maintained a distinction between *Trichoderma*, species of which produce conidia in dry heads, and *Gliocladium*, which has slimy heads of conidia. Arx (1987) did not accept this as the only distinction between *Gliocladium* and *Trichoderma* and included *G. virens* in *Trichoderma* but did not explain the placement. By this act he blurred the boundary between *Trichoderma* and *Gliocladium*.

The inclusion of *G. virens* in *Trichoderma* - as *T. virens* - has not been adopted by applied mycologists, but is supported by overwhelming evidence, including cultural and morphological characters as well as molecular data. *Trichoderma virens* shares characters of habit, rapid colony growth, conidial color, chlamydospore production and branching pattern with more typical *Trichoderma* species (Bissett, 1991a). Although the conidiophores of *T. virens* appear somewhat more penicillate than the typical *Trichoderma*, a complete range of intermediates is found amongst the *Trichoderma* anamorphs of some *Hypocrea* species (Doi, 1972), and in *Trichoderma* sect. *Pachybasium* (Bissett, 1991b).

Trichoderma virens bears little resemblance to *Gliocladium roseum*, also often used in biological control studies, or to *G. penicillioides*, the type species of *Gliocladium*. *Gliocladium roseum* and *G. penicillioides* are the anamorphs of distantly related genera of the ascomycete order Hypocreales. While no teleomorph is known for *T. virens*, the species is similar to the anamorph of *Hypocrea gelatinosa* (Webster, 1964). Direct sequence analysis of polymerase chain reaction (PCR) products from ribosomal large subunit RNA (Rehner & Samuels, 1992) clearly placed *T. virens* with *H. gelatinosa* among other *Trichoderma* species and widely separated from *G. penicillioides* and *G. roseum*. In addition, *G. penicillioides* and *G. roseum* were shown to be generically distinct from each other.

Given the success in using *T. virens* in biological control of soilborne plant pathogens, it would be useful to identify closely related species in the possibility that they, too, might have commercial potential. Thus the question is where *T. virens* fits in *Trichoderma*. Following Bissett (1991b), one would survey species of *Trichoderma* sect. *Pachybasium*. A second hypothesis is that the relatives of *T. virens* are to be found among the species of *Hypocrea* sect. *Creopus*, including *H. gelatinosa*. The sequence analysis (Rehner & Samuels, 1992) was not variable enough to show interrelationships among *Trichoderma* and *Hypocrea* species. However, preliminary restriction analysis of the internal transcribed spacers of the nuclear ribosomal repeat indicate that *T. virens* is more similar to species in *Hypocrea* sect. *Creopus* than to other species of *Hypocrea* and *Trichoderma*, including species in sect. *Pachybasium*. This observation suggests that efficient biological control species may be found in the former.

Concept of Species

Before the early 1960's a common view was that any green spored *Trichoderma* was *T. viride* (see review in Rifai 1969). Rifai (1969) proposed nine "aggregate species," each of which comprised more than one species but individual species could not be recognized. Bissett (1984, 1991a, b) subdivided *Trichoderma* into five sections and revised two of them.

The basic problem is one of recognizing the limits of individual species. All taxonomic work with *Trichoderma* has attempted to characterize species mainly on morphological characters, but there may not be enough morphological characters to resolve all the species. We lack an objective measure of the variability of any *Trichoderma* species. In *Trichoderma* no assumptions can be made about the genetic relatedness of isolates because of the apparent absence of sexual reproduction. Any research using non-morphological characters must bear in mind the fact that morphological similarity of two *Trichoderma* isolates does not necessarily mean that the isolates are the same species.

Macromolecular approaches, either isozyme or nucleic acid analyses, can be useful in indicating relationships, but few have been applied to *Trichoderma*. Stasz et al. (1989) studied allozyme polymorphisms in a collection of *Trichoderma* isolates distributed among five of Rifai's aggregate species and found little correlation between the distribution of alleles and morphological species.

Most *Hypocrea* species have *Trichoderma* anamorphs and variation within an individual *Hypocrea* anamorph can be used as an objective guide to the amount of variation that can be expected in any *Trichoderma* species. Samuels et al. (1991), using isozyme polymorphism, found that geographically isolated collections of *Hypocrea schweinitzii* clustered independently of each other, suggesting that they are different taxa. Reexamination of the morphology of the respective teleomorphs and their anamorphs supported the taxonomic separation. Within each group there was high isozyme similarity and low morphological and cultural variability. The results of this limited study of *H. schweinitzii* suggest that individual species of *Trichoderma* should be viewed narrowly and that there may be far more species than previously thought.

Taxonomic reevaluation of *Trichoderma* must combine morphological and non-morphological characters. Morphological subtlties may assume a new importance. Meyer & Plaskowitz (1989) observed differences in conidial surface ornamentation as seen in scanning electron microscopy and divided the *T. viride* aggregate species into two groups. Meyer (1991) correlated mitochondrial DNA restriction fragments to each of the two groups. In spite of the near morphological and cultural identity between the "type" isolates of *T. reesei* and *T. longibrachiatum* (Bissett, 1984), they were genetically distinct when isozyme (Samuels et al., 1991) and DNA fingerprint analyses (Meyer et al., 1992) were applied. There are examples in the Hypocreales, to which *Hypocrea* and ultimately *Trichoderma* belong, of anamorphs belonging to distinct but closely related teleomorphs that are not morphologically separable and this may be the case with *T. longibrachiatum* and *T. reesei* even though neither species has been conclusively linked to a teleomorph.

Conclusions and Future Prospects

Taxonomic stability is urgently needed in the face of the large amount of biological and biotechnological research on *Trichoderma*. The recognition that *Trichoderma* species produce mycotoxins (see Taylor, 1986) emphasizes that urgency if these species are to be used in food production.

Identification of species of *Trichoderma* today exists in a kind of "fool's paradise" in the sense that, as was clearly stated by Rifai in 1969, the names we use refer to aggregates of species. Because of the difficulty in defining individual species and in finding characters that can be reliably used to identify them, the probability is high that there are far more species of *Trichoderma* than are currently recognized. Names that are used in the literature will have to be evaluated in taxonomic studies. As unsettling as it may be, frequently cited names such as *T. longibrachiatum* or *T. harzianum* will no doubt be much more restricted in their application than at present and new names will be introduced. The introduction of new names is a genuine attempt to ensure the repeatability of results from laboratory to laboratory.

Classical techniques of morphology and cultural characteristics must be combined with non-morphological methods. Vegetative compatibility should be explored in an effort to recognize heterokaryosis. Macromolecular techniques, especially gene sequencing, should be undertaken especially in studies that include isolates derived from discrete teleomorphs as well as morphologically similar, named anamorphs. Correlating metabolically interesting isolates of *Trichoderma* with unexplored isolates of *Hypocrea* species may help resolve the problem of genetically manipulating asexual fungi to achieve more favorable activity. At least a few species of *Hypocrea* have been induced to undergo sexual reproduction in the laboratory thereby giving the possibility of obtaining recombinants (Mathiesen, 1952; Webster, 1964).

References

Arx, J. A. "Plant pathogenic fungi." *Nova Hedwigia Beih.*, **1987**, *87*, 1-288.

Bissett, J. "A revision of the genus *Trichoderma*. I. Section *Longibrachiatum* sect. nov." *Canad. J. Bot.*, **1984**, *69*, 924-931.

Bissett, J. "A revision of the genus *Trichoderma*. II. Infrageneric classification." *Canad. J. Bot.*, **1991a**, *69*, 2357-2372.

Bissett, J. "A revision of the genus *Trichoderma*. III. Section *Pachybasium*." *Canad. J. Bot.*, **1991b**, *69*, 2372-2417.

Doi, Y. "Revision of the Hypocreales with cultural observations IV. The genus *Hypocrea* and its allies in Japan (2) Enumeration of the species." *Bull. Nat. Sci. Mus.*, **1972**, *15*, 649-751.

Mathieson, M. J. "Ascospore dimorphism and mating type in *Chromocrea spinulosa* (Fuckel) Petch n. comb." *Ann. Bot. (London)*, **1952**, *16*, 449-466.

Meyer, R. J.; Plaskowitz, J. S. "Scanning electron microscopy of conidia and conidial matrix of *Trichoderma*." *Mycologia*, **1989**, *81*, 312-317.

Meyer, R. J. "Mitochondrial DNAs and plasmids as taxonomic characteristics in *Trichoderma viride*." *Appl. Environ. Microbiol.*, **1991**, *57*, 2269-2276.

Meyer, W.; Morawetz, R.; Börner, T.; Kubicek, C. P. "The use of DNA-fingerprint analysis in the classification of some species of the *Trichoderma* aggregate." *Curr. Genet.*, **1992**, *21*, 27-30.

Papavizas, G. C. "*Trichoderma* and *Gliocladium*: biology, ecology, and potential for biocontrol." *Ann. Rev. Phytopathol.*, **1985**, *23*, 23-54.

Rehner, S. A.; Samuels, G. J. "Molecular phylogenetics and systematics of *Gliocladium*." *1992 APS/MSA Joint Meeting, Portland, Abstracts of Presentations*, **1992**, M350.

Rifai, M. A. "A revision of the genus *Trichoderma*." *Mycol. Pap.*, **1969**, *116*, 1-56.

Samuels, G. J.; Manguin-Gagarine, S.; Meyer, R.; Petrini, O. "Morphological and macromolecular characterization of *Hypocrea schweinitzii* and its *Trichoderma* anamorph (Abstract)." *Petria*, **1991**, *1*, 121-122.

Stasz, T. E.; Nixon, K.; Harman, G. E.; Weeden, N. F.; Kuter, G. A. "Evaluation of phenetic species and phylogenetic relationships in the genus *Trichoderma* by cladistic analysis of isozyme polymorphism." *Mycologia*, **1989**, *81*, 391-403.

Taylor, A. "Some aspects of the chemistry and biology of the genus *Hypocrea* and its anamorphs, *Trichoderma* and *Gliocladium*." *Proc. Nova Scotia Inst. Sci.*, **1986**, *36*, 27-58.

Webster, J. "Culture studies on *Hypocrea* and *Trichoderma* I. Comparison of perfect and imperfect states of *H. gelatinosa*, *H. rufa* and *Hypocrea* sp. 1." *Trans. Brit. Mycol. Soc.*, **1964**, *47*, 75-97.

Dynamics of *Sporidesmium*, a Naturally Occurring Fungal Mycoparasite

P. B. Adams and D. R. Fravel, Research Plant Pathologists, Biocontrol of Plant Diseases Laboratory, Plant Sciences Institute, Agricultural Research Service, U.S. Department of Agriculture, Beltsville, Maryland 20705

Lettuce drop, caused by *Sclerotinia minor*, was controlled in the field by the soil-inhabiting fungus *Sporidesmium sclerotivorum* applied at 0.2 kg/ha. *S. sclerotivorum* is an obligate mycoparasite of sclerotia primarily of species of the Sclerotiniaceae and can survive in soil for at least five years without hosts. Macroconidia of *S. sclerotivorum* germinate in response to host sclerotia. Extra-sclerotial growth and sporulation disseminate *S. sclerotivorum* in soil. Dissemination of the mycoparasite is aided by the aggregate distribution of pathogen sclerotia in soil. Since most sclerotia of *S. minor* are produced above-ground at the end of the season, spraying *S. sclerotivorum* on lettuce debris before tilling results in contact of the mycoparasite with pathogen sclerotia.

By taking a holistic view of the disease setting, various features of the biocontrol agent, plant pathogen and crop can be exploited so that the system as a whole is nudged toward higher agricultural productivity. An example of this approach is the use of *Sporidesmium sclerotivorum* to control lettuce drop caused by *Sclerotinia minor*. This paper discusses how the natural attributes of the mycoparasite, pathogen and crop are utilized to minimize disease loss. In addition, guidelines are given for applying these concepts to other mycoparasites, pathogens and crops.

Distribution and Biology of *Sporidesmium*

In nature, *S. sclerotivorum* behaves as an obligate parasite (biotroph) of sclerotia of the plant pathogens *Sclerotium cepivorum*, *Botrytis cinerea* (and probably other *Botrytis* species), *Sclerotinia minor*, *S. sclerotiorum* and *S. trifoliorum*. This mycoparasite was first isolated from a sclerotium of *S. minor* from soil in a field in Beltsville, MD (Uecker *et al.*, 1978). *Sporidesmium sclerotivorum* has since been detected in soils from 15 additional states, as well as in soils from Australia, Canada, Finland, Hungary, Japan, and Norway (Adams and Ayers, 1985; Litkei, 1988). *Sporidesmium sclerotivorum* has been implicated in the natural biological control of onion white rot (*S. cepivorum*) in New Jersey, stem rot of potato (*S. sclerotiorum*) in

Washington (Adams and Ayers, 1981) and Sclerotinia stem rot of sunflowers in Minnesota and North Dakota (P.B.Adams and T. Gulya, unpublished data). *Sporidesmium sclerotivorum* probably is involved in the natural regulation of other diseases caused by fungi in the Sclerotiniaceae.

Sporidesmium sclerotivorum is a dematiaceous hyphomycete that produces mycelium, microconidia, macroconidia (Uecker, *et al.*, 1978), chlamydospores (Ayers and Adams, 1979), and microsclerotia. The function of the microconidia is unknown while the chlamydospores, macroconidia and microsclerotia are thought to be involved in survival of the fungus in soil. Macroconidia have thick walls that contain phenolic compounds in their outer layers, features commonly found in resting structures (Bullock, *et al.*, 1989). Germination rates of macroconidia vary widely among isolates. Fifty percent of the macroconidia of two isolates germinated in 24 hours. Five other isolates reached 50 % germination between 35 and 50 hours, while another isolate took almost 80 hours to reach 50 % germination (P.B. Adams, unpublished data). *S. sclerotivorum* isolates also vary in growth habit in liquid culture. Growth of 30 isolates varied from production of a thin mat to a dense mycelial ball and ranged from 7 to 57 mg dry wt after 4 weeks of incubation (Adams, 1987b). However, when two "fast" growing isolates were compared to two "slow" growing isolates, all isolates were equally effective in infecting and destroying sclerotia of *S. minor* in soil (Adams, 1987b).

Sporidesmium sclerotivorum survives best in the field at depths greater than 2 cm (Adams, 1987a). In soil infested with sclerotia of *Sclerotinia* species, infection of sclerotia by *S. sclerotivorum* is optimal at soil temperatures of 20-25 C, a soil pH of 5.5-7.5 and a soil matric potential of -800 kPa or wetter (Adams and Ayers, 1980). These conditions are encountered in most agricultural fields. Macroconidia can survive in water at 20 C or in moist soil in the laboratory for about 5 years (P.B. Adams, unpublished data). In soil in the laboratory, macroconidia of *S. sclerotivorum* infect 90-100% of the sclerotia of *S. minor* within 5-6 weeks. Within an additional 4-6 weeks incubation, 90% of the pathogen's sclerotia are decayed and are no longer detectable in the soil (Ayers and Adams, 1979).

Life Cycle of *S. sclerotivorum*

In the presence of nutrients, macroconidia of *S. sclerotivorum* germinate readily and infect sclerotia. Glucose, mannose, maltose, or glycerol stimulate germination of macroconidia. However, none of these sugars stimulates germination as well as an extract of sclerotia (Ayers *et al.*, 1981). Further, live sclerotia stimulate germination far better than dead sclerotia, extracts from live sclerotia or chemicals (P.B. Adams, unpublished data).

After penetrating between the rind cells of a sclerotium, hyphae of *S. sclerotivorum* remain intercellular (Adams and Ayers, 1983) while producing haustoria in cells of the sclerotia (Bullock *et al.*, 1986). During this process, the sclerotial glucanase activity is doubled, yielding sufficient glucose for *S. sclerotivorum* to continue its parasitic activity (Adams and Ayers, 1983). After infection is well established within the sclerotium, hyphae of the mycoparasite grow to the surface of the sclerotium where *S. sclerotivorum* sporulates profusely. In soil, about 1.5×10^4 macroconidia are produced from each sclerotium of *S. minor* infected by *S. sclerotivorum* (Adams *et al.*, 1984). The mycoparasite uses approximately 80% of the energy (calories) in the pathogen sclerotia for development within the sclerotium and for subsequent mycelial growth in soil. The mycoparasite uses the remaining 20% of the energy stored in the sclerotium for production of macroconidia (Adams *et al.*, 1985).

During sporulation, mycelium of *S. sclerotivorum* can grow up to 3 cm from an infected sclerotium and infect healthy sclerotia (P.B. Adams and W. A. Ayers, unpublished data). Since aggregates of pathogen sclerotia generally contain up to 12 sclerotia/g of soil, vast amounts of mycelium of *S. sclerotivorum* are produced. Many spores will not germinate until they have been dispersed by tillage of the field because of the "staling products", presumably carbon dioxide, produced in soil as a result of the extensive mycelial growth (Adams, unpublished data). Macroconidia that are produced as a result of infection and destruction of sclerotia can germinate and infect healthy sclerotia or they can remain dormant for at least 5 years. Thus, *S. sclerotivorum* has three characteristics useful for biocontrol: i) secondary infection of pathogen sclerotia, ii) lateral spread of the mycoparasite within the soil mass and iii) a tremendous increase in population of the mycoparasite which can reduce the pathogen population in future seasons.

Production and Formulation of *S. sclerotivorum*

Biotrophs grow very slowly or not at all on highly specialized media. The difficulty in culturing biotrophs should not be a reason to dismiss biotrophs from consideration as potential commercial biocontrol products. Biotrophs are difficult to grow only because of our ignorance. Sufficient knowledge of the organism should result in capability for production in commercial quantities.

In the case of *S. sclerotivorum*, much research was required to develop a satisfactory synthetic medium for growth of the biotroph. Glutamine and glucose are the preferred nitrogen and carbon sources (Barnett and Ayers, 1981). Mannitol, mannose, and cellobiose also supported adequate growth. A liquid medium composed of mineral salts, glucose, glutamine, thiamine, biotin, succinic acid, KH_2PO_4, $CaCl_2$, minor elements, and ferric-potassium salt of

EDTA provides substantial growth and sporulation of the biocontrol agent (Ayers and Adams, 1983). The mycoparasite can be grown on vermiculite saturated with the synthetic liquid medium. Vermiculite provides ample surface area for sporulation after 6 to 12 weeks incubation. From such cultures a product can be formulated which contains 10^6 macroconidia / ml. Only 100 ml of this formulation would be required to treat 1 ha.

Pathogen Distribution

Any biocontrol agent must be used in a manner that is consistent with the characteristics of the biocontrol agent, plant pest, crop culture and epidemiology of the plant disease. In the case of *S. sclerotivorum*, the target pest is *S. minor* which causes severe losses on lettuce and peanuts.

Lettuce is grown in a nearly uniform distribution with approximately 5.4×10^4 plants / ha. Crops require 40-80 days to reach maturity, depending on the weather and cultivar. Incidence of lettuce drop is positively correlated with soil populations of sclerotia of *S. minor* (Adams and Tate, 1975). All types of lettuce are susceptible to *S. minor* throughout their life cycles. However, because of the growth habit of lettuce, most infections of the lettuce occur within 2-3 weeks of harvest when the outer leaves expand and expose this leaf tissue to the soil surface (Dillard and Grogan, 1985).

A single diseased lettuce plant supports production of an average of 3.5×10^3 sclerotia of *S. minor* / plant with a range of about 0 to 1.2×10^4 sclerotia / plant (Adams, 1986). After harvest, the remaining lettuce debris is disked into the soil, resulting in an aggregate distribution both vertically and horizontally of the pathogen sclerotia. Inoculum densities within the aggregates range from 0 to over 12 sclerotia / g of soil (Adams, 1986). With subsequent disking and plowing the sclerotia become more dispersed but still remain in an aggregate distribution. The aggregate distribution of the pathogen results in an aggregate distribution of lettuce drop (Marois and Adams, 1985). The aggregate distribution of the pathogen has important implications for biocontrol. Because *S. sclerotivorum* can spread from one sclerotium to another, the aggregate distribution of the pathogen facilitates the epidemic destruction of sclerotia.

Delivery and Epidemiology

In the case of lettuce drop, what is the most efficient way to apply *S. sclerotivorum* to the field to control the disease caused by *S. minor*? In an early field test, the mycoparasite was applied to soil at a rate of about 10^3 macroconidia / g of soil (Adams and Ayers, 1982). At this application rate, it would take nearly 1 ha of Petri dish surface area to produce the 2.24×10^{12} macroconidia needed to treat one ha to a depth of 6 cm. A more efficient delivery system was clearly needed. The natural attributes of the biocontrol agent, pathogen and crop were used to improve delivery with reduced rates of the biocontrol agent.

Macroconidia of *S. sclerotivorum* will infect sclerotia of *S. minor* only when the spores are in close proximity to the sclerotia. Hence, the most efficient way to apply the biocontrol agent is directly on the sclerotia. Since sclerotia are produced on the above-ground portion of lettuce plants, spraying *S. sclerotivorum* onto the diseased crop refuse after harvest, but before disking the field, insures that a high percentage of macroconidia of the mycoparasite will contact sclerotia formed on the diseased lettuce tissue. These sclerotia will become infected after the crop refuse is plowed and/or disked into the soil. The reduction in inoculum density of *S. minor* by *S. sclerotivorum* was sufficient to reduce the incidence of lettuce drop in the field (Adams and Ayers, 1982). The rate of infection of sclerotia of *S. minor* by *S. sclerotivorum* is density dependent (Adams *et al.*, 1984). That is, the higher the population of the mycoparasite the quicker the sclerotia decay in soil. Conversely, with a constant population of *S. sclerotivorum* the more sclerotia that are added to the soil, the quicker the sclerotia decay in soil. This density dependent relationship is extremely important in both natural and imposed biological control of plant diseases. Since sclerotia of the pathogen are found in the field as aggregates, the mycoparasite can spread from sclerotium to sclerotium within an aggregate and sometimes from aggregate to aggregate. One consequence of this spread is that the pathogen population may be reduced below an economic threshold for disease development. By applying *Sporidesmium* to the crop debris before disking and waiting for the proliferation of the mycoparasite, control of lettuce drop in the field was obtained with a single application of *S. sclerotivorum* applied at rates as low as 0.2 kg (fresh wt) / ha (Adams and Fravel, 1990).

Analysis of disease progress curves indicated that within a growing season, *S. sclerotivorum* delays the onset of the epidemic but does not change the rate of disease increase (Fravel, Adams and Potts, 1992). After five successive croppings of lettuce (three fall and two spring crops), disease incidence was reduced to a negligible level with an initial *S. sclerotivorum* density of less than 10 macroconidia / g soil. Since pesticides used to control lettuce drop are not harmful to *S. sclerotivorum* in soil (Adams and Wong, 1991), it should be possible to apply the biocontrol agent followed by the currently recommended two to three fungicide applications per crop until *S. sclerotivorum* reduces the population of pathogen sclerotia below an economic threshold. Alternatively, *S. sclerotivorum* could be applied at a higher rate to achieve disease control sooner.

Unconventional Biocontrol

Another way to increase the population of *S. sclerotivorum* and decrease the population of *S. minor* is to add sclerotia of *S. minor* to the field. About 50% of the agricultural fields assayed for mycoparasites of *Sclerotinia* spp. have contained either *S. sclerotivorum* and/or *Teratosperma oligocladum* (another mycoparasite of *Sclerotinia* spp.). The assay is a baiting technique using sclerotia of *S. minor* and has an estimated sensitivity of approximately 10 macroconidia / g of soil. Before selecting a field to conduct a field test for control of *S. minor*, soil samples were collected and assayed for mycoparasites. Only fields testing negative for mycoparasites of *S. minor* were used. Two fields that tested negative for *S. sclerotivorum* were subsequently found to contain the mycoparasite after a large number of sclerotia were added to the field. In fact, the population of *S. sclerotivorum* increased

to the point where the mycoparasite caused a decrease in the inoculum density of *S. minor* to about 10% of the original inoculum level resulting in a substantial reduction in the incidence of lettuce drop. Thus, just adding a sufficient quantity of the plant pathogen to the field was sufficient to affect biological control, even when the fungus was present in the field at populations of less than 10 macroconidia / g of soil.

Adding sclerotia of *S. minor* to a field may have unwanted repercussions (in addition to increased disease!). Just as growth in the population of *S. minor* results in proliferation of *S. sclerotivorum*, increases in *S. sclerotivorum* can stimulate population increases in fungi parasitic on *S. sclerotivorum*. For example, increases in *S. sclerotivorum* in a field test resulted in proliferation in *Laterispora brevirama*, a mycoparasite of *S. sclerotivorum* (Uecker *et al.*, 1982). *Laterispora brevirama* did not affect infection of pathogen sclerotia by *S. sclerotivorum* but did reduce production of *S. sclerotivorum* macroconidia in soil (Ayers and Adams 1985). This increase of an undesirable mycoparasite in response to population increases of a beneficial mycoparasite underscores the need for a basic understanding of microbial interactions in soil.

Characteristics of the Mycoparasitic Biocontrol Agent and Pathogen

Some of the attributes that make *S. sclerotivorum* a good mycoparasite for biocontrol include i) it is an obligate mycoparasite in nature (which makes it target specific), ii) it utilizes its host as a source of nutrients for growth and sporulation, iii) its resting propagules are good survival structures, and iv) it can grow out into the soil from an infected pathogen propagule and infect a healthy pathogen propagule.

Disappointing performance by mycoparasites that attack the mycelium of plant pathogens has been noted (Boosalis and Mankau 1970). Pathogen mycelium may lack sufficient nutrients to sustain the necessary growth of the mycoparasite through the soil matrix to achieve biological control. Sclerotia provide an ample substrate for sporulation, growth and spread of the mycoparasite. In addition, many sclerotial pathogens, such as *S. minor*, do not produce secondary inoculum. The lack of secondary cycles makes it possible to achieve control through destruction of the primary inoculum.

Screening for Mycoparasites

Is *S. sclerotivorum* unique or are there other mycoparasites in nature that are potential biocontrol agents for plant diseases? There may be other potentially useful mycoparasites that have not been discovered due to the methods commonly used to look for new biocontrol agents. Most screening for biocontrol agents depends on isolation on agar medium. This method would not recover obligate mycoparasites. Plating pathogen propagules recovered from soil onto moist sand, filter paper or directly onto pathogen propagules increases the chance of finding a mycoparasite. Since some obligate mycoparasites grow slowly, the pathogen propagules plated onto sand, filter paper or fresh pathogen propagules should be observed periodically for 4 weeks or more. This technique was used to discover *T. oligocladum*, a parasite of *Sclerotinia* spp. (Ayers and Adams, 1981; Parfitt and Coley-Smith, 1983; Uecker, *et al.* 1980) and *Laterispora brevirama*, a parasite of *S. sclerotivorum* and *T. oligocladum* (Ayers and Adams, 1985). If sclerotia of *S. minor* infected with either *S. sclerotivorum* or *T. oligocladum* had been plated onto agar containing nutrients, saprophytic fungi would have grown out and the biotrophs would never have been discovered.

Summary

Biological control of plant diseases can be thought of as a form of biological warfare in which the plant pathogen is the enemy. This thinking leads us to ask what can be done to insure, or increase the probability, that a particular biocontrol agent will successfully attack a particular plant pathogen resulting in a reduction of disease incidence or severity. The chances of successful biocontrol are greatly enhanced if the natural habitat, abilities and functions of the agent, pathogen and crop are employed in the control scheme. Too often,

attempts are made to force organisms to do what they do not do normally. Working against the organisms, rather than with them, is a prescription for failure at biocontrol. Initial field tests used 2×10^3 to 2×10^5 kg/ha of *S. sclerotivorum* to control lettuce drop (Adams and Ayers, 1982). Using *S. sclerotivorum* correctly, in an ecological sense, resulted in successful control with only 0.2 kg/ha (Adams and Fravel, 1990).

The control system described in this chapter exploits features of *S. sclerotivorum*, *S. minor* and lettuce to achieve control of lettuce drop at reasonable application rates of the biocontrol agent. Characteristics of the biocontrol agent that were utilized include: i) ability to parasitize sclerotia, ii) ability to grow out from infected sclerotia for up to 3 cm through soil and infect healthy sclerotia, iii) ability to reproduce abundantly in soil. Useful features of the pathogen include i) production of most sclerotia occurs on the above-ground portion of the plant (which facilitates delivery of the biocontrol agent at practical rates), and ii) sclerotia are distributed in aggregates in the soil (which facilitates the epidemic destruction of sclerotia). Finally, close and uniform spacing of plants enhances aggregate to aggregate spread of *S. sclerotivorum*. Careful examination of other pathosystems should reveal other opportunities for biocontrol.

References

Adams, P.B. "Production of sclerotia of *Sclerotinia minor* on lettuce in the field and their distribution in soil after disking." *Plant Dis.* **1986**, *70*, 1043-1046.

Adams, P.B. "Effects of soil temperature, moisture, and depth on survival and activity of *Sclerotinia minor*, *Sclerotium cepivorum*, and *Sporidesmium sclerotivorum*." *Plant Dis.* **1987a**, *71*, 170-174.

Adams, P.B. "Comparison of isolates of *Sporidesmium sclerotivorum* in vitro and in soil for potential as active agents in microbial pesticides." *Phytopathology* **1987b**, *77*, 575-578.

Adams, P.B., Ayers, W.A. "Factors affecting parasitic activity of *Sporidesmium sclerotivorum* on sclerotia of *Sclerotinia minor* in soil." *Phytopathology* **1980**, *0*, 366-368.

Adams, P.B., Ayers, W.A. *"Sporidesmium sclerotivorum*: Distribution and function in natural biological control of sclerotial fungi." *Phytopathology* **1981**, *71*, 90-93.

Adams, P.B., Ayers, W.A. "Biological control of Sclerotinia lettuce drop in the field by *Sporidesmium sclerotivorum.*" *Phytopathology* **1982**, *72*, 485-488.

Adams, P.B., Ayers, W.A. "Histological and physiological aspects of infection of sclerotia of two *Sclerotinia* species by two mycoparasites." *Phytopathology* **1983**, *73*, 1072-1076.

Adams, P.B., Ayers, W.A. "The world distribution of the mycoparasites *Sporidesmium sclerotivorum, Teratosperma oligocladum* and *Laterispora brevarama.*" *Soil Biol. Biochem.* **1985**, *17*, 583-584.

Adams, P.B., Ayers, W.A., Marois, J.J. "Energy efficiency of the mycoparasite *Sporidesmium sclerotivorum in vitro* and in soil. *Soil Biol. Biochem.*" **1985**, *17*, 155-158.

Adams, P.B., Fravel, D.R. "Economical biological control of Sclerotinia lettuce drop by *Sporidesmium sclerotivorum.*" *Phytopathology* **1990**, *80*, 1120-1124.

Adams, P.B., Marois, J.J., Ayers, W.A. "Population dynamics of the mycoparasite, *Sporidesmium sclerotivorum*, and its host, *Sclerotinia minor*, in soil." *Soil Biol. Biochem.* **1984**, *16*, 627-633.

Adams, P.B., Tate, C.J. "Factors affecting lettuce drop caused by *Sclerotinia sclerotiorum.*" *Plant Dis. Reptr.* **1975**, *59*, 140-143.

Adams, P.B., Wong, J.A.-L. "The effect of chemical pesticides on the infection of sclerotia of *Sclerotinia minor* by the biocontrol agent *Sporidesmium sclerotivorum.*" *Phytopathology* **1991**, *81*, 1340-1343.

Ayers, W.A., Adams, P.B. "Mycoparasitism of sclerotia of *Sclerotinia* and *Sclerotium* species by *Sporidesmium sclerotivorum.*" *Can. J. Microbiol.* **1979**, *25*, 17-23.

Ayers, W.A., Adams, P.B. "Mycoparasitism of sclerotial fungi by *Teratosperma oligocladum.*" *Can. J. Microbiol.* **1981**, *27*, 886-892.

Ayers, W.A., Adams, P.B. "Improved media for growth and sporulation of *Sporidesmium sclerotivorum.*" *Can. J. Microbiol.* **1983**, *29*, 325-330.

Ayers, W.A., Adams, P.B. "Interaction of *Laterispora brevirama* and the mycoparasites *Sporidesmium sclerotivorum* and *Teratosperma oligocladum.*" *Can. J. Microbiol.* **1985**, *31*, 786-792.

Ayers, W.A., Barnett, E.A., Adams, P.B. "Germination of macroconidia and growth of *Sporidesmium sclerotivorum* in vitro." *Can. J. Microbiol.* **1981**, *27*, 664-669.

Barnett, E.A., Ayers, W.A. "Nutrition and environmental factors affecting growth and sporulation of *Sporidesmium sclerotivorum.*" *Can. J. Microbiol.* **1981**, *27*, 685-691.

Boosalis, M.G., Mankau, R. In *Ecology of Soilborne Plant Pathogens*; Baker, K.F., Snyder, W.C., Ed.; Univ. Calif. Press: Berkeley, 1970, pp 374-389.

Bullock, S., Adams, P.B., Willetts, H.J., Ayers, W.A. "Production of haustoria by *Sporidesmium sclerotivorum* in sclerotia of *Sclerotinia minor.*" *Phytopathology* **1986**, *76*, 101-103.

Bullock, S., Willetts, H.J., Adams, P.B. "Morphology, histochemistry, and germination of conidia of *Sporidesmium sclerotivorum.*" *Can. J. Microbiol.* **1989**, *67*, 313-317.

Dillard, H.R., Grogan, R.G. "Relationship between sclerotial spatial pattern and density of *Sclerotinia minor* and the incidence of lettuce drop." *Phytopathology* **1985**, *75*, 90-94.

Fravel, D.R., Adams, P.B., Potts, W.A. "Use of disease progress curves to study the effects of the biocontrol agent *Sporidesmium sclerotivorum* on lettuce drop." *Biocontrol Sci. Technol.* **1992**, 2, 339-348.

Litkei, J. "Occurrence of *Sporidesmium sclerotivorum* Uecker, Ayers and Adams in Hungary. A hyperparasitic fungus on sclerotia of *Sclerotinia sclerotiorum* Lib. deBary." *Acta Phytopathologica et Entomologica Hungarica* **1988**, *23*, 115-118.

Marois, J.J., Adams, P.B. "Frequency distribution analysis of lettuce drop caused by *Sclerotinia minor.* " *Phytopathology* **1985**, *75*, 957-961.

Parfitt, D., Coley-Smith, J.R. "*Teratosperma oligocladum*, a mycoparasite of fungal sclerotia." *Plant Pathol.* **1983**, *32*, 495-460.

Uecker, F.A., Ayers, W.A., Adams, P.B. "A new hyphomycete on sclerotia of *Sclerotia sclerotiorum.*" *Mycotaxon* **1978**, *7*, 275-282.

Uecker, F.A., Ayers, W.A., Adams, P.B. "*Teratosperma oligocladum*, a new hyphomycetous mycoparasite on sclerotia of *Sclerotinia sclerotium*, *S. trifoliorum*, and *S. minor*." *Mycotaxon* **1980**, *10*, 421-427.

Uecker, F.,A., Ayers, W.A., Adams, P.B. "*Laterispora brevirama*, A new hypomycete on Sclerotia of *Sclerotinia minor*." *Mycotaxon* **1982**, *14*, 491-496.

Managing Soilborne Plant Pathogens with Fungal Antagonists

Robert D. Lumsden, Jack A. Lewis and James C. Locke, U. S. Department of Agriculture, Agricultural Research Service, Beltsville Agricultural Research Center, Beltsville, MD 20705

Agricultural use of commercially available biocontrol products to manage soilborne plant pathogens is an emerging technology. Research, since the 1930's has demonstrated the potential for fungi to parasitize plant pathogens, form antibiotic metabolites, or compete spatially and nutritionally for ecological niches. Factors affecting discovery, efficacy, formulation, product acceptance, and registration, especially with regard to the first product developed in the U.S., "GlioGard™", are indicated. Problems of practical application in the past include variability and inconsistency in product performance. These difficulties most likely can be overcome by proper strain selection, genetic improvement, and improved formulation and delivery systems for biocontrol fungi.

The first definitive indications that saprophytic soil fungi (*Trichoderma* spp., *Gliocladium* spp.) were responsible for the inhibition and destruction of plant pathogenic fungi (*Rhizoctonia solani*, *Pythium* spp.) were made by Weindling in the early 1930's (Cook and Baker, 1983). These studies remained of academic interest until the publication of Rachel Carson's Silent Spring in 1962 in which the dangers of pesticide application were brought to public notice. Shortly thereafter, plant pathologists focused on approaches to achieve biological control with reduced dependence on synthetic fungicides: 1) soil management practices (use of organic amendments, crop rotation, and sublethal chemical and physical treatments), and 2) mass introduction of biological agents to soils and seeds (Papavizas, 1981).

Current Status

This chapter addresses the mass introduction of biocontrol agents in agricultural applications for which the application technology is more amenable to commercialization than is the soil management approach, and conditions are more easily defined. In addition, only research involving field application since the last BARC Symposium on biocontrol in 1980 (Papavizas, 1981) will be presented. Since that time, several general reviews on the subject have appeared, including those by Cook and Baker (1983), Deacon (1988), Fravel and Keinath (1991), Howell (1990), Lumsden and Lewis (1987), and Nelson (1991). *Trichoderma* and *Gliocladium* species are the most thoroughly and widely studied biocontrol fungi (Papavizas, 1985). They are easy to isolate and identify, they grow rapidly, colonize treated soils readily, and produce a wide range of lytic enzymes and inhibitory metabolites.

In spite of the many years of research with several biocontrol fungi, there are only a few fungal formulations which have been commercialized. Of these, only three (BINAB-T™, GlioGard™, and *T. harzianum* AG2) are registered for use in the United States. BINAB-T™ (Binab Bio-Innovation AB, Sweden) is a mixture of *Trichoderma harzianum* and *Trichoderma polysporum* registered for use as a mycofungicide to control tree wound decay and wood rot (Ricard, 1981). Specifically, it is claimed to control *Heterobasidion annosum* and *Chondrostereum purpureum* in wounds of ornamentals, shade trees, and forest trees. GlioGard™ (W. R. Grace & Co.-Conn.) was the first registered fungal formulation developed in the United States. It is a solid prill composed of fermentor-produced biomass of isolate Gl-21 of *G. virens*, powdered wheat bran, and sodium alginate suspension which is gelled, dried, and applied to greenhouse horticultural soilless growing media to prevent damping-off of seedlings caused by species of *Pythium* and

Rhizoctonia (Lewis and Papavizas, 1985; Lumsden and Locke, 1989). It is marketed by Grace-Sierra Horticultural Products in the U.S. A formulation of *T. harzianum* AG2 (Cornell Research Foundation), a biotype prepared by protoplast fusion, protects against a wide range of soilborne plant pathogenic fungi on several crops. The fungus is delivered as a seed treatment, broadcast application to soil, or aerial application (Harman, 1990; and *personal communication*).

Other fungal antagonists are currently formulated commercially in other parts of the world. *Pythium oligandrum* for the suppression of damping-off diseases caused by *P. ultimum* is produced in the Czech Republic (Vesely, 1987) and is called 'Polygandron'. A suspension of mycelia and oospores is coated onto sugar beet seed before sowing. Alternatively, *P. oligandrum* is grown on vermiculite/cornmeal and pelleted onto cress and sugar beet seed (Mc Quilken et al., 1990). PG-Suspension (Ecological Labs, England) consists of dehydrated tablets containing spores of *Phlebia gigantea*, which are rehydrated in water, and poured or painted over tree stumps to prevent the spread of root rot and heart decay caused by the pathogenic basidiomycete, *Heterobasidion annosum* (Rishbeth, 1975). *Coniothyrium minitans*, a mycoparasite on sclerotia of pathogens such as *Sclerotinia* and *Sclerotium* is used to produce Micon and Coniothyrin, in Russia and Hungary, respectively. These formulations reduce damping-off diseases in greenhouses and stalk rot and wilt of sunflowers caused by *Sclerotinia sclerotiorum*. In a 3-year field study in Canada, seed furrow application of *C. minitans* reduced wilt of sunflower and increased yield (Huang, 1981). In Russia, a biological fungicide, Trichodermin-3, is produced by growing *T. lignorum* on sterilized turf for use against *R. solani*, *Fusarium* spp., and *Colletotrichum* spp. which cause root and hypocotyl rot of cucumbers and tomatoes (Lipa 1985).

Since 1980, several fungal antagonists have shown effectiveness in field applications or pilot tests. This suggests that commercialization may be imminent if industry is willing to invest in the development of products based on liquid or semi-solid fermentation and applied to soil broadcast, in furrow, or as a seed treatment.

One of the most likely candidates is *Sporidesmium sclerotivorum*, effective against lettuce drop caused by *Sclerotinia sclerotiorum* in field applications (Adams, 1990 and chapter in this volume). The product consists of the antagonist grown on nutrient-supplemented vermiculite which is broadcast then rototilled into infested soils. Other interesting and novel approaches to the prevention or reduction of plant diseases include the use of non-pathogenic isolates of pathogenic species of fungi. Fungal antagonists used in field studies which have potential for development are indicated in Table 1.

Factors Considered for Commercialization

Commercialization of a new product requires novel approaches to marketing. Successful completion of the scientific stages of product development, including discovery, demonstration of efficacy, formulation, and product stability, is one part of the process. Economic considerations such as production costs, market size and potential value, consumer acceptance, and producer marketing philosophy are critical to develop appropriate marketing strategies. Regulation of biocontrol agents, in which product safety, environmental impact, and product protection must be addressed, also interjects new requirements for product commercialization. Improved efficiencies at any stage in the process will improve chances of success.

In the development of GlioGard™, problems were anticipated and addressed for bringing a new biological control product into the marketplace. Before initiating the discovery phase of research, specific criteria were developed: 1) the target pathogens (*Pythium ultimum* and *Rhizoctonia solani*) were chosen for their short-term, damping-off disease activity and their importance in seedling production, 2) indigenous candidate organisms, including various genera of fungi, actinomycetes, and bacteria were chosen for testing, 3) a single agent was selected rather than a mixture of agents, 4) a cropping system was used where environmental variables such as pH, moisture, and temperature were relatively consistent and the growing medium was relatively uniform, and

Table 1. Biological control fungi used in the field with potential for formulation

Biological control	Pathogen and crop	Reference
Epicoccum purpurescens	*Sclerotinia sclerotiorum* on beans (white mold)	Zhou and Reeleder (1989)
Fusarium oxysporum	*F. oxysporum* on various crops (wilt)	Alabouvette et al. (1987)
Gliocladium virens	*Rhizoctonia solani* *Pythium ultimum* on cotton (damping-off)	Howell (1982, 1991)
Penicillium oxalicum	*P. ultimum* on chickpea (damping-off)	Kaiser and Hannan (1990)
Rhizoctonia spp.	*R. solani* on sugar beets (crown and root rot)	Herr (1987)
Sporidesmium sclerotivorum	*S. sclerotiorum* on lettuce (lettuce drop)	Adams (1990)
Talaromyces flavus	*Verticillium dahliae* on eggplant (wilt)	Marois et al. (1982)
Trichoderma spp.	*R. solani, Sclerotium rolfsii* on various crops (damping-off, root rot, blight)	Chet (1990), Harman (1990)
	S. cepivorum on onion (white rot)	Abd-El Moity and Shatla (1981)
	P. ultimum on peas (damping-off)	Harman (1990)
	R. solani on potatoes (black scurf, stem canker)	Beagle-Ristaino and Papavizas (1985)
Verticillium biguttatum	*R. solani* on potatoes (black scurf, stem canker)	Jeger and Velvus (1986)

5) a market niche was targeted that was of sufficient value and size to interest commercial development.

Successful development of GlioGard™ was largely the result of a close working relationship between the USDA's Beltsville Agricultural Research Center (BARC), W. R. Grace's Washington Research Center (WRC), and Grace's Horticultural Products Division. The close geographical proximity of the BARC to the WRC was a key component of the agreement. Although the initial screening, fermentation, formulation, and efficacy studies were conducted at BARC, the engineering aspects of fermentation and formulation scale-up were carried out at WRC. The on-going efficacy and quality control bioassays were accomplished at BARC, and formulation refinement and product stability (shelf life) studies were done at WRC. This cooperation, was facilitated by the Technology Transfer Act of 1986, and was administered through a Cooperative Research and Development Agreement (CRADA) in which scientists at all locations were able to exchange ideas and information during the development process. The W. R. Grace & Co., through its Horticultural Products Division, contributed market analysis and commercialization expertise. The test marketing of GlioGard™ was ultimately achieved by Grace-Sierra Horticultural Products Co., an organization resulting from the merger of Grace's Horticultural Products Division and the Sierra Chemical Co. of Milpitas, California. Product uses, including range of pathogens controlled, crop species safety, and compatibility with established horticultural practices such as pesticide and growth regulator usage, were investigated by all three cooperating groups at locations throughout the USA. The development process included the following:

A) *Discovery/screening* - Naturally occurring, indigenous microorganisms with previously identified biocontrol potential were screened by a simple, easily repeatable zinnia damping-off bioassay with both *Pythium ultimum* and *Rhizoctonia solani* (Lumsden and Locke, 1989). *Gliocladium virens* isolates showed the best activity against these pathogens.

B) *Fermentation* - Efficient production of chlamydospores by *Gliocladium virens* was achieved in liquid culture (Lewis and Papavizas, 1983) using inexpensive molasses and brewer's yeast in deep tank fermentation. Modification of the fermentation medium and process resulted in increased shelf-life of the formulated biomass.

C) *Formulation* - Technology protected by three USDA patents on formulation of biocontrol agent biomass (Fravel et al., 1985, Lewis and Papavizas, 1985) provided a readily applied product which was compatible with current horticultural production systems, tested under simulated growers' conditions.

D) *Efficacy* - Evaluation of the granulated *G. virens* product at various locations, utilizing several bedding plant crops (Lumsden et al. 1990), demonstrated product efficacy under a range of environmental conditions. Evaluations in the presence of commonly used soil plant protection products showed no reduction in efficacy (Locke and Lumsden, 1989).

E) *Test Marketing* - Initial sales to commercial growers in four states provided an encouraging response on product performance and indicated no problems related to use of the product with a variety of crops. Grower acceptance of this microbial product approach to control damping-off pathogens was established (Knauss, 1992).

F) *Registration* - The U.S. Environmental Protection Agency (EPA) requires registration of materials intended for control of plant diseases which includes a multitiered system for evaluating toxicological effects on mammals and ecological effects on wildlife. Origin, biological characteristics and use patterns determine the extent of enforcement of these requirements and allowable exemptions. Final approval limited usage of GlioGard™ to greenhouse production of crop seedlings, with certain safety considerations appended.

Future Approaches to Performance Improvement

Variability in performance of biocontrol agents in the greenhouse and field can be a major source of inconsistency in practical applications (Lumsden and Lewis, 1987). Attempts to overcome this obstacle include optimal strain selection, improved formulation and delivery systems, and genetic improvement. Genetic improvement by mutagenic methods, protoplast fusions, and molecular transformations is the topic of a separate chapter in this book (see chapter by Harman), and therefore, will not be covered here.

Improved performance by nongenetic means may be ineffective with a blind, empirical approach based on random isolation of strains from nature. Selection of the best strains, based on accumulated scientific knowledge of important antagonist characteristics and behavioral attributes offers the best approach.

Elucidating the mechanism of action of biocontrol agents against plant pathogens and studying their interactions *in situ* will provide clues for developing means to improve performance. One or more mechanisms of action may be responsible for the control responses observed. The most commonly associated disease control mechanisms are 1) mycoparasitism; 2) competition for space and nutrients; 3) production of antibiotics, metabolites or lytic enzymes; and 4) induction of host plant resistance. These systems may influence pathogen survival in soil, ability to colonize host plants and organic substrates, and potential to germinate and proliferate in soil (Cook and Baker, 1983; Papavizas and Lumsden, 1980).

This chapter will not review the large number of contributions made to developing an understanding of mechanisms of biocontrol. Instead, emphasis will be placed on strain Gl-21 recently commercialized as GlioGard™.

Mycoparasitism has little or no role in antagonism of *Pythium ultimum* by *G. virens* (Howell, 1987) and a minor role in control of *R. solani* (Howell, 1982). We have recently shown that *Gliocladium virens* did not directly attack *P. ultimum*, but it caused lysis of hyphal tips of *R. solani* in a nonsterile potting soil system, and it subsequently colonized and sporulated on the remains of *R. solani* mycelia (Harris and Lumsden, unpublished). This suggests a necrotrophic type of mycoparasitism (Lumsden, 1992), in contrast to active mycoparasitism as demonstrated by *Sporidesmium sclerotivorum* (Adams, 1990).

Competition for space and nutrient resources between *G. virens* and plant pathogens in the soil is indirect. Saprophytic colonization of organic substrates by *G. virens* results in production of microbially inhibitory metabolites such as the epipolythiodiketopiperazine

antibiotic, gliotoxin. This compound has been isolated from soil and soilless media containing GlioGard™ (Lumsden et al., 1992a). Gliotoxin is produced by *G. virens* isolate GL-21 along with other metabolites in soil preparations (Lumsden et al., 1992b). Another related antibiotic, gliovirin, is produced by other strains of *G. virens* (Howell and Stipanovic, 1983). Gliotoxin is active against *P. ultimum* and *R. solani*, with resultant inhibition of germination of propagules of these fungi at concentrations as low as 1.0 μg/ml (1 ppm). Although gliotoxin may not be the only antibiotic effective in the antagonism, additional evidence strongly suggests its implication. *G. virens* causes leakage from mycelium of *R. solani*, which is at least partially caused by gliotoxin (Lewis et al., 1991), and it affects sporangial germination and mycelial growth of *P. ultimum* (Lumsden et al., 1992b; Roberts and Lumsden, 1990).

Gliotoxin is produced most abundantly early in the growth cycle of *G. virens*, with a peak of activity within four days of initiation of mycelial growth. It remains active for two weeks or more, but it is converted to a biologically inactive derivative, dimethylgliotoxin, after peak production (Lumsden et al., 1992b and unpublished). We are attempting to manipulate the production of gliotoxin to enhance the amount and duration of its existence in the plant rhizosphere and to prolong activity. Prevention of conversion of gliotoxin to dimethylgliotoxin would also be advantageous. In preliminary experiments, we have almost doubled gliotoxin production and simultaneously reduced its conversion to dimethylgliotoxin several-fold by formulaton of *G. virens* in alginate prills containing an inorganic source of nitrogen (unpublished results).

In addition to enhancing activity by manipulation of the environment, it may also be possible to select strains of *G. virens* that produce more gliotoxin. Biocontrol by several strains of *G. virens* against damping-off of seedlings by *R. solani* varied from poor to excellent (Lumsden and Locke, 1989). The production of gliotoxin by these same strains also was varied from strain to strain (unpublished results). The ability to genetically alter *G. virens* to produce more gliotoxin than present strains through gene deregulation, or prevent production of undesireable metabolites by blocking gene activity is an attractive approach. We have isolated proteins from gliotoxin-producing strains of *G. virens* that are different from proteins from non-producing strains (Ridout and Lumsden, 1992). Antibodies to some of these proteins have been produced and will be used to determine the function of the proteins by probing genomic libraries of *G. virens* for eventual identification of genes controlling production or action of gliotoxin.

Induced resistance, another possible mechanism of antagonism, has not been associated with *G. virens*. This mechanism is most notably associated with saprophytic *Fusarium* spp., the topic of another chapter in this volume dealing with naturally-occurring disease suppressive soils (see Alabouvette).

Conclusions

Fungal biocontrol products are proving effective for control of several soilborne plant pathogens. Fungi produced in liquid fermentation systems allow formulation of the biomass into delivery systems that provide ease of application, prolonged shelf life, and effective biocontrol activities. Successful results are especially evident in management of seed and seedling diseases in greenhouse production of bedding plants. Partnerships between private and public research laboratories and private industry will make available more biocontrol systems for future agricultural implementation.

A successful partnership between the U.S. Department of Agriculture, Agricultural Research Service, the W. R. Grace & Co.-Conn. and Grace/Sierra Horticultural Products has resulted in the introduction of the first fungal biocontrol formulation developed in the U.S., GlioGard™, for use against damping-off diseases in greenhouse production of seedling plants. Future use of *G. virens* against pathogens of field grown crops most likely will require improved efficacy through better understanding of the mechanisms of disease control, selection of improved strains of biocontrol fungi, and improvements in formulations. If current trends continue, fungal biocontrol agents could, in large measure, replace or reduce the use of synthetic chemical fungicides and perhaps even perform disease control in environments or situations

where chemicals have not been effective in the past. At the very least, we expect biocontrol fungi to be available for control of specific plant pathogens in important "niche" markets in horticultural production systems.

References

Abd-El-Moity, T.H., Shatla, M.N. "Biological control of white rot disease of onion (*Sclerotium cepivorum*) by *Trichoderma harzianum*." *Phytopathol. Z.* **1981**. *100*, 29-35.

Adams, P.B. "The potential of mycoparasites for biological control of plant diseases." *Annu. Rev. Phytopathol.* **1990**, *28*, 59-72.

Alabouvette, C., de la Broise, D, Lemanceau, P., Couteaudier, Y., Louvet, J. "Utilisation de souches non pathogenes de *Fusarium* pour lutter contre les fusarioses." *Bulletin OEPP/EPPO* **1987**, *17*, 665-674.

Beagle-Ristaino, J.E., Papavizas, G.C. "Biological control of Rhizoctonia stem canker and black scurf of potato." *Phytopathology* **1985**, *75*, 560-564.

Carson, R. *Silent Spring*, Houghton Mifflin, Boston, MA, 1962.

Chet, I. "Biological control of soilborne plant pathogens with fungal antagonists." In *Biological Control of Soilborne Plant Pathogens*; Hornby, D., Ed.; CAB International, Oxon, UK **1990**, pp. 15-26.

Cook, R.J., Baker, K.F. *The Nature and Practice of Biological Control of Plant Pathogens*, American Phytopathological Society, St. Paul, MN, 1983.

Deacon, J.W. "Biocontrol of soil-borne plant pathogens with introduced inocula." *Phil. Trans. R. Soc. London B* **1988**, *318*,249-264.

Fravel, D.R., Keinath, A.P. "Biocontrol of soil borne plant pathogens with fungi." In *The Rhizosphere and Plant Growth*; Keister, D.L. and Cregan, P.B. Eds.; Kluver Academic Publisher, Dortrecht, Netherlands, 1991, pp. 237-243.

Fravel, D.R., Marois, J.J., Lumsden, R.D., Connick, W.J., Jr. "Encapsulation of potential biocontrol agents in an alginate-clay matrix." *Phytopathology* **1985**, *75*, 774-777.

Harman, G.E. "Development tactics for biocontrol fungi in plant pathology." In *New*

Directions in Biological Control; Baker, R.R. and Dunn, P.E., Eds.; A.R. Liss, New York, NY, 1990, pp. 779-792.

Herr, L. "Biocontrol of Rhizoctonia crown and root rot of sugar beet by binucleate *Rhizoctonia* spp. and *Laetisaria arvalis*." *Ann. Appl. Biol.* **1988**, *113*, 107-118.

Howell, C.R. "Effect of *Gliocladium virens* on *Pythium ultimum*, *Rhizoctonia solani*, and damping-off of cotton seedlings." *Phytopathology* **1982**, *72*, 496-498.

Howell, C.R. "Relevance of mycoparasitism in the biological control of *Rhizoctonia solani* by *Gliocladium virens*." *Phytopathology* **1987**, *77*, 992-994.

Howell, C.R "Fungi as biological control agents." In *Biotechnology of Plant-Microbe Interactions*; Nakas, J.P. and Hagedorn, C. Ed; McGraw-Hill Pub. Co. New York, 1990 pp. 257-279.

Howell, C.R. "Biological control of Pythium damping-off of cotton with seed-coating preparations of *Gliocladium virens*." *Phytopathology* **1991**, *81*, 738-741.

Howell, C.R., Stipanovic, R.D. "Gliovirin, a new antibiotic from *Gliocladium virens*, and its role in the biological control of *Pythium ultimum*." *Can. J. Microbiol.* **1983**, *29*, 321-324.

Huang, H.C. "Distribution of *Coniothyrium minitans* in Manitoba sunflower fields." *Can. J. Plant Pathol.* **1981**, *3*, 219-222.

Jeger, G., Velvus, H. "Biological control of *Rhizoctonia solani* on potatoes by antagonists." *Netherlands J. Plant Pathol.* **1986**, *92*, 231-238.

Kaiser, W.J., Hannon, R.M. "Biological control of seed rot and preemergence damping-off of chickpea with *Penicillium oxalicum*." *Plant Dis.* **1984**, *68*, 806-811.

Knauss, J.F. "*Gliocladium virens*, a new microbial for control of *Pythium* and *Rhizoctonia*." *Florida Foliage* **1992**, *18*, 6-7.

Lewis, J.A., Papavizas, G.C. "Production of chlamydospores and conidia by *Trichoderma* spp. in liquid and solid growth media." *Soil Biol. Biochem.* **1983**, *15*, 351-357.

Lewis, J.A., Papavizas, G.C. "Characteristics of alginate pellets formulated with *Trichoderma* and *Gliocladium* and their effect on the proliferation of the fungi in soil." *Plant Pathol.* **1985**, *34*, 571-577.

Lewis, J.A., Papavizas, G.C. "Application of *Trichoderma* and *Gliocladium* in alginate pellets for control of Rhizoctonia damping-off." *Plant Pathol.* **1987**, *36*, 438-446.

Lewis, J.A., Roberts, D.P., Hollenbeck, M.D. "Induction of cytoplasmic leakage from *Rhizoctonia solani* hyphae by *Gliocladium virens* and partial characterization of a leakage factor." *Biocontr. Sci. Technol.* **1991**, *1*, 21-29.

Lipa, J.J. "History of biological control in protected culture in Eastern Europe." In *Biological Pest Control: The Glasshouse Experience*; Hussey, N.W. and Scopes, N., Eds.; Blandford Press, Pools, UK, 1985, pp. 23-29.

Locke, J.C., Lumsden, R.D. "Compatibility of some commonly used soil drench fungicides and insecticides with the biocontrol agent *Gliocladium virens*." *Phytopathology* **1989**, *79*, 1152.

Lumsden, R.D. "Mycoparasitism of soilborne plant pathogens." In *The Fungal Community*: Carroll, G.C. and Wicklow, D.T., Eds.; Marcel Dekker, Inc., New York, NY, 1992, pp. 275-293.

Lumsden, R.D., Lewis, J.A. "Problems and progress in the selection, production, formulation and commercial use of plant disease control fungi." In *Biotechnology of Fungi for Improving Plant Growth*, Whipps, J.M. and Lumsden, R.D. Eds.; Cambridge Univ. Press, Cambridge, UK, 1987, pp. 171-190.

Lumsden, R.D., Locke, J.C. "Biological control of damping-off caused by *Pythium ultimum* and *Rhizoctonia solani* in soilless mix." *Phytopathology* **1989**, *79*, 361-366.

Lumsden, R.D., Locke, J.C., Adkins, S.T., Walter, J.F., Ridout, C.J. "Isolation and localization of the antibiotic gliotoxin produced by *Gliocladium virens* from alginate prill in soil and soilless media." *Phytopathology* **1992a**, *82*, 230-235.

Lumsden, R.D., Locke, J.C., Lewis, J.A., Johnston, S.A., Peterson, J.L., Ristaino, J.B. "Evaluation of *Gliocladium virens* for biocontrol of Pythium and Rhizoctonia damping-off of bedding plants at four greenhouse locations." *Biol. Cult. Control Tests* **1990**, *5*, 90.

Lumsden, R.D., Ridout, C.J., Vendemia, M.E.,

Harrison, D.J., Waters, R.M., Walter, J.F. "Characterization of major secondary metabolites produced in soilless mix by a formulated strain of the biocontrol fungus *Gliocladium virens.*" *Can. J. Microbiol.* **1992b**, 1274-1280.

Marois, J.J., Johnston, S.A., Dunn, M.T. and Papavizas, G.C. "Biological control of Verticillium wilt of eggplant in the field." *Plant Dis.* **1982**, *66*, 1166-1168.

Mc Quilken, M.P., Whipps, J.M., Cooke, R.C. "Control of damping-off in cress and sugar beet by commercial seed-coating with *Pythium oligandrum.*" *Plant Pathol.* **1990**, *39*, 452-462.

Nelson, E.B. "Current limits to biological control of fungal phytopathogens." In *Handbook of Applied Mycology*; Arora, D.K. Ed.; Marcel Dekker, New York, NY, **1991**, pp. 327-356.

Papavizas, G.C. *Biological Control in Crop Protection*, BARC Symposium, Allanheld, Osmun and Co., Totowa, NJ, 1981.

Papavizas, G.C. "*Trichoderma* and *Gliocladium*: Biology, ecology, and potential for biocontrol." *Ann. Rev. Phytopathol.* **1985**, *23*, 23-54.

Papavizas, G.C., Lumsden, R.D. "Biological control of soilborne fungal propagules." *Annu. Rev. Phytopathol.* **1980**, *18*, 389-413.

Ricard, J.L. "Commercialization of a *Trichoderma*-based mycofungicide - Some problems and solutions." *Biocont. News Inf.* **1981**, *2*, 95-98.

Ridout, C.J., Lumsden, R.D., Hruschka, W.R. "Identification of mycelial polypeptides associated with gliotoxin-producing strains of the biocontrol fungus *Gliocladium virens.*" *Phytopathology* **1992**, *82*, 479-484.

Rishbeth, J. "Stump inoculation: A biological control of *Fomes annosus.*" In *Biology and Control of Soil-Borne Plant Pathogens*; Bruehl, G.W., Ed.; American Phytopathological Society, St. Paul, MN, 1975, pp. 158-162.

Roberts, D.P., Lumsden, R.D. "Effect of extracellular metabolites from *Gliocladium virens* on germination of sporangia and mycelial growth of *Pythium ultimum.*" *Phytopathology* **1990**, *80*, 461-465.

Vesely', D. "Germinating power of oospores of *Pythium oligandrum* in a powder preparation." *Folia Microbiol.* **1987**, *32*, 502-506.

Zhou, T., Reeleder, R.D. "Application of *Epicoccum purpurascens* spores to control white mold of snap bean." *Plant Dis.* **1989**, *73*, 639-642.

Naturally Occurring Disease-Suppressive Soils

Claude Alabouvette, I.N.R.A., Laboratoire de Recherches sur la Flore Pathogène du Sol, BV 1540, F 21034 DIJON

The incidence of diseases due to soilborne plant pathogens can be limited by different mechanisms naturally occurring in suppressive soils. Suppressiveness is fundamentally based on microbial interactions between the pathogen and all or a part of the saprophytic microflora. Studies of soils suppressive to fusarium wilt have shown the involvement of populations of fluorescent *Pseudomonas* spp. and non-pathogenic *Fusarium oxysporum*. Some of these microorganisms are being developed as biocontrol agents. Another approach toward biological control is the enhancement of natural suppressiveness that exists in every soil.

Diseases caused by soilborne plant pathogens are among the most difficult to control and they are responsible for severe losses in many crops of economical importance around the world. Because of the importance of soilborne plant pathogens, many attempts have been made to utilize naturally occurring suppressive soils to control these diseases.

In nature, suppressive soils can be detected by the observation that disease incidence remains low despite the presence of a susceptible host plant, climatic conditions favourable to disease expression and ample opportunity for the pathogen to have been introduced. The hypothesis that such a soil is suppressive to the disease is easy to test in greenhouse experiments. The pathogen is artificially produced and introduced into the soil, and disease incidence on a susceptible host is compared to that produced by the same inoculum in a conducive control soil. All experimental conditions being comparable, differences in disease incidence have to be attributed to differences in the soil environment.

Disease suppression may be due to a direct effect of the soil on the pathogen or to an indirect effect through the host plant. Therefore, Cook and Baker (1983) distinguished pathogen suppressive-soils from disease-suppressive soils. Only studies on the mechanism of suppressiveness enable the distinction between the two types of suppressive soils.

Although variations of disease incidence related to soil types have been recognized for many years (Stover, 1962), it is only during the last thirty years that most of the mechanisms involved in disease suppression have been discovered (Cook, 1990). Progress made in the understanding of microbial interactions responsible for disease suppression may lead to the development of biological control procedures for commercial applications.

Because many recent review papers on disease-suppressive soils and biological control are available (Hornby, 1990 ; Whipps, 1992), this paper will summarize the most important facts dealing with fusarium wilt-suppressive soils as an example of the different approaches followed to develop biological control procedures.

Basic Knowledge on Disease-Suppressive Soils

Soils suppressive to some of the most important soilborne pathogens have been described, showing that soil suppressiveness to diseases is not an exceptional phenomenon (Cook and Baker, 1983 ; Schneider, 1982 ; Schippers, 1992). Three examples of suppressive soils that have been studied extensively are : the take-all disease of wheat caused by *Gaeumannomyces graminis* var. *tritici*, fusarium wilts due to different *formae speciales* of *Fusarium oxysporum* and root rot of tobacco caused by *Thielaviopsis basicola* (Thomashow and Weller, 1990 ; Alabouvette, 1990 ; Defago et al., 1990). These few examples show a great diversity of mechanisms responsible for disease suppression. It is therefore difficult to make general statements about disease-suppressive soils. However most of the suppressive soils have several properties in common demonstrating that suppressiveness is fundamentally microbial in nature. Disease suppression results from more or less complex microbial interactions between the pathogen and all or a part of the saprophytic microflora. Indeed, the suppressive effect disappears upon destruction of organisms by biocidal treatments and can be restored by mixing a small quantity of suppressive soil into a previously heat-treated soil (Louvet et al., 1976 ; Scher and Baker, 1980 ; Shipton et al., 1973).

Properties that account for the microbiological nature of disease suppression do not demonstrate that soil physio-chemical properties play any role in the mechanisms of suppressiveness. On the contrary, early studies on fusarium wilt suppressive soils established correlations between soil type, presence of smectite clays and soil suppressiveness to fusarium wilts in Central America (Stover, 1962 ; Stotzky and Martin, 1963). But, the most documented example of the role of abiotic factors is that of Swiss soils suppressive to black root rot of tabacco ; Stutz et al. (1989) established that only soils derived from moraine and containing vermiculitic clay minerals are suppressive to *Thielaviopsis basicola*. Many other abiotic factors such as soil texture, water potential, aeration, pH, organic matter content, cation availability (Al, Fe, Mn) are indirectly involved in the mechanisms of disease suppression, but it is difficult to generalize the findings from one soil to another.

Finally, cultural practices, such as monoculture as opposed to crop rotation, may play an important role in the establishment of soil suppressiveness to soilborne plant pathogens. Several examples demonstrate that monoculture of a susceptible host plant is responsible for the build-up of soil suppressiveness (Baker and Chet, 1982), and the well known phenomenon of take-all decline is the best example of "acquired" suppressiveness. Conversely, suppressiveness to fusarium wilts appears stable with time and has been described as constitutive but it may be not always independant from the cropping history of the soils. Sneh et al. (1987) reported an observation that successive plantings of susceptible muskmelons induced suppressiveness of a soil in Israël. In fact, involvement of abiotic factors or cultural practices is difficult to demonstrate and most of the recent progress concerns the role of microbial populations inhibiting the growth and activity of the pathogen in suppressive soils.

Mechanisms and Microorganisms Responsible for Suppression of Fusarium Wilts in Naturally Occurring Suppressive Soils

Two complementary approaches have been followed to detect and prove the involvement of specific mechanisms and particular microorganisms in the suppression of fusarium wilts in suppressive soils. First, manipulation of the nutrient status of the soil enabled the demonstration of the role of competition for nutrients in the mechanisms of suppression. Both carbon and iron have been proved to be limiting nutrients. Scher and Baker (1982) modified iron availability in the suppressive soil from the Salinas Valley and demonstrated that the level of suppressiveness was either decreased or increased by addition of FeEDTA or FeEDDHA that respectively make iron more or less available for *F. oxysporum* growth. The same results were obtained by Lemanceau et al. (1988) who also demonstrated that addition of glucose increased disease incidence in the suppressive soil from Châteaurenard. In that case, competition for carbon was considered to be most important because addition of glucose resulted in greater disease incidence when the amount of available iron was reduced by addition of EDDHA. Secondly, manipulation of the microbial balance of suppressive soils enabled the demonstration of the role of specific populations of microorganisms antagonistic to the pathogen. Rouxel et al. (1979) established that suppressiveness of the soil from Châteaurenard disappeared after elimination of the populations of nonpathogenic *Fusarium* by heat-treatment and reappeared after their reintroduction into the heat-treated soil. Similarly Scher and Baker (1980) isolated strains of fluorescent *Pseudomonas* spp. from a suppressive soil from the Salinas Valley and demonstrated that introduction of these bacteria into a conducive soil made it suppressive. Other antagonistic microorganisms have been proposed as having a role in the suppression of fusarium wilts (Alabouvette, 1990) but only populations of nonpathogenic *F. oxysporum* and fluorescent *Pseudomonas* spp. have been repeatedly proved to be involved in the mechanisms of suppression of fusarium wilts, in naturally occuring disease-suppressive soils (Alabouvette, 1990 ; Schippers, 1992).

These two experimental approaches have provided good evidence that competition for nutrients on one hand, and populations of fluorescent *Pseudomonas* spp. and nonpathogenic *F. oxysporum* on the other hand are involved in the mechanisms of suppression. However, these approaches do not demonstrate that the specific antagonistic populations associated with suppression are responsible for the competitive effects demonstrated in the suppressive soils. Kloepper et al. (1980) demonstrated that competition for iron resulting from the production of siderophores by fluorescent *Pseudomonas* spp. could be a mechanism of disease suppressiveness. The use of mutant strains of *Pseudomonas* spp. which have lost the capacity to chelate iron, inhibit the growth of *Fusarium* spp. in vitro, and induce suppressiveness in conducive soils, led to the conclusion that production of siderophores by fluorescent *Pseudomonas* spp. is responsible, at least in part, for competition for iron that is more intense in suppressive than in conducive soils (Schippers, 1992)

Since the soil microbial biomass in a suppressive soil was more responsive to glucose amendment than the biomass of a conducive soil, competition for carbon was attributed to the activity of the total soil biomass (Alabouvette et al., 1985). Subsequently, it was also established that pathogenic and nonpathogenic *F. oxysporum* are competing for carbon in soil (Couteaudier, 1992). Indeed, the most effective strains of nonpathogenic *F. oxysporum* in biocontrol experiments are those which are the most efficient in glucose consumption (Alabouvette and Couteaudier, 1992).

Recently, Lemanceau et al. (1993) demonstrated that competition for carbon and competition for iron are not independant from each other. The intensity of the intraspecific competition for carbon between pathogenic and nonpathogenic *F. oxysporum* depends on iron availability controlled by the activity of siderophore producing *Pseudomonas* spp. In vitro, the pathogenic *F. oxysporum* f.sp. *dianthi* was more sensitive to competition for carbon than the nonpathogenic strain Fo47, and its growth yield (defined as the germ-tube length per unit of glucose consumed) was reduced when the availability of iron was decreased by addition of the pseudobactin produced by the strain WCS358 of *Pseudomonas putida*. These results provide a good explanation for the synergistic effect of the coinoculation of nonpathogenic *F. oxysporum* with fluorescent *Pseudomonas* spp. to control fusarium-wilt incidence of several crops (Lemanceau and Alabouvette, 1991 ; Lemanceau et al., 1992).

Interaction of competition for carbon and iron involving populations of fluorescent *Pseudomonas* spp. and nonpathogenic *F. oxysporum*, show the great complexity of mechanisms responsible for natural suppressiveness of soils to fusarium wilts. Moreover, other mechanisms involving the host-plant, such as systemic induced resistance, may also play a role as suggested recently by Van Peer et al. (1991) and Mandeel and Baker (1991). It is evident that stable suppressiveness such as suppressiveness of the soil from the Salinas Valley or Châteaurenard is based on a cooperation among several microorganisms and mechanisms (Schippers, 1992). Although our understanding of the mechanisms of soil suppressiveness is still incomplete, present knowledge allows applications towards biological control of soilborne plant pathogens.

Applications to Biological Control of Soilborne Plant Pathogens

Two types of strategies can be used to utilize suppressive soils to control soilborne plant pathogens. One strategy is based on enhancement of the level of suppressiveness that exists in any soil. The other strategy consists of the selection of efficient strains of antagonistic microorganisms isolated from suppressive soils and their utilization as biocontrol agents.

1. Enhancement of the natural level of suppressiveness.
Every soil has a potential for disease suppression. Soil receptiveness to disease may be considered as a continuum from strongly suppressive to highly conducive (Alabouvette et al., 1982, Linderman et al., 1983).

Mechanisms of competition for nutrients, populations of nonpathogenic *Fusarium oxysporum* and populations of fluorescent *Pseudomonas* spp. exist in every soil. However depending on the level and activity of the populations of antagonists and on the intensity of competition, the same types of interactions will result in either conduciveness or suppressiveness of the soil to fusarium wilts. Biocontrol could be achieved by managing the soil environment in order to enhance populations and mechanisms responsible for suppression.

Most attempts to control fusarium wilts by manipulations of the abiotic environment have failed (Alabouvette, 1989). From a theoretical point of view, iron availability is related to pH, therefore application of lime should increase the level of suppressiveness of soils to fusarium wilts. Indeed, most of the suppressive soils have a pH value greater than 7 and, in Florida, liming soils effectively reduces severity of wilts when associated with ammonium nitrogen (Jones and Woltz, 1981). In the soilless culture of tomatoes the use of FeEDDHA as the source of iron in the nutrient solution decreases fusarium wilt incidence in comparison with FeEDTA (Lemanceau and Alabouvette, unpublished).

A correlation has been established between suppressiveness and presence of smectite type clays in suppressive soils. Indeed, introduction of 25% of montmorillonite in a conducive sandy soil made it suppressive to fusarium wilt of flax (Amir and Alabouvette, 1993). Clearly, it is not possible to add clays on a pratical scale to induce suppressiveness in conducive soils.

A more realistic approach would be the choice of cultural practices favoring the enhancement of the suppressive potential of the soil. Jones et al. (1989) reviewed the roles of different macro and micro-elements in relation to fusarium wilts of vegetables and ornamentals. Fertilization sometimes has a clear effect on the reduction of disease incidence, but this generalization is not always true. Most studies involving nutrients did not allow the distinction between the direct effect of the elements on the plant physiology and an indirect effect on the level of soil suppressiveness to fusarium wilts.

The only well documented example of the influence of cultural practices on disease incidence and soil suppressiveness is that of take-all decline due to the monoculture of wheat (Cook and Veseth 1991). The crop itself, as well as weeds or cover crops, affects the suppressive level of the soil. Field observations in oil-palm groves suggested that soil covered with *Calopogonium* could be more conducive to fusarium wilts than the soil planted with *Puraria* (Abadie and Alabouvette, unpublished).

Organic amendments have been proposed to control soilborne diseases (Lumsden et al., 1983) although the effect of organic amendments have not been studied in relation to induction of suppressiveness in soil. One exception is the Ashburner system to control *Phytophthora* root rot of Avocado in Australia based on the incorporation of large amounts of organic matter to reproduce the environment of naturally suppressive soils that exist in the rain forest (Cook, 1982).

Clearly, more knowledge is needed before biological procedures based on modifications of soil factors could be proposed confidently.

2.   Application of Selected Strains of Antagonistic Microorganisms.

Most of the studies of disease suppressive soils have led to the selection of antagonistic microorganisms that are now evaluated for use as biocontrol agents.

The selection of effective strains may be based on two different approaches : i) *in vitro* tests designed to reveal a specific antagonistic activity such as the production of antibiotics or siderophores by *Pseudomonas* spp., and ii) *in vivo* tests designed to reveal the efficiency in bioassays with the pathogen and the host in the soil environment. Such a method was developed in Dijon, and a statistical analysis enables the ranking of the strains for their biocontrol activity (Corman et al., 1986). This procedure led to the selection of a strain of nonpathogenic *F. oxysporum* (Fo47) and a fluorescent *Pseudomonas* sp. (C7) used in further experiments for biological control.

Production of large amounts of antagonist inoculum is necessary for biocontrol experiments. Fortunately, the nonpathogenic Fo47 is easily cultured in fermentors. The resulting microchlamydospores produced can be mixed with talc and stored for several months at 4°C. Shelf life of this formulation would be sufficient to commercialize the product (Alabouvette and Couteaudier, 1992).

Under laboratory conditions, where inoculum of the nonpathogen can be mixed thoroughly into the entire volume of soil, strain Fo47 gave significant control of fusarium wilts in natural soils. However, this strain failed to control fusarium wilt of muskmelons in open fields. Lack of control in the field may be due to various factors including poor distribution of the antagonist throughout the soil and insufficient colonization of root tips. Subsequent experiments were conducted in soil fumigated with methylbromide or in soilless cultures, in order to allow the nonpathogenic strain the time and opportunity to colonize the free space before introduction of the pathogen.

Under laboratory conditions, where pathogen and antagonist population can be adjusted, precolonization of the rooting medium by Fo47 at 1×10^5 CFU ml^{-1} resulted in control of the disease after introduction of the pathogen at 5×10^2 CFU ml^{-1}. The antagonist reduced disease incidence and increased yield compared to the infested control (Alabouvette et al., 1993).

The association of the strain C7 of *Pseudomonas* with Fo47 gave almost total control of the crown and root rot of tomato due to *Fusarium oxysporum* f.sp. *radicis lycopersici*. The yield in the protected plots was not significantly different from the yield in the non infested control (Lemanceau and Alabouvette, 1991, Lemanceau et al., 1992).

Since the nonpathogenic strain Fo47 satisfied toxicity tests required by the French government, this antagonist may be tested in commercial crops. Strain Fo47 has been used in comparison with the fungicide hymexazol to control Fusarium crown and root rot of tomato in soilless cultures. Both treatments may give good control of this disease but the results were not consistent from one location to another. In fact, disease control depends on the inoculum pressure and on environmental factors (Jeannequin et al., 1991). Failure of the chemical treatment does not affect grower confidence in chemical control as does failure of biological control. Efficacy of strain Fo47 is currently being assessed under different conditions to control fusarium wilt of cyclamen, carnation, basil and asparagus.

Prospects

This brief review of studies dealing with

naturally occurring disease-suppressive soils emphasizes the complexity of mechanisms and the roles of several populations of antagonistic microorganisms responsible for disease suppression. Due to the cooperation among several populations of microorganisms having different modes of action, naturally occuring suppressiveness can limit disease incidence efficiently and consistently (Baker, 1991). Biological control based on the application of a single antagonistic strain is often inconsistent. Biological control may be improved by mimicking the complexity of the mechanisms operating in nature, learning how to use several populations of microorganisms together, and managing environmental factors to enhance microbial antagonism in soil. Results already obtained with the association of nonpathogenic *F. oxysporum* with fluorescent *Pseudomonas* spp. are encouraging. It is likely that registration of a product containing a mixture of microorganisms will be more difficult to obtain than registration of a product with a single strain. In fact, the reason for having conducted our experiments with Fo47 alone is that the strain C7 of fluorescent *Pseudomonas* sp. has not yet satisfied the requirements of the EEC for experimentation under commercial crop conditions. The coinoculation approach was used by Pierson and Weller (1991) to demonstrate the beneficial effect of adding several strains of fluorescent *Pseudomonas* spp. for control of take-all of wheat.

Even an efficient association of antagonists may require specific environmental conditions to be effective in controlling a disease. It will be necessary to adapt the formulation of the biological product according to the mode of application and soil environmental conditions. Several review papers have addressed all the questions related to mass production, formulation and delivery of antagonists to seeds and soil (Lewis and Papavizas, 1991 ; Harman, 1991). It is clear that the biotechnological problems are as important as the study of the mode of action of the biological control agents to assure the success of biological control.

Even when practical solutions have been found and biological control agents are available to growers, biological control of soilborne plant pathogens will probably be limited to specific crops under well defined conditions. Vegetable and flower crops under greenhouses which are often cultivated on artificial substrates offer opportunities to use biological control. The recent evolution of agriculture linked to the concept of sustainable agriculture will require more research to promote biological control based not only on the application of selected strains of antagonists but also based on the enhancement of the low

level of suppressiveness that naturally exists in every soil.

References

Alabouvette, C. "Manipulation of soil environment to create suppressiveness in soils", 457-478. In *Vascular Wilt diseases of Plants - Basic studies and control*; Tjamos E.C., Beckman C.H., Eds.; Springer Verlag: Berlin, DE, 1989, NATO ASI Series, H28.

Alabouvette, C. "Biological control of fusarium wilt pathogens in suppressive soils", 27-43. In *Biological control of soilborne plant pathogens*, Hornby D., Ed.; C.A.B. International: Wallingford, UK, 1990.

Alabouvette, C., Couteaudier, Y. "Biological control of fusarium wilts with nonpathogenic *Fusaria*", 415-426. In Biological control of plant diseases, Tjamos E.C., Papavizas G.C. and Cook R.J., Eds.; Plenum Press: New York, US, 1992.

Alabouvette, C., Couteaudier, Y., Louvet, J. "Comparaison de la réceptivité de différents sols et substrats de culture aux fusarioses vasculaires". *Agronomie* **1982**, 2, 1-6.

Alabouvette, C., Couteaudier, Y., Louvet, J. "Recherches sur la résistance des sols aux maladies. XII - Activité respiratoire dans un sol résistant et un sol sensible aux fusarioses vasculaires enrichis en glucose". *Agronomie* **1985**, 5, 69-72.

Alabouvette, C., Lemanceau, P., Steinberg, C. "Recent advances in the biological control of fusarium wilts". *Pestic. Sci.* **1993** (in press)

Amir, H., Alabouvette, C. "Involvement of soil abiotic factors in the mechanisms of soil suppressiveness to fusarium wilt". *Soil Biol. Biochem.* **1993**, 25, 157-164.

Baker, R., Chet, I. "Induction of suppressiveness", 35-50. In *Suppressive soils and plant disease*; Schneider, R.W., Ed.; Am. Phytopathol. Soc.: St Paul, Minnesota, US, 1982.

Baker, R. "Diversity in biological control". *Crop Protect.* **1991**, 10, 85-94.

Cook, R.J. "Use of pathogen-suppressive soils for disease control", 51-65. In *Suppressive soils and plant disease*; Schneider, R.W., Ed.; Am. Phytopathol. Soc.: St Paul, Minnesota, US, 1982.

Cook, R.J. "Twenty-five years of progress towards biological control", 1-14. In *Biological control of soilborne plant pathogens*; Hornby D., Ed.; C.A.B. International: Wallingford, UK, 1990.

Cook, R.J., Baker, K.F. *The nature and practice of biological control of plant pathogens*; Am. Phytopathol. Soc.: St Paul, Minnesota, US, 1983, 533p.

Cook, R.J., Veseth, R.J. *Wheat Health Management*, Am. Phytopathol. Soc.: St Paul, Minnesota, US, 1991, 152p.

Corman, A., Couteaudier, Y., Zegerman, M., Alabouvette, C. "Réceptivité des sols aux fusarioses vasculaires : méthode statistique d'analyse des résultats". *Agronomie* **1986**, 6, 751-757.

Couteaudier, Y. "Competition for carbon in soil and rhizosphere, a mechanism involved in biological control of fusarium diseases", 99-104. In *Biological control of plant diseases*, Tjamos E.C., Papavizas G.C., Cook, R.J., Eds.; Plenum Press: New York, US, 1992.

Defago, G., Berling, C.H., Burger, U., Haas, D., Kahr, G., Keel, C., Voisard, C., Wirthner, P., Wuthrich, B. "Suppression of black root rot of tobacco and other root diseases by strains of *Pseudomonas fluorescens* : Potential Applications and Mechanisms", 93-108. In *Biological control of soilborne plant pathogens*; Hornby D., Ed.; C.A.B. International: Wallingford, UK, 1990.

Harman, G.E. "Seed treatments for biological control of plant disease". *Crop Prot.* **1991**, 10, 166-171.

Hornby, D. Ed.; Biological control of soilborne plant pathogens; C.A.B. International: Wallingford, UK, 1990, 479p.

Jeannequin, B., Martin, C., Alabouvette, C. "Lutte contre la fusariose du collet de la tomate", Vol. I/III, 321-329. In *Troisième Conférence internationale sur les maladies des plantes*; ANPP: Paris, FR, 1991.

Jones, J.P., Engelhard, A.W., Woltz, S.S. "Management of fusarium wilt of vegetables and ornementals by macro -and-microelement nutrition", 18-32. In Soilborne plant pathogens : *Management of diseases with macro- and microelements*; Engelhard, A.W., Ed.; Am. Phytopathol. Soc., St Paul, Minnesota, U.S., 1989.

Jones, J.P., Woltz, S.S. "Fusarium-incited diseases of tomato and patato and their control", 157-168. In *Fusarium : Disease, biology and taxonomy*; Nelson P.E., Toussoun T.A., Cook R.J., Eds.; Pennsylvania, US, 1981.

Kloepper, J.W., Leong, J., Teintze, M., Schroth, M.N. "Pseudomonas siderophores : a mechanism explaining disease-suppressive soils". *Curr. Microbiol.* **1980**, 4, 317-320.

Lemanceau, P., Alabouvette, C. "Biological control of fusarium diseases by fluorescent *Pseudomonas* and non-pathogenic *Fusarium*". *Crop Protect.* **1991**, 10, 279-286.

Lemanceau, P., Alabouvette, C., Couteaudier, Y. "Recherches sur la résistance des sols aux maladies. XIV - Modification du niveau de réceptivité d'un sol résistant et d'un sol sensible aux fusarioses vasculaires en réponse à des apports de fer ou de glucose". *Agronomie* **1988**, 8, 155-162.

Lemanceau, P., Bakker, P.A.H.M., De Kogel, W.J., Alabouvette, C., Schippers, B. "Effect of pseudobactin 358 production by *Pseudomonas putida* WCS358 on suppression of fusarium wilt of carnations by nonpathogenic *Fusarium oxysporum* Fo47". *Appl. Environ. Microbiol.* **1992**, 58, 2978-2982.

Lemanceau, P., Bakker, P.A.H.M., De Kogel, W.J., Alabouvette, C., Schippers, B. "Antagonistic effect of nonpathogenic *Fusarium oxysporum* strain Fo47 and pseudobactin 358 upon pathogenic *Fusarium oxysporum* f.sp. *dianthi*". *Appl. Environ. Microbiol.* **1993**, 59, 74-82.

Lewis, J.A., Papavizas, G.C. "Biocontrol of plant diseases : the approach for tomorrow". *Crop. Prot.*" **1991**, 10, 95-105.

Linderman, R.G., Moore, L.W., Baker, K.F., Cooksey, D.A. "Strategies for detecting and characterizing systems". *Plant Dis.* **1983**, 67, 1058-1064.

Louvet, J., Rouxel, F., Alabouvette, C. "Recherches sur la résistance des sols aux maladies. I - Mise en évidence de la nature microbiologique de la résistance d'un sol au développement de la fusariose vasculaire du melon". *Ann. Phytopathol.* **1976**, 8, 425-436.

Lumsden, R.D., Lewis, J.A., Papavizas, G.C. "Effect of organic amendments on soilborne plant diseases and pathogen antagonists", 51-70. In *Environmentally Sound Agriculture*, Lockeretz W., Ed. Praeger Press, New York, US, 1983.

Mandeel, Q., Baker, R. "Mechanisms involved in biological control of fusarium wilt of cucumber with strains of nonpathogenic *Fusarium oxysporum*". *Phytopathology* **1991**, 81, 462-469.

Pierson, E.A., Weller, D.M. "Recent work on control of take-all of wheat by fluorescent pseudomonads", 96-97. In *Plant growth-promoting rhizobacteria, Progress and Prospects*; Keel, C.; Koller, B. ; Defago, G., Eds; IOBC/WPRS; 1991, 14/8.

Rouxel, F., Alabouvette, C., Louvet, J. "Recherches sur la résistance des sols aux maladies. IV - Mise en évidence du rôle des *Fusarium* autochtones dans la résistance d'un sol à la Fusariose vasculaire du melon". *Ann. Phytopathol.* **1979**, 11, 199-207.

Scher, F.M., Baker, R. "Mechanism of biological control in a *Fusarium*-suppressive soil". *Phytopathology* **1980**, 70, 412-417.

Scher, F.M, Baker, R. "Effect of *Pseudomonas putida* and a synthetic iron chelator on induction of soil suppressiveness to fusarium wilt pathogens". *Phytopathology* **1982**, 72, 1567-1573.

Schippers, B. "Prospects for management of natural suppressiveness to control soilborne pathogens", 21-34. In *Biological Control of Plant Diseases*; Tjamos E.C., Papavizas G.C., Cook R.J., Eds. Plenum Press: New York, US, 1992.

Schneider, R.W. *Suppressive Soils and Plant Disease*; Am. Phytopathol. Soc.: St. Paul, Minnesota, US, 1982, 88p.

Shipton, P.J., Cook, R.J., Sitton, J.W. "Occurrence and transfer of a biological factor in soil that suppresses take-all of wheat in eastern Washington". *Phytopathology* **1973**, 63, 511-517.

Sneh, B., Pozniak, D., Salomon, D. "Soil suppressiveness to fusarium wilt of melon, induced by repeated croppings of resistant varieties of melons". *J. Phytopathol.* **1987**, 120, 347-354.

Stotzky, G., Martin, T. "Soil mineralogy in relation to the spread of fusarium wilt of banana in Central America". *Plant Soil* **1963**, 18, 317-337.

Stover, R.H. *Fusarial wilt of bananas and other Musa species*. Commonw. mycol. Inst. Phytopat. Pap. **1962**, 4, 117p.

Stutz, E.W., Kahr, G., Defago, G. "Clays involved in suppression of tobacco black root rot by a strain of *Pseudomonas fluorescens*". *Soil Biol. Biochem.* **1989**, 21, 361-366.

Thomashow, L.S., Weller, D.M. "Role of antibiotics and siderophores in biocontrol of take-all disease of wheat", 109-122. In *Biological Control of Soilborne Plant Pathogens*; Hornby D., Ed.; C.A.B. International: Wallingford, UK, 1990.

Van Peer, R., Niemann, G.J., Schippers, B. "Induced resistance and phytoalexine accumulation in biological control of fusarium wilt of carnation by *Pseudomonas* sp.strain WCS417r". *Phytopathology* **1991**, 81, 728-734.

Whipps, W.M. "Status of biological disease control in horticulture". *Biocontrol Sci. Technol.* **1992**, 2, 3-24.

Integrated Biological and Chemical Control of *Sclerotium rolfsii*

C.D. Hoynes, Botany Department, University of Maryland, College Park, MD 20742
J.A. Lewis and **R.D. Lumsden,** Biocontrol of Plant Diseases Laboratory, BARC-West, ARS, USDA, Beltsville, MD 20705.

Interactions occurred between preparations of the biocontrol fungus *Gliocladium virens*(GL3) and nitrogen (N) fertilizers in the control of damping-off and blight of snap bean (*Phaseolus vulgaris* L.) caused by *Sclerotium rolfsii* (Sr1). In a loamy sand, low rates of ammonium sulfate and GL3 biomass applied together resulted in suppressed disease. The effect of interaction between the soil fumigant metham and chlamydospores (fermentor biomass) and conidia of several isolates of *Trichoderma* spp. and *G. virens* resulted in reduced sclerotial germination. However, the magnitude of reduction in germination depended on the propagules of the biocontrol fungus used (conidia vs. chlamydospores) and time of biocontrol application with respect to fumigation.

It is unlikely that biological control alone used in various natural ecosystems will effectively prevent soilborne diseases. In addition, the current systems for disease control and for plant growth stimulation require amounts of fungicides and fertilizers which are often expensive, ineffective, and contribute to groundwater pollution and commodity contamination. This brief report describes the effect of biological control fungi (*Gliocladium virens, Trichoderma* spp.) applied to soil with N-fertilizers or the fumigant metham on the pathogen *Sclerotium rolfsii* and the diseases it incites on snap beans (*Phaseolus vulgaris* L.).

Materials and Methods

To test the feasibility of applying chlamydospores in fermentor biomass of *G. virens* (GL3) together with fertilizers on damping-off and blight of snap beans, greenhouse experiments using two loamy sands (Sassafras, Rumford) were conducted. Sclerotia of *S. rolfsii* (Sr1, Sr3) produced on potato dextrose agar (PDA) were added to soils at rates estimated to cause 80% damping off of Blue Lake 274 snap beans. N-fertilizers (urea, ammonium nitrate, ammonium bicarbonate, ammonium sulfate) were added at recommended rates ($4\text{-}8.7 \times 10^{-5}$g N/g soil) alone or together with fermentor biomass of

GL3, which was added at rates estimated to provide 20% disease control. Amended soils were incubated in plastic bags for 7-10 days and then placed in pots and planted with bean seed (Papavizas and Lewis 1989). After 3 weeks, a disease severity index (DSI) was determined with 0 = no disease to 6 = all dead. Experiments on sclerotial survival also were conducted using a modification of the method of Rodriguez-Kabana et al. (1980).

Conidia and chlamydospores (fermentor biomass) of biocontrol fungi were also studied in association with the fumigant metham (Vapam®, ICI America, Inc., Wilmington, DE.) Dosage response of sclerotia of *S. rolfsii* (Sr3) as well as of conidia and chlamydospores of various *Trichoderma* spp. and GL3 was determined. In these trials, 200-g portions of moist soil were mixed with fresh sclerotia from PDA and placed in plastic pots covered with glass petri dish lids. After 1 week, soils were drenched with metham solutions of various concentrations (0-40 μg/g). After an additional week, pots were uncovered. The soil was allowed to aerate 2 days, then it was remoistened and incubated an additional 9 days. Fermentor biomass or conidial suspensions of biocontrol fungi were added at different times with respect to fumigation. Sclerotial survival was determined by the method of Rodriguez-Kabana et al. (1980).

Results and Discussion

In Sassafras loamy sand (pH 5.9), a low rate of ammonium sulfate (4×10^{-5} g N/g soil) applied together with a low rate of GL3 biomass (0.10 mg/g soil) resulted in a DSI of 1.6. These rates of fertilizer and biomass, applied individually, gave DSI values of 3.6 and 3.4, respectively. The DSI of beans in infested soil with no additives was 4.1. Results with combined treatments were significantly different from those with individual treatments ($P = \leq 0.05$). In addition, effective combined treatments reduced sclerotial germination (>50%) compared to that of individual treatments. Integrated control of disease was as effective as that achieved with high rates of ammonium sulfate (8.7×10^{-5} g N/g soil) and GL3 biomass used alone. In contrast to results with ammonium sulfate, agronomic rates of other fertilizers such as urea and ammonium nitrate increased disease, whereas ammonium bicarbonate did not affect disease.

In experiments with Rumford loamy sand (pH 6.5) a similar effect to that found with Sassafras loamy sand was achieved with low rates of ammonium sulfate and GL3 biomass. However, neither in Rumford loamy sand nor in Sassafras loamy sand limed to pH 7.0, was germination of Sr1 sclerotia reduced.

Generally, these data suggest that disease control may be accomplished with low rates of an effective biocontrol fungus applied to soil simultaneously with low or normal rates of a compatible fertilizer. However, in a review by Punja (1989), it was reported that direct control of diseases caused by *S. rolfsii* with various N-fertilizers was inconsistent and variable, but Maiti and Sen (1985) reported successful interation with fertilizer and microbial antagonist combinations.

Disease reduction that occurred in loamy sands of pH >6.5 without an appreciable reduction in germination of sclerotia of Sr1 may be the result of sclerotial "weakening" rather than of sclerotial destruction. Papavizas and Collins (1990) reported that sclerotia of *S. rolfsii* may somehow be conditioned to remain viable but still not incite diseases. It may be possible that GL3 (or a microflora stimulated by the fertilizer) attacked hyphae that developed in soil from sclerotia that germinated in response to root bean exudates.

Other parameters which affect an integrated control approach include timing of treatment application as well as soil type and pH. For example, our data suggest that time is necessary for the interaction between biocontrol fungus and pathogen propagules to occur before planting. Since it is common practice to apply fertilizers to the soil about a week before planting, it would be appropriate if biocontrol preparations could be applied and disked in at the same time. We are investigating the biocompatibility of other fertilizers using rates that provide little or no direct control of *S. rolfsii*. Also, many fertilizers applied at normal use rates enhance the development of *S. rolfsii* (Punja 1989). Therefore, it is necessary to determine if fertilization will provide a relative benefit to the antagonist rather than the pathogen.

In the application of biocontrol fungi in association with soil fumigation, it was observed that conidia of several *Trichoderma* spp. and Gl3 in soil, tolerated 10X more metham than did chlamydospores found in the fermentor biomass. However, investigations were continued using chlamydospores because these are the propagules most likely produced during commercial fermentation. For example, it was shown that the biocontrol fungus *T. viride* (TV1) added as chlamydospores (fermentor biomass) to Rumford loamy sand fumigated with 4 μg metham/g soil decreased survival of sclerotia from 92% in the fumigated control to 71%, 34% and 24% when biomass was added before fumigation, at aeration and 2 days after aeration, respectively ($P = \leq 0.05$). Similar results were achieved with biomass of GL3.

Survival and germination of conidia of various isolates of *Trichoderma* spp. and *G. virens* after treatment with metham were tolerant to the fumigant (Lewis and Papavizas 1984). The current investigation suggested that the type of antagonist propagules (conidia, chlamydospores) as well as time of fumigant application were important in the reduction of survival of *S. rolfsii* (Sr3) sclerotia.

The data also indicated that the best biocontrol was obtained when the antagonist was added after the fumigation period (either shortly before or after aeration). However, application of antagonist immediately before

212

fumigation or several days or weeks after aeration was less effective.

Since sclerotia which survive fumigation can aggressively recolonize the soil, it may be useful to include an antagonist that degrades hyphae growing from sclerotia that escaped fumigation in a practical integrated control program.

References

Lewis, J.A.; Papavizas, G.C. "Effect of the Fumigant Metham on *Trichoderma* spp." *Can. J. Microbiol.*, **1984**, *30*, 739-745.

Maiti, D.; Sen, C. "Integrated Biocontrol of *Sclerotium rolfsii* with Nitrogenous Fertilizers and *Trichoderma harzianum*." *Ind. J. Agric. Sci.*, **1985**, *55*, 464-468.

Papavizas, G.C.; Collins, D.J.; "Influence of *Gliocladium virens* on Germination and Infectivity of *Sclerotium rolfsii*."Phytopathology, **1990**, *80*, 627-630.

Papavizas, G.C.; Lewis, J.A. "Effect of *Gliocladium* and *Trichoderma* on Damping-off and Blight of Snapbean Caused by *Sclerotium rolfsii* in the Greenhouse." *Plant Pathology*, **1989**, *38*, 277-286.

Punja, Z.K. In *Soilborne Plant Pathogens: Management of Diseases with Macro- and Microelements*, Engelhard, A.W., Ed.; APS Press: St. Paul, MN, 1989; pp 75-89.

Rodriguez-Kabana, R.; Beute, M.K.; Backman, P.A., "A Method for Estimating Numbers of Viable Sclerotia of *Sclerotium rolfsii* in Soil." *Phytopathology*, **1980**, *70*, 917-919.

Fungi and Fungus/Bioregulator Combinations for Control of Plant-Parasitic Nematodes

Susan L. F. Meyer and Robin N. Huettel, USDA ARS, Nematology Laboratory, Beltsville Agricultural Research Center-West, Beltsville, MD 20705

Development of new management strategies for control of plant-parasitic nematodes has become increasingly important as many effective chemical nematicides have been removed from the market or restricted in use. Biological control organisms could be applied as part of integrated pest management programs, but few commercial formulations have been developed, and use is limited. Bioregulatory compounds such as pheromones have been utilized for insect control and may also prove useful for reduction of soilborne nematode populations. Studies on a mutant fungus strain with enhanced biocontrol potential, on bioregulatory compounds, and on combinations of these agents are being conducted as part of the search for novel means of nematode management.

Application of commercial nematicides has been severely restricted because of problems associated with groundwater contamination, food safety, and worker protection. Approximately 13 chemical nematicides are currently registered in the United States (Richard Michell, EPA, personal communication), and about half of these are very limited in use. Methyl bromide may be removed from this list, further impacting chemical control of nematodes. Although a highly efficient soil fumigant, it has been implicated in ozone layer depletion, leading to a proposal to remove it from use by the year 2000 (Anonymous, 1993). The continuing loss of nematicides from the marketplace emphasizes the immediate need for replacement management agents for nematode pests. Unfortunately, nematology research has not been able to provide economically viable alternatives to the chemical control measures on which nematode management has relied for so many years (Thomason, 1987; Stirling, 1991). Although some nematodes are managed with resistant cultivars, crop rotation, and other cultural techniques, problems can arise. These include shifts in nematode races or species, resistance to limited numbers of nematode species, lower crop yield or quality, and limitations on crop selection by growers (Young, 1992).

Because integrated pest management systems for nematodes would benefit greatly from the introduction of new techniques, research on biological control organisms has accelerated in recent years. The use of biocontrol agents to decrease nematode populations shows promise, but few organisms are commercially available (e.g., Stirling, 1991; Kerry, 1990). Long-term, intensive study is necessary in this area to learn more about the organisms and to develop useful products.

Another alternative to conventional chemical control is application of naturally occurring bioregulatory compounds, such as pheromones. Bioregulators have been successfully utilized against insect pests, but studies have only recently begun on their effectiveness in reducing nematode populations (DeMilo and Huettel, 1991; Meyer and Huettel, 1991, 1992a). Such compounds could interfere with some stage of a nematode life cycle (Stern et al., 1988), thereby decreasing numbers of nematodes in the soil. The above approaches are being united in a research program that encompasses use of antagonistic fungi, bioregulatory compounds, and combinations of these agents as novel controls for plant-parasitic nematodes (Huettel and Meyer, 1992; Meyer and Huettel, 1991, 1992a).

Fungal Biological Control

Fungi that act as nematode antagonists fall

into three broad categories: endoparasitic species that attack vermiform nematodes, nematode-trapping (or predacious) fungi, and fungi that colonize stationary females, cysts, and eggs (Morgan-Jones and Rodríguez-Kábana, 1987; Stirling, 1991). Studies have been conducted on species in all of these groups, but only a small number of biological control fungi have been applied by growers, and even these fungi are not in widespread use (Stirling, 1991). In the search for biocontrol fungi, potential antagonists are often isolated from nematode populations. The identification of such fungi, even from nematodes in suppressive soils, does not necessarily lead to the production of successful control agents. Because each fungus strain may affect viability of only a limited number of nematodes (Kerry, 1990), the impact of any one fungus species can be minor. It is often difficult to determine which fungal species are the most pathogenic. Some fungi are mainly saprophytic, and rarely infest nematodes, or only affect weakened individuals. Other fungi may colonize dead nematodes and exhibit no antagonism to live eggs or adults. Fungi that kill nematodes via toxin or enzyme production before colonization may have more effect than would be indicated from bioassays that measure parasitism of live specimens.

Beyond the problems in determining fungal virulence, difficulties are present at every subsequent stage in development of a biocontrol fungus. Competitiveness with other soil organisms may need to be enhanced before a nematode antagonist can function well as a management agent. Problems are often encountered with mass production, formulation, and delivery of beneficial fungi. Such obstacles are being addressed in numerous studies, with much of the effort in nematode biocontrol directed to cyst nematodes (*Heterodera* and *Globodera* spp.) and root-knot nematodes (*Meloidogyne* spp.) (Kerry, 1988; Kerry and de Leij, 1992). These groups include significant pests of many crops grown throughout the world (Noel, 1986; Kerry and de Leij, 1992).

The research on fungi and bioregulators discussed in this chapter focuses on *Heterodera glycines*, the soybean cyst nematode. In the United States, this nematode alone was estimated to cause more than $250 million in yield losses during the 1991 growing season (Agricultural Research Service Quarterly Report, 1992). At present, there are few registered nematicides available for control of this pest (Rodríguez-Kábana, 1992). Common control methods currently employed include use of resistant soybean cultivars and crop rotation. In addition, cultural practices such as no-till planting, late planting, and irrigation may increase yields (Wrather et al., 1992). Crop losses demonstrate that these control measures are not always effective. The introduction of a biocontrol agent would provide another useful tool for management of this pest.

As part of the research on biocontrol organisms for *H. glycines*, many fungi have been isolated from females, cysts, and eggs (Meyer et. al., 1990; Carris and Glawe, 1989; Rodríguez-Kábana and Morgan-Jones, 1988; Kim and Riggs, 1991). To aid in selecting fungus strains with biocontrol potential, a petri dish bioassay was developed to determine whether or not fungi actually killed nematode eggs (Meyer et al., 1988; Meyer et al., 1990). Of the fungi evaluated, one strain of *Verticillium lecanii* that significantly decreased egg viability was selected for additional research. This species was chosen for several reasons. First, *Verticillium lecanii* had already been investigated as a biocontrol organism for fungi and insects, and was commercially produced for control of aphid and whitefly in the greenhouse (Hussey, 1984; Harper and Huang, 1986; Uma and Taylor, 1987; Heintz and Blaich, 1990). The fungus had also exhibited antagonism to nematodes in previous studies (Gintis et al., 1983; Rodríguez-Kábana and Morgan-Jones, 1988; Hänssler, 1990; Hänssler and Hermanns, 1981). In addition, other researchers had produced mutant strains with changes in several characteristics, including altered enzyme activity, pigmentation, nutritional requirements, and spore density and release (Jackson, 1984; Heale, 1987). This mutagenesis research was of particular interest, because part of the projected research program included efforts to enhance biocontrol capabilities of *V. lecanii* (Meyer, 1992). The planned enhancement process followed the approach of earlier studies, in which mutagenesis led to production of fungus strains with increased tolerance to fungicides and greater antagonism to certain plant pathogens (Papavizas, 1985). Consequently, a fungus species amenable to mutagenesis was a useful research tool.

To determine whether mutagenesis would be useful for nematode biocontrol work with *V. lecanii*, mutants were induced with ultraviolet light. Strains with increased tolerance to benomyl were selected (Meyer, 1992), and greenhouse studies were conducted on the mutant and wild type strains. When data from experiments in steamed and unsteamed soil were combined, the most effective mutant strain caused a significant decrease in soybean cyst nematode numbers compared to controls. The wild type did not (Meyer, unpublished). In experiments conducted in steamed soil, a difference was noted between effects of the mutant strain and those of the wild type strain. At application rates greater than ca. 0.6 g (dry weight fungus) per 20-cm diameter pot, both the wild type and the mutant strain caused similar reductions in nematode populations compared to untreated controls. At lower application rates (approximately 0.05-0.15 g per pot), the mutant strain caused greater reductions in cyst populations than the wild type (Meyer and Huettel, 1991). The rate of 0.6 g per pot is about equivalent to 200 lb/acre, while 0.05-0.15 is roughly 18-60 lb/acre. These results were of particular interest because lower application rates are preferred for economical delivery to the field. In unsteamed soil, application of the mutant resulted in population reductions compared to untreated controls; significant reductions were recorded at 0.1-0.15 g fungus per 20-cm diameter pot (Meyer, unpublished). Application of the wild type strain to unsteamed soil at this lower rate did not result in significant population decreases. These greenhouse experiments demonstrated that the mutant strain could reduce nematode populations when applied at rates feasible for field use. Studies on formulations and application rates are continuing. A patent application was filed by Meyer (1990) for use of the mutants in nematode management.

Pheromone and Pheromone Analogs

Bioregulators are endogenous or exogenous substances that influence various life processes of an organism. Semiochemicals, a type of bioregulator, are "chemicals that mediate interactions between organisms" (Nordlund, 1981). Because of their function, semiochemicals provide opportunities to interfere with the life cycles of plant pests. There are two categories of semiochemicals. One group, the allelochemicals, includes compounds that influence interspecific interactions. For example, they can function in such areas as host recognition, inhibition of competitors, and predator avoidance. Compounds in the other category, called pheromones, mediate chemical communications between members of the same species. Since the first research on a nematode pheromone by Greet (1964), nematologists have primarily studied a category known as attractive sex pheromones, which play a role in interactions between sexes (Bone et al., 1979; Rende et al., 1982; Huettel, 1986). Attractive sex pheromonal activities of nematodes generally have been demonstrated through bioassays that measured movement of males toward females. Most bioassays are relatively simple; for example, nematodes are often placed on petri plates which contain a support medium (such as agar) and a point source of the attractant (Huettel and Rebois, 1986). However, plant-parasitic nematodes are attracted to or repelled by numerous chemicals (Riddle and Bird, 1985; Huettel and Jaffe, 1987; Papademetriou and Bone, 1983), and many assays for sex pheromone activity failed to recognize this phenomenon. The recent development of a sophisticated bioassay that allows for controlled release of the compounds being tested and more precise tracking of the nematodes may eliminate some problems (Castro et al., 1990).

Another approach was described by Huettel and Rebois (1986), who developed a different behaviorally based assay. They determined that the specific coiling behavior of male soybean cyst nematodes occurred only in the presence of females or crude extracts from females. Males were attracted to several other compounds, but did not coil in the presence of those compounds. Because of this specific behavior, purified fractions of crude extracts from females could be

216

tested for pheromonal activity. Chemical purification and behavioral bioassay of components from female extracts were used to identify the *H. glycines* sex pheromone as vanillic acid (Jaffe et al., 1989). Even though it has been shown in many other nematode species that males are attracted to females, few attempts have been made to chemically identify pheromones. Consequently, vanillic acid is the only nematode sex pheromone that has been structurally identified (Jaffe et al., 1989).

Following identification of the sex pheromone, vanillic acid and fifteen pheromone analogs were studied for effects on nematodes propagated in root explant cultures (Stern et al., 1988; Huettel, DeMilo and Stern, unpublished). The numbers of soybean cyst nematode females that developed were counted 10-12 days post infection and again at five weeks post infection. The former time period allowed for development of the first generation and the latter for growth of the second generation. At 10-12 days, vanillic acid and 14 of the analogs did not cause a significant decrease in the number of females as compared with untreated controls. Presumably, these compounds did not inhibit root penetration. However, treatment with one pheromone analog significantly decreased the number of females developing on the root explants. At five weeks, treatment with some of the analogs reduced second generation females by up to 100%. Vanillic acid reduced the number of second generation females by 25%.

These studies suggest that at least four analogs of the soybean cyst nematode pheromone are parapheromones; i.e., compounds that act similarly to naturally produced pheromones (Inscoe et al., 1990). These possible parapheromones may have reduced the second generation by preventing males from locating females. Preliminary *in vitro* assays have indicated that tested compounds are not ovicides or larvacides, and their mode of action is not known at this time (unpublished data). Sex pheromone activity is a possibility. The one analog that decreased the number of first generation females may have inhibited the juveniles from finding the host plant, thus mimicking the effect of an allelochemical.

To determine if the pheromone and selected analogs have potential as management agents for *H. glycines* on soybean, greenhouse and field studies were conducted. In the greenhouse, application of either the pheromone or any one of several analogs significantly reduced nematode numbers relative to untreated controls. In a 1991 field study, the pheromone and four analogs were applied individually to a susceptible soybean variety. With each bioregulator treatment, yields obtained were greater than those recorded from untreated susceptible controls (Meyer and Huettel, 1992a). While none of the increases were significant, the trend was encouraging, especially because treatment with one analog resulted in a yield increase of 57%, similar to the 43% increase obtained with a resistant variety (unpublished). In 1992, a significant yield increase (29%) resulted from treatment with one of the analogs. Although not as high as the 36% yield increase recorded from plants treated with this analog the previous year, the 29% increase was significantly higher ($P<0.05$) than the yield from untreated susceptible controls. A patent application has been filed for use of the pheromone and its analogs as management agents of the soybean cyst nematode (DeMilo and Huettel, 1991).

Fungus/Bioregulator Combinations

Research has shown that biological agent/bioregulator combinations can reduce nematode populations. In greenhouse studies of soybean cyst nematode on soybean, two combination control agents were tested. These were a mutant strain of *V. lecanii* incorporated with the sex pheromone, and the same strain with an analog. Both combinations caused significant decreases in cyst numbers relative to controls (Meyer and Huettel, unpublished), and a patent application was filed on use of fungus/bioregulator agents for nematode management (Meyer and Huettel, 1992b).

Difficulties with formulations have affected microplot and field tests. For these studies, fungus/bioregulator combinations were incorporated into alginate-bran prills (formulation after Fravel et al., 1985). Placement of these prills immediately adjacent to soybean seeds resulted in decreased soybean stand counts. Consequently, new formulations are currently being explored. However, there were indications that beneficial results may be obtained with more efficacious delivery systems. In 1991, the mutant strain of *V. lecanii*, vanillic acid, an analog, fungus/vanillic acid, and fungus/analog combinations were applied to field microplots. All applications resulted in increased yields compared to untreated control plants and to plants treated with autoclaved fungus prills (unpublished). None of the increases were significant, however. In 1992, significant yield increases were not recorded from plants in treated microplots, but the fungus/pheromone combination applied to two small-scale field plots did result in a 22% yield increase (P<0.10) relative to untreated susceptible controls. This was the second highest yield obtained from any of the fungus, bioregulator, or combination treatments, comparing favorably with the 29% yield increase (P<0.05) recorded from the most effective analog treatment that year. Yields recorded after treatment with *V. lecanii* mutant alone and vanillic acid alone were 19% and 12% higher (respectively) than yields from untreated controls, but these values were not significantly different from the controls. The highest yield compared to the untreated susceptible variety was recorded with aldicarb treatment (40% increase, P<0.01), and the lowest was the 11% increase obtained with a resistant variety (unpublished). Further tests need to be conducted over a spectrum of field conditions to determine the efficacy of fungus/bioregulator combinations.

This work is clearly in its infancy, but it presents the intriguing possibility that fungi and bioregulators used in tandem may affect more stages of the nematode life cycle than when used individually. Several scenarios can be suggested: these would obviously depend upon the nematode, the fungus, and the bioregulator. These concepts have been summarized in diagrammatic form (Huettel and Meyer, 1992), and are outlined here, with *H. glycines* as an example.

A sex pheromone used alone could attract males away from the females, or habituation of males might occur, preventing the males from following the pheromone gradient to the females. The consequence would be a reduction in fertilization. A chemical that inhibits juveniles from penetrating roots could reduce the population at a different stage. A fungus that attacks vermiform stages could trap or otherwise infect males and juveniles; alternatively, a fungus that primarily affects egg stages could be selected.

If a bioregulator and antagonistic fungus were used together, the vermiform stages that were attracted to the bioregulator would have an increased chance of being parasitized by the fungus. This would further lower the possibility that the vermiform stages could complete their function in the life cycle. Vermiform nematodes not attracted to the compound might still be infected by the fungus. A formulation might include a bioregulator with a fungus that decreases viability of stationary eggs, so that vermiform stages would be inhibited by the bioregulator while mycelium grew out to kill the eggs. This combination could severely limit nematode populations even though males or juveniles were not directly affected by the biocontrol organism. It might prove beneficial to utilize a fungus with adhesive conidia. Males attracted to pheromone in the delivery system, if not killed by the fungus, might later transport the conidia closer to the future cyst location for later action against eggs. A sex pheromone, an allelochemical, a nematode-trapping fungus, and a fungus that affects egg viability could all be utilized in a single product. Many combinations are possible, and the selection of the agents would depend upon the nematode to be managed, the stages of the life cycle targeted, the compatibility of the agents, and the cost-effectiveness of the combinations.

Although bacteria are not the focus of this paper, they or other living organisms could be incorporated into a control formulation. Fungicides and food sources could be manipulated to encourage growth of the biocontrol organisms. In addition, low levels of chemical nematicides could be applied at rates that had minimal impact on the environment, and yet were still destructive to the nematodes attracted to the delivery systems.

Live agents can be combined with various bioregulators in almost limitless combinations. Certainly, more work is needed on the ecology of selected organisms and on the modes of action of the bioregulators. Semiochemicals from other plant-parasitic nematodes have not yet been identified. Numerous formulations, including granules, liquids, dusts, pastes, and gels (Fravel and Lewis, 1992), are possible for biocontrol organisms and need to be investigated. Eventually, this research should lead to information that can be applied to development of new management strategies for plant-parasitic nematodes.

References

Anonymous. "Methyl bromide as an ozone depletor; a NAPPO position statement." *NAPPO (North American Plant Protection Organization) Newsletter* **January 1993**, *13*, 3, 6.

Agricultural Research Service. "Into the marketplace. Patent licenses." *USDA, ARS Quarterly Report* **Oct.-Dec. 1992**, 1.

Bone, L. W.; Gaston, L. K.; Hammock, B. D.; Shorey, H. H. "Chromatographic fractionation of aggregation and sex pheromones of *Nippostrongylus brasiliensis* (Nematoda)." *J. Exper. Zool.* **1979**, *208*, 311-318.

Carris, L. M.; Glawe, D. A. Fungi Colonizing Cysts of *Heterodera glycines*; Bulletin 786; United States Department of Agriculture, University of Illinois at Urbana-Champaign, College of Agriculture, Agricultural Experiment Station, IL, **1989**.

Castro, C. E.; Belser, N. O.; McKinney, H. E.; Thomason, I. J. "Strong repellency of the root knot nematode, *Meloidogyne incognita* by specific inorganic ions." *J. Chem. Ecol.* **1990**, *16*, 1199-1205.

DeMilo, A.; Huettel, R. *Method and Composition for Controlling the Soybean Cyst Nematode with a Sex Pheromone and Analogs Thereof.* Serial No. 07/645,438. Patent application filed 1991.

Fravel, D. R.; Marois, J. J.; Lumsden, R. D.; Connick, W. J., Jr. "Encapsulation of potential biocontrol agents in an alginate-clay matrix." *Phytopathology* **1985**, *75*, 774-777.

Fravel, D. R.; Lewis, J. A. In *Pesticide Formulations and Application Systems;* Bode, L. E.; Chasin, D. G., Eds.; American Society

for Testing and Materials: Philadelphia, PA, 1992, Vol. 11; pp 173-179.

Gintis, B. O.; Morgan-Jones, G.; Rodríguez-Kábana, R. "Fungi associated with several developmental stages of *Heterodera glycines* from an Alabama soybean field soil." *Nematropica* **1983**, *13*, 181-200.

Greet, D. N. "Observations on sexual attraction and copulation in the nematode *Panagrolaimus rigidus* (Schneider). *Nature* **1964**, *204*, 96-97.

Hänssler, G. "Parasitism of *Verticillium lecanii* on cysts of *Heterodera schachtii*." *Z. PflKrankh. PflSchutz.* **1990**, *97*, 194-201.

Hänssler,. G.; Hermanns, M. "*Verticillium lecanii* as a parasite on cysts of *Heterodera schachtii*." *Z. PflKrankh. PflSchutz.* **1981**, *88*, 678-681.

Harper, A. M.; Huang, H. C. "Evaluation of the entomophagous fungus *Verticillium lecanii* (Moniliales: Moniliaceae) as a control agent for insects." *Environ. Entomol.* **1986**, *15*, 281-284.

Heale, J. B. In *Fungi in Biological Control Systems;* Burger, M. N., Ed.; University Press: Manchester, 1987; pp 211-234.

Heintz, C.; Blaich, R. "*Verticillium lecanii* as a hyperparasite of grapevine powdery mildew (*Uncinula necator*)." *Vitis* **1990**, *29*, 229-232.

Huettel, R. N. "Chemical communicators in nematodes." *J. Nematol.* **1986**, *18*, 3-8.

Huettel R. N.; Jaffe, H. "Attraction and behavior of *Heterodera glycines*, the soybean cyst nematode, to some biological and inorganic compounds." *Proc. Helminthol. Soc. Wash.* **1987**, *54*, 122-125.

Huettel, R. N.; Meyer, S. L. F. In *Biological Control of Plant Diseases, Progress and Challenges for the Future;* Tjamos, E. C.; Papavizas G. C.; Cook R. J., Eds.; Plenum Press: New York, NY, 1992; pp 273-276.

Huettel, R. N., Rebois, R. V. "Bioassay comparisons for pheromone detection in *Heterodera glycines*, the soybean cyst nematode." *Proc. Helminthol. Soc. Wash.* **1986**, *53*, 63-68.

Hussey, N. W. In *The Role of Biological Control in Pest Management;* Allen, G.; Rada, A., Eds.; University of Ottawa Press: Ottawa, Ontario, 1984; pp 128-136.

Inscoe, M. N.; Leonhardt, B. A.; Ridgeway, R.

L. In *Behavior-modifying Chemicals for Insect Management;* Ridgeway, R. L.; Silverstein, R. M.; Inscoe, M. N., Eds.; Marcel Dekker: New York, NY, l990; pp 631-715.

Jackson, C. W. Genetical Studies on the Entomopathogenic Fungus *Verticillium lecanii* (Zimm.) Viégas; Ph.D. Dissertation; University of London: London, 1984.

Jaffe, H.; Huettel, R. N.; DeMilo, A. B.; Hayes, D. K.; Rebois, R. V. "Isolation and identification of a compound from soybean cyst nematode, *Heterodera glycines,* with sex pheromone activity." *J. Chem. Ecol.* **1989,** *15,* 2031-2043.

Kerry, B. R. In *Agriculture, Ecosystems, and Environment; Biological Interactions in Soil;* Edwards, C. A.; Stinner, B. R.; Stinner, D.; Rabatin, S., Eds; Elsevier: New York, NY, l988; pp 293-305.

Kerry, B. R. "An assessment of progress toward microbial control of plant-parasitic nematodes." *J. Nematol. Suppl.* **1990,** *22,* 621-631.

Kerry, B. R.; Leij, de, F. A. A. M. In *Biological Control of Plant Diseases, Progress and Challenges for the Future;* Tjamos, E. C.; Papavizas, G. C.; Cook, R. J., Eds.; Plenum Press: New York, NY, 1992; pp 139-144.

Kim, D. G.; Riggs, R. D. "Characteristics and efficacy of a sterile hyphomycete (ARF18), a new biocontrol agent for *Heterodera glycines* and other nematodes." *J. Nematol.* **1991,** *23,* 275-282.

Meyer, S. L. F. *Benomyl Tolerant Strains of the Fungus Verticillium lecanii and Methods of Use for Biocontrol.* Serial No.: 07/633,815. Patent application filed 1990.

Meyer, S. L. F. "Induction of increased benomyl tolerance in *Verticillium lecanii,* a fungus antagonistic to plant-parasitic nematodes." *J. Helminthol. Soc. Wash.* **1992,** *59,* 237-239.

Meyer, S. L. F.; Huettel, R. N. "Comparisons of fungi and fungus-bioregulator combinations for control of *Heterodera glycines,* the soybean cyst nematode." *J. Nematol.* **1991,** *23,* 540.

Meyer, S. L. F.; Huettel, R. N. "Field evaluation of potential control agents for soybean cyst nematode on soybean." *Phytopathology* **l992a,** *82,* 1155.

Meyer, S. L. F.; Huettel, R. N.

Fungus-bioregulator Compositions and Methods for Control of Plant-parasitic Nematodes. Serial No.: 07/937,764. Patent application filed 1992 (b).

Meyer, S. L. F.; Huettel, R. N.; Sayre, R. M. "Isolation of fungi from *Heterodera glycines* and in vitro bioassays for their antagonism to eggs." *J. Nematol.* **1990,** *22,* 532-537.

Meyer, S. L. F.; Sayre, R. M.; Huettel, R. N. "Comparisons of selected stains for distinguishing between live and dead eggs of the plant-parasitic nematode *Heterodera glycines.* " *Proc. Helminthol. Soc. Wash.* **1988,** *55,* 132-139.

Morgan-Jones, G.; Rodríguez-Kábana, R. In *Vistas on Nematology;* Veech, J. A.; Dickson, D. W., Eds.; Society of Nematologists, Inc.: Hyattsville, MD, 1987; pp 94-99.

Noel, G. R. In *Cyst Nematodes;* Lamberti, F.; Taylor, C. E., Eds.; Plenum Press: New York, NY, 1986; pp 257-268.

Nordlund, D. A. In *Semiochemicals: Their Role in Pest Control;* Nordlund, D. A.; Jones, R. L.; Lewis, W. J., Eds.; John Wiley and Sons: New York, NY, 1981; pp 13-28.

Papademetriou, M. K.; Bone, L. W. "Chemotaxis of larval soybean cyst nematode, *Heterodera glycines* race 3, to root leachates and ions." *J. Chem. Ecol.* **1983,** *9,* 387-396.

Papavizas, G. C. In *Annual Review of Phytopathology;* Cook, R. J.; Zentmyer, G. A.; Cowling, E. B., Eds.; Annual Reviews, Inc.: Palo Alto, CA, 1985, Vol. 23; pp 23-54.

Rende, J. F.; Tefft, P. M.; Bone, L. W. "Pheromone attraction in the soybean cyst nematode *Heterodera glycines* Race 3." *J. Chem. Ecol.* **1982,** *8,* 981-991.

Riddle, D. L; Bird, A. F. "Responses of the plant parasitic nematodes *Rotylenchulus reniformis, Anguina agrostis,* and *Meloidogyne javanica* to chemical attractants." *Parasitology* **1985,** *91,* 185-195.

Rodríguez-Kábana, R. In *Biology and Management of the Soybean Cyst Nematode;* Riggs, R. D.; Wrather, J. A., Eds.; The American Phytopathological Society: St. Paul, MN, 1992; pp 115-123.

Rodríguez-Kábana, R.; Morgan-Jones, G. "Potential for nematode control by mycofloras endemic in the tropics." *J. Nematol.* **1988,** *20,* 191-203.

Stern, S.; Jaffe, H.; DeMilo, A.; Huettel, R. N. "Disruption of mate finding in soybean cyst nematodes with analogs of the nematode sex attractant." *J. Nematol.* **1988,** *20,* 661.

Stirling, G. R. *Biological Control of Plant Parasitic Nematodes: Progress, Problems and Prospects;* CAB International: Wallingford, U.K., 1991.

Thomason, I. J. In *Vistas on Nematology;* Veech, J. A.; Dickson, D. W., Eds.; Society of Nematologists, Inc.: Hyattsville, MD, 1987; pp 469-476.

Uma, N. U.; Taylor, G. S. "Parasitism of leek rust urediniospores by four fungi." *Trans. Br. Mycol. Soc.* **1987,** *88,* 335-340.

Wrather, J. A.; Anand, S. C.; Koenning, S. R. In *Biology and Management of the Soybean Cyst Nematode;* Riggs, R. D.; Wrather, J. A., Eds.; The American Phytopathological Society: St. Paul, MN, 1992; pp 125-131.

Young, L. D. "Problems and strategies associated with long-term use of nematode resistant cultivars." *J. Nematol.* **1992,** *24,* 228-233.

BIOCONTROL AGENTS FOR SUPPRESSION
OF WEEDS

Foreign Plant Pathogens for Environmentally Safe Biological Control of Weeds

William L. Bruckart, USDA-ARS-FDWSRU, Ft. Detrick, Bldg. 1301, Frederick, MD 21702
Nina Shishkoff, Dynamac Corp., 2275 Research Blvd., Rockville, MD 20805

Pathogens are developed for weed control if they are virulent and environmentally safe. Using criteria for risk assessment, we present evidence that, through proper selection of the pathogen and use of scientific evaluation, foreign pathogens for weed control pose no greater risk than endemic pathogens. Evaluations of the autoecious rust fungi *Puccinia carduorum* and *P. jaceae* are used to illustrate the process of risk assessment of foreign pathogens in containment and field experiments. Concepts of infection, injury and damage by fungi are considered as part of risk assessment.

Use of plant pathogens for biological control of weeds is a rapidly developing part of the weed management arsenal. Despite a number of successful applications and the considerable potential for plant pathogens in weed control, challenges abound in the areas of discovery, development, risk assessment, and technologies for mass-production, formulation, application, and strain improvement. Development of plant pathogens for use in weed biocontrol is a very complex process. Aspects of this complexity include: Diversity of plant pathogens; Strategies for using pathogens in weed management; Options for improved utilization of pathogens; Opportunities for creating desirable traits in pathogens; Variability of target weeds; and Ecological and agricultural settings where weed management is desired. Although a brief general perspective is provided in this paper, it is neither possible nor the intention of this paper to provide a complete review. Readers who desire to more fully experience the challenges and excitement of this field are encouraged to read papers by Charudattan, 1991; TeBeest, et al., 1992; and Watson, 1991.

There are two attributes that every candidate weed biocontrol agent should have. The first is efficacy or virulence. Efficacy is of most concern to those who develop and benefit directly from the technology; either the pathogen works or it doesn't. The second is safety, which unlike efficacy, is important to everyone. As a consequence, aspects of safety comprise major parts of current USDA, Animal and Plant Health Inspection Service (APHIS) and Environmental Protection Agency (EPA) regulation and registration processes.

The focus of this paper is safety in the context of risk assessment, with particular emphasis on safe use of foreign (imported) plant pathogens. Using concepts of infection and disease development, the safety of foreign pathogens is discussed and illustrated by case studies from evaluations of *Puccinia carduorum* and *P. jaceae* for biological control of musk thistle (*Carduus thoermeri*) and yellow starthistle (*Centaurea solstitialis*), respectively.

Overview

Fungi have been the object of most research and development. However, pathogenic microorganisms from all major groups have been considered. Of 83 weed control projects listed by Templeton (1982), 71 involved fungi, six involved viruses, and three each involved bacteria and nematodes. Charudattan (1991) recently listed 109 mycoherbicide projects, including 73 not listed by Templeton (1982). Two nematodes, *Subanguina picridis* for control of *Acroptilon repens*, Russian knapweed, and *Orrina phyllobia* for control of *Solanum elaeagnifolium*, silverleaf nightshade, have been field evaluated for weed biocontrol (Parker, 1991). Recent investigations of bacteria in the genera *Erwinia*, *Pseudomonas* and *Xanthomonas* have been reported (Haygood, 1992; Kennedy, et al., 1991; Kremer et al., 1990).

The use of plant pathogens for weed control has been considered in almost every ecosystem where there are weeds, including agricultural (Charudattan, 1991; Watson, 1991), aquatic (Joye, 1990), forest, public, recreational, and industrial settings (Fowler, et al., 1991; Gardner, 1990).

Three major strategies are involved in using plant pathogens for weed control: 1) bioherbicide, 2) augmentative, and 3) classic (inoculative). They represent a continuum from regular and intensive to a limited number of applications of pathogens to achieve the desired level of weed control. These strategies are based on general principles of ecology and epidemiology; each is a way to initiate a disease epidemic within the target population (Shrum, 1982). Augmentation combines features of the bioherbicide and classical strategies.

All of these strategies involve plant pests, so interstate and international shipments are regulated by APHIS. These organisms also are pesticides according to the Federal Insecticide, Fungicide, and Rodenticide Act (FIFRA), so research and registration of weed control pathogens are regulated by the EPA. The regulation of plant pathogens for weed control is summarized by Charudattan (1990).

What About Risk?

Pest control involves risk, regardless of pesticide or strategy for use. Risk is the combination of hazard (danger) and exposure. It follows that if there is no hazard, there is no risk. Also, if there is no exposure, then there is no risk (Alexander, 1990). Risk, ideally, can be expressed in terms of probability, either as precise ratios (1 chance in 100) or in general terms (high or low).

The risk of using any plant pathogen for weed control must be determined, because each is a biologically active, self-replicating microbe that can damage plants. Where they are native, indigenous and non-indigenous candidate weed control pathogens do not damage valuable plant species at their naturally-occurring levels. The potential for risk increases when pathogens are used deliberately for weed control.

The general assumption is that risk is greater for foreign pathogens in a new location than for endemic pathogens, because the act of introduction is an irreversible process and ostensibly there is likely to be exposure of native or crop plants that may lack resistance to an introduced pathogen. These are valid concerns and worthy of attention. As Wilson (1969) pointed out, however, these concerns "should influence but not deter future work in this area." With proper procedure and careful, scientific testing, foreign pathogens approved for release will pose no greater risk than endemic pathogens for weed control. Supportive evidence comes from the successful use of foreign plant pathogens in Australia, Chile, South Africa, and the United States (Watson, 1991).

The objective of any evaluation is to discover candidate pathogens that damage the target weed without damaging other plant species. The process of evaluation has three components: 1) selection of the biocontrol agent, 2) scientific evaluation of potential weed control pathogens, and 3) determination of the risk. Safety is almost always associated with host specificity

and there have been several cases where the number of plants susceptible to a pathogen increases in an experimental situation (Watson, 1985). In discussing "equivocal results" from host specificity testing for insect weed-control agents, Cullen (1990) identified two problems: 1) measuring host specificity comprehensively under experimental conditions, and 2) interpreting results in a way that predicts effects on nontarget species in nature. These also are recognized problems in the evaluation of plant pathogens for weed control (Watson, 1985), but both Cullen (1990) and Watson (1985) feel that valid decisions can be made from good research and proper interpretation of the results. Their recommendations include testing under "realistic conditions" and the use of field evaluations (Watson, 1985), and reliance on "a certain amount of experimental improvisation" and careful interpretation of results (Cullen 1990).

The Pathogen. Many perceived risks can be eliminated by selecting appropriate pathogens. Pathogens vary in a number of attributes, including general host specificity, expected heterogeneity, and ability to disperse. These attributes, in and of themselves, are neither good nor bad but may be useful in the selection and evaluation of candidate pathogens (Leonard, 1982). Pathogens known to have broad host ranges are not considered for introduction, while obligate parasites, intimately adapted to one or a few plant species, are among the best candidates for classical biocontrol of weeds. Both *P. carduorum* and *P. jaceae* are obligate parasites of one or a few plant species (autoecious), and evidence in the literature suggests these pathogens are host specific (Gaumann, 1959; Savile, 1970). Evidence for host specificity of an heteroecious rust fungus has been reported, as well. *Uromyces rumicis* requires both *Rumex crispus* (curly dock) and *Ranunculus ficaria* (lesser celandine) to complete its life cycle. Tests of both the uredinial and telial stages of the pathogen suggest the life cycle is limited only to these two species (Inman, 1971; Schubiger et al., 1985). This evaluation was more complicated because it required host range determination for both forms of the pathogen.

Several pathogens that are facultative saprophytes have acceptable levels of host specificity for applications in biological weed control, although they may have broader host ranges or greater heterogeneity than obligate pathogens (Leonard, 1982). Use of foreign pathogens of this type may be more difficult to justify as bioherbicides on the basis of host specificity alone. Other attributes also may be important. Most facultative fungi spread only a few meters in a season (Agrios, 1980), and if host range is sufficiently narrow, it is conceivable that certain foreign pathogens could be used effectively as bioherbicides against introduced weeds in cultivated agriculture. Standards used in evaluating *Phytophthora palmivora* for biocontrol of *Morrenia odorata*, stranglervine (Ridings, 1986) would be sufficient for evaluating foreign pathogens of this nature. The potential nontarget effects from *P. palmivora* were reduced by specifying localities for use on the label (Kenney, 1986).

Testing, general. The main focus of risk assessment is host specificity, and a valuable perspective about the potential risk is developed from these tests. But this information is only part of the story, because a foreign pathogen for classical or bioherbicide application will be used in nature where conditions for disease are usually less than optimal. Other aspects of the host and pathogen interaction must be considered to make the best judgment about risk in nature (Cullen, 1990). Reliance strictly on host range data is insufficient.

Testing pathogens for efficacy and safety involves at least two phases. The first phase is the host range determination conducted under optimal conditions for disease. The focus here is on individual plant susceptibility, which is necessary to identify potential hazards. The second phase consists of detailed study of potentially susceptible plants when nontarget infections are not perceived to be damaging. Additional information from these studies may clarify the decision-making process.

Testing, initial. Generally, host specificity is the best way to insure the safe use of any plant pathogen. This process has been described for evaluations of foreign plant pathogens (Bruckart and Dowler, 1986; Watson, 1985). Host range plant lists are traditionally organized using the centrifugal phylogenetic testing sequence (Wapshere, 1975; Watson. 1985), based on plant taxonomic relationships and the assumption that the most likely susceptible species will be close relatives of the target. Weidemann (1991) suggests that this is reasonable for obligate pathogens but not for facultative pathogens. We have found that Wapshere's approach provides a good framework for organizing and interpreting data from the initial stages of host range determinations for rust fungi (Bruckart, 1989, Politis, et al., 1984).

For host specificity tests, it is important to understand concepts of infection and disease development as they relate to plant damage. Infection, according to Agrios (1988) is "establishment of a parasite within a host plant". Nutter, et al. (1993) differentiate between infection (which they call 'injury') and damage caused by plant pathogens. Infection may result in "visible or measurable symptoms and/or signs caused by pathogens or pests" (Nutter, et al. 1993). Infection may be beneficial, as in the cases for mycohorrizal fungi and nitrogen-fixing bacteria (Wilson 1977). Many plants also tolerate disease and perform acceptably while providing necessary habitat for pathogen growth and reproduction (Mussell, 1980). Infection may lead to damage, which is a "reduction in the quantity and/or quality of yield" (Nutter, et al. 1993). It is possible to describe the effect of a pathogen on target and nontarget species as a continuum from immunity (no infection) through infection to damage.

Screening of nontarget plant species under optimal conditions eliminates many species that are clearly not susceptible. Although negative results are valuable, positive results (i.e., symptomatic infection) are not justification to discount a pathogen for biological weed control (Cullen, 1990). Additional testing is warranted if the infection is less severe than that on the target species (Cullen, 1990).

The optimal conditions for infection of yellow starthistle and musk thistle are 12-16 hr dew at 20 C (Politis and Bruckart, 1986; Bennett et al., 1991). Under these conditions, disease severity was high and both *P. carduorum* from musk thistle and *P. jaceae* from yellow starthistle caused measurable damage to their respective hosts. Data also indicated strong preference of *P. carduorum* from musk thistle and *P. jaceae* from yellow starthistle for the species from which they were isolated (Bruckart, 1989; Politis et al., 1984; Bruckart, unpublished). Host range determinations also suggested that neither *P. carduorum* nor *P. jaceae* infected species outside of the tribe Cynareae in the Asteraceae under optimal conditions. Positive reactions were noted on *Cirsium* spp. and artichoke (*Cynara scolymus*) by *P. carduorum;* infection of two *Cirsium* spp., safflower (*Carthamus tinctorius*), and bachelor's button (*Centaurea cyanus*) resulted from inoculations with *P. jaceae*. Both disease incidence (proportion of infected plants)· and severity (amount of disease per infected plant) were much less on nontarget species than on target species, except for *C. cyanus* inoculated with *P. jaceae* (Bruckart, 1989; Politis et al., 1984). Additional tests were justified.

Testing, Secondary. Host range determinations often involve a second step, which includes those species that developed symptoms in the first series. It may involve testing under more realistic greenhouse or laboratory conditions, (Watson, 1985), but other approaches should be considered. Examples include whole-leaf microscopic examination (Bouzzese and Hasan, 1986), histological comparisons (Watson, 1986), study of latent infections (Cerkanskus, 1988), and inclusion of endemic pathogens in greenhouse evaluations of candidates (Bruckart 1989, Bruckart and Peterson, 1991).

We conducted studies with *P. carduorum* and *P. jaceae* under optimal conditions to quantify disease severity and damage to selected plant species. Because disease development or damage is expected to be much less in nature where conditions for disease are less than optimal, insignificant reactions under optimal greenhouse conditions will be of no consequence in nature. Of 24 *Cirsium* spp. inoculated with *P. carduorum* from musk thistle 4-5 wk after planting, no macroscopic symptoms developed on individuals in 11 of the species. Of the *Cirsium* spp. where susceptible individuals were noted, only 25% of these plants were symptomatic and reinoculation of these individuals 2 wk later resulted in a very few new susceptible reactions (Politis et al., 1984). *P. carduorum* could not be maintained on any of these nontarget species (Bruckart, unpublished). Similar results occurred from inoculation of artichoke with *P. carduorum* (Bruckart et al., 1985) and two of nine susceptible *Cirsium* spp. with *P. jaceae* (Bruckart, 1989).

Comparative studies with related pathogens or pathogens of related plant species in North America have been used to provide perspectives for results from greenhouse evaluations of candidate pathogens. For *P. carduorum* the comparison involved a California strain of the rust occurring on slenderflower thistle, *Carduus tenuiflorus* (Watson and Brunetti, 1984). Although rust-infected slenderflower thistles are common near artichokes in California, no rust disease of artichokes has been reported in California (or in the world). The slenderflower strain infected artichokes (but not musk thistle) in the greenhouse. Strains of *P. carduorum* from musk and slenderflower thistles also were distinguished by isozyme analysis (Bruckart and Peterson, 1991).

For *P. jaceae* the comparison involved *P. carthami*, the cause of safflower rust worldwide. A series of side-by-side comparisons of safflower susceptibility to *P. jaceae* and *P. carthami* indicated that significant differences occur in susceptibility of safflower to each of these pathogens under optimal conditions (Bruckart, 1989). Also, the evidence suggested that safflower cultivar susceptibility to *P. jaceae* may correlate at a much lower level of disease severity with susceptibility to *P. carthami* (Bruckart, 1989). We observed that safflower was so susceptible to *P. carthami* as a foliar pathogen in the greenhouse that safflower plants in the controls and *P. jaceae* treatments frequently were contaminated by *P. carthami* without a deliberate artificial dew period.

We also attempted to maintain *P. jaceae* on four safflower cultivars for 10 "generations" and select a "safflower strain" of *P. jaceae*. It was only possible to maintain *P. jaceae* on the most susceptible cultivar, "Pacific 1"; the other cultivars never produced enough inoculum for the experiment. Values for expected number of infected leaves per plant and the number of pustules per leaf at the tenth generation were derived by regression analysis. Estimates for infected leaves per plant and the slope are compared for four safflower cultivars (Table 1). "Pacific 1" was the most susceptible of the cultivars, and estimates of slope suggested either negative or neutral trends at the tenth "generation". Results were similar for pustule count data.

A detailed comparison of damage (loss in root dry weight) caused by strains of *P. jaceae* from three *Centaurea* spp. did not always relate to the number of pustules on the infected plants (Tables 2 and 3). The number of pustules was highly variable and did not differ for each combination of host and pathogen; cornflower has smaller leaves and therefore lower pustule numbers (Table 2). Similarities in pustule counts did not reflect data on damage. Significant reduction of root biomass occurred only with yellow starthistle inoculated with the yellow starthistle strain of *P. jaceae;* inoculation of cornflower by any of the strains did not result in damage as measured by root biomass (Table 3).

Testing, field. Field tests are part of the experimental process and are not entirely isolated from influences of the research process (Cullen, 1990). The more closely an experiment mimics the natural situation, the more useful the information and the more realistic the interpretation of data (Cullen, 1990). Two field studies conducted in Europe as part of North American evaluations of *P. carduorum* and *P. jaceae* included artificial inoculation of test plants when favorable conditions for infection were expected. Symptoms developed on nontarget species as a result, but they were similar to those from the greenhouse studies; nontarget species were much less susceptible than the targets (Hasan et al., 1990; Defago, personal communication).

Field tests are best if natural spread of inoculum to both target and test species can be arranged. It is necessary to verify that the test species receive adequate exposure if data are to be valid. Recently, *P. carduorum* was field tested in Virginia for potential to infect musk thistle and selected species of *Cynara* and *Cirsium* (Baudoin et al., 1993). In this study,

Table 1. Disease severity of safflower inoculated ten generations (Gen.) with *Puccinia jaceae* from safflower

Host Cultivar	Infected Leaves/Plant 10th Gen. (Predicted)	95% CI[a] (Upper)	Slope (Estimated)	(pr > f)[b]
CH 353	0	0.01	0.002	0.16
C/W 74	0	n/e[c]	0	n/e
UC-41	0	0.25	-0.170	0.01
PAC-1	9.1	14.0	-0.400	0.68

[a]CI = Confidence interval.
[b](pr > f), probability that slope is equal to "0".
[c]n/e = not estimated, none of the plants were infected and without variability, the upper confidence interval could not be estimated.

Table 2. Pustule counts on cornflower, diffuse knapweed (DK), purple starthistle (PS), and yellow starthistle (YS) inoculated with three strains of *Puccinia jaceae*

P. jaceae[a] (Strain)	Pustules/leaf[b] On host	On cornflower
DK	109 ± 48	19 ± 10
PS	70 ± 36	25 ± 12
YS	59 ± 68	24 ± 8

SOURCE: Adapted from Shishkoff and Bruckart, 1993.
[a]Each strain of *P. jaceae* was used to inoculate both the host from which it came and cornflower, a nontarget species universally susceptible to *P. jaceae*.
[b]Mean ± standard deviation for all infected leaves.

Table 3. Root dry weight (g) of cornflower, diffuse knapweed (DK), purple starthistle (PS), and yellow starthistle (YS) inoculated with three strains of *Puccinia jaceae*

Treatment[b]	Dry weight (g) of host roots Strain of *P. jaceae*[a] On host DK	PS	YS	On cornflower DK	PS	YS
C	1.6a	2.9a	2.3a	1.1a	0.3a	0.3a
2 + 3	1.6a	2.5a	1.8b	1.1a	0.3a	0.2a
5 + 6	1.4a	2.8a	1.7b	1.0a	0.2a	0.2a

SOURCE: Adapted from Shishkoff and Bruckart, 1993.
[a]Each strain of *P. jaceae* was used to inoculate both the host from which it came and cornflower, a nontarget species universally susceptible to *P. jaceae*.
[b]Treatments are: C = uninoculated control, 2 + 3 = plants inoculated twice at 2 and 3 wk after planting, 5 + 6 = plants inoculated twice at 5 and 6 wk after planting.

musk thistle transplants in plots were inoculated and observed for infection. Transplants of the test species were placed between the plots and received inoculum from infected musk thistle in the plots. Musk thistle monitor plants were placed around the plots or transplanted with test plants between plots. Good infection occurred on inoculated musk thistle in the plots and on the uninoculated monitor plants between or outside the plots; no infection occurred on any of the *Cirsium* spp. and only one very small pustule developed on one of 32 artichokes late in the second year (Baudoin et al., 1993).

Judgment of Results. Even with careful planning, results always require interpretation, and the final step is to make a judgment about risk based primarily on data and the literature (Cullen, 1990). Researchers do not submit proposals for high risk candidates. In proposals for use of *P. carduorum* and *P. jaceae*, we started with two obligate, autoecious parasites, evaluated them at optimal conditions for disease, compared them with known pathogens in the United States, considered information in the literature, and made the judgment that the risk was very acceptable.

Alexander (1990) suggests that proper assessment of risk should involve individuals who are not specialists in the technology. It is up to the scientist to develop an adequate case for these specialists, but those who make the decisions need to understand both the process the scientist follows and the rationale for submitting the proposal. This is likely to require communications between scientist and decision-makers. Such should be encouraged.

References

Agrios, G. N. "Escape from disease." In *Plant Disease: An Advanced Treatise*; Horsefall, J. G.; Cowling, E. B., Eds; Academic Press: New York, NY, 1980, Vol. 5; pp 17-37.

Agrios, G. N. *Plant Pathology*, Third Ed.; Academic Press: New York, NY, 1988; p 779.

Alexander, M. "Potential impacts on community function." in *Risk Assessment in Agricultural Biotechnology*; Marois, J. J., Bruening, G.; Eds.; Univ. Calif.: Oakland, CA, 1990; pp 121-125.

Baudoin, A. B. A. M.; Abad, R. G.; Kok, L. T.; Bruckart, W. L. "Field evaluation of *Puccinia carduorum* for biological control of musk thistle." *Biol. Contr.* **1993**, *3*, (In Press).

Bennett, A. R., Bruckart, W. L., Shishkoff, N. "Effects of dew, plant age, and leaf position on the susceptibility of yellow starthistle to *Puccinia jaceae*." *Plant Dis.* **1991**, *75*, 499-501.

Bruckart, W. L. "Host range determination of *Puccinia jaceae* from yellow starthistle." *Plant Dis.* **1989**, *73*, 155-160.

Bruckart, W. L., Dowler, W. M. "Evaluation of exotic rust fungi in the United States for classical biological control of weeds." *Weed Sci.* **1986**, *34* (Supplement 1), 11-14.

Bruckart, W. L.; Peterson, G. L. "Phenotypic comparison of *Puccinia carduorum* from *Carduus thoermeri*, *C. tenuiflorus*, and *C. pycnocephalus*." *Phytopathology* **1991**, *81*, 192-197.

Bruckart, W. L.; Politis, D. J.; Sutker, E. M. "Susceptibility of *Cynara scolymus* (artichoke) to *Puccinia carduorum* observed under greenhouse conditions." In *Proc. VI Int. Symp. Biol. Contr. Weeds*; Delfosse, E. S., Ed.; Agric. Can.: Ottawa, Canada, 1985; pp 603-607.

Bruzzese, E.; Hasan, S. "Host specificity of the rust *Phragmidium violaceum*, a potential biological control agent of European blackberry." *Ann. Appl. Biol.* **1986**, *108*, 585-596.

Cerkauskas, R. F. "Latent colonization by *Colletotrichum* spp.: Epidemiological considerations and implications for mycoherbicides." *Can. J. Plant Pathol.* **1988**, *10*, 297-310.

Charudattan, R. "Release of fungi: Large-scale use of fungi as biological weed control agents." *In Risk Assessment in Agricultural Biotechnology*; Marois, J. J., Bruening, G.; Eds.; Univ. Calif.: Oakland, CA, 1990; pp 70-84.

Charudattan, R. "The mycoherbicide approach with plant pathogens." In *Microbial Control of Weeds*; TeBeest, D. O., Ed.; Chapman and Hall: New York, NY, 1991; pp 24-57.

Cullen, J. M. "Current problems in host-specificity testing" In *Proc. VII Int. Symp. Biol. Contr. Weeds*; Delfosse, E. S., Ed.; Inst. Sper. Patol. Veg. (MAF): Rome, Italy, 1990; pp 27-36.

Fowler, S. V.; Holden, A. G. N.; Schroeder, D. "The possibilities for classical biological control of weeds of industrial and amenity land in the U.K. using introduced insect herbivores or plant pathogens." *Proc. Brighton Crop Prot. Conf., Weeds.* **1991**, *3*, 1173-1180.

Gardner, D. E. *Role of biological control as a management tool in national parks and other natural areas;* Technical Report NPS/NRUH/NRTR-90/01; U.S. Dept. Int.: Washington, DC, **1990**; 40 pp.

Gaumann, E. *Die Rostpilze Mitteleuropas.* Buchler and Co.: Bern, Switzerland, **1959**; 1407 pp.

Hasan, S.; Chaboudez, P.; Mortensen, K. "Field experiment with European knapweed rust (*Puccinia jaceae*) on safflower, sweet sultan and bachelor's button." In *Proc. VII Int. Symp. Biol. Contr. Weeds*; Inst. Sper. Patol. Veg. (MAF): Rome, Italy, 1990; pp 499-509.

Haygood, R. A. "Field control of annual bluegrass (*Poa annua* L.) using a unique pathovar of the bacterium *Xanthomonas campestris*." *WSSA Abstracts* 1992, 28.

Inman, R. E. "A preliminary evaluation of *Rumex* rust as a biological control agent for curly dock." *Phytopathology* 1971, *61*, 102-107.

Joye, G. F. "Biological control of aquatic weeds with plant pathogens." ACS Symposium Series: No 439; American Chemical Society: Washington, DC, 1990; pp 155-174.

Kennedy, A. C., Elliott, L. F., Young, F. L., Douglas, C. L. "Rhizobacteria suppressive to the weed downy brome." *Soil Sci. Soc. Am. J.* 1991, *55*, 722-727.

Kenney, D. S. "DeVine - the way it was developed - An industrialist's view." *Weed Sci.* 1986, *34* (Supplement 1), 15-16.

Kremer, R. J., Begonia, M. F. T., Stanley, L., Lanham, E. T. "Characterization of rhizobacteria associated with weed seedlings." *Appl. Environ. Microbiol.* 1990, *56*, 1649-1655.

Leonard, K. J. "The benefits and potential hazards of genetic heterogeneity in plant pathogens." In Charudattan, R.; Walker, H. L.; Eds.; *Biological Control of Weeds with Plant Pathogens*; John Wiley and Sons: New York, NY, 1982; pp 99-112.

Mussell, H. "Tolerance to Disease." In *Plant Disease: An Advanced Treatise*; Horsefall, J. G.; Cowling, E. B., Eds; Academic Press: New York, NY, 1980, Vol. 5; pp 39-52.

Nutter, F. W.; Teng, P. S.; Royer, M. H. "Terms and concepts for yield, crop loss, and disease thresholds." *Plant Dis.* 1993, *77*, 211-215.

Parker, P. E. "Nematodes as biological weed control agents." In *Microbial Control of Weeds*; TeBeest, D. O., Ed.; Chapman and Hall: New York, NY, 1991; pp 58-68.

Politis, D. J.; Bruckart, W. L. "Infection of musk thistle by *Puccinia carduorum* influenced by conditions of dew and plant age." *Plant Dis.* 1986, *70*, 288-290.

Politis, D. J.; Watson, A. K.; Bruckart, W. L. "Susceptibility of musk thistle and related composites to *Puccinia carduorum*." *Phytopathology* 1984, *74*, 687-691.

Ridings, W. H. "Biological control of stranglervine in citrus - A researcher's view." *Weed Sci.* 1986, *34* (Supplement 1), 31-32.

Savile, D. B. O. "Some eurasian *Puccinia* species attacking Cardueae." *Can. J. Bot.* 1970, *48*, 1553-1556.

Schubiger, F. X.; Defago, G.; Sedlar, L.; Kern, H. "Host range of haplontic phase of *Uromyces rumicis*." In *Proc. VI Int. Symp. Biol. Contr. Weeds*; Delfosse, E. S., Ed.; Agric. Can.: Ottawa, Canada, 1985; pp 653-659.

Shishkoff, N.; Bruckart, W. L. "Evaluating infection of target and nontarget hosts by isolates of the potential biocontrol agent, *Puccinia jaceae*, that infect *Centaurea* spp." *Phytopathology* 1993, *83*: (In Press).

Shrum, R. D. "Creating epiphytotics." In Charudattan, R.; Walker, H. L.; Eds.; *Biological Control of Weeds with Plant Pathogens*; John Wiley and Sons: New York, NY, 1982; pp 113-136.

TeBeest, D. O., Yang, X. B., Cisar, C. R. "The status of biological control of weeds with fungal pathogens." *Annu. Rev. Phytopath.* 1992, *30*, 637-657.

Templeton, G. E. "Status of weed control with plant pathogens." In Charudattan, R.; Walker, H. L.; Eds.; *Biological Control of Weeds with Plant Pathogens*; John Wiley and Sons: New York, NY, 1982; pp 29-44.

Wapshere, A. J. "A protocol for programmes for biological control of weeds." *PANS* 1975, *21*, 295-303.

Watson, A. K. "Host specificity of plant pathogens in biological weed control." In *Proc. VI Int. Symp. Biol. Contr. Weeds*; Delfosse, E. S., Ed.; Agric. Can.: Ottawa, Canada, 1985; pp 577-586.

Watson, A. K. "Host range of, and plant reaction to, *Subanguina picridis*." *J. Nematol.* 1986, *18*, 112-120.

Watson, A. K. "The classical approach with plant pathogens." in *Microbial Control of Weeds*; TeBeest, D. O., Ed.; Chapman and Hall: New York, NY, 1991; pp 3-23.

Watson, A. K.; Brunetti, K. "*Puccinia carduorum* on *Carduus tenuiflorus* in California." *Plant Dis.* 1984, *68*, 1003-1005.

Weidemann, G. L. "Host-range testing: safety and science." in *Microbial Control of Weeds*; TeBeest, D. O., Ed.; Chapman and Hall: New York, NY, 1991; pp 83-96.

Wilson, C. L. "Use of plant pathogens in weed control." *Annu. Rev. Phytopath.* 1969, *7*, 411-434.

Arthropods for Suppression of Terrestrial Weeds

Richard D. Goeden, Department of Entomology, University of California, Riverside, CA 92521

Abstract. The biological control of terrestrial weeds with arthropods in the continental United States and Canada is reviewed from its inception in California during the late 1930's to date. A total of 165 species of natural enemies (insects, mites, nematodes, and phytopathogens) has been released for the biological control of 75 species or congeneric species groupings of weeds. Of these, 117 species have been arthropods, including 114 species of insects and three species of mites. The overwhelming majority of these releases comprised exotic species of insects intentionally introduced for classical biological control of weeds. Major developments, trends, and actions involved in the growth and maturation of biological control of weeds with insects are highlighted and discussed.

The use of phytophagous arthropods to reduce or suppress weed population densities (known throughout the world as "biological control" in all but the "USDA-speak" used in the titles in these Proceedings) has undergone considerable changes in North America since its beginnings in California during the late 1930's, 1940's, and early 1950's (Goeden 1978, 1988). In this paper I present a broad overview of these trends and events relative to the biological control of terrestrial weeds with insects, lauding some trends and actions and deploring others, then assess briefly the state of the art, while itemizing some of the things learned along the way.

Where We Have Been

Biological control of weeds in North America began as a joint-USDA/University of California venture by J. K. Holloway (USDA) and H. S. Smith (UC) with the project on biological control of prickly pear cacti (*Opuntia* spp.) on Santa Cruz Island off the coast of southern California during the late 1930's (Goeden et al. 1967). At the close of WWII and ably assisted by C. B. Huffaker, they next undertook the spectacularly successful project on biological control of Klamath weed (*Hypericum perforatum* L.) in northern and central California (Holloway and Huffaker 1949). This latter project started in the mid-1940's and other early projects started in the United States during the 1950's and early 1960's, i.e., ragwort (*Senecio jacobaea* L.) and gorse (*Ulex europeaus* L.), relied initially on European insects previously studied and imported by Australian and New Zealand workers, who pioneered biological control of weeds with insects during the first half of the 20th Century (Goeden 1978, 1988; Julien 1992).

Canada also began its biological control of weeds program with a "transfer project" by subsequently introducing the insects successfully used on Klamath weed (= St. Johnswort) from California in the early 1950's, but soon took an independent course with respect to weeds targeted, agencies involved, and means of obtaining natural enemies. Thus, by the late 1950's, the Canada Department of Agriculture (now Agriculture Canada) was introducing insects for biological control of Canadian weeds that were provided under contract by the European Station of the Commonwealth Institute of Biological Control (CIBC) [now CAB International Institute of Biological Control (IIBC)] (Harris 1971; Julien 1992). Early Canadian projects involving this approach (which continues today, cf., Schroeder and Goeden 1986) began introducing insects on common toadflax (*Linaria vulgaris* Miller) in 1952, and on Canada thistle [*Cirsium arvense* (L.) Scopoli] in 1963 (Harris 1971).

Meanwhile, biological control of terrestrial weeds research in the United States continued largely as a joint USDA-Agricultural Research Service (ARS)/UC venture during the 1960's. Domestic research was centered at Albany and Riverside, California, and overseas in Rome, Italy. The first projects unique to the continental United States were on Scotch broom (*Cytisus scoparius* (L.) Link (Frick 1964) and puncturevine (*Tribulus terrestris* L.) (Andres and Angelet 1963, Maddox 1976). Additional joint projects begun during the mid-1960's targeted the Mediterranean weeds, yellow starthistle (*Centaurea solstitialis* L.), Russian thistle (*Salsola australis* R. Brown), milk thistle [*Silybum marianum* (L.) Gaertner], and Italian thistle (*Carduus pycnocephalus* L.), and the predominantly temperate field bindweed (*Convolvulus arvensis* L.) (Julien 1992). Concurrently, projects undertaken solely by the USDA-ARS were begun on halogeton

[*Halogeton glomeratus* (M. von Bieberstein) C. Meyer] (Goeden et al. 1974) and Mediterranean sage (*Salvia aethiopis* L.) (Andres and Rizza 1965). Each of these projects on terrestrial weeds resulted in the introduction of one or more insect species obtained mainly from southern Europe, but also from Pakistan and Turkey during the 1960's and early 1970's, with immediate and considerable benefits documented for puncturevine (Kirkland and Goeden 1978, Maddox 1981, Huffaker et al. 1983).

Another project on alligatorweed [*Alternanthera phylloxeroides* (C. Martius) Grisebach] set precedent by targeting an aquatic weed. This project also redirected and expanded federal research activities away from California and adjacent western States into the southeastern United States, where aquatic weeds were major problems (Andres and Bennett 1975), and ultimately into South America to obtain natural enemies (Coulson 1977).

The 1960's and 1970's saw expansion of biological control of weeds activities in both Canada and the United States, mainly within the USDA-ARS, but also at selected land-grant universities, and in Hawaii by the State Department of Agriculture (Goeden et al. 1974). Canada began importations of insects on spurges, *Euphorbia* spp., in 1965 and on *Carduus* thistles in 1968 (Harris 1971). Shortly thereafter, USDA-ARS entomologists and cooperators at selected eastern and western universities in the United States also targeted these weeds and began transferring the flower-head weevil, *Rhinocyllus conicus* Froelich, from Canada in 1969 (Kok and Surles 1975). They later imported additional material from Europe, including biotypes of *R. conicus* selectively attacking different *Carduus* spp. or milk thistle (Goeden et al.1985). Biological control projects on *Carduus, Cirsium,* and *Silybum* thistles, and subsequently, spurges and knapweeds (*Centaurea* spp.) in the United States have benefited greatly from these and subsequent natural enemy transfers and from enabling research on weeds sponsored in Europe by Canada, much more so, in my opinion, than Canada has benefited reciprocally.

The 1980's was a decade of enhanced ecological or environmental awareness, but contrary to what otherwise might be expected, research and development activities in biological control of terrestrial weeds in the United States did not proportionally benefit from this movement. Instead, natural enemy importations languished as USDA-ARS and university researchers studied natural systems, partly in response to ecological concerns generated by passage of the Endangered Species Act and similar state legislation for protection of endangered and threatened native plant species

(Andres 1985, Turner 1985, Goeden et al. 1985, Goeden and Ricker 1987, Turner et al. 1987). For example, up to that time native congeners or close relatives of alien weeds largely had been ignored or their intrinsic worth discounted relative to losses caused by target weeds in procedures for screening the host specificities of arthropods for biological control of weeds (Zwölfer and Harris 1971, Andres 1985, Goeden and Ricker 1987, Harris 1988).

Other events symptomatic of the growth of the subdiscipline of biological control of weeds during the 1980's presaged subsequent parallel developments in the subdiscipline of biological control of arthropod pests with imported parasites and predators. For example, guidelines for introducing foreign organisms into the United States for biological control of weeds were formulated in consultation with federal and university biological control scientists. These guidelines were widely published (Klingman and Coulson 1982a, 1982b, 1983) in an attempt to standardize procedures for U.S. biological control scientists. Another symptom of growth was the new, state-of-the-art, quarantine greenhouse and laboratory for USDA-ARS research on biological control of terrestrial weeds built at Albany, California, during the mid-1980's. Unfortunately, this magnificent facility was all but abandoned by administrative decree in 1987 and most research personnel retired or transferred to another new, state/university laboratory in Bozeman, Montana. At last word, only one full-time research scientist and a technician now occupy the multi-million dollar Albany facility (Turner et al. 1992). In its favor, the facility in Montana is located in the Northwest, the region of current extensive research activity in biological control of rangeland weeds (Nowierski 1985), where several States support active programs in biological control with insects (Story 1985, Piper 1985). A research unit on biological control of native and other rangeland weeds was established at the USDA-ARS Grasslands, Soil, and Water Research Laboratory in Temple, Texas. Most terrestrial weeds targeted for biological control with insects worldwide have been introduced perennial rangeland weeds (Julien 1992).

Cashing-in on newfound and widespread public willingness to support environmentally responsible alternatives to pesticides, other federal agencies (e.g., USDA-APHIS), various states, even county agencies and commercial companies were created, or independently assumed responsibility, to *implement* biological control, usually under the guise of integrated pest management (IPM). Justifications advanced for formation of these units were multitudinous, e.g., the implication that available imported natural enemies were underutilized; the promise that new natural enemies could more rapidly be put into widespread use by these new agencies following their importation by biological control scientists; and a host of other rationales exploiting the good-will and positive environmental, economic, and efficaceous attributes of biological control. Thus, with suitable fanfare and maximum publicity, these implementation agencies began to spread whatever agents had been introduced by USDA-ARS, Agriculture Canada, and university workers up to that time, taking care to document their efforts with photographs, news releases, videotapes, and as a token gesture to science, counts of agents released. Sometimes these redistributions took place *ad absurdum,* whereby every isolated, minor weed colony received its "dosage" (some of these IPM or newly recruited, biological control workers formerly had dispensed chemical pesticides with equal fervor!) of long-established natural enemies, and until further blanket releases like these became indefensible.

The result of these activities was that the total effort in biological control of terrestrial weeds with insects was greatly expanded during the late 1970's and 1980's, but this expansion largely came at the expense of the supply side of biological control--and ultimately, these implementation agencies ran out of imported agents to implement (or shortly will)! Public funds spent on implementation are funds not spent on research and development! One remedy devised for keeping these implementation agencies in business after they ran through their initial supply of readily available, already imported insect agents was to embrace the augmentative approach. This approach further required building, staffing, and maintaining expensive insectary facilities to rear selected species of newly-acquired agents for release in ever greater numbers at ever decreasing intervals. Alternatively, some implementation agencies also have become disbursement agencies that identify targets, set priorities, and dole out research grants and contracts, and not incidently, thereby assure their own survival. Another recourse to this depletion of "implementable" natural enemies is for the implementation agency to enter the research and development arena and undertake foreign research independently, or to contract for same. This latter approach indeed has happened in the case of the recently developed project on the biological control of purple loosestrife (*Lythrum salicaria* L.), an aquatic weed. At a total cost of $543,000, an IIBC research project recently was completed within 7 years, from start to final screening reports, that included environmental assessments for six control

agents. The first surveys in Europe during 1986-87 were initiated and funded by the USDA-ARS, and the following 5 years were supported by the U.S. Department of Interior (USDI), Fish and Wildlife Service, with additional funding during the last 2 years provided by Washington State Department of Agriculture and Wildlife. The first three insect agents were released just last year in seven States during the seventh year of this project (D. Schroeder, pers. commun. 1993). Consortia like this provide a partial answer to the perennial problem of underfunding of classical biological control of weeds research and development (Schroeder and Goeden 1986, Harris 1991).

Another recent use made of biological control has been to sell high technology as an aid to taxonomic study and precise identification of target hosts and natural enemies. Improvement of existing biological control agents through genetic engineering and proposals to use designer, arthropod "anatural enemies" for terrestrial weed control has found few buyers to date (Harris 1991, Goeden 1992, Turner et al. 1992).

Where We Are Now

To date, as determined from published and unpublished sources (see Acknowledgments), 165 species of natural enemies (insects, mites, nematodes, and phytopathogens) have been released in the continental United States and Canada for biological control of 75 species or congeneric species groupings, e.g., *Cuscuta* spp., of terrestrial weeds. This calculation essentially followed the criteria of Julien (1992), who catalogued only initial field releases of a natural enemy in a particular country. However, the above calculation did not count as separate projects any subsequent transfers between the United States and Canada of the same species of natural enemy initially introduced into the other country on the same weed. The three species each of exotic or native nematodes and mites used in biological control of weeds in North America to date were included in the above calculation. The 75 species of exotic insects and pathogens released for biological control of 18 Hawaiian weeds also were not counted, as they warrant separate analysis involving unique plant species largely addressed independently during Hawaii's long history of participation in biological control of weeds (Waterhouse and Norris 1987, Goeden 1988, Davis et al. 1993, Julien 1992). I also acknowledge only in passing and mainly for purposes of comparison that about 50 species of phytopathogens, mostly indigenous fungi, have been evaluated in the United States and Canada as bioherbicides involving at least some field trials (A. Watson, pers. commun. 1992). Most of this research activity with phytopathogens has occurred since Goeden et al. (1974) listed agents introduced for biological control of weeds in the United States and Canada.

To date, 114 species of phytophagous insects have been released in the continental United States or Canada for biological control of 35 species or congeneric species groups of terrestrial weeds. This calculation includes an overwhelming majority of imported species of insects and the few native or naturalized species that have been mass-reared and used on an experimental basis in inundative releases against yellow and purple nutsedges, *Cyperus* spp. (Keeley et al. 1970, Frick and Chandler 1978, Frick et al. 1983) and hedge bindweed, *Convolvulus sepium* L. (Parrella and Kok 1979). The remaining 111 releases of insect agents have involved the so-called "classical" approach to biological control, i.e., the introduction of agents into new areas, but not simple redistributions of native or accidentally and intentionally introduced insects, e.g., interstate or interprovincial transfers. However, some insect species were counted more than once if released on different weed species, e.g., *Rhinocyllus conicus*.

As already noted, the United States has benefited greatly from the Canadian program on biological control of terrestrial weeds with insects. About 30 insect species introduced by Canada on 11 target weeds subsequently were transferred or directly introduced from Eurasia into the United States for biological control of the same weed. This is about half the number of insect species subsequently transferred from the United States to Canada on a reciprocal basis. On the other hand, 43 additional releases of insect species have been made in the United States independently of Canada; whereas, 24 insect species released in Canada were not subsequently transfered to the United States. These latter figures reflect, among other factors, agents that were not established, differences in weed floras, differing priorities among mutually-targeted weeds, localized projects on weeds absent in the other country, climatological differences, numbers of researchers in the continental 48 States versus Canada as well as initial insect introductions too recent to allow transfers to have taken place to the other country. The number of initial releases of natural enemies are one measure of research activity, as these agents represent the main stock in trade of classical biological control of weeds (Julien 1992). However, their introduction represents just one phase of any classical biological control program (Harris

1971, 1991; Schroeder and Goeden 1986; Turner et al. 1992). Too often, natural enemy releases and redistributions are highly publicized, but their subsequent effectiveness is given little mention, and unfortunately, may remain unstudied, especially if their effect is subtle or minimal (Schroeder and Goeden 1986).

Harris (1991) proposed four graded steps for evaluating the "success" of biological control of weeds programs, which recognized progress in terms of natural enemy establishment, "biological success," host-plant impact, and control achieved. Stepwise, measured progress on biological control projects can and should be recognized. Unfortunately, research on natural enemy effectiveness is largely ignored or grossly underfunded (Schroeder and Goeden 1986). Most current projects on terrestrial weeds in North America would register as at least partial successes under Harris' (1991) scheme (Julien 1992). However, aside from early successes with prickly pear cacti, Klamath weed, and puncturevine in California, only the biological control of musk thistle in Canada (Harris 1984) and the northeastern United States (Kok and Surles 1975), plumeless thistle (*Carduus acanthoides* L.) in Virginia (Kok and Mays 1991), and, perhaps, tansy ragwort in northern California (Hawkes and Johnson 1978) and Oregon (McEvoy et al. 1991), qualify as substantial to complete successes among projects on terrestrial weeds in North America to date. Given the long development times for some projects (Harris 1979), this record is not surprising, considering the relatively recent starting dates of some projects. However, some projects of long standing clearly have been unsuccessful, e.g., field bindweed and nutsedges (Julien 1992).

The many federal and State agencies and laboratories and State universities in the United States and agencies and universities in Canada currently involved with different degrees in research of various persuasion on biological control of weeds were listed in Julien (1992). The USDA-ARS and Agriculture Canada currently are by far the main practitioners of biological control of terrestrial weeds with insects in North America.

The Future

What form biological control of terrestrial weeds with insects in North America will take in the 21st Century is difficult to predict in light of today's fast-changing politico-economic context. Teaching and research and development activities in classical biological control at the University of

California have been sharply curtailed and USDA-ARS laboratories at Albany and Riverside closed or underutilized in recent years, despite continued introductions of new foreign pests. Some suggest that the IPM researchers will conduct whatever biological control research might be needed in the future. Evidence that part-time dabbling by novices in biological control will be effective so far is lacking. Instead, we continue to see glib pronouncements that biological control is but one of several tools available for use in IPM programs. In other words, "Because one says, 'I use it', one can, and does, do it!" I suggest, instead, that there will continue to be a need for well-trained, experienced researchers specializing in one of the major subdisciplines of biological control, such as the biological control of weeds, which for the foreseeable future substantially will remain an art that requires a full-time philosophical committment.

Full recognition, adoption, and integration of alternatives to herbicides such as biological control by the field of Weed Science unfortunately also remains a future undertaking (Zimdahl 1991, Turner et al. 1992). This action probably awaits an environmentally healthy dose of herbicidal resistance among weeds, which as a tonic similar to insecticidal resistance among insect pests, the field of Economic Entomology reluctantly had to swallow two decades ago.

Recommendations

1. The United States should be at the forefront of biological weed control, but presently is not, in large part because of duplicative, uncoordinated programs and wasteful competition among the USDA-ARS, APHIS, universities, and State biological control agencies. Multidisciplinary, multi-agency teams of applied and theoretical researchers and implementers assigned and funded long-term to work on the biological control of specific target weeds (or congeneric weed species) may provide one answer.

2. The IIBC now is clearing agents for release in North America faster than the USDA-ARS. Why not use present federal resources to contract enhanced programs through the IIBC?

3. The cost of screening candidate agents is sufficiently high that whenever possible, it should be financed as a cooperative international effort between the United States and Canada or Mexico (and other interested countries).

4. The APHIS agent distribution program is expensive and could be better done by States. APHIS currently merely transfers insects, not the essential technology, and feed-back from their release efforts via effectiveness studies necessary to improve future success, currently an ARS responsibility, has been minimal.

5. There is a continuing need and place for private firms to distribute biological control agents and information. At present, these firms range from highly responsible to "fast-buck" operations. Why not use licensing of firms to eliminate the worst offenders?

6. Biological control of weeds is partly constrained by the costs and time needed for agent approval. Why not use experts in biological control on the Technical Advisory Committee (TAG) (Coulson 1992), including university authorities, not just representatives of federal stake-holder agencies, to advise and help regulators to expedite their risk-benefit judgments?

Acknowledgments

I especially thank Lloyd Andres, Peter Harris, Loke Kok, Cliff Moran, Dieter Schroeder, and Helmut Zwölfer, and many other colleagues too numerous to single out, for their stimulating and provocative discussions and correspondence during the past quarter of a century which helped to form some of the opinions expressed in this paper as we worked together in our respective countries on biological control of weeds with insects. However, the opinions expressed and the calculations offered in this paper are my sole responsibility. I also note with gratitude the access to unpublished tabulations or manuscripts provided by Lloyd Andres, Jack Coulson, Peter Harris, and Alan Watson, on which this paper was based in part. I sincerely thank Tom Bellows, Jack Coulson, David Headrick, Peter Harris, Loke Kok, Dieter Schroeder, and Jeff Teerink for their thoughtful and helpful comments on early drafts of this paper.

References

Andres, L. A. In *Proceedings of the VI International Symposium on Biological Control of Weeds;* Delfosse, E. S., Ed., Agriculture Canada, Ottawa, **1985**, pp 235-239.

Andres, L. A.; Angelet, G. W. "Notes on the Ecology and Host Specificity of *Microlarinus lareynii* and *M. lypriformis* (Coleoptera: Curculionidae) and the Biological Control of Puncture Vine,

Tribulus terristris." *J. Econ. Entomol.* **1963**, *56*, 333-340.

Andres, L. A.; Bennett, F. D. "Biological Control of Aquatic Weeds." *Ann. Rev. Entomol.* **1975**, *20*, 31-46.

Andres, L. A.; Rizza, A. "Life history of *Phrydiuchus topiarus* (Coleoptera: Curculionidae) on *Salvia verbenacea* (Labiatae)." *Ann. Entomol. Soc. Am.* **1965**, *58*, 314-319.

Coulson, J. R. "Biological Control of Alligatorweed, 1959-1972. A Review and Evaluation." *U.S. Dept. Agric. Tech. Bull.* **1977**, *1547*, 1-98.

Coulson, J. R. In *Proceedings of a USDA/CRS Workshop on Regulations and Guidelines: Critical Issues in Biological Control;* Charudattan, R.;Browning, H. W., Eds.; Instituted of Food and Agricultural Sciences, University of Florida, Gainesville, 1992, pp 53-60.

Davis, C. J.; Yoshioka, E.; Kageler, D. In *Biocontrol of Lantana camara, Opuntia spp., and Ageratina riparia in Hawaii; a Review and Update;* Smith, C.; Tunison, T.; Stone, C. P., Eds; University of Hawaii Press: Honolulu, HI, 1993, in press.

Frick, K. E. *Leucoptera spartifoliella*, an Introduced Enemy of Scotch Broom in the Western United States." *J. Econ. Entomol.* **1964**, *57*, 589-591.

Frick, K. E.; Chandler, J. M. "Augmenting the Moth *(Bactra verutana)* in Field Plots for Early-Season Suppression of Purple Nutsedge *(Cyperus rotundus). Weed Sci.* **1978**,*26*, 703-710.

Frick, K. E.; Hartley, G. G.; King, E. G. "Large Scale Production of *Bactra verutana* (Lep.: Tortricidae) for the Biological Control of Nutsedge." *Entomophaga.* **1983**, *28*, 107-115.

Goeden, R. D. In *Introduced Parasites and Predators of Arthropod Pests and Weeds: A World Review;* Clausen, C. P., Ed.; U. S. Department of Agriculture Agricultural Handbook 480, Washington, D.C., 1978, Part II; pp 357-545.

Goeden, R. D. "A Capsule History of Biological Control of Weeds."*Biocontr. News Info.* **1988**, *9*, 55-61.

Goeden, R. D. In *Proceedings of a USDA/CRS Workshop on Regulations and Guidelines: Critical Issues in Biological Control;* Charudattan, R.; Browning, H. W., Eds.; Institute of Food and Agricultural Sciences, University of Florida, Gainesville, 1992, pp 107-114.

Goeden, R. D.; Andres, L. A.; Freeman, T. E.; Harris, P.; Pienkowski, R. L.; Walker, C. R. "Present Status of Projects on the Biological Control of Weeds with Insects and Plant Pathogens in the United States and Canada." *Weed Sci.,* **1974**, *22*, 490-495.

Goeden, R. D.; Fleschner, C. A.; Ricker, D. W. "Biological control of prickly pear cacti on Santa Cruz Island, California." *Hilgardia.* **1967**, *38*, 509-606.

Goeden, R. D.; Ricker, D. W. "Phytophagous Insect Faunas of Native *Cirsium* Thistles, *C. mohavense, C. neomexicanum,* and *C. nidulum,* in the Mojave Desert of Southern California." *Ann. Entomol. Soc. Am.* **1987**, *80*, 161-175.

Goeden, R. D.; Ricker, D. W.; Hawkins, B. A. In *Proceedings of the VI International Symposium on Biological Control of Weeds;* Delfosse, E. S., Ed.; Agriculture Canada, Ottawa, 1985, pp 181-189.

Harris, P. In *Biological Control Programmes Against Insects and Weeds in Canada 1959-1968;* Simmonds, F. J., Ed.; Commonwealth Institute of Biological Control Technical Communication No. 4, Commonwealth Agricultural Bureaux, Farnham Royal, Slough, England, 1971, pp 67-76.

Harris, P. "Cost of Biological Control of Weeds by Insects in Canada." *Weed Sci.* **1979**, *27*, 242-250.

Harris, P. In *Biological Control Programmes Against Insects and Weeds in Canada 1969-1980;* Kelleher, J. S; Hume, M. A., Eds., Commonwealth Agricultural Bureaux, Farnham Royal, Slough, England, 1984, pp 159-169.

Harris, P. "Environmental Impact of Weed-control Insects." *Bioscience* **1988**, *38*, 542-548.

Harris, P. "Invitation Paper (C.P. Alexander Fund): Classical Biocontrol of Weeds: Its Definition, Selection of Effective Agents, and Administrative-Political Problems." *Can. Entomol.* **1991**, *123*, 827-849.

Hawkes, R. B.; Johnson, G. R. In *Proceedings of the IV International Symposium on Biological Control of Weeds,* Freeman, T. E., Ed., University of Florida, Gainesville, 1987, pp 193-196.

Holloway, J. K.; Huffaker, C. B. "Klamath Weed Beetles." *Calif. Agric.* **1949**, *3*, 3-10.

Huffaker, C. B.; Hamai, J.; Nowierski, R. M. "Biological Control of Puncturevine, *Tribulus terrestris* in California After Twenty Years of Activity of Introduced Weevils." *Entomophaga* **1983**, *28*, 387-400.

Julien, M. H., Ed.; *Biological Control of Weeds: A World Catalogue of Agents and Their Target Weeds;* C.A.B. International, Wallingford, U.K., 1992.

Keeley, P. E.; Thullen, R. J.; Miller, J. H. "Biological Control Studies on Yellow Nutsedge with *Bactra verutana* Zeller." *Weed Sci.* **1970**, *18*, 393-395.

236

Kirkland, R. L.; Goeden, R. D. "An Insectical-Check Study of the Biological Control of Puncturevine (*Tribulus terrestris*) by Imported Weevils, *Microlarinus lareynii* and *M. lypriformis* (Col.: Curculionidae)." *Environ. Entomol.* **1978,** *7*, 349-354.

Klingman, D. L.; Coulson, J. R. "Guidelines for Introducing Foreign Organisms Into the United States for Biological Control of Weeds." *Weed Sci.* **1982a,** 661-667. *Plant Disease* **1982b,** 1205-1209. *Bull. Entomol. Soc. Am.* **1983,** 55-61.

Kok, L. T.; Mays, W. T. "Successful Biological Control of Plumeless Thistle, *Carduus acanthoides* L.[Campanulatae: Asteraceae (= Compositae)], by *Trichosirocalus horridus* (Panzer) (Coleoptera: Curculionidae) in Virginia. Biol. Control. **1991,** *1*, 197-202.

Kok, L. T.; Surles, W. W. "Successful Biological Control of Musk Thistle by an Introduced Weevil, *Rhinocyllus conicus.*" *Environ. Entomol.* **1975,** *4*, 1025-1027.

Maddox, D. M. "History of Weevils of Puncturevine in and Near the United States." *Weed Sci.* **1976,** *24*, 414-416.

Maddox, D. M. In *Proceedings of the V International Symposium on Biological Control of Weeds;* Delfosse, E. S., Ed.; Commonwealth Scientific and Industrial Research Organization, Melbourne, Australia, 1981, pp 447-467.

McEvoy, P.; Cox, C.; Coombs, E. "Successful Biological Control of Ragwort, *Senecio jacobaea,* by Introduced Insects in Oregon."*Ecol. Applications* **1991,** *1*, 443-452.

Nowierski, R. M. In *Proceedings of the VI International Symposium on Biological Control of Weeds;* Delfosse, E. S., Ed., Agriculture Canada, Ottawa, 1985, pp 811-815.

Parrella, M. P.; Kok, L. T. "*Oidaematophorus monodactylus* as a Biocontrol Agent of Hedge Bindweed: Development of a Rearing Program and Cost Analysis." *J. Econ. Entomol.* **1979,** *72*, 590-592.

Piper, G. L. In *Proceeedings of the V International Symposium on Biological Control of Weeds;* Delfosse, E. S., Ed.; Agriculture Canada, Ottawa, 1985, pp 817-826.

Schroeder, D.; Goeden, R. D. "The Search for Arthropod Natural Enemies of Introduced Weeds for Biological Control--In Theory and Practice." *Biocontr. News Info.* **1986,** *7*, 147-155.

Story, J. M. In *Proceedings of the VI International Symposium on Biological Control of Weeds;* Delfosse, E. S., Ed.; Agriculture Canada, Ottawa, 1985, pp 837-842.

Turner, C. E. In *Proceedings of the VI International Symposium on Biological Control of Weeds;* Delfosse, E. S., Ed., Agriculture Canada, Ottawa, 1985, pp 203-225.

Turner, C. E.; Anderson, L. W.; Foley, P.; Goeden, R. D.; Lanini, W. T.; Lindow, S. E.; Qualset. In *Beyond Pesticides: Biological Approaches to Pest Management in California;* Madden, J. P.; Schroth, M. N., Eds; University of California, Division of Agriculture and Natural Sciences, Oakland, California, 1992, pp 33-67.

Turner, C. E.; Pemberton, R. W.; Rosenthal, S. S. "Host Utilization of Native *Cirsium* Thistles (Asteraceae) by the Introduced Weevil *Rhinocyllus conicus* (Coleoptera: Curculionidae) in California." *Environ. Entomol.* **1987,** *16*, 111-115.

Waterhouse, D. F.; Norris, K. R. *Biological Control: Pacific Prospects;* Inkata Press: Melbourne, Australia, 1987; pp 332-341.

Zimdahl, R. L. *Weed Science A Plea for Thought;* Symposium Preprint, U. S. Department of Agriculture, Cooperative State Research Service, Washington, DC; 1991, pp 1-34.

Zwölfer, H.; Harris, P. "Host Specificity Determination of Insects for Biological Control of Weeds." *Ann. Rev. Entomol.* **1971,** *16*, 159-178.

Wheat Flour Granules Containing Mycoherbicides and Entomogenous Nematodes

William J. Connick, Jr., Southern Regional Research Center, ARS, USDA, New Orleans, LA 70179
William R. Nickle, Nematology Laboratory, USDA, ARS, BARC-W, Beltsville, MD 20705
C. Douglas Boyette, Southern Weed Science Laboratory, ARS, USDA, Stoneville, MS 38776

Versatile new products called "Pesta" consist of granules containing fungal weed pathogens (*Fusarium oxysporum* and *Colletotrichum truncatum*) or entomogenous nematodes (*Steinernema carpocapsae* strain All). Wheat flour, fillers, adjuvants, and biocontrol agent are blended to make a dough that is rolled into a sheet, dried, and ground into granules. After exposure to moisture, Pesta granules containing entrapped mycoherbicide agents become covered with new fungal growth, resulting in sustained production and release of newly-formed spores. Pesta granules with entrapped nematodes soften when wet, and the nematodes escape to seek insect pests. Weeds or insects were controlled by these products.

New formulation technology is needed to expand the use of biocontrol agents in agriculture. Granular products are often desired because they can be applied easily to soil, and entrapment or encapsulation in a solid matrix can protect delicate biocontrol organisms from rapid desiccation, UV light, and other adverse environmental stresses. Versatile, pasta-like, wheat flour-based granular formulations ("Pesta") are being developed for use with biocontrol agents. This paper reports results obtained with Pesta granules containing fungal weed pathogens (mycoherbicides) and entomogenous nematodes.

Experimental Methods

Mycoherbicide/Pesta granules were prepared as in Boyette et al. (in press) and Connick et al. (1991). Briefly, fungus grown in liquid fermentation was homogenized, added to a mixture of semolina (durum wheat flour) and kaolin (80:20 w/w), and kneaded to make a dough that was rolled into a sheet 1.0-1.5 mm thick. After drying, the sheet was ground or broken and then sieved to obtain 1-2 mm granules. Two mycoherbicide agents were studied: *Fusarium oxysporum* (NRRL #18279) which is pathogenic to sicklepod (*Cassia obtusifolia*), hemp sesbania (*Sesbania exaltata*), and coffee senna (*Cassia occidentalis*); and

Colletotrichum truncatum (COLTRU), pathogenic to hemp sesbania.

Weed control (percent stand emergence or mortality) with Pesta/*F. oxysporum* was evaluated on sicklepod, hemp sesbania, and coffee senna seedlings in the greenhouse 4 weeks after inoculation. Granules contained 10^7 infective fungal propagules/g.

Sustained conidial production with Pesta/COLTRU was quantified by placing granules on wet sand and incubating at 25 °C (12 h photoperiod). New spores growing on the granules were harvested at 3, 5, 7, and 10 days by washing with water. Effect of inoculum concentration (10^3-10^6 conidia/g) on viability of COLTRU in Pesta was determined by incubating on wet sand at 25 °C (12 h photoperiod) for 10 days, followed by washing off and counting the conidia produced. Samples were tested at 13-week intervals for 52 weeks.

In soybean field tests conducted in 1992 at Stoneville, MS, Pesta/COLTRU granules were applied by tractor-mounted equipment at rates of 13-130 lb/A preplant incorporated, pre-emergence, and post-emergence. Hemp sesbania control 4 weeks later was determined by percent stand emergence. Soybean yields were taken at the end of the growing season.

Nematode/Pesta granules were prepared as in Connick et al. (1993) and Nickle et al. (submitted to Journal of Nematology). Briefly, an aqueous suspension containing about 700,000

Steinernema carpocapsae (All) nematodes/ml (Biosys, Palo Alto, CA) was blended with 32 g semolina, 6 g kaolin, and 2 g peat moss to make a dough that was dried overnight (20-24% moisture is desirable), processed into granules as described above, and stored at 5 °C.

In a greenhouse test, two rows of corn were grown in potting soil in caged 51 cm x 36 cm flats containing 200 second instar Western corn rootworms (WCR). Pesta granules were added to deliver 50,000-200,000 nematodes. Adult insects emerging into the cages were collected and counted. Data from each of 4 trials were expressed as an average of 2 replicates of each of 4 nematode dosage rates.

Results and Discussion

In the greenhouse, Pesta/*F. oxysporum* granules controlled three weeds. Weed control ratings of coffee senna, sicklepod, and hemp sesbania were 95%, 98%, and 80%, respectively, when the granules were preplant incorporated in the soil. Conidial spray application under the same conditions gave 45% or less control (Boyette et al., in press).

On moist sand, Pesta/COLTRU granules rapidly became covered with new fungal growth that continuously produced conidia. In addition to forming the gluten matrix, the wheat flour component provides nutrition for the entrapped fungus. New conidia that were formed on granules inoculated at 10^6 conidia/g were harvested at 3, 5, 7, and 10 days. The total cumulative harvest was 10^8 conidia/g after 10 days. This demonstrates that, unlike conidial spray formulations, Pesta granules can provide a net gain in active ingredient with time. Storage viability of COLTRU in Pesta improved with increasing inoculum level (Fig. 1). At 10^6 conidia/g, the fungus was viable for at least one year when the granules were stored refrigerated.

Pesta/COLTRU granules applied preplant incorporated, pre-emergence, and post-emergence at 130 lb/A to hemp sesbania in the soybean field gave 84%, 70%, and 85% control, respectively. The corresponding soybean yields were 2520, 2090, and 2315 Kg/ha, compared with 1090 Kg/ha for the untreated control.

Infective stage *Steinernema carpocapsae* nematodes and their associated bacteria survived the Pesta process and emerged in large numbers in soil to kill WCR larvae and prepupae of the Colorado potato beetle (CPB) (Nickle et al., submitted to Journal of Nematology). Treatment with 50,000 nematodes killed 68% of the WCR, and treatments with 100,000-200,000 nematodes killed 90-95% of the insects (Fig. 2). Against CPB, there were indications from greenhouse tests (data not shown) that Pesta granules were twice as effective as a suspension drench application. Shelf life at 21 °C has been extended recently from 4 weeks to about 20 weeks by a formulation additive.

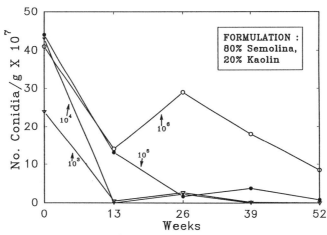

Fig. 1: Effect of inoculum concentration on conidial production by *Colletotrichum truncatum* incorporated in Pesta and stored at 4 °C.

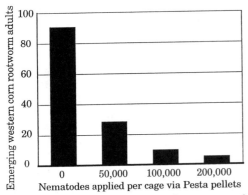

Fig. 2: Effect of application rate of Pesta/*Steinernema carpocapsae* (All) nematode-containing granules on 200 2nd instar Western corn rootworms in soil.

References

Boyette, C. D.; Abbas, H. K.; Connick, W. J., Jr. "Evaluation of *Fusarium oxysporum* as a potential bioherbicide for sicklepod (*Cassia obtusifolia*), coffee senna (*C. occidentalis*), and hemp sesbania (*Sesbania exaltata*)." *Weed Sci.*, in press.

Connick, W. J., Jr.; Boyette, C. D.; McAlpine, J. H. "Formulation of mycoherbicides using a pasta-like process." *Biol. Control* **1991**, *1*, 281-287.

Connick, W. J., Jr.; Nickle, W. R.; Vinyard, B. T. " 'Pesta': New granular formulations for *Steinernema carpocapsae*." *J. Nematol.*, **1993**, *25*, 198-203.

Progress and Promise in Management of Aquatic-Site Vegetation Using Biological and Biotechnological Approaches

Lars W. J. Anderson, USDA-ARS Aquatic Weed Control Research Laboratory, Botany Department, University of California, Davis, CA 95616

Highly selective, non-native biological control agents have reduced impacts of alligatorweed, waterhyacinth, salvinia, and waterlettuce. Less selective agents such as the grass carp, have also provided control of native and exotic aquatic weeds. However, efficacy has not been achieved for mycoherbicides, even though several have shown good potential in small scale studies. Infection and virulence may be enhanced by various stressors, but this approach has been examined only sporadically and needs better focus. Long-term management also will require identification and selection of macrophytes (possibly via transgenic plants) having desirable establishment, growth and canopy characteristics.

Aquatic vegetation management presents unique and interesting challenges for both the research community and the end-users for three basic reasons: (1) water, with its lateral and vertical mobility coupled with its myriad uses, constitutes some of the most environmentally sensitive ecosystems in nature; (2) it is the *use* of these systems that is directly affected by aquatic weeds rather than a specific commodity such as soybeans, corn or wheat; and (3) the goal of modern aquatic weed management is usually not complete removal of particular species, but rather, the attainment of a "livable" and appropriate mix of plant biomass and species diversity. The consequences of these important features necessitates an approach to aquatic weed control that focuses on the ecosystem level (Anderson, 1987).

Trend Toward Biological Control in Aquatic Sites

These management goals and their inherent constraints have driven solutions to aquatic weed problems progressively toward increasing emphasis on biological control strategies and away from reliance on herbicides and mechanical devices.

As evidence of this trend, in the last 15 years just two new aquatic herbicide active ingredients have been registered in U.S. products (glyphosate as Rodeo® and fluridone as Sonar®). However, during this same period, nearly a dozen different biological control agents (insects, fish, and fungi) have been released on target species in the U.S. (Julien, 1992; Creed and Sheldon, 1992; Cofrancesco, 1992). The wide breadth of target species and agents is clear from Tables 1 and 2, which list worldwide releases of agents for biocontrol of aquatic weeds. Although several agents have been established in the U.S. since the early 1970's (e.g. *Neochetina* spp., *Agasicles hygrophila*, and *Vogtia malloi*), others have been released in the last five years.

In addition to these classical introductions, interest is increasing in identifying and augmenting native insects which might provide moderate control of *Myriophyllum spicatum* (Macrae et al., 1990; Creed and Sheldon, 1992; Creed and Sheldon, 1993).

Public Acceptance of Biological Control

The selectivity and general unobtrusiveness of most herbivorous insects have led to generally good public acceptance for even high-use recreational sites (Center, 1992). This response is in contrast to sporadic but continuing objections to the use of various aquatic

Table 1. Exotic and Native Invertebrates/Vertebrates Released For Control Of Aquatic Weeds[1]

Target Weed	Biological Control Agent	Target Weed	Biological Control Agent
Alternanthera philoxeroides (alligator weed)	*Agasicles hygrophila*	*Salvinia molesta*	*Cyrtobagous salviniae*
	Amynothrips andersoni		*Cyrtobagous singularis*
	Disonycha argentinensis	*Salvinia minima*	*Paulinia acuminata*
	Vogtia malloi	(=*Salvinia rotundifolia*) (water fern)	*Samea multiplicalis*
		Pistia stratiotes (water lettuce)	*Neohydronomous affinis*
Hydrilla verticillata (hydrilla)	*Bagous affinis*	Various Aquatic Weeds and Algae	*Aristichthys nobilis*
	Hydrellia balciunasi		*Ctenopharyngodon idella*
	Hydrellia pakistanae		*Hypophthalmichthys molitrix*
	Parapoynx diminutalis		*Oreochromis aureus*
	(an immigrant species)		*Sartherodon aureus*
			Oreochromis mossambicus
Mimosa pigra	*Acanthoscelides puniceus*		*Oreochromis niloticus*
	Acanthoscelides quadridentatus		*Tilapia nilotica*
	Carmenta mimosa		*Osphronemus goramy*
	Chlamisus mimosae		*Puntius javanicus*
	Neurostrota gunniella		*Tilapia macrochir*
	Scamurius sp		*Tilapia melanopleura*
			Tilapia zillii
Eichhornia crassipes (waterhyacinth)	*Acigona infusella*		
	Neochetina bruchi		
	Neochetina eichhorniae		
	Sameodes albiguttalis	*Myriophyllum spicatum* (Eurasian watermilfoil)	*Phytobius leucogaster*
	Bellura densa		*Cricotopus myriophylli*[2]
	(= *Arzama densa*)		*Euhrychiopsis lecontei*[3]
	Orthogalumna terebrantis (=*Leptogalumna* sp.)		

[1]SOURCE: Adapted from Julien, 1992
[2]Macrae et al., 1990, (NOTE: Not released; this midge found as possible native herbivore in British Columbia)
[3]Creed and Sheldon, 1993, (NOTE: Not released; this weevil found as native herbivore in northeast U.S.)

herbicides, even when such uses are generally considered safe and acceptable by state and federal regulatory agencies. The perception that herbicide use is a threat to water resources appears to be a predictable result of efforts to initiate management or "eradication" programs, even for rapidly spreading exotic weeds such as hydrilla, waterhyacinth and eurasian watermilfoil. Successful programs, such as the waterhyacinth control program for the Sacramento Delta often have had to incorporate a herbicide residue monitoring component, which adds to costs and requires periodic reporting and documentation (Anderson, 1990).

For less selective biocontrol agents such as the grass carp (*Ctenopharygodon idella*), acceptance by the public and regulatory agencies has been mixed. For example, until the (sterile) triploid grass carp was available in the late 1980's, most states in the U.S. had prohibited its use. However, during the past four years progressively more states have implemented permit systems whereby either the diploid or triploid grass carp can be released for control of a variety of submersed aquatic weeds. At present, only a few states still ban, or severely restrict the use of this non-native fish (Fig.1). But it is clear that without a non-reproducing form, there would not be widespread use of the grass carp in the U.S.

Operational-Level Efficacy, or What Has Really Worked?

Notwithstanding its status as an exotic, generalist herbivore, without doubt, the grass carp has been the most consistently efficacious, biocontrol agent yet released in the U.S. and worldwide (van der Zweerde, 1990). Reductions in aquatic weed impacts from use of this single species alone have matched or

Table 2. Pathogens and Targeted Aquatic Weeds[1]

Target Weed Species	Pathogen
Eichhornia crassipes (waterhyacinth)	*Acremonium zonatum*
	Alternaria eichhorniae
	Cercospora piaropi
	Cercospora rodmanii
	Myrothecium roridum
	Rhizoctonia sp. (*Aquathanatephorus pendulus*)
Eleocharis kuroguwaii (water chestnut)	*Epicoccosorus nematosporus* gen. et. sp. nov.
Hydrilla verticillata (hydrilla)	*Fusarium roseum* 'Culmorum'
	Fusarium solani
	Sclerotium sp.
	Macrophomina phaseolina[2]
Lemna spp. (duckweed)	*Pythium aphanidermatum*
	P. myriotylum
Myriophyllum aquaticum (parrotfeather)	*Pythium carolinianum*
M. spicatum (eurasian watermilfoil)	*Colletotrichum gloeosporioides*
	Mycoleptodiscus terrestris
Nymphaea odorata (fragrant waterlily)	*Dichotomophthoropsis nymphaearum*
Nymphoides orbiculata	Various fungi

[1]SOURCE: Adapted from Charudattan, 1990, 1991.
[2]SOURCE: Joye and Paul, 1991.

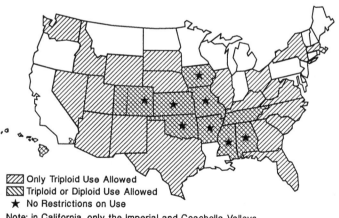

Only Triploid Use Allowed

Triploid or Diploid Use Allowed

★ No Restrictions on Use

Note: in California, only the Imperial and Coachella Valleys

(Note: Permits and other restrictions are usually in effect)

Fig. 1 Status of permitting for use of the grass carp (*Ctenopharyngodon idella*) in the United States as of 1993. (Reproduced and modified with permission from Anderson, 1991.)

exceeded results from herbicide applications in a variety of sites. In the Imperial Valley of California, for example, a massive infestation of *Hydrilla verticillata* was virtually eliminated from nearly 500 miles of irrigation supply canals within three years after initial stocking (Stocker, 1987, 1992). Similar results are typical for small lakes, ponds and reservoirs not only in the U.S., but in many other countries. In fact, the main problem associated with the use of the grass carp is a general lack of accurate stocking models that provide reasonable prediction of efficacy. As a result, over-stocking has occurred, and, due to non-selective feeding, resulted in denuding of some sites. However, better models are in development which more correctly account for net plant growth rates and fish consumption rates. Though this is still an imprecise biological control method, the adoption of this approach throughout the U.S. in just the past five years has been remarkable.

A second group of highly effective agents has been the *Neochetina* spp. (the water-hyacinth weevils) and a suite of insects targeted against alligatorweed, namely, *Agasicles hygrophila, Amyothrips andersoni,* and *Vogtia malloi.* Both target species have declined over several year spans within many of the release zones, and suppressions of biomass has been sustained.

Perhaps the most impressive recent examples are the highly effective control of *Salvinia molesta* by *Cyrtobagous salviniae* in several countries outside the U.S. (Room, 1981, 1986), and apparent successes in control of *Pistia stratiotes* (waterlettuce) with the weevil *Neohydronomous affinis* (Dray and Center, 1992; Chikwenhere and Forno, 1991). Several infestations of *S. molesta* have been controlled within one to two years after release of the insect. (One reason for the rapid reduction in target plant may be the relatively small size of this water fern compared to the weevil. In contrast to this, the ratio of waterhyacinth plant size to *Neochetina* spp. is orders of magnitude larger.) Similarly, waterlettuce populations in two sites in Florida declined dramatically within three years of introductions of *Neohydronomous.*

The Mycoherbicide/Pathogen Problem

Excellent discussions on research activity and releases of mycoherbicides in a variety of target weeds have been published recently (Charudattan, 1990; 1991). Table 2 summarizes target weeds and potential pathogens. As Charrudattan points out (1991), in the current regulatory environment, mycoherbicides are those pathogens whose production and release (i.e. application) are considered similar to herbicides and fall under the purview of the U.S. EPA. Dozens of fungal pathogens as potential mycoherbicides have been evaluated, and perhaps 10 to 12 have shown promise in secondary testing. However, consistent field-scale efficacy has not been demonstrated, particularly in submersed aquatic weed species. Yet there is reason for optimism; very recently, the U.S. Environmental Protection Agency granted an Experimental Use Permit for *Mycoleptodiscus terrestris* ("MT") for field testing against *Myriophyllum spicatum* (Cofrancesco 1992). In addition, R. Charudattan (Pers. communication, 1993) recently announced that *Cercospora rodmanii*, a fungus isolated in Florida, has been authorized for release in South Africa for control of water-hyacinth. Charudattan states that this is "...the first deliberate and authorized release of a foreign pathogen against waterhyacinth anywhere in the world".

In spite of many years of laboratory and field evaluations, why is there as yet no effective, commercially available mycoherbicide for aquatic weeds? There are probably two main problems associated with the use of fungal pathogens and mycoherbicides in aquatic systems: (1) most organisms have been isolated from declining, or senescent populations of target species and therefore tend to be opportunistic and not highly virulent; (2) releases, or application of propagules (e.g. spores) into the aqueous environment are fraught with uncertainty of propagule viability, proper conditions for infection and poor host contact. This, however, points to areas where additional basic research may be extremely fruitful, especially when coupled with development of improved formulations and delivery systems.

Optimizing Conditions for Microbial Biological Control Agents

Several features of aquatic macrophytes and aquatic sites make these environments particularly difficult ones in which to achieve effective control from microbial agents. For example, water quality varies greatly (pH, hardness, dissolved oxygen, dissolved organic material); vertical temperature gradients can be very steep; wind-driven and convection flows can move inocula away from target sites; most target species reproduce clonally at very rapid rates and thus can "out-grow" pathogens; epidermal access (for hyphal penetration, or uptake of any toxins) may be restricted by epiphytes such as diatoms and filamentous algae; human activities such as boating and fishing can reduce physical access and contact time. Some of these problems might be mitigated through preconditioning of target weeds and other measures (Table 3). Recently, the role of lectins in pathogen binding and penetration has been examined in some aquatic plants (Theriot, 1992). Perhaps this avenue of research will lead to improvement in field efficacy through better formulations.

Some of these approaches have been used with varying success (Center et al. 1982; Haag and Habeck, 1991; Van, 1988). Due to the current lack of registered growth regulators, the most practical methods at present are sub-lethal herbicide doses and mechanical disruptions. With the increased availability of the sterile (triploid) grass carp, combination or sequential use of this herbivore and myco-herbicides may prove useful. This approach will require judicious use of stocking rates and sizes so that herbivory creates only stress in the target weed.

In an extensive review of senescence in aquatic plants Rejmankova (1989) pointed out that a multitude of environmental conditions lead to plant stress and may lead to plant death (Fig. 2). But until there is a more deliberate and systematic examination of these interactions, the use of "stressors" borders closer to an art than a science.

Table 3. Potential Approaches To Enhance Efficacy of Microbial Biocontrol Agents

Action:	Effect:
Use sub-lethal herbicide dose	Slow growth Disrupt epidermis
Apply growth regulators inhibitors (paclobutrizole, ABA, ethylene, cytokinins)	Reduces biomass & Block plant defense Mimic senescence
Alter water level to expose plants	Water stress, heat stress; prevents removal of inocula by water currents
Cutting, chopping	Induces wounding, reduces biomass
Herbivorous insects and fish	Induces wounding, reduces biomass

Limitations and Potential for Improvement of Grass Carp Utility

The burgeoning use of the grass carp, an exotic and quite non-selective herbivore, has resulted from the convergence of several technological developments. First, production of triploidy by heat-shock and now, more commonly, by hydrostatic pressure on newly fertilized eggs has overcome concerns that these fish may become too numerous in a given site. Secondly, the ability to differentiate diploid and triploid fish at immature stages has provided for quality control of ploidy. This relies on sensitive cell-volume analyzers or on direct measurement of cellular DNA (Pine and Anderson, 1990). These methods enable excellent quality control at the production end as well as rapid pre- and post-stocking inspection and certification. Without this ability, state and federal regulatory activity would be extremely costly and time-consuming. Third, development of fish-stocking and plant-growth models has begun to provide better forecasting of feeding effects.

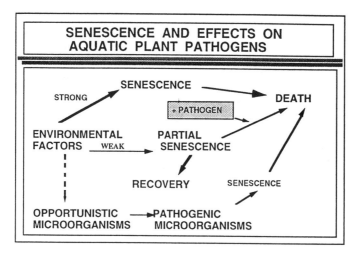

Fig. 2 Interactions and relationships between environmentally induced senescence and pathogens. (Modified from Rejmankova, 1989).

Yet, in spite of these advances, little effort has been made in the U.S. to actually improve the grass carp. There are several potential areas where application of current technology could increase the breadth and safety of using this fish. Although the triploid fish cannot reproduce, its longevity is probably 10 to 15 years. If its life-span could be reduced, there would probably be less reluctance to release it into natural waters. Some of the most pressing aquatic weed problems are in natural riverine and static systems such as the Columbia River (Washington state) and associated water bodies. The insertion of age-dependent "lethal" genes may be a worthy long-term research goal. A fish with a "guaranteed" longevity of say 5 +/- 2 years could find acceptance in natural river systems. Likewise, breeding and selection for particular pharyngeal structures might impart more feeding selectivity so that beneficial aquatic plants are not consumed as readily as target plants.

Beneficial Plants

A serious impediment to sustained aquatic vegetation management has been an almost complete lack of suitable "desirable" species that can fill niches opened-up after initial control is obtained. This is in part due to a kind of "Catch 22" as follows: competition for light in aquatic systems is keen, and plants which are successful tend to produce extensive canopies with high biomass. Yet it is precisely this large biomass, distributed generally in the upper one-half to one-third of the water column that makes certain plants "weeds". Other "weedy"characteristics include high rate of clonal growth and production of numerous, long-lived propagules. In contrast, desirable plants may have several different growth habits, physiological adaptations and canopy architecture (Table 4.).

The only extensive studies of purposeful introduction of beneficial plants to suppress submersed aquatic weeds were conducted on two species of spikerushes (*Eleocharis acicularis* and *E. coloradoensis*) (Yeo and Thurston, 1984; Ashton et al. 1984). Though this plant has some of the aforementioned characteristics, it was found to be difficult to establish and requires relatively high light levels. Other candidates might include the alga *Chara* sp., some diminutive *Potamogeton* species, *Limnosella* spp. ("mudwarts"), and some *Najas* species.

Though there have been a few studies on allelopathy in aquatic plants, including *Eleocharis,* unequivocal proof of action under field conditions is lacking (Frank and Dechoretz, 1980; Elakovich and Wooten, 1989; Sutton and Portier, 1989; Cheng and Riemer, 1989). The ability to obtain active, inhibitory extracts

Table 4. Some Characteristics of Potentially Beneficial Aquatic Plants

Character/Trait	Effect
Short stature	Low biomass & open surface-water
Open branching	Open gaps in canopy
Majority of biomass near bottom	Blocks invasion & open upper water column
Rapid lateral spread	Early establishment
Low light compensation point	Competes well for low light
Low temperature-adapted	Comptetes well in cooler depths
Allelopathy	Blocks invasion

from plant tissue should not be construed as evidence for allelopathy; rather it is only evidence of a potential source of allelochemicals. However, this does not diminish the importance of identifying naturally occurring inhibitory compounds since these might be excellent herbicides or stressors for use in conjunction with other biological control agents.

The traits listed in Table 4 could be a good starting point for classical breeding and selection, but to date this effort has not been made. Likewise, with the ability to produce transgenic plants from direct insertions into genomic DNA, the time may be ripe for a well-focused attempt to transfer transcripts for specific enzymes to provide low light compensation point, low-temperature affinity and other traits to plants that have inherently short stature. Similarly, desirable canopy-structure traits might be moved into plants that already have desirable physiological competence. Whether such transgenic aquatic plants could gain public and regulatory acceptance is another question.

Other Related Approaches

The use of microbial inocula to alter and amend aquatic sediments could target such critical nutrients as iron, or could disrupt nitrification processes. Since it is clear that many of the most problematic aquatic weeds derive most of their nutrients from sediments, small shifts in availability could, over time, lead to reduced biomass. With the current technology and successes in selecting bacterial and fungal strains to decompose various wastes, this approach probably has merit. But, once again, the complexities of aquatic sediments and their associated microflora present a much more difficult system than simple batch-cultures of organisms through which waste is recycled.

Summary and Conclusions

Tremendous progress has been made in the use of classical biological control introductions targeted for aquatic weeds. Considering the relatively small "commercial market" compared to terrestrial weeds, these releases represent a very high proportion of agents released for all types of weeds (Charudattan, 1991). It is no doubt the highly sensitive nature of the aquatic environment that has, on the one hand dissuaded continued herbicide discovery and registration efforts by corporations, while at the same time presented opportunities for investment in foreign explorations for biocontrol agents. For control of native aquatic weeds, other approaches, primarily augmentative, are receiving more attention now. New methodology in gene transfers as well as improved methods for mycoherbicide production and formulation could have as much of an impact on native weed management as the last 15 years of classical introductions have had on exotic weeds. But this will require a serious and long-term investment in basic and applied research and probably will not be successful without state & federal multi-agency collaboration. Another serious question is whether sufficient changes in the research proposal review process will take place so that increasing funds through such programs as the National Research Initiative and "sustainable agriculture" monies can flow to laudable biotechnology-based aquatic weed control research.

References

Anderson, L.W.J. "Sterile fish offers hope for weedy water "Hazards"". North. Calif. Golf Assoc. Newsletter, **1991**, 11, 18-19.

Anderson, L.W.J. "Recent developments and future trends in aquatic weed management." Proc. EWRS/AAB 7th Symposium on Aquatic Weeds. 1987, 9-16.

Anderson, L.W.J. "Aquatic weed problems and management in the western United States and Canada." In *Aquatic Weeds: The Ecology and Management of Nuisance Aquatic Vegetation*; Pieterse, H. and Murphy, K.J. Eds.; Oxford Univ. Press, N.Y., **1990**; 371-391.

Ashton, F. M.; Bissell, S.R.; DiTomaso, J. M.; Wach, M. J. "Research on biological control of aquatic weeds with competitive species of spikerush (*Eleocharis* spp.)," Final Report to USDA, 1984, 187 pp.

Center, T. D.; Steward, K. K.; Bruner, M.C. "Control of waterhyacinth (*Eichhornia crassipes*) with *Neochetina eichhorniae* (Coleoptera: Curculionidae) and a growth retardant." *Weed Sci.* **1982**, 30, 453-457.

Center, T.D. "Biological control of weeds in waterways and on public lands in the southeastern United States of America." Proc. First Internatl. Weed Control Congress, Melbourne, Australia, 1992, 256-263.

Confrancesco, A. F. "Biological control overview." Proc. 26th Annual Meeting, Aquatic Plant Control Research Prog. Miscl. paper A-92-2, 1992, 203-204.

Charudattan, R. "The mycoherbicide approach with plant pathogens." In TeBeest, D.O. Ed. *Microbial Control of Weeds*. Chapman and Hall, N.Y., **1991**; pp. 24-57.

Charudattan, R. "Biological control of aquatic weeds by means of fungi." In Pieterse, H. and Murphy, K.J. Eds. *Aquatic Weeds: The ecology and Management of Nuisance Aquatic Vegetation*; Oxford Univ. Press: N.Y., **1990**; pp. 186-201.

Cheng, T.S.; Riemer, D.N. "Characterization of allelochemicals in American eelgrass." *J. Aquat. Plant Manage.* **1989**, 27, 84-89.

Chikwenhere, G.P.; Forno, I.W. "Introduction of *Neohydronomus affinis* for biological control of *Pistia stratiotes* in Zimababwe." *J. Aquat. Plant Manage.* **1991**, 29, 53-55.

Creed, R. P. Jr.; Sheldon, S.P. "Further investigations into the effect of herbivores on Eurasian watermilfoil (*Myriophyllum spicatum*)." Proc. 26th Annual Meeting, Aquatic Plant Control Research Program. **1992**, Miscl. paper A-92-2, 244-252.

Creed, R.P. Jr.; Sheldon, S.P. "The effect of feeding by a North American weevil, *Euhrychiopsis lecontei*, on eurasian watermilfoil (*Myriophyllum spicatum*)." *Aquatic Botany* **1993**, 45, 245-256.

Dray, F. A., Jr.; Center, T.D. "Biological control of *Pistia stratiotes* L. (waterlettuce) using *Neohydronomus affinis* Hustache (Coleoptera: Curculionidae)." Technical Report A-92-1, U.S. Army Corps of Engineers Waterways Experiment Station, Vicksburg, MS, 1992, 62p.

Elakovich, S.D.; Wooten, J. W. "Allelopathic potential of sixteen aquatic and wetland plants." *J. Aquat. Plant Manage.* **1989**, 27, 78-84.

Frank, P.A., Dechoretz, N. "Allelopathy in dwarf spikerush (*Eleocharis coloradoensis*)." *J. Aquat. Plant Manage.* **1980**, 28, 499-505.

Haag, K. J.; Habeck, D. H. "Enhanced biological control of waterhyacinth following limited herbicide application." *J. Aquat. Plant Manage.* **1991**, 29, 24-28.

Joye, G. F. Paul; R. "Histology of infection of hydrilla by *Macrophomina phaseolina*." Technical Report A-91-6, US Army Corps of Engineer Waterways Experiment Station, Vicksburg MS., 1991, 21p.

Julien, M. H., Ed. "*Biological Control of Weeds: A world Catalogue of Agents and their Target Weeds.*", Third Edition. CAB International: Brisbane, Queensland, **1992**, 186 pp.

Macrae, I. V.; Winchester, R. A.; Ring, R. A. "Feeding activity and host preference of the milfoil midge, *Cricotopus myrophylli* Oliver (Diptera: Chironomidae)". *J. Aquat. Plant Manage.* **1990**, 28, 89-92.

Pine, R.; Anderson, L.W.J. "Blood preparation for flow cytometry to identify triploidy in grass carp." Prog. Fish Cult. **1990**, 52, 266-268.

Rejmankova, E. "Review of senescence as an important factor determining the relationship among aquatic plants, their epiphytes, and pathogens." Miscl. Paper A-89-3, US Army Corps of Engineer Waterways Experiment Station, Vicksburg, MS. 107p.

Room, P. M.; Harley, K.S.S.; Forno, I.W.; Sands, D.P. A. "Successful control of the floating weed salvinia." *Nature* (London), **1981**, 294, 78-80.

Room, P. M. "Biological control is solving the world's *Salvinia molesta* problems." Proc. EWRS/AAB 7th Symposium on Aquatic Weeds, 1986, 271-276.

Stocker, R. 1986 "Annual Report to the Hydrilla Technical Advisory Committee." Report: Imperial Irrigation District, Imperial California, 1987. 34 pp.

Stocker, R. K. "Water related projects of the Imperial Irrigation District." 32nd Annual Meeting Aquatic Plant Management Soc., Daytona Beach FL, 1992, Abstract.#92, 32.

Sutton, D. L.; Portier, K. M. "Influence of allelochemicals and aquatic plant extracts on growth of duckweed." *J. Aquat. Plant Manage.*, **1989**, 27, 90-95.

Theriot, E.A. "Attachment and infection of fungal pathogens on submersed aquatic plants." 32nd Annual Meeting Aquatic Plant Management Soc., Daytona Beach, FL, 1992, Abstract.#48, 17.

van der Zweerde, W. "Biological control of aquatic weeds by means of phytophagous fish." In *Aquatic Weeds: The Ecology and Management of Nuisance Aquatic Vegetation*; Pieterse, H. and Murphy, K.J. Eds.; Oxford Univ. Press, N.Y., **1990**; 201-221.

Van, T.K. "Integrated control of waterhyacinth with *Neochetina* and paclobutrazol." *J. Aquat. Plant Manage.*, **1988**, 26, 59-61.

Yeo, R.R.; Thurston, J.R. "The effect of dwarf spikerush (*Eleocharis coloradoensis*) on several submersed aquatic weeds." *J. Aquat. Plant Manage.*, **1984**, 22, 52-56.

NATURAL COMPOUNDS IN PEST MANAGEMENT

Phytochemical Contributions to Pest Management

W. S. Bowers, Laboratory of Chemical Ecology, Department of Entomology, The University of Arizona, Tucson, AZ 85721

The earliest insecticides, composed of phytochemicals, soon gave way to the use of highly toxic, broad-spectrum synthetic insecticides. The realization that a few persistent synthetic pesticides threaten planetary ecology generated an intensive search seeking environmentally pacific strategies for plant and public health protection. Studies of insect biology and chemical ecology have revealed numerous points of attack that promise environmental compatibility. Included in this developing arsenal are insect growth, feeding and behavior regulators optimized from insect hormones and phytochemicals.

Nicotine, rotenone, ryanodine alkaloids and pyrethrum were important phytochemnical insecticides until the discovery of DDT initiated the dawn of synthetic pesticides. The superior efficacy and cost effectiveness of synthetics dramatically increased agricultural yields and for the first time in history embarrassing surpluses of food were experienced. Their impact on the control of insect disease vectors stimulated an expansion in world population that continues out of control today. No other discovery in the history of science, not excluding the antibiotics or immunization techniques, so sensationally improved the quality of life. The use of insecticides for control of malaria vectors alone is estimated to have saved over 2 billion lives (Bruce-Chwatt, 1971). The impact of insecticides, fungicides and herbicides on agricultural production and human nutrition is inestimable. Yet, improved chemical analytical techniques and the evolution of the disciplines of toxicology combined with our increasing awareness of the linkages in planetary ecology revealed disturbing accumulations of some of the more persistent pesticides and their metabolites. Evidence that certain pesticides accumulated in animals at the top of the trophic pyramid and killed or disabled certain non-target species generated calls for restriction and elimination of many pesticides (Carson, 1962). These demands provoked consideration, for the first time, of the importance of pesticides in our society and the realization that we could not return to the more pastoral, less efficient agricultural practices of yesteryear. Attempts to limit the use of insecticides in vector control revealed that many devastating diseases like malaria, filariasis, plague and yellow fever rapidly returned with a vengeance.

Environmental concern coupled with the rapid development of resistance by many pests to synthetic pesticides prompted an outcry for the discovery of safe and effective control measures and increased support for investigations targeted to the discovery of environmentally friendly chemistry for plant and public health protection. Clearly, the rapid development of analytical and synthetic organic chemistry had outpaced those elements of natural product chemistry that might have revealed new biological agents for pest control. Yet, a remarkable fund of basic biological information had accumulated on the hormonal control of insect growth, development, reproduction and diapause as well as an understanding that much of insect behavior is regulated by volatile secretions called pheromones. As in no previous peacetime endeavor, government, industry and academia initiated collaborations to pool intellectual and monetary resources to seek new modalities for plant and public health protection. New vitality was given to research in novel non-toxic chemicals, biological control efforts with parasites, predators and diseases and the developing arts of genetic engineering were marshaled to target improved insect control. One of the most promising discoveries has been the realization that many plants possess non-toxic defensive chemistry (Bowers et al., 1966).

Phytochemical Resources

Efforts in drug discovery have yielded approximately 120 useful compounds from only 90 plant species. About three quarters of these compounds were discovered as a result of native ethnomedical practices. Eighty-eight percent of people in developing countries (3-4 billion people) use plants as their primary source of medicine (Sjamsul, 1992).

The high value placed on drugs for human medicine has always guaranteed continued searching among cultural traditions for plants used in disease management. The principal natural resources that have contributed to the discovery of new drugs are microorganisms, plants and invertebrates. Since the development of a new drug in the U. S. is estimated to cost about $ 200 million this searching modality is not inexpensive. Nevertheless, twenty-five percent of new drugs depend on leads from natural products.

In addition to drugs, plants are used for clothing, cosmetics, perfumes, food, shelter, and as biological agents such as insecticides, repellents etc. Despite the investigation of plants for useful chemicals less than 10 per cent of terrestrial plants have been evaluated for insect control other than as direct toxicants.

Basic research in insect biology and their chemical ecology demonstrate that plants possess a vast store of secondary chemicals many of which appear to be used in defense against insect predation. Indeed, many toxic phytochemicals are well known, but some defensive secondary compounds of plants appear to focus on the disruption of insect feeding, development, reproduction and behavior. Thus, plants are clearly capable of defending themselves, not only by poisoning hungry insects, but by using their chemistry to subtly perturb discrete aspects of insect life. In answer to the question "what is a weed" the 19th century nature philosopher, Ralph Waldo Emerson replied "A plant whose virtue has not been discovered". In a similar vein, insect scientists asked "can natural phytochemical defensive strategies against insects be optimized for modern pest control" ?

Insect Growth Regulators

The discovery of the juvenile hormone in insects (Wigglesworth, 1934) and the prediction (Williams, 1956) that this hormone could be used as a selective method for insect control set the stage for the development of the first bio-rational chemicals employed in insect pest management. Normally present in immature insects and reproducing adults, the juvenile hormone (JH) must be absent during those brief periods when insects differentiate into mature adults. During these sensitive periods, insects contacted with JH cannot mature, and die without further development or reproduction. The synthesis of the JH and its' isolation from insects (Bowers, 1965, Judy et al., 1973) stimulated industrial efforts to optimize its' chemistry for commercial use (Fig. 1) and provided several products with outstanding safety and efficacy. When JH activity was discovered in a plant, i.e., the balsam fir, (Slama and Williams 1965) and chemically characterized (Bowers, et al., 1966) a new mode of plant defense utilizing mimics of insects' own hormones was revealed (Fig. 2). Additional discoveries of phytochemicals with JH activity from the plant *Sesamin indicum* (Bowers, 1968,

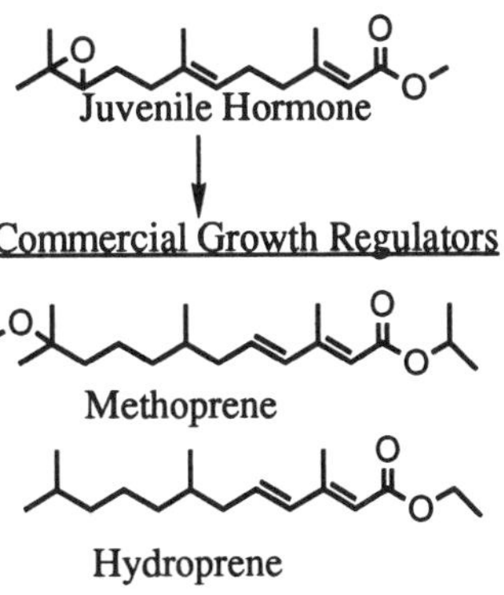

Fig. 1. Synthetic optimization of the natural juvenile hormone led to the first biorational products for insect control called insect growth regulators.

1969) yielded novel chemistry that stimulated the synthesis of aromatic hormonal analogs with extraordinarily high activity. Finally, the isolation and identification of extremely active hormone analogs from the sweet basil plant, *Ocimum basilicum* (Bowers and Nishida, 1980) suggested the synthesis of polyaromatic structures which ultimately yielded a second generation of commercial hormonal products (Fig. 3). Viewed alone the new hormonal products would appear to bear scant resemblance to the natural hormone, yet every biological response to the natural hormone can be completely duplicated by the optimized products. Only by following the step by step synthetic alterations prompted by discoveries of new phytochemical models can the evolution of

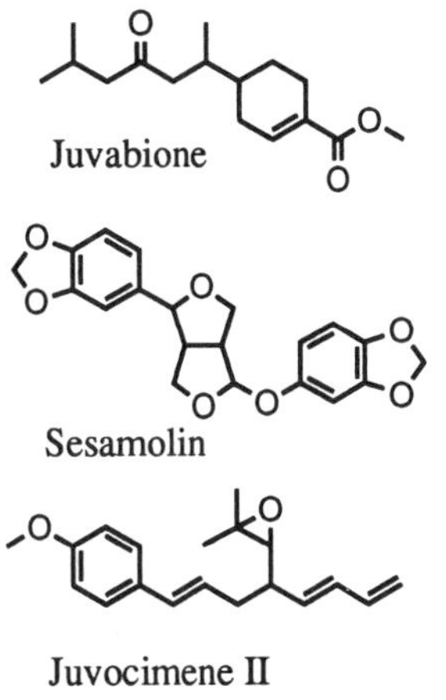

Fig. 2. Phytochemicals with insect juvenile hormone activity were discovered in the plants *Abies balsamea* (juvabione - balsam fir), *Sesamum indicum* (sesamolin - sesame oil) and *Ocimum basilicum* (juvocimene II - sweet basil).

these new products be comprehended. Thus, the lead chemistry derived from plants was vital to the development of these first bio-rational products for which the term insect growth regulator was originally coined. Other compounds with alternative kinds of insect growth regulant activities are known, however, lacking a natural origin, they will be omitted from this discussion.

Anti-juvenile Hormones

Since JH interferes with the formation of the mature stages the first commercial products were targeted to control insects that inflict their damage in the adult stage, i.e., mosquitoes, flies, fleas etc. In agriculture the growth regulatory action of JH comes too late to interfere with the larval feeding damage to crops. Consequently, a method of limiting immature growth and development and/or preventing adult reproduction would be very beneficial for the control of plant feeding insects. From classical studies it can be visualized that if an excess of JH cannot disturb the normal course of immature stages, who are

full of JH anyway, it might be possible to damage their development if somehow JH secretion or action could be prevented. Surgical extirpation of the corpora allata, the glands responsible for JH biosynthesis and secretion results in termination of the immature phase, promoting instead precocious maturation into non-viable, sterile diminutive adults. An anti-JH would be an ideal form of growth regulator for immature insect pests as well as for other stages. A search among plants revealed a variety of phytochemicals with anti-JH activity including the precocenes from *Ageratum houstonianum*, (Bowers, 1976) and *Nama rothrockii*, (Binder et al., 1991), and a polyacetylenic sulfoxide isolated from *Chrysanthemum coronarium*, (Bowers and Aregullin, 1987)(Fig. 4). Synthetic studies to maximize the biological activity of the precocenes yielded several active compounds, in particular, 7-ethoxy and 7-isopropoxy analogs of precocene 2 were found to be superior to the natural products, but none were sufficiently active against important plant pests to encourage commercial development. Industrial synthetic efforts provided a number of anti-JH active compounds, but none with commercial potential (for review see Staal, 1986). An anti-JH, effective against the immature stages of plant feeding insects, and possessing the desired safety of other growth regulators remains an elusive target.

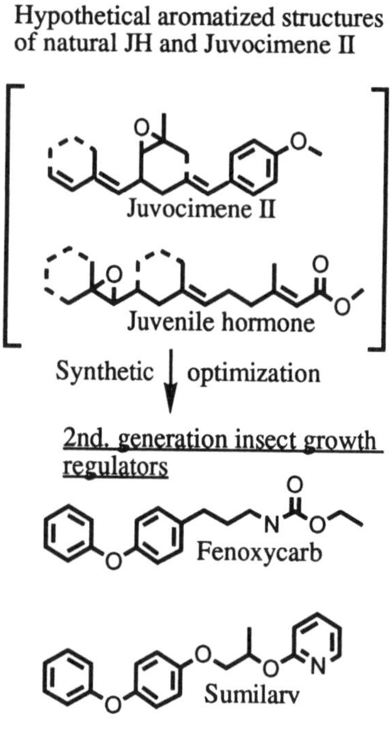

Pheromones

Insects are dependent on many chemicals with communicative significance. Called pheromones these chemicals are used to signal for sexual receptivity, courtship, resource marking, colonization, alarm, social recognition, orientation etc. (for reviews see Bell and Carde, 1984). The first pheromone, identified after a 30 year effort, was that of the commercial silkworm *Bombyx mori* (Butenandt et al., 1959). However, the greatest stimulus to pheromone research followed the development of the electroantennegram method for identifying the sex pheromones of Lepidoptera (Roelofs, 1979). Highly species specific, the sex pheromones are used to great

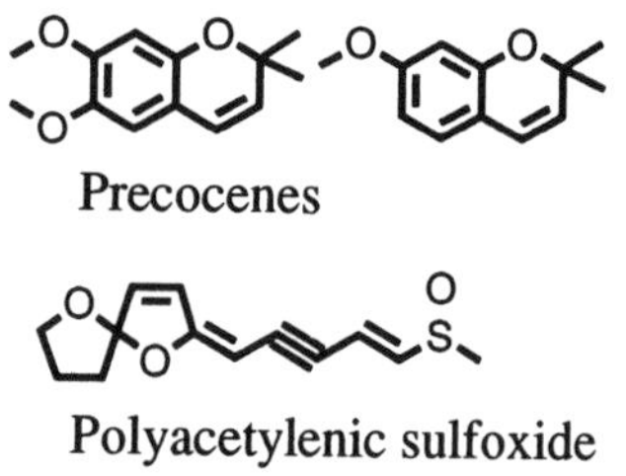

Hypothetical aromatized structures of natural JH and Juvocimene II

Juvocimene II

Juvenile hormone

Synthetic optimization

2nd. generation insect growth regulators

Fenoxycarb

Sumilarv

Fig. 3. Theoretical aromatization of elements of the natural juvenile hormone and of the phytojuvenoid juvocimene II inspired the development of a second generation of insect growth regulators.

Precocenes

Polyacetylenic sulfoxide

Fig. 4. Phytochemicals with anti-juvenile hormonal activity.

advantage to monitor the presence of reproductively active insects in the vicinity of susceptible crops. Successful application of pheromones in monitoring for pests in apple orchards can reduce insecticide and miticide applications by up to 50% (Roelofs, 1981). Pheromones have also been used to permeate the atmosphere and have been demonstrated to prevent successful mating of certain pests including the cabbage looper and pink bollworm (Roelofs, 1981). Although lengthy efforts were made to isolate and identify the sex pheromones of the American cockroach, phytochemicals in Verbena and spruce oil were found to precipitate the identical attraction and mating display of these insects (Bowers and Bodenstein, 1971, Nishino et al., 1977, Tahara et al., 1975, Kitamura et al. 1976) long before the isolation and identification of the two natural pheromones was accomplished (Persoons et al. 1979, Nishino et al., 1988).The phytopheromonal sex attractants were (+)-bornyl acetate, (+)-E-verbenyl acetate and germacrene D.

Repellents

The alarm pheromones of aphids (Dahl, 1971) were isolated and identified (Bowers et al., 1972, Bowers et al., 1977, Nishino et al., 1977) as (E)-ß-farnesene and (-)-germacrene A (Fig.5). Acting as natural repellents, the alarm pheromones can be used to drive aphids off of their host plants, but due to their extreme lability in air and light no useful methods for their commercial applications have been reported. (E)-ß-farnesene has been shown to be a constituent released from the trichomes of aphid resistant potato cultivars (Griffiths and Pickett, 1980, Gibson and Pickett, 1983). This demonstrated, for the first time, the use of a natural insect pheromone, by a plant for defense. Slow release formulations for (E)-ß-farnesene have been developed (Dawson et al., 1982) but field efficacy remains to be demonstrated.

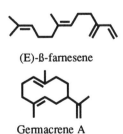

(E)-ß-farnesene

Germacrene A

Fig. 5. Aphid alarm pheromones - natural insect repellents.

The origins and use of personal repellents are doubtless lost in prehistory but the use of volatile chemicals to repel biting arthropods are universal to all cultures. Repellents are commonly secreted by animals for protection against parasites and predators. Thus, insect defensive secretions contain a striking array of simple hydrocarbons, esters, aldehydes and alcohols employed to fend off attack by other arthropods (Blum, 1981). Benzoquinones, naphthoquinones and phenolics are common repellent components of arthropods (Tschinkel, 1969). Insects are the principal vectors of all of the ancient plagues and seem to conspire constantly to keep man in an endless state of ill health. The reproductive potential of humans appears to be a constant, but historically malignant diseases like malaria, yellow fever, typhus and trypanosomiasis savagely checked population growth. Prior to World War II most personal repellents were derived from plant volatiles and included the well known plant secondary chemicals such as geraniol, citronellol, camphor and menthol. The need for a long-lasting repellent during the war supported a massive screening effort of industrial chemicals that ultimately led to the adoption of the chemical diethyl m-toluamide or DEET. DEET was quickly moved into production and can be credited with the protection of millions of soldiers from vector borne disease. Following the war, DEET entered the civilian market and largely displaced natural repellents. It should be noted that DEET was a patented repellent preparation and could be sold for a premium price whereas the natural repellents lacked commercial advantage and were soon pushed out of the market. Recently, re-examination of DEET has revealed that it is not an innocuous chemical and possesses significant toxicity to humans (Knowles, 1990).

Our improved understanding of insect biology and of the natural chemical defenses of plants opens the possibility of discovering new natural repellents lacking the toxicity of DEET and similar synthetic chemicals. The exciting discipline of chemical ecology investigates how plants and animals interact in nature. Some of their defenses are clearly based on the deployment of repellent chemicals. We have discovered several plants that possess potentially useful natural repellents. Among the plant secondary chemical repellents, we find that simple aromatics like coumarin, piperonal, piperitone and linalool are highly effective space repellents for ants, mosquitoes and blackflies (Bowers, 1991, unpub. data). Their volatility is a limitation when a long period of efficacy is required, but modern controlled-release formulation technology and special release media promise to dramatically improve the persistence of more volatile compounds . Beyond the use of natural repellents for personal protection it must be recognized that repellents serve their host plants well and constitute a natural mechanism to fend off herbivores. Problems of high volatility and sensitivity to light and air can now be overcome by suitable formulation.

Increasingly, investigations focus on the potential application of natural repellents for the direct protection of crops.

Anti-feedants

Numerous phytochemicals have been demonstrated to deter insect feeding and it is assumed that this constitutes an important natural strategy against herbivory. Successful field trials have demonstrated the efficacy of the phytochemical antifeedants ajugarin I and (-)-polygodial (Pickett et al. 1987). The extracts of the neem and chinaberry trees have long been known to contain a variety of tetranortriterpenoid antifeedants. A commercial product extracted from the neem tree seed, composed mainly of the antifeedant compound azidirachten, is currently approved for the protection of non-food plants. A serious deficiency with the anticipation of using natural antifeedants is that few plants occur in sufficient abundance in nature to provide an adequate supply for large scale pest control. This drawback might be overcome through synthetic optimization efforts which have been undertaken to simplify the chemistry of the phytochemical antifeedants and make them suitable for commercial production (Ley, 1987).

Summary

Counteradaptations that have developed from plant / insect interactions have generated in plants many secondary chemical defensive strategies that target insect specific elements of biology other than simple intoxication. Defenses that interfere with insect growth, development, reproduction and behavior may have no counterpart receptors in higher animals. Thus, some of these subtle defenses offer the opportunity to develop safe and environmentally pacific methods for pest control.

References

Bell, W. J., Carde, R.T., Eds.; Chemical Ecology of Insects; Sinauer Associates, Inc.: Sunderland, MA, 1984.

Binder, B. F., Bowers, W. S., Evans, P.H. "Insect anti-juvenile hormone and juvenile *Nama.*"Experientia, **1991**, 47, 199-201.

Blum, M. S. Chemical Defenses of Arthropods; Academic Press: Orlando, FL, 1981.

Bowers, W. S., Thompson, M. C., Uebel, E. C., "Juvenile and gonadotropic hormone activity of l0,ll-epoxyfarnesenic acid methyl ester" Life Sci. **1965**, 4: 2323-2331.

Bowers, W. S., Fales, H. M, Thompson, M. J., Uebel, E.C."Juvenile Hormone: Identification of an active compound from balsam fir". Science, **1966,** 54: 1020-1022.

Bowers, W. S. "Juvenile Hormone: Activity of natural and synthetic synergists". Science, **1968,** 161: 895-897.

Bowers, W. S. "Juvenile Hormone: Activity of aromatic terpenoid ethers". Science, **1969**, 164: 323-325.

Bowers W. S., Bodenstein, W. G. "Sex pheromone mimics of the American cockroach". Nature, **1971**, 232, 59-261.

Bowers, W. S., Nault, L.R., Webb, R.E., Dutky, S.R. "Aphid alarm pheromone: I solation, identification, synthesis". Science, **1972,** 177: 1121-1122.

Bowers, W. S. Discovery of insect antiallatotropins. In, The Juvenile Hormones. Ed.; L. I. Gilbert. Plenum Press, New York. 1976.

Bowers, W. S., Nishino, C., Montgomery, M. E., Nault, L. R. Nielsen., M.W.," Sesquiterpene progenitor, Germacrene A: An alarm pheromone in aphids". Science, **1977**, 196: 680-681.

Bowers, W. S., Nishida, R. "Juvocimenes: Potent juvenile hormone mimics from sweet basil". Science, **1980,** 209: 1030-1032.

Bowers, W. S. Aregullin, M. "Discovery and dentification of an antijuvenile hormone from *Chrysanthemum cornarium.";* Memoria do Instituto Oswaldo Cruz, Special Issue, suppl. III, vol. 82). 1987.

Bruce-Chwatt, L. J. "Insecticides and the control of vector borne diseases". Bull. Wld. Hlth. Org. No. 1-2-3; **1971**, 44; 419-424.

Butenandt, A. R., Beckmann, C., Stamm, D., Hecker, E. "Uber den sexuallockstoff des seidenspinners*Bombyx mori*". Naturforschung, **1959**, 146: 283-284.

Carson, R. Silent Spring; Houghton Mifflin Co, Boston. The Riverside Press, Cambridge, MA, 1962

Dahl, M. L. "Uber einen schreckstoff bei aphidehn". Dtsch. Ent. Z, **1971,** 18: 121-128.

Dawson, G. W., Griffiths, D. C., Pickett, J. A., Smith , M. C. Woodcock, C. M. "Improved preparation of (E)-ß-farnesene and its activity with economically important aphids". J. Chem. Ecology, **1982**, 1111-1117.

Gibson, R. W., Pickett, J. A. "Wild potato repels aphids by release of aphid alarm pheromone". Nature, **1983**, 608-609.

Griffiths D. C., Pickett, J. A. "A potential application of aphid alarm pheromones". Entomologia Experimentalis et Applicata, **1980**, 27 : 199-201.

Henrick, C. A., Staal, G. B., Siddall, J. B. "3,7,11-trimethyl-2,4-dodecadienoates, a new class of potent insect growth regulators with juvenile hormone activity". Agri. Food Chem., **1973**, 21:354-359.

Judy, K. J., Schooley, D. H. Dunham, L.L., Hall, M. S. Bergot. B. J., Siddall, J. B. "Isolation, structure and absolute configuration of a new natural insect juvenile hormone from *Manduca sexta*". Proc. Nat. Acad. Sci. U. S. A., **1973**, 70: 1509-1513.

Kitamura C., Takahashi, S. Tahara, S. Mizutani, J., A sex stimulant to the male American cockroach in plants. J. Agr. Biol. Chem., **1976**, 40: 1965-1969.

Knowles, C. O. In Encyclopedia of Pesticide Toxicology; Misouri AgriculturalStation Journal; Hayes, W. J., Laws, E. R., Eds.; series No. 10308: Columbia, MO, 1990.

Ley, S. V. In Pest. Sci. and Biotech.Greenhalgh, R., Roberts, T. R. Eds.; Blackwell Sci. Publ. Oxford, 1987.

Nishino, C., Bowers, W. S., Montgomery, M. E., Nault, L. R. "Aphid alarm pheromone mimics: sesquiterpene hydrocarbons". Agr. Biol. Chem., **1976**, 40 : 2303-2304.

Nishino, C., Bowers, W. S., Montgomery, M. E., Nault, L. R., Nielson, M. W. "Alarm pheromone of the spotted alfalfa aphid, *Therioaphis maculata*". J. Chem. Ecol., **1977**, 3 : 349-357.

Nishino, C., Kobayashi, K. Fukushima,K. Imanari, M. Nojima, K. Kohno, S. "Structure and receptor participation of periplanone A, the sex pheromone of the American cockroach", Chemistry Letters, **1988**, 517-520.

Persoons, C. J., Verweil, P. E. Ritter, J., Talman, E., Nooijen, P. J. F., Nooijen, W. J. "Sex pheromones of the American cockroach, *Periplaneta americana*: A tentative structure of periplanone B".Tetrahedron Lett., **1976**, 2055-2058.

Pickett, J. A., Dawson, G. W., Griffiths, D.C., Hassanali, A., Merritt, A., Mudd, M. C., Smith, L. J., Wadhams, C. M., Zhang Z. In Pest. Sci. and Biotech. Greenhalgh, R., Roberts, T. R. Eds.; Blackwell Sci. Publ. Oxford. 1987.

Roelofs, W. L. "Electroantennograms". Chemtech , **1979**, 9, 222-227.

Roelofs, W. L. In, Semiochemicals : Their Role in Pest Control. Nordlund, D.A., Jones, R. L., Lewis, W. J. Eds.; John Wiley & Sons Pub. NY, 1981.

Sjamsul, A. A. Ed.; Chemistry of Rainforest Plants and their Utilization for Development. Institut Teknologi, Bandung, Indonesia, 1992.

Slama, K., Williams, C. M. "Juvenile hormone activity for the bug *Pyrrhocoris apterus*.". Proc. Natl. Acad. Sci. U.S.A., **1965**, 54: 411-414.

Staal, G. B. Anti juvenile hormone agents. In Ann. Rev. Entomol. Mittler, T. E. Radovsky, F. J., Resh, V. H. Eds.; Annual Review of Entomology. Annual Reviews Inc. Palo Alto CA., 1986.

Tahara, S., Yoshida, M. , Mizutani, J., Kitamura, C., Takahashi, S. "A sex stimulant to the male American cockroach in Compositae plants". Agri. Biol. Chem., **1975**, 39: 1517- 1518.

Tschinkel, W. R. "Phenols and quinones from the defensive secretions of the tenebrionid beetle, *Zophobas rugipes*". J. Insect Physiol., **1969**, 15 : 191-195.

Williams, C. M. "The juvenile hormone of insects". Nature, **1956**,178: 212-213.

Wigglesworth, V.B. "The physiology of ecdysis in *Rhodnius prolixus* (Hemiptera). II Factors controlling moulting and metamorphosis". Quart. J. Mic. Sci. **1934**, 77:191-222.

Attracticides for the Control of Diabroticite Rootworms

Robert L. Metcalf, Lesley Deem-Dickson, and Richard L. Lampman, Department of Entomology, University of Illinois, Urbana-Champaign, IL, 61801

An effective integrated pest management (IPM) alternative to the prophylactic application of broad spectrum, highly persistent soil insecticides for the control of corn rootworms is the use of formulated bait granulars or sprays incorporating cucurbitacin arrestants and phagostimulants with minimal dosages of carbamate or organophosphorus insecticides. Such semiochemical-baits applied at 10 kg per ha have given 90 to 100% control of adult Diabroticite rootworms with quantities of insecticides as low as 10 g per ha. The addition of volatile attractants and structural analogues derived from *Cucurbita* blossoms (eg., indole, cinnamaldehyde, cinnamyl alcohol, 4-methoxycinnamaldehyde, and 4-methoxyphenethanol) improved the efficacy of the cucurbitacin-based baits by at least 3-fold.

Rootworms including *Diabrotica barberi*, the northern corn rootworm (NCR), *D. undecimpunctata howardi*, the southern corn rootworm (SCR), *D. virgifera virgifera*, the western corn rootworm (WCR), *D. balteata*, the banded cucumber beetle (BCB), and *Acalymma vittatum*, the striped cucumber beetle, (SCB) cause an estimated annual loss to U.S. agriculture of $1 billion in damage and control costs on corn, soybeans, and cucurbits. The conventional control practice over the past half-century has been preemergent application of soil insecticides and these have been used on as much as 60% of the corn acreage, about 18 million ha. Total amounts of rootworm insecticides applied ranged from 11 million kg in 1971 to 14 million kg in 1982 (Adkisson, 1986).

Corn rootworm resistance, accelerated microbial degradation of insecticides in soils, and pollution of groundwater have resulted in a shift in soil insecticides from the relatively inexpensive organochlorines to the increasingly expensive carbamates, organophosphates, and pyrethroids. Consequently, control costs over the period of 1950 to 1990 have increased from $2.50 per acre to as much as $20 per acre, despite relatively static crop prices. It is apparent that reliance on soil insecticides for corn rootworm control is not a viable control technology for the 21st century.

Cucurbitacins as Kairomones

Diabroticite beetles have coevolved with plants of the family Cucurbitaceae where they were originally pollinators of the open, bowl-shaped flowers (Avila et al., 1989). There is abundant evidence that the bitter oxygenated tetracyclic triterpenoid cucurbitacins (Cucs) (Fig.1) characteristic of this family developed as allomones to deter herbivores and that the rootworm beetles, after evolutionary development

of appropriate detoxification systems for these highly toxic phytochemicals, now recognize the Cucs as kairomones for host selection (Metcalf, 1986; Metcalf and Metcalf, 1992). Although the Cucs are essentially non-volatile, they are locomotory arrestants and phagostimulants for Diabroticite beetles, producing behavioral modification in minute quantities through perception by specific receptors on the maxillary palpi. Diabroticites exposed to Cucs were immobilized and fed compulsively whether the source was in plant tissue or on inert filter paper, silica gel TLC plate, or corn grit granule. For example, the parent Cucs B and E were arrestants in cage tests with *D. balteata*, *D. undecimpunctata howardi*, and *D. v. virgifera* at concentrations as low as 5×10^{-12} g and phagostimulants at 1×10^{-9} g (Metcalf et al., 1980; Peterson and Schalk, 1985).

Bitter Cucurbits as Bait Sources

The behavioral responses of the rootworm beetles to minute quantities of Cucs suggested the use of these phytochemicals in poison baits for control (Metcalf et al. 1981). For this purpose, it was necessary to secure a dependable plant source for the biosynthesis of the Cucs. This problem was solved by the production of hybrid cultivars of *Cucurbita andreana* X *C. maxima* (AND X MAX) whose fruit contained about 1 mg per g Cucs B and D, and of *C. texana* X *C. pepo* (TEX X PEP) whose fruit contained about 0.5 mg per g Cuc E and its glycoside (Rhodes et al., 1980). When the Cuc-containing fruits were dried and ground, the resulting 2 mm granules contained about 4.5 mg per g of Cucs B, D and glycosides (AND X MAX) or 5.5 mg per g of Cucs E, I, and glycosides (TEX X PEP) (Metcalf et al., 1987). The Cucs in the dried, ground baits are very stable and baits stored for more than 5 years

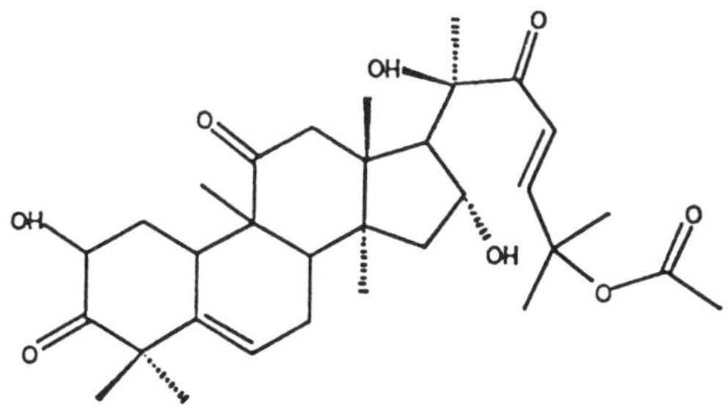

Fig. 1 Cucurbitacin B

remained fully effective. The ground roots of *C. foetidissima*, containing 3 mg per g of Cucs E and I and glycosides in a starch matrix, provide an additional source of Cucs for bait formulations (Metcalf et al., 1987).

The Cuc-containing, granular preparations of bitter *Cucurbita* fruits and roots were formulated as toxic-baits by impregnating them with insecticides from acetone slurries. A variety of small plot field evaluations in corn and cucurbits infested with SCR and WCR beetles showed that Cuc-baits containing methomyl and carbaryl at 0.01 to 0.1% were fed upon readily by the Diabroticites and produced intoxication within minutes, rapidly followed by paralysis and death. Dimethoate, isofenphos, and malathion used at the same concentrations were also effective but appreciably slower in action. Permethrin, cypermethrin, and flucythrinate at 0.01% were appreciably less effective and were somewhat repellent to the Diabroticites (Metcalf et al., 1987).

Several seasons of field experiments applying these Cuc-baits to hybrid field corn infested with Diabroticites, showed that when broadcast at 1.1 to 11 kg per ha, with concentrations of insecticide at about 10 to 100 g per ha and of Cucs at about 40 g per ha; 90 to 100% beetle mortality was obtained (Metcalf et al., 1987).

In these *Cucurbita* baits, most of the Cucs are in the interior of the bait granules and are unavailable to stimulate Diabroticite feeding. A marked improvement in bait efficiency was obtained by impregnating corn grits carrier with Cucs extracted from *Cucurbita* fruit so that the total Cuc concentration was on the exterior of the granules, and ranged from 0.0036 to 0.12% of Cucs. In field tests, such formulated baits containing 0.1% carbaryl produced significant Diabroticite mortalities at Cuc concentrations as low as 0.012% (Metcalf et al., 1987). This concentration, extrapolated to the application of 10 kg of bait per ha, is equivalent to a Cuc dosage of 1 to 2 g per ha.

Structure-Activity of Volatile Kairomones

The blossoms of Cucurbitaceae are well known to be highly attractive to Luperini beetles (Coleoptera: Chrysomelidae; Galerucinae) and most host plant records for the Aulacophorites of the Old World (480 species) and the Diabroticites of the New World (900 species) refer to cucurbit blossoms. Two North American species, *Diabrotica barberi* (=*longicornis*) (NCR) and the economically important *D. v. virgifera* were first collected from the blossoms of the buffalo gourd, *C. foetidissima* in southeastern Colorado in 1824 and 1868 respectively (Smith and Lawrence, 1967).

Cucurbit floral volatiles are long range Diabroticite attractants and 30 g of shredded blossoms of *C. maxima* in cheesecloth-covered cylindrical sticky traps had mean (±SD) catches of 87 (±31) *D. v. virgifera* adults and 12 (±5) *A. vittatum* adults after 1 hour field exposure as compared to 7 (±7) and 2 (±1) respectively for unbaited control traps (Metcalf and Lampman, 1989a). GC-fractionation of blossom volatiles showed the presence of numerous constitutents: 40 from *C. maxima*, 16 from *C. moschata*, and 12 from *C. pepo* (Andersen and Metcalf, 1987; Andersen, 1987). Rootworm beetle preference for *C. maxima* blossoms was correlated with high release rates of 1,2,4-trimethoxybenzene, indole, cinnamaldehyde, cinnamyl alcohol, and ß-ionone. The remarkable aggregations of Diabroticite beetles found in *Cucurbita* blossoms were attributed to the increased arrival rate of the beetles due to long-range attraction to blossom volatiles, while the arrestant and phagostimulant effects of the Cucs in the blossoms delayed beetle departure (Andersen and Metcalf, 1987).

Systematic field evaluations of these volatile kairomones singly and in mixtures, provided the rationale for the development of practical lures (Andersen and Metcalf, 1986; Lampman et al., 1987; Lampman and Metcalf, 1987, 1988; Metcalf and Lampman 1989a,b,c; Lewis et al., 1990). Each Diabroticite species investigated displayed a distinctive pattern of lure responses when exposed to a broad spectrum of the blossom volatile kairomones. Thus *D. v. virgifera* (WCR) responds specifically to indole, cinnamaldehyde, and ß-ionone. *D. barberi* (NCR) responds to cinnamyl alcohol and the closely related *D. cristata*. responds intermediately between WCR and NCR. *D. undecimpunctata howardi* (SCR) responds specifically to cinnamaldehyde and the related *Acalymma vittatum* (SCB) to indole. All of the species investigated show chemosensory affinities to cucurbit blossom volatiles. Indole, which is responsible for the musky aroma of *Cucurbita* blossoms, was shown to have a fundamental role in attraction of several species, and when combined with phenylacetaldehyde and

1,2-dimethoxybenzene (veratrole) (VIP mixture) exhibited a marked degree of olfactory synergism to SCR beetles but not to WCR beetles. However, the presence of relatively large amounts of cinnamaldehyde and 1,2,4-trimethoxybenezene in *C. maxima* blossoms and the structural analogies of these to phenylacetaldehyde and to veratrole, led to the simplified synthetic blossom mixture of trimethoxybenzene, indole and cinnamaldehyde (TIC). This mixture was attractive to all the Diabroticite species and showed olfactory synergism of about 2-fold to *D. barberi, D. cristata, D. undecimpunctata howardi, D virgifera virgifera*, and *A. vittatum* (Lampman and Metcalf, 1987, 1988; Lampman et al., 1991; and Lewis et al., 1990). The TIC mixture has proven to be a useful general purpose lure for Diabroticite beetles and when applied to cylindrical sticky traps at doses as low as 0.1 mg, resulted in a linear response between mean number of beetles trapped and log dose of attractant (Lewis et al.,1990).

The relatively specific attraction of beetle species to structurally related compounds isolated from blossom volatiles (e.g. cinnamaldehye for SCR and cinnamyl alcohol for NCR) focused attention on the role of small changes in the molecular structure of these phenylpropanoids. The frequently dramatic shift in species-specific response to these kairomones is associated with changes in the terminal groups of the unsaturated side chains and in the substituents of the phenyl rings. This was demonstrated by comparing the attractiveness of pairs of phenylpropanoids and structural analogues of blossom volatiles (parakairomones) with differing side chains, one of each pair with a 4-methoxyphenyl substitution (Table 1) (Metcalf and Lampman, 1989a). The

ratio of SCR to WCR captured on baited traps varied by a factor of more than 10 between cinnamaldehyde and 4-methoxycinnamaldehyde or cinnamonitrile and 4-methoxycinnamonitrile . As shown in Table 1, cinnamaldehyde and its bioisostere cinnamonitrile were excellent lures for SCR but were unattractive to WCR. In contrast, 4-methoxycinnamaldehyde and 4-methoxycinnamonitrile were excellent lures for WCR but were unattractive to SCR. The 4-methoxycinnamaldehyde proved to be the most effective lure yet found for WCR and produced significant attraction in field studies at doses as low as 0.03 mg per sticky trap (Metcalf and Lampman, 1989b). Traps baited with 100 mg of 4-methoxycinnamaldehyde were 50-fold more attractive over an 18 day period than unbaited control traps. This semiochemical was also effective for *D. cristata* but was not attractive to NCR (Table 1).

Investigations of analogues of cinnamyl alcohol showed that at equivalent dosages, the bioisostere 3-phenylpropanol (phenpropanol) was about 0.8 X as attractive and 2-phenylethanol (phenethanol) substantially less attractive (Metcalf and Lampman, 1989c). Incorporation of 4-methoxy-groups into these molecules produced the very effective and persistent NCR attractants 4-methoxyphenpropanol and 4-methoxy-phenethanol (Table 1). The latter was attractive in field tests at dosages as low as 0.1 mg per sticky trap (Metcalf and Lampman, 1991) and is the most effective NCR attractant yet discovered. Neither compound, however, is appreciably attractive to WCR and SCR (Table 1). The relative chemosensory specificity of these phenylpropanoid attractants is dependent both on the degree of fit between the attractant molecule

Table 1. Mean *Diabrotica* beetle capture on sticky traps (24 hour, n = 4, controls subtracted) baited with volatile kairomones (100 mg)[1].

Volatile	NCR	WCR	SCR
Date: 8-5-86			
$C_6H_5CH=CHC(O)H$	-	3a	413d
$4\text{-}CH_3OC_6H_4CH=CHC(O)H$	-	143c	25b
$C_6H_5CH=CHCCN$	-	1a	139c
$4\text{-}CH_3OC_6H_4CH=CHCCN$	-	58b	0a
Date: 8-7-87			
$C_6H_5CH=CHC(O)H$	0a	51b	58c
$C_6H_5CH=CHCH_2OH$	9b	3a	20b
$4\text{-}CH_3OC_6H_4CH=CHC(O)H$	0a	365c	0a
Date: 8-12-91			
$4\text{-}CH_3OC_6H_4CH=CHC(O)H$	2a	12b	-
$4\text{-}CH_3OC_6H_4CH_2CH_2OH$	84b	1a	-

[1]Means for a given species and date followed by different letters are significantly different at p < 0.05. Data from Metcalf and Lampman 1989b and unpublished.

and the beetle's antennal receptor and on the "release rate" (volatility) of the attractant at ambient temperatures. Thus, 4-methoxy-phenethanol for NCR and 4-methoxy-cinnamaldehyde for WCR have the lowest thresholds and the lowest release rates for the respective species (Metcalf and Lampman, 1991).

Volatile Lures for Attracticide Baits

Knowledge of an array of volatile kairomone attractants for Diabroticites, suggested their use as additional ingredients in "attracticide baits" (Metcalf and Lampman, 1991). A basic formulation of 95% corn grits and 5% *Cucurbita foetidissima* root powder (for a final concentration of approximately 0.01% Cucs) was impregnated with 0.3% carbaryl. Solvent (acetone) impregnation was used to incorporate a spectrum of one to several kairomone volatiles, typically at 0.1% each. Such bait combinations were compared in a variety of field experiments to determine the most effective formulations for Diabroticite pest control (Table 2). The most complex formulation contained 1,2,4-tri-methoxybenzene, indole, cinnamaldehyde, cinnamyl alcohol, phenpropanol, 4-methoxy-cinnamaldehyde, and 4-methoxyphenethanol ("Slam" granular, formerly called "Nemesis", Microflo Company, Lakeland,FLA). From these and other field evaluations during 1990-1992, it appears that a Cuc-containing bait with 4-

methoxycinnamaldehyde, 4-methoxyphenethanol, cinnamaldehyde,and indole (Table 2) provides an optimum formulation for adult Diabroticite control.

Preliminary experiments were conducted in a hybrid field corn averaging 0.07 WCR beetles per plant (n=100 plants). The whorls of 20 individual corn plants, each 10 m apart, were treated with 200 mg of granular bait with or without volatile attractants. Each treated row was separated by 10 rows of untreated plants. The rate of application was approximately equivalent to 10 kg per ha for a field of 50,000 plants per ha. After 1 day, the mean number ($\pm$SD) of dead and moribund WCR beetles per treated corn plant were: bait lacking attractants 1.4($\pm$0.28), bait with 4-methoxy-cinnamaldehyde 3.29($\pm$0.29), and bait with 7 attractants 4.44($\pm$0.32). ANOVA showed that both baits with attractants were significantly more effective than the bait lacking attractants (P<0.0001) (unpublished data).

Similar experiments with 200 mg of the various baits exposed in cardboard lids wired to corn plants at ear height also showed the greater efficiency of the baits with volatile attractants. After 1 day, the mean number of dead and moribund WCR ($\pm$S.E) for bait lacking attractants was 0.0 ($\pm$0.0), for bait with 4-methoxy-cinnamaldehyde was 1.9 ($\pm$0.35), and for bait with the 7 attractants was 5.3 ($\pm$1.1). ANOVA showed that the baits with volatile attractants were significantly more effective than the bait lacking

Table 2. Mean *Diabrotica* beetle capture on sticky traps (24 hour, n = 4) baited with granulars containing 0.2 mg volatile kairmones[1].

Volatile	NCR	WCR
none	5a	19a
cinnamaldehyde, indole (CI)	8a	160b
trimethoxybenzene, indole, cinnamaldehyde (TIC)	8a	212bc
methoxycinnamaldehyde (MCA)	4a	266bc
MCA/CI	4a	409c
methoxyphenethanol (MPE)	42b	18a
MPE/CI	46b	133b
MCA/MPE	25b	231bc
MCA/MPE/CI	32b	360c
TIC/MCA/MPE/cinnamyl alcohol/phenpropanol	28b	367c

[1]Means for each columnfollowed by different letters are significantly different at p < 0.05. Unpublished data (8-28-90).

volatile attractants (P<0.0001) (unpublished data). These results suggest that incorporation of volatile attractants can increase attracticide effectiveness by 3-fold or more.

Attractant Dosage vs Diabroticite Response

It has also been demonstrated that the number of Diabroticites caught by cylindrical sticky traps baited with volatile attractants is directly proportional to the log dose of lure, in conformity with the Weber-Fechner Law (Metcalf and Lampman, 1989a, 1991; Lewis et al., 1990). This fundamental relationship was explored further by applying various quantities of the formulated granular baits described above, to the tops of cylindrical sticky traps. Results with 4-methoxycinnamaldehyde attractants for WCR and with 4-methoxyphenethanol for NCR beetles are shown in Fig. 2. As expected there is a highly significant (P<0.001) relationship between the mean numbers of beetles trapped and the log dose of volatile attractant used (unpublished data). This technique provides a practical and realistic way to evaluate various formulations of "attracticide" baits.

Other investigators (Lance et al., 1992) have reported inconsistent results with dose-response and preference studies using diabroticite attractants. These inconsistencies are largely attributable to the randomized block design of the experiments, which would generate large areas of plume overlap. For this reason, we arrange dose-response experiments and preference tests in linear randomized complete blocks or latin-square designs which are perpendicular to prevailing winds in order to minimize plume interactions.

Attracticide Field Evaluation

With the cooperation of the U.S. Department of Agriculture, Northern Grain Insects Research Laboratory, Brookings S.D., a commercially prepared granular bait ("Slam" formerly "Nemesis", Microflo Company, Lakeland, FLA) was applied by air to approximately 20 ha (44 acre) of hybrid seed corn near Pesotum, IL (8-9-90). An equal sized contiguous area was left as an untreated reference field. Before application, 400 corn plants in each corner of the field were sampled and showed populations of WCR 0.7 (± 0.05) [mean (±SE)]and NCR 0.1 (±0.005)per corn plant. The formulated bait, applied at 10 kg per ha, consisted of 94% corn grits, 0.3% carbaryl, 5% of *Cucurbita foetidissima* root powder (containing about 0.2 mg per g Cucs), and 0.1% each of cinnamaldehyde, cinnamyl alcohol, indole, trimethoxybenzene, 3-phenyl-propanol, 4-methoxyphenethanol, and 4-methoxycinnamaldehyde. The ingredients applied per ha were 30g carbaryl, 1.0 g Cucs, and 10 g of each attractant.

Within 1 to 2 minutes after application, corn rootworm beetles were observed feeding on bait granules adhering to the corn plants, and within 15 minutes, thousands of moribund beetles littered the ground. The overall effects of the attracticide treatment were determined after 1,2,7, and 14 days by comparing the number of NCR and WCR adults in untreated and treated portions of the field using: (a) mean whole plant counts (n = 400) in each corner, and (b) mean 24 hour counts on baited (TIC) cylindrical sticky traps (n = 4) 20 m apart in each corner (Levine and Metcalf, 1989). The results shown in Table 3 indicated a rapid kill of rootworm beetles after 1

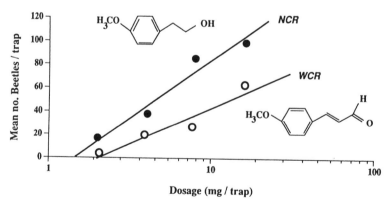

Fig. 2. Attractancy of granular "attracticide baits" as a function of trap dose (n = 4).for: *D. v. virgifera* (WCR); o, 4-Methoxycinnamaldehyde (y = 61.46 log (x)- 18.00, R2 = 0.88); *D. barberi* (NCR); •, 4-Methoxyphenethanol (y = 99.89 log (x)- 14.80, R2 = 0.96)

Table 3. Mean *Diabrotica* beetle counts in hybrid seed corn following aerial application of attracticide bait at 10 kg per ha.[1]

Field treatment		days following application			
		1	2	7	14
		Plant count (n = 400)			
untreated acreage	NCR	0.1±0.01	0.14±0.11	0.25±0.03	0.28±0.04
	WCR	0.7±0.05	0.72±0.18	0.94±0.06	0.89±0.06
treated acreage	NCR	0.0	0.0	0.005	0.035±0.005
	WCR	0.09±0.02	0.1±0.04	0.1±0.01	0.330±0.073
		Sticky trap count (n = 16)			
untreated acreage	NCR	17±1.9	36±02.0	72±5.4	72±6.2
	WCR	503±40	256±22	249±22	457±22
treated acreage	NCR	0.1±0.1	0.4±0.2	9.5±1.1	31.6±5.6
	WCR	33±4.8	96±11	81±11	237±16

[1]Pesotum, IL, 8-9-90.

day: NCR 100% (plant count), 97% (trap count); WCR 86% (plant count), 93% (trap count). Mortality remained high after 7 days: NCR 98% (plant count), 87% (trap count); WCR 90% (plant count), 67% (trap count), despite evidence of beetle migrations into the treated area.

Summary

"Attracticide-baits" containing cucurbitacin arrestants and feeding stimulants, and carbamate or organophosphorous insecticides can provide very high mortalities of Diabroticite rootworm beetles when applied at about 10 kg per ha. The minimum effective concentrations of cucurbitacins are about 1 g per ha and insecticide about 10 g per ha. The addition of volatile kairomones to these Cuc-baits improved the performance by several fold. The most effective long-range attractants were 4 - methoxycinnamaldehyde for WCR and 4 - methoxyphenethanol for NCR.

The use of such baits has promise of providing highly selective Diabroticite control with doses of insecticides of about one one-hundredth those used in conventional aerial application. Much remains to be learned about the proper use of the "attracticide-bait" technology but this alternative insect control method has considerable potential in IPM, e.g. for use as "artificial trap crops" for many insect pests such as tephritid fruit flies, Japanese beetle, and tse-tse flies; and for the dissemination of insect growth regulators and pathogens.

Acknowledgements

This research has been supported in part by the U.S. Department of Agriculture Competitive Research Grants Office, AG 87-CRCR-1-2373 and by a grant from the LISA program; and by the National Science Foundation Grant PCM-8312778.

References

Adkisson, P.L. Integrated Pest Management. *Bull. Entomol. Soc. Amer.* **1986**, *32*, 136.

Andersen, J.F. Composition of the floral odor of *Cucurbita maxima* (Duchesne: Cucurbitaceae). *J. Agr. Food Chem.* **1987**, *35*, 60.

Andersen, J.F.; Metcalf, R.L. Identification of a volatile attractant for *Diabrotica* and *Acalymma* species from the blossoms of *Cucurbita maxima* Duchesne. *J. Chem. Ecol.* **1986**, *12*, 687.

Andersen, J.F.; Metcalf, R.L. Factors influencing the distribution of *Diabrotica* spp in the blossoms of cultivated *Cucurbita* spp. *J. Chem. Ecol.* **1987**, *14*, 681.

Avila, C.J.; Martinho, M.R.; Campos, J. J.P. de. Pollination and pollinators in the production of fruits and hybrid seeds of squash (*Cucurbita pepo* var *melopepo*) *Ann. Soc. Entomol. Brazil* **1989**, *18*, 13.

Ladd, T.L.; Stinner, B.L.; Kruger, H.R. Eugenol, a new attractant for the northern corn rootworm (Coleoptera: Chrysomelidae) *J. Econ. Entomol.* **1983**, *76*, 1049.

Lampman, R.L.; Metcalf, R.L. Multicomponent kairomone lures for southern and western corn rootworms (Coleoptera: Chrysomelidae: *Diabrotica* spp.). *J. Econ. Entomol.* **1987**, *80*, 1137.

Lampman, R.L.; Metcalf, R.L. The comparative response of *Diabrotica* species (Coleoptera: Chrysomelidae) to volatile attractants. *Environ. Entomol.* **1988**, *17*, 644.

Lampman, R.L.; Metcalf, R.L.; Deem-Dickson, L.; Reid, C.D. Attraction of *Diabrotica* species (Coleoptera: Chrysomelidae) to a multicomponent lure and implications for corn rootworm baits. In Press, *J. Econ. Entomol.*. 1993.

Lance, D.R.; Scholtz, W.; Stewart, J.W.; Fergen, J.K. Non-pheromonal attractants for Mexican corn rootworm beetles *Diabrotica virgifera zeae* (Coleoptera: Chrysomelidae). *J. Kansas Entomol Soc.*, **1992**, *65*, 10.

Levine, E.; Metcalf, R.L. Sticky attractant traps for monitoring corn rootworm beetles. *Ill. Natr. History Sur. Rept.* **1988**, 279.

Lewis, P.A.; Lampman, R.L.; Metcalf, R.L. Kairomonal attractants for *Acalymma vittatum*. *Environ. Entomol.* **1990**, *19*, 9.

Metcalf, R.L. Coevolutionary adaptations of rootworm beetles. (Coleoptera: Chrysomelidae) *J. Chem. Ecol.* **1986**, *12*, 1109.

Metcalf, R.L.; Ferguson, J.E.; Lampman, R.L.; Andersen, J.F. Dry cucurbitacin containing baits for controlling diabroticite beetles (Coleoptera: Chrysomelidae). *J. Econ. Entomol.* **1987**, *80*, 870.

Metcalf, R.L.; Lampman, R.L. Chemical ecology of Diabroticites and Cucurbitaceae. *Experientia* **1989a**, *45*, 240.

Metcalf, R.L.; Lampman, R.L. Estragole analogues as attractants of *Diabrotica* species (Coleoptera: Chrysomelidae) corn rootworms. *J. Econ. Entomol.* **1989b**, *82*, 123.

Metcalf, R.L.; Lampman, R.L. Cinnamyl alcohol and analogues as attractants for the adult northern corn rootworm *Diabrotica barberi* (Coleoptera: Chrysomelidae). *J. Econ. Entomol.* **1989c**, *82*, 1620.

Metcalf, R.L.; Lampman, R.L. Evolution of Diabroticite (Coleoptera: Chrysomelidae) receptors for *Cucurbita* blossom volatiles. *Proc. Nat. Acad. Sci.* (U.S.A.) **1991**, *88*, 123.

Metcalf, R.L.; Metcalf, E.R. *Plant Kairomones in Insect Ecology and Control*, ; Chapman and Hall: New York, NY, **1992**

Metcalf, R.L.; Lampman, R.L. Volatile attractants for *Diabrotica* species. U.S. PAT. 4880624, **1989**.

Metcalf, R.L.; Rhodes, A.M. Diabroticite pest control. Can. Pat. 1195922, **1985**.

Metcalf, R.L.; Rhodes, A.M.; Metcalf, E.R. Monitoring and controlling corn rootworm beetles with baits of dried, bitter *Cucurbita* hybrids. *Cucurbita Genetics Coop.* **1981**, *4*, 37.

Metcalf, R.L.; Metcalf, R.A.; Rhodes, A.M. Cucurbitacins as kairomones for diabroticite beetles. *Proc. Nat. Acad. Sci.* (U.S.A.) **1980**, *77*, 3769.

Metcalf, R.L.; Rhodes, A.M.; Metcalf, R.A.; Ferguson, J.; Metcalf, E.R.; Lu, P-y. Cucurbitacin contents and Diabroticites (Coleoptera: Chrysomelidae) feeding upon *Cucurbita* spp. *Environ. Entomol.*, **1982**, *11*, 931.

Petersen, K.K., Schalk, J.M. Semiquantitative bioassay for levels of cucurbitacins using the banded cucumber beetle (Coleoptera: Chrysomelidae)*J. Econ. Entomol.*, **1985**, *78*, 738.

Rhodes, A.M., Metcalf, R.L., Metcalf, E.R. Diabroticite response to cucurbitacin kairomones. *J. Amer. Soc. Hort. Sci.*, **1980**, *105*, 838.

Smith, R.F., Lawrence, J.F. *Classification of the Status of Type Specimens of Diabroticites (Coleoptera: Chrysomelidae, Galerucini)*. Univ. Cal. Press, Berkeley, **1967**.

Strategies and Tactics for the Use of Semiochemicals Against Forest Insect Pests in North America

John H. Borden, Centre for Pest Management, Department of Biological Sciences, Simon Fraser University, Burnaby, B.C. V5A 1S6 Canada

Five strategies and 19 tactics for the management of forest insect pests with semiochemicals are described. Pheromone-based survey and detection tactics are now used to monitor many defoliators, shoot and tip feeders and bark and timber beetles. Disruption of communication through wide-scale deployment of lepidopteran sex pheromones is operational for three species, but for others it is suboptimally effective, possibly because of missing pheromone components. The strategy of removing the pest has been successfully applied using a number of tactics, including: semiochemical-based containment and concentration of bark beetle infestations followed by logging; mass-trapping to suppress populations; and the use of semiochemical-baited trap trees and logs. Modification of behavior using the tactic of wide-scale deployment of antiaggregation pheromones has been operationally successful for only one bark beetle species.   Improvement in tactics may require blends of semiochemicals or integration with silvicultural treatments. Two promising tactics are to use epideictic pheromones produced by lepidopteran eggs and larvae to deter oviposition and feeding, respectively.   The tactic of semiochemical-induced competitive displacement of bark beetles is a promising means for using semiochemicals to enhance natural controls. Future commercial use of semiochemicals in forest environments will demand accommodating regulatory policies and an industry willing to grow and survive through the development and marketing of numerous minor species-specific products and services.

The development of semiochemicals (message-bearing chemicals) (Nordlund 1981) as tools for the management of forest insect pests in the past three decades has been encouraging. Rather than documenting the many and varied applications by taxa, regions, or forest types, I prefer to view them in a conceptual manner and to categorize them by the strategies and tactics involved. I define a strategy as a broad plan and a tactic as a specific action or maneuver that alone or in combination with other tactics can be used to satisfy a strategic objective. There are five strategies and 19 tactics that encompass current or anticipated applications of semiochemicals against forest insect pests in North America (Table 1).

Monitoring

The most readily implemented strategy has been to exploit the attraction of forest insect pests to pheromones to monitor populations. The various monitoring tactics usually employ either sex or aggregation pheromones in baited traps.

Sex pheromones for lepidoptera are used alone, mimicking the attraction of males to females in nature (Fig. 1). Aggregation phero-mones for bark or timber beetles are usually supplemented by the addition of attractive host kairomones, reflecting the role of semio-chemical blends that in nature induce attraction to trees or logs (Fig. 2) that provide food, shelter and the opportunity to mate and reproduce (Borden 1985).

Table 1.   Semiochemical-based strategies and tactics used or considered for use against forest insect pests in North America

Strategies and tactics	Status[a]	Examples of target species with selected reference[b]
Monitoring		
Detection and survey of exotic pests using semio-chemical-baited traps as decision-making aids	Implemented	Gypsy moth, *Lymantria dispar* (Kolodny-Hirsch and Schwalbe, 1990); European pine shoot moth, *Ryacionia buoliana* (Borden 1990); smaller European elm bark beetle, *Scolytus multistriatus* (Lanier et al. 1976)
Survey and detection of indigenous pests using pheromone-baited traps as decision-making aids	Implemented	Douglas-fir tussock moth, *Orygia pseudotsugata*; Nantucket pine tip moth, *Rhyacionia frustrana* (Daterman, 1990); cranberry girdler, *Chrysoteuchia topiaria* (Borden, 1990); forest tent caterpillar, *Malacosoma disstria* (Grant, 1991);
	Operational	Black army cutworm, *Actebia fennica* (Grant, 1991); coneworms, *Dioryctria* spp. (Daterman, 1990); eastern spruce budworm, *Choristoneura fumiferana* (Sanders, 1990a)
	Developmental	Jack pine budworm, *Choristoneura pinus* pinus; western spruce budworm, *Choristoneura occidentalis*; larch casebearer, *Coleophora laricella*; larch bud moth, *Zeiraphera improbana*; large aspen tortrix, *Choristoneura conflictana*; oak leaf shredder, *Croesia semipurpurana*; oak olethreutid leafroller, *Pseudoxentera spoliana*; spruce budmoth, *Zeiraphera unfortunana*; Douglas-fir cone moth, *Barbara colfaxiana*; spruce seed moth, *Cydia strobilella*; spruce coneworm, *Dioryctria reniculelloides*; variegated cutworm, *Peridroma saucia* (Grant, 1991); eastern and western hemlock loopers, *Lambdina fiscellaria fiscellaria* and *L. f. lugubrosa*, respectively (J.H. Borden, unpubl.; R.J. West, Forestry Canada, St. John's, Nfld., pers. comm.); cone beetles, *Conophthorus* spp. (P. deGroot, Forestry Canada, Sault St. Marie, Ont., pers. comm.)
Survey and detection using semiochemical-baited traps as basis for design of mass-trapping program	Implemented	Ambrosia beetles, *Trypodendron lineatum*, *Gnathotrichus sulcatus* and *G. retusus* (Borden, 1990)

[a] Implemented: routinely used for some time, usually over a wide area. Operational: efficacy experimentally demonstrated, ready for or in early stages of implementation. Developmental: potential efficacy demonstrated experimentally, applied research in progress, possibly delayed because of technological or scientific limitations. Research: promising idea under scientific investigation.

[b] References selected for ease of access to information, e.g. review papers containing information on as many species and tactics as possible chosen over original papers with information on only one subject.

Table 1. (continued) Semiochemical-based strategies and tactics used or considered for use against forest insect pests in North America

Strategies and tactics	Status[a]	Examples of target species with selected reference[b]
Detection of onset and peak flight using semio-chemical-baited traps and trees as basis for regulatory measures or post-flight direct control	Implemented	Mountain pine beetle, *Dendroctonus ponderosae* (Stock, 1984; Hall, 1989)
Determination of predator: ratio using semiochemical-baited traps as a measure of risk	Implemented	Southern pine beetle, *Dendroctonus frontalis* prey (Billings 1988)
Disruption of communication		
Suppression of mating by wide-scale deployment of sex pheromone	Operational	Western pine shoot borer, *Eucosma sonomana* (Daterman, 1990); *O. pseudotsugata* (Sower et al., 1990); *L. dispar* (Kolodny-Hirsch and Schwalbe, 1990)
	Research	*C. occidentalis* (Daterman et al., 1985); *C. fumiferana* (Sanders 1990a)
Removal of pest		
Containment and concen-tration of infestations using semiochemical-baited trees prior to logging	Implemented	*D. ponderosae* (Borden 1990); Spruce beetle, *Dendroctonus rufipennis* (D.G. Holland, U.S.D.A., For. Serv., Ogden, UT, pers. comm.)
	Operational	Douglas-fir beetle, *Dendroctonus pseudotsugae* (Ringold et al., 1975)
	Research	Western balsam bark beetle, *Dryocoetes confusus* (Stock 1991)
Mass-trapping of both sexes using semiochemical-baited traps	Implemented	*T. lineatum, G. sulcatus, G. retusus* (Borden 1990)

[a] Implemented: routinely used for some time, usually over a wide area. Operational: efficacy experimentally demonstrated, ready for or in early stages of implementation. Developmental: potential efficacy demonstrated experimentally, applied research in progress, possibly delayed because of technological or scientific limitations. Research: promising idea under scientific investigation.

[b] References selected for ease of access to information, e.g. review papers containing information on as many species and tactics as possible chosen over original papers with information on only one subject.

Continued on next page.

Table 1. (continued) Semiochemical-based strategies and tactics used or considered for use against forest insect pests in North America

Strategies and tactics	Status[a]	Examples of target species with selected reference[b]
	Operational	California fivespined ips, *Ips paraconfusus* (P.J. Shea, U.S.D.A., For. Serv., Davis, CA, pers. comm.); *S. multistriatus* (Lanier et al., 1976); pine engraver, *Ips pini* (Phero Tech 1988)
	Developmental	Western pine beetle, *Dendroctonus brevicomis* (DeMars et al., 1980); *I. pini*; *D. pseudotsugae* (B.S. Lindgren, Phero Tech Inc., Delta, B.C., pers. comm.)
Mass-trapping of males in pheromone-baited traps	Operational	*L. dispar* (Kolodny-Hirsch and Schwalbe, 1990)
Trap trees or logs baited with semiochemicals and treated with insecticide	Implemented	*S. multistriatus* (Lanier, 1988); *D. ponderosae* (Hall, 1989)
	Operational	*D. brevicomis* (Smith, 1986)
	Research	*T. lineatum, G. sulcatus, G. retusus* (Lindgren et al., 1982)
Modification of behavior		
Prevention of attack by wide-scale deployment of antiaggregation pheromone	Operational	*D. pseudotsugae* (Furniss et al., 1981)
	Developmental	*D. ponderosae* (Amman, 1993); *D. rufipennis* (Lindgren et al., 1989);
As above, with silviculture treatment	Operational	*D. frontalis* (Payne et al., 1992)
As above, with competitor synomone	Research	*D. brevicomis* (Paine and Hanlon, 1991)
As above, with competitor synomone plus non-host kairomones	Research	*D. ponderosae* (J.H. Borden, unpubl.)
As above, with semiochemical-baited traps or trees in adjacent area ("push-pull" tactic)	Operational	*I. paraconfusus* (Berson 1992)
	Developmental	*D. ponderosae* (Lindgren and Borden 1993)

[a] Implemented: routinely used for some time, usually over a wide area. Operational: efficacy experimentally demonstrated, ready for or in early stages of implementation. Developmental: potential efficacy demonstrated experimentally, applied research in progress, possibly delayed because of technological or scientific limitations. Research: promising idea under scientific investigation.

[b] References selected for ease of access to information, e.g. review papers containing information on as many species and tactics as possible chosen over original papers with information on only one subject.

Table 1. (continued) Semiochemical-based strategies and tactics used or considered for use against forest insect pests in North America

Strategies and tactics	Status[a]	Examples of target species with selected reference[b]
Prevention of attack by wide-scale deployment of competitor synomones	Operational	*I. pini* (Miller et al., 1993)
Inhibition of oviposition by epideitic pheromone in eggs	Research	Obliquebanded leafroller, *Choristoneura rosaceana* (Poirier and Borden, 1991)
Inhibition of feeding by epideictic pheromone in larval oral exudate	Research	*C. fumiferana, C. occidentalis* (L.M. Poirier and J.H. Borden, unpubl.)
Enhancement of natural control		
Competitive displacement of primary species by baiting hosts with secondary species pheromone	Research	*D. ponderosae* (Rankin and Borden 1991)
	Research	*D. frontalis* (Payne and Richerson 1985)

[a] Implemented: routinely used for some time, usually over a wide area. Operational: efficacy experimentally demonstrated, ready for or in early stages of implementation. Developmental: potential efficacy demonstrated experimentally, applied research in progress, possibly delayed because of technological or scientific limitations. Research: promising idea under scientific investigation.

[b] References selected for ease of access to information, e.g. review papers containing information on as many species and tactics as possible chosen over original papers with information on only one subject.

The use of semiochemicals in monitoring is relatively simple when compared to other strategies, and thus its importance tends to be overlooked. Nonetheless, monitoring tactics are applied with increasing frequency and precision, and they are beginning to supplement or replace other survey methods.

Detection of the introduction and spread of exotic pests with pheromone-baited traps has been widely applied. Citizens throughout much of the U.S.A. and Canada recognize the stan-dard delta trap for gypsy moths as a reassuring signal that urban forests are monitored with care by vigilant public servants. If disparlure-baited traps had not been deployed in 1991 on the west coast of North America, the introduction and establishment of Asian gypsy moths might have escaped detection until it was too late for eradication to be successful. Other, less spectacular, programs continue to monitor introduced populations of such pests as smaller European elm bark beetles and pine shoot moths.

International travel and commerce will maintain the threat of introduction of exotic

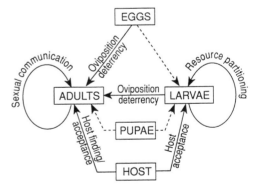

Fig. 1. Pathways and functions for known (solid lines) or possible (broken lines) semiochemical-based communication in lepi-doptera. Interspecific communication between lepidoptera and associated insect species not shown.

269

pests, e.g. in timber imported into the U.S.A. from Siberia (U.S.D.A. 1991). The alarming detection of the southern pine beetle in semio-chemical-baited traps in Israel (Mendel and Argman 1986) suggests that traps baited with appropriate semiochemicals could be used in U.S. and Canadian ports for early detection of unwanted intruders from Siberia or elsewhere.

In a few cases, programs have been imple-mented in which traps baited with synthetic sex pheromones have been used to evaluate popula-tions of native lepidopteran pests as the basis for management decisions (Table 1). This tactic is commonly based on the establishment of a con-sistent relationship over time between catches in traps and levels of populations determined by some other sampling method, usually egg, larval or pupal counts.

Catches in traps may be used directly as a basis for management decisions, e.g. for cran-berry girdlers in conifer nurseries, where almost no damage is tolerable and scheduling of insect-icide sprays may be more important than deter-mining population levels precisely. In other cases, e.g. for Douglas-fir tussock

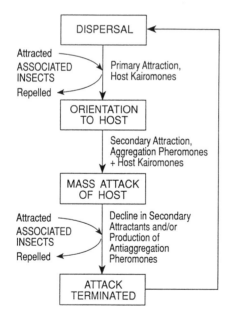

Fig. 2. Generalized sequence of host selection and attack by bark and timber beetles, showing role of semiochemical-mediated behavior.

moths, catches of males in pheromone-baited traps may be used to determine the initial release of populations from natural controls, some years ahead of the need for direct intervention. Some other method of more accurate population assessment, e.g. sequential sampling for egg masses, is then applied until a management decision is made and direct controls are applied if necessary (Daterman 1990).

Progress in the identification of sex pheromones has led to a large number of lepi-doptera for which surveys are listed as "under development" (Table 1). It may take several years before a consistent relationship between trap catches and population levels is established (Sanders 1990a). One means of decreasing the development time is to test pheromone-baited traps against numerous pest populations of different levels, disclosing the predictive capability of pheromone-baited traps in one or two years (Sanders 1990a).

Three tactics have been implemented in which attractive semiochemicals are used to survey for populations of bark and timber beetles (Table 1). For ambrosia beetles, a semiochemical-based survey is set up in a timber processing area during the first year of a full-service mass-trapping program (Borden 1990). The focus of the mass-trapping effort in subsequent years is directed where the highest catches were achieved in the survey year, with changes in focus made if there are shifts in the spatial distributions of captured beetles. In western North America peaks in catches of mountain pine beetles in multiple funnel traps can be used as a basis for restricting log hauling when there is a high risk of transporting emergent beetles (Stock 1984). Catches in traps or observations of attacks on semiochemical-baited trees are used to establish the three-week post-attack window of efficacy during which treatments of monosodium methane arsonate will be effective in killing beetles in newly-attacked trees (Hall 1989).

In the southern U.S.A. semio-chemical-baited traps are used in an ingenious method to determine the risk of southern pine beetle infestations based upon the prey:predator ratio (Billings 1988). An upward shift in the ratio of southern pine beetles to clerid predators signals the potential onset of an outbreak.

Disruption of Communication

Disruption of communication conceptually could be achieved through tactics in two broad categories: those that use some unnatural agent, e.g. halogenated phero-mones, to inhibit sensory receptors, or those that allow perception to occur, but inhibit natural responses. No operational-ly-promising disruption of communication by inhibiting perception of pheromones has been demonstrated.

In practice, disruption employs natural pheromone (usually a sex pheromone) or pheromone analogues that act through one or more of the following mechanisms (Cardé 1990): 1) neural inhibition through sensory adaptation or habituation, 2) false trail following caused by competition between multiple point sources of synthetic pheromone and pheromone-emitting females, 3) inability to perceive "camouflaged" natural pheromone sources in the presence of high levels of uniformly dispersed synthetic pheromone, and 4) use of a partial pheromone blend to disrupt response to a complex natural blend, e.g. by overwhelming the natural pheromone sources by a massive release of one or two components.

The success of disruption against forest lepidoptera has been mixed. The greatest success has been achieved against the western pine shoot borer, Eucosma sonomana (Daterman 1990). The population levels of terminal shoot borers are limited by the numbers of trees; halving their damage may provide adequate control. Hence this species is an ideal target for disruption. Large plot experiments with aerial application of pheromone-impregnated hollow fibres or hand-hung polyvinylchloride lures resulted in respective reductions in infested terminals in the following year of 68-81% (Sower et al. 1982) and 45-62% (Sartwell et al. 1983). The wide spacing of the hand-hung polyvinyl-chloride lures suggests strongly that disruption in this case occurred through false trail following. The pheromone has been registered as a pesticide but use has been minimal, in part due to cost and in part to a lack of recognition by forest managers of the damage caused by the pest (Daterman 1990).

For Douglas-fir tussock moths, eastern spruce budworms, and Gypsy moths, all devas-tating defoliators, there is no problem recognizing their impact. However, compared to the efficacy of chemical or microbial insecticides, or to pheromone-based disruption for some agricultural pests, disruption trials for these insects have been much less effective. Even at low, preoutbreak densities of Douglas-fir tussock moths, aerial applications of (Z)-6-henicosen-11-one in hollow fibres at 25 g/ha resulted in respective reductions of eggs and larvae in the next generation of only 74 and 68% (Sower et al. 1990). Because of environmental concerns with conventional pesticides, there has been immense interest in developing disparlure, cis-7,8-epoxy-2-methyloctadecane, for disruption of Gypsy moth populations (Cameron, 1981). However, lack of data showing adequate population suppression using various application techniques against high or low populations is a principal factor impeding implementation (Kolodny-Hirsch and Schwalbe 1990). The case is similar for eastern spruce budworms. Aerial applications of pheromone (Z)-and (E)-tetradecanal, in capsules, fibers or flakes have yielded inconclusive evidence for reductions in mating or production of fertile eggs (Sanders 1985).

Attempts to suppress mating through dis-ruption of communication may fail for a number of technological or environmental causes. Another reason for a lack of efficacy may be that there are unknown pheromone components yet to be discovered. Spruce budworm females significantly outcompete the known sex pheromone components in attracting males (Sanders 1984, 1990b). Males of at least five species respond to the known pheromone com-ponent in Douglas-fir tussock moths. This lack of species specificity, as well as evidence that the synthetic pheromone does not match the attractive capability of a female, again suggests a missing component (Daterman et al. 1976).

The incomplete efficacy of disruption of communication in spruce budworms and

271

Douglas-fir tussock moths argues against the existence of a partial blend disruption mechanism (Cardé 1990) in these species. Moreover, it suggests that the quest for new semiochemicals should be at least partially directed toward species already studied. Coupled gas chromatography-electroantennogram detection (GC-EAD) techniques now make it possible for antennae to detect minute amounts of active pheromone components that are not detectable by GC (Gries et al. 1991). Once the complete pheromone blends are known, it is likely that there will be renewed interest in the disruption strategy.

Removal of the Pest

Semiochemicals can be used very effectively in a number of tactics designed to remove the pest from the forest (Table 1). Most of these tactics apply to bark and timber beetles, and exploit the phenomenon of semiochemical-mediated, secondary attraction leading to mass attack of new hosts (Fig. 1).

One of the easiest and most logical tactics is to induce mass attack of trees prior to harvest within a confined infested area using a grid of attractive baits, usually applied to trees at 50 m centres (Borden 1992). This containment and concentration tactic prevents infestations from spreading, attracts beetles from about 75 m outside of the baited grid into the infested area, and allows the offending infestation to be removed by clearcutting. It is simple and cheap, accommodates logging that would occur any-way, and demands only that cutting be rapid and directed at the infestation. Moreover, it alters nature only to the extent that the beetles attack the baited trees and those around them where mankind directs. The full implementation of this tactic is disclosed by the fact that in 1991 there were 46,186, 6,735 and 1,562 attractive semio-chemical baits used in B.C. against the mountain pine beetle, the spruce beetle, and the Douglas-fir beetle, respectively (Hall 1992). The principal application of these baits was to contain and concentrate infestations prior to harvest, saving millions of

dollars in lost timber and the cost of salvaging it, as well as greatly reducing the size and environmental impact of harvested areas.

The next most logical tactic is to suppress the pest population by mass-trapping. This tactic has been developed to the greatest extent for ambrosia beetles on the West Coast of North America (Borden 1990). Attack by ambrosia beetles in the forest is best avoided by removing cut logs before they are attacked. However, mass trapping is feasible at timber processing areas, mainly dryland sorts where logs are sorted by species and grade, and temporarily stored. These field "warehouses" often harbor large populations of ambrosia beetles in a limited area filled with an expensive inventory. It is thus feasible and cost effective to ring these areas with a small number of semiochemical-baited traps (e.g. 100), which are often supplemented with piles of baited trap logs (a modification of the trap tree tactic).

Trapping is directed at three species, the striped ambrosia beetle, Trypodendron lineatum, and two Gnathotrichus spp. Separate trapping and bait systems are required, and seasonal adjustment of emphasis must occur, as spring to summer populations of all three species give way to a large late summer to fall flight of G. sulcatus. The traps and sort operations are monitored periodically by ambrosia beetle managers, who advise industrial personnel with regard to risk of attack, inventory management and when to remove infested trap piles. In 1991, in B.C. 8,991 semiochemical baits were used for ambrosia beetle management (Hall 1992). Applied judiciously, year after year, mass trapping for ambrosia beetles can be very effective (Lindgren, 1990; Lindgren and Fraser, 1993).

I contend that mass trapping for insects other than ambrosia beetles has acquired an unjustifiably bad reputation. For bark beetles it is more difficult than tree-baiting, because of the need to out compete natural sources of attraction. Yet when it was tried on one generation of the western pine beetle in a small infestation in southern California, there were encouraging signs of success (Bedard and Wood 1974; DeMars et al. 1980). Landscape-wide mass trapping for any insect may not be feasible, but as in all the examples in Table 1, it should work against small populations in limited areas.

The need to mass trap in limited areas is especially true for lepidoptera, such as the gypsy moth (Table 1), which may not be considered as candidates for mass trapping because the sex pheromones of moths attract only males. Yet, small isolated populations of gypsy moths can be eradicated by persistent mass trapping (Kolodny-Hirsch and Schwalbe 1990).

The ultimate development of mass trapping is dependent on three factors. Firstly, basic research directed at identifying all of the semiochemicals used by the target species must ideally precede serious efforts to make the tactic work. Without the potential capacity to out-compete natural sources of attraction, mass trapping will probably fail. Elucidation of complete semiochemical blends may justify resurrection of previously inadequate programs.

Secondly, there must be a commitment to apply the tactic intensively for as long as is necessary. Had the pilot study on an isolated infestation of the western pine beetle in southern California (Bedard and Wood 1974) been continued beyond a single generation of beetles, the outbreak might have been terminated.

Lastly, spillover attacks or some degree of defoliation must not be viewed as negative side effects that make mass trapping undesirable. Rather, they must be viewed as opportunities for integrating other tactics with mass trapping. For example, infestations that develop in the vicinity of traps for bark beetles could be used to modify stand management by the application of group selection or mini clearcuts, in a transition to an unevenaged silvicultural system. Similarly, spot applications of an environmentally-benign microbial pesticide might be used to reduce moth populations to levels at which mass trapping can be effective or more rapid (Kolodny-Hirsch and Schwalbe 1990).

The tactic of integrating the use of lethal pesticides with attractive semio-chemicals has also been applied to bark and timber beetles. The most widespread variation of this tactic involves baiting disposable trees with semiochemicals to induce attack, and then introducing a translocating arsenical herbicide into the sapstream. Attacking beetles or their brood are killed in the inner bark, either by the toxic action of the arsenical (Maclauchlan et al. 1988) or by drying of the moribund phloem caused by invading fungi (Webber 1981). The late G.N. Lanier developed and refined this procedure in a highly successful program against the smaller European elm bark beetle, Scolytus multistriatus (Lanier 1989).

Another variation of the pesticide plus semiochemical tactic has been to treat the surface of semiochemical-baited trees or logs with a toxic pesticide. This technique is effective for limited use against bark beetles (Smith 1986), but met with less success for ambrosia beetles (Lindgren et al. 1982). Efficacy is increased if the insecticide is slow-acting, allowing attacking beetles to supplement the semiochemical baits with their own pheromone before they die.

The "attracticide" procedure in which sex pheromones of moths are used to attract males to an insecticide-treated substrate (Haynes et al. 1986) could be effective against some forest lepidoptera. However, unless the insecticide can be transmitted by contact with other insects, as might be expected for pathogens or insect growth regulators, this technique offers little advantage over mass trapping.

Modification of Behavior

A family of closely-related tactics exploits the natural role of antiaggregation (epideictic) pheromones (Fig. 1) to prevent or deter attack by bark beetles (Table 1). Numerous researchers have used antiaggre-gation pheromones to prevent or disperse infestations. These agents are relatively expensive and labor-intensive unless applied aerially. Thus they will probably be of greatest utility in addressing small, isolated infestations, and in deterring attack in limited areas where harvesting or use of conventional chemical pesticides is inappropriate or prohibited. Such areas include riparian zones, game corridors, commercial recreational facilities, parks and wilderness areas, and rural residential zones.

While results with the aerially-dispersed antiaggregation pheromone, 3-methylcyclohex-2-en-1-one, for the Douglas-fir beetle have been encouraging (Furniss et al. 1981, 1982), results with aerially or ground-applied verbenone for the mountain pine beetle have been dis-couraging (Amman 1993). Tactics using ver-benone, against D. ponderosae have been incon-sistently effective among doses (Borden and Lindgren 1988, Gibson et al. 1991), years (Shea et al. 1992, Amman 1993), host trees (Gibson et al. 1991, Amman 1993) and geographic locations (Amman 1993). In addition to technological and environmental factors that could cause verbenone appli-cations to fail (Amman 1993), verbenone in the vapor phase photo-isomerizes readily to chrysanthenone, an inactive material (Kostyk et al. 1993).

Recently, there have been some exciting results indicating that blends of antiaggregants can be much more effective than a single component (Table 1). A combination of verbenone and ipsdienol, an ips aggregation pheromone, was very effective at deterring the response of western pine beetles to traps baited with attractive semiochemicals (Paine and Hanlon 1991). Similarly, attacks by pine engravers were almost completely deterred in the presence of verbenone and ipsenol, synomones of two competing species (Borden et al. 1992; Miller et al. 1993). The discovery that green leaf volatiles characteristic of hardwood forests can deter responses of conifer-infesting bark beetles to attractive semiochemicals (Dickens et al. 1992) suggests that these compounds could be incorporated into anti-aggregant blends. These "supernatural" blends could become the operationally effective dis-ruptants that we have long sought.

Integrating tactics to prevent attack also shows great promise. Payne et al. (1992) found that felling infested trees combined with apply-ing verbenone to trees at the advancing head of a southern pine beetle infestation was more effective than verbenone treatment alone. In another approach "push-pull" tactics have integrated the application of antiaggregants in one area with attractive sources in another. This technique is promising for the mountain pine beetle (Lindgren and Borden 1993). A remarkable "push-pull" program devised by P.J. Shea of the U.S. Forest Service has saved a small stand of Pinus torreyana, an endangered species, from devastation by the California fivespined ips (Berson 1992).

Almost all research on lepidopteran semiochemicals has been devoted to the study of sex pheromones. However, there is a growing body of evidence indicating an epideictic (spacing) function in which lepidopteran eggs (Schoonhoven, 1990; Poirier and Borden, 1991; Thiery and LeQuére, 1991; Thiery et al., 1992) and larvae (Corbet, 1973; Dittrick et al., 1983; Hilker and Klein, 1989) can employ pheromones to deter oviposition by adults. Larvae of spruce budworms apparently ward off intruders into their feeding tunnel territories by releasing a drop of oral exudate that has both repellent and antifeedant functions (L.M. Poirier and J.H. Borden (unpublished). This complex of pheromone-based communication coupled with additional influences of host tree semiochemicals, e.g. as oviposition stimulants (Städler, 1974) (Fig. 2) suggests that there may be a multitude of unexploited opportunities for the use of semiochemicals against forest lepidopteran pests. Two research projects are exploring these possibilities on Choristoneura spp. (Table 1).

Enhancement of Natural Control

In a visionary but unsuccessful experiment, Chatelain and Schenk (1984) attempted to use semiochemicals to enhance predation by clerid beetles on the mountain pine beetle. Any success of this tactic with either predators or parasitoids must await a better understanding of the role of semio-chemicals in their selection of hosts.

Another type of natural control of tree-killing bark beetles occurs through competition from secondary species that can occupy the inner bark habitat so rapidly that they competitively displace the primary species (Rankin and Borden 1991).

Preliminary research suggests that semiochemical-induced attack by Ips spp. can greatly reduce brood production in trees attacked by southern pine beetles (Watterson et al. 1982, Payne and Richerson 1985) or mountain pine beetles (Rankin and Borden 1991). When the weakly aggressive ips emerge from a tree they will pose relatively little threat to the remaining standing trees. Research is underway to develop semiochemical-induced competitive displacement as an operational tactic against the mountain pine beetle.

Because success of competitive displacement will be limited by the population sizes of secondary species, it will probably be integrated with other tactics and will be used only in restricted areas, e.g. within 10 m of fish-bearing streams, where conventional pesticides cannot be used or clearcutting is prohibited.

Constraints and Opportunities

Management of forest insect pests with semiochemicals is an increasingly attractive option. Semiochemicals are viewed by the informed public as natural and safe (although this is not necessarily a wise assumption). When there is sufficient knowledge of the blend of semiochemicals used to convey a particular message, and when the natural context in which the message is used is well understood (Borden 1989), semiochemicals can be very effective pest management tools. In some cases they can be competitive in cost with other pest management agents.

Despite the current and predicted successes (Table 1) there are constraints to the development of semiochemicals for use in forests. By nature they are species-specific. This is good from an environmental perspective, for there will be little impact on non-target organisms. But species specificity means that a separate product and the technology to use it must be available for every species of importance. Unlike the situation in agricultural and urban environments, there are few persistently acute pests in forest environments. Most occur in intermittent outbreaks over limited areas. In integrated forest pest management programs, semiochemicals may be essential tools but may be sparingly and judiciously used. For example, Safranyik and Hall (1990) identified six strategies and 25 tactics that could be used in IPM of the mountain pine beetle. Only three of the tactics involved the use of semiochemicals. These constraints mean that it would be fool-hardy for a conventional industry to invest in development and marketing of semiochemicals for most forest insect pests. The potential to produce substantial revenue is just not there.

If semiochemicals are to be produced commercially for use in forest environments, there must be a special kind of industry willing to grow in small, incremental (species-specific) steps, with overall profits based on cumulative sales and services. Economies of scale will rarely be achieved, costs of research and development, production, and marketing will be high, and the total earning potential may or may not be enough to justify investment (Kydonieus and Beroza, 1982). There will be little room for competition, because the prize is too small. Therefore, only a few companies with relative monopolies over limited product lines will emerge as stable and permanent.

There may also be a need for government to subsidize pesticide development and registration costs, unless these costs can be reduced to levels consistent with anticipated profit margins. I believe that wise politicians and regulators will ensure that receptive and workable regulatory systems are adopted.

References

Amman, G.D. In *Proc. Symp. on Management of Western Bark Beetles with Pheromones-Recent Research and Development.* Shea, P.J., Ed; U.S. Dept. Agric., For. Serv. Gen. Tech. Rep. PSW (in press).

Bedard, W.D.; Wood, D.L. In *Pheromones.* Birch, M.C., Ed.; North Holland, Amsterdam, 1974; pp 441-449.

Berson, D.R. Pheromones Help Save Torrey Pines. U.S. Dept. Agric., For. Serv., Res. West., 1992, Sept., 17-20.

Billings, R.F. In *Integrated Control of Scolytid Bark Beetles.* Payne, T.L.; Saarenmaa, H., Eds.; Virginia Polytech. Inst. and State Univ., Blacksburg, Virginia, 1988; pp 295-306.

Borden, J.H. In *Behaviour.* Kerkut, J.A., Ed.; Pergammon Press, Oxford, 1985; pp 257-285.

Borden, J.H. "Semiochemicals and Bark Beetle Populations: Exploitation of Natural Phenomena by Pest Management Strategists." *Holarctic Ecol.,* 1989, 12, 501-510.

Borden, J.H. "Two Tree Baiting Tactics for the Management of Bark Beetles with Semiochemicals." *J. App. Entomol.* 1992, 114, 201-207.

Borden, J.H. In *Behavior-Modifying Chemicals for Insect Management.* Ridgway, R.L.; Silverstein, R.M.; Inscoe, M.N., Eds.; Marcel Dekker, N.Y., 1990; pp 281-315.

Borden, J.H.; Lindgren, B.S. In *Integrated Control of Scolytid Bark Beetles.* Payne, T.L.; Saarenmaa, H., Eds.; Virginia Polytech. Inst. and State Univ., Blacksburg, Virginia, 1988; pp 247-255.

Borden, J.H.; Devlin, D.R.; Miller, D.R. Synomones of Two Sympatric Species Deter Attack by the Pine Engraver, Ips pini (Say). *Can. J. For. Res.* 1992, 22, 381-387.

Cameron, E.A. In *The Gypsy Moth: Research Toward Integrated Pest Management.* Doane, C.C.; McManus, M.L., Eds.; U.S. Dept. Agric., For. Serv. Tech. Bull. 1584, 1981; pp 554-560.

Cardé, R.T. In *Behavior-Modifying Chemicals for Insect Management.* Ridgway, R.L.; Silverstein, R.M.; Inscoe, M.N., Eds.; Marcel Dekker, N.Y., 1990; pp 47-71.

Chatelain, M.P.; Schenk, J.A. Evaluation of Frontalin and Exo-Brevicomin as Kairomones to Control Mountain Pine Beetle (Coleoptera: Scolytidae) in Lodgepole Pine. *Environ. Entomol.,* 1984, 13, 1666-1674.

Corbet, S.A. Oviposition Pheromone in Larval Mandibular Glands of Ephestia kuehniella. *Nature,* 1973, 243, 537-538.

Daterman, G.E. In *Behavior-Modifying Chemicals for Insect Management.* Ridgway, R.L.; Silverstein, R.M.; Inscoe, M.N., Eds.; Marcel Dekker, N.Y., 1990; pp 317-343.

Daterman, G.E.; Peterson, L.J.; Robbins, R.G.; Sower, L.L.; Daves, G.D.; Jr.; Smith, R.G. Laboratory and Field Bioassay of the Douglas-fir Tussock Moth Pheromone, (Z)-6-Heneicosen-11-One. *Environ. Entomol.,* 1976, 5, 1187-1190.

Daterman, G.E.; Sower, L.L.; Sartwell, C. In *Recent Advances in Spruce Budworms Research.* Sanders, C.J.; Stark, R.W.; Mullins, E.J.; Murphy, J., Eds.; Can. For. Serv., Ottawa, Ontario, 1985; pp 386-387.

DeMars, C.J.; Slaughter, G.W.; Bedard, W.D.; Norick, N.X.; Roettgering, B. Estimating Western Pine Beetle-Caused Mortality for Evaluating an Attractive Pheromone Treatment. *J. Chem. Ecol.,* 1980, 6, 853-866.

Dickens, J.C.; Billings, R.F.; Payne, T.L. Green Leaf Volatiles Interrupt Aggregation Pheromone Response in Bark Beetles Infesting Southern Pines. *Experientia,* 1992, 48, 523-524.

Dittrick, L.E.; Jones, R.L.; Chiang, H.C. An Oviposition Deterrent for the European Corn Borer, Ostrinia nubialis (Lepidoptera: Pyralidae), Extracted from Larval Frass. *J. Insect Physiol.,* 1983, 29, 119-121.

Furniss, M.M.; Clausen, R.W.; Markin, G.P.; McGregor, M.D.; Livingston, R.L. Effectiveness of Douglas-fir Beetle Antiaggregation Pheromone Applied by Helicopter. U.S. Dept. Agric., For. Serv. Gen. Tech. Rep. INT-101, 1981.

Furniss, M.M.; Markin, G.P.; Hager, V.J. Aerial Application of Douglas-fir Beetle Antiaggregative Pheromone: Equipment and Evaluation. U.S. Dept. Agric., For. Serv. Gen. Tech. Rep. INT-137, 1982.

Gibson, K.E.; Schmitz, R.F.; Amman, G.D.; Oakes, R.D. Mountain Pine Beetle Response to Different Verbenone Doses in Pine Stands of Western Montana. U.S. Dept. Agric., For. Serv. Res. Pap. INT-444, 1991.

Grant, G.G. Development and Use of Pheromones for Monitoring Lepidopteran

Forest Defoliators in North America. *For. Ecol. Manage.,* **1991,** 39, 153-162.

Gries, G.; Gries, R.; Borden, J.H.; Li, J.; Slessor, K.N.; King, G.G.S.; Bowers, W.W.; West, R.J.; Underhill, E.W. 5,11-Dimethyl-heptadecane and 2,5-Dimethylheptadecane: Sex Pheromone Components of the Geometrid Moth, Lambdina fiscellaria fiscellaria. *Naturwiss.,* **1991,** 78, 315-317.

Hall, P.M. In *Proc. Symp. on the Management of Lodgepole Pine to Minimize Losses to the Mountain Pine Beetle.* Amman, G.D., Ed.; U.S. Dept. Agric., For. Serv. Gen. Tech. Rep. INT-262, 1989; pp 101-107.

Hall, P.M. Reporting of Bait Use. Pest Management Progress, B.C. Ministry of Forests, 1992, 11(1), 19.

Haynes, K.F.; Li, W.-G.; Baker; T.C. Control of Pink Bollworm Moth (Lepidoptera: Gelechiidae) with Insecticides and Pheromones (Attracticide): Lethal and Sublethal Effects. *J. Econ. Entomol.*; **1986,** 79, 1466-1471.

Hilker, M.; Klein, B. Investigations of Oviposition Deterrent in Larval Frass of Spodoptera littoralis (Boisd.). *J. Chem. Ecol.,* **1989,** 15, 929-938.

Kolodny-Hirsch, D.M.; Schwalbe, C.D. In *Behavior-Modifying Chemicals for Insect Management.* Ridgway, R.L.; Silverstein, R.M.; Inscoe, M.N., Eds.; Marcel Dekker, N.Y., 1990; pp 363-385.

Kostyk, B.C.; Borden, J.H.; Gries, G. Photoisomerism of the Antiaggregation Pheromone Verbenone: Biological and Practical Implications with Respect to the Mountain Pine Beetle, Dendroctonus ponderosae Hopkins (Coleoptera: Scolytidae). *J. Chem. Ecol.,* **1993** (in press).

Kydonieus, A.F.; Beroza, M. In *Insect Suppression with Controlled Release Pheromone Systems.* Kydonieus, A.F., Beroza, M., Eds.; CRC Press, Boca Raton, Florida, 1982, Vol II; pp 187-199.

Lanier, G.N. Trap Trees for Control of Dutch Elm Disease. *J. Arboriculture,* **1989,** 15, 105-111.

Lanier, G.N.; Silverstein, R.M.; Peacock, J.W. In *Perspectives in Forest Entomology.* Anderson, J.F.; Kaya, H.K., Eds ·

Academic Press, N.Y., 1976; pp 149-175.

Lindgren, B.S. Ambrosia Beetles. *J. For.,* **1990,** 88, 8-11.

Lindgren, B.S.; Borden, J.H. Displacement and Aggregation of Mountain Pine Beetles, Dendroctonus ponderosae (Coleoptera: Scolytidae), in Response to their Antiaggregation and Aggregation Pheromones. *Can. J. For. Res.,* **1993** (in press).

Lindgren, B.S.; Fraser, R.G. Ambrosia Beetles: More than a Decade of Mass Trapping at a Dryland Sorting Area. *For. Chron.,* **1993** (in press).

Lindgren, B.S.; Borden, J.H.; Gray, D.R.; Lee, P.C.; Palmer, D.A.; Chong, L. Evaluation of Two Trap Log Techniques for Ambrosia Beetles (Coleoptera: Scolytidae) in Timber Processing Areas. *J. Econ. Entomol.,* **1982,** 75, 577-586.

Lindgren, B.S.; McGregor, M.D.; Oakes, R.D.; Meyer, H.E. Suppression of Spruce Beetle Attacks by MCH Released from Bubble Caps. *Western J. Appl. For.,* **1989,** 4, 49-52.

Maclauchlan, L.E.; Borden, J.H.; D'Auria, J.M.; Wheeler, L.A. Distribution of Arsenic in MSMA-Treated Lodgepole Pines Infested by the Mountain Pine Beetle, Dendroctonus ponderosae (Coleoptera: Scolytidae), and its Relationship to Beetle Mortality. *J. Econ. Entomol.,* **1988,** 81, 274-280.

Mendel, Z.; Argman, Q. Discovery of the Southern Pine Beetle, Dendroctonus frontalis, in Israel. *Phytoparasit.,* **1986,** 14, 319.

Miller, D.R.; Devlin, D.R.; Borden, J.H. In *Proc. Symp. on Management of Western Bark Beetles with Pheromones-Recent Research and Development.* Shea, P.J., Ed.; U.S. Dept. Agric., For. Serv. Gen. Tech. Rep. PSW (in press).

Nordlund, D.A. In *Semiochemicals. Their Role in Pest Control.* Nordlund, D.A.; Jones, R.L.; Lewis, W.J., Eds.; Wiley, N.Y., 1981; pp 13-28.

Paine, T.D.; Hanlon, C.C. Response of Dendroctonus brevicomis and Ips paraconfusus (Coleoptera: Scolytidae) to Combinations of Synthetic Pheromone Attractants and Inhibitors Verbenone and

Ipsdienol. *J. Chem. Ecol.,* **1991**, 17, 2163-2176.

Payne, T.L.; Richerson, J.V. Pheromone-Mediated Competitive Displacement Between Two Bark Beetle Populations: Influence on Infestation Suppression. *Z. angew. Entomol.,* **1985**, 99, 131-138.

Payne, T.D.; Billings, R.F.; Berisford, C.W.; Salom, S.M.; Grossman, D.M.; Dalusky, M.J.; Upton, W.W. Disruption of Dendroctonus frontalis (Coleoptera: Scolytidae) Infestations with an Inhibitor Pheromone. *J. Appl. Entomol.,* **1992**, 114, 341-347.

Phero Tech Inc. Pine Engraver Beetle Trapping with the Lindgren Funnel Trap. Phero Tech Inc. Tech. Bull., Delta, B.C., Canada, 1988.

Poirier, L.M.; Borden, J.H. Recognition and Avoidance of Previously Laid Egg Masses by the Obliquebanded Leafroller (Lepidoptera: Tortricidae). *J. Ins. Behav.,* **1991**, 4, 501-508.

Rankin, L.J.; Borden, J.H. Competitive Interactions between the Mountain Pine Beetle and the Pine Engraver. *Can. J. For. Res.,* **1991**, 21, 1029-1036.

Ringold, G.B.; Gravelle, P.J.; Miller, D.; Furniss, M.M.; McGregor, M.D. Characteristics of Douglas-fir Beetle Infestation in Northern Idaho Resulting From Treatment with Douglure. U.S. Dept. Agric., For. Serv. Res. Note INT-189, 1975.

Safranyik, L.; Hall, P.M. Strategies and Tactics for Mountain Pine Beetle Management. B.C. For. Serv. Unpubl. Rep., 1990.

Sanders, C.J. Sex Pheromone of the Spruce Budworm (Lepidoptera: Tortricidae): Evidence for a Missing Component. *Can. Entomol.,* **1984**, 116, 93-100.

Sanders, C.J. In Recent Advances in Spruce Budworms Research. Sanders, C.J., Stark, R.W., Mullins, E.J., Murphy, J., Eds.; Can. For. Serv., Ottawa, Ontario, 1985; p 397.

Sanders, C.J. In Behavior-Modifying Chemicals for Insect Management. Ridgway, R.L., Silverstein, R.M., Inscoe, M.N., Eds.; Marcel Dekker, N.Y., 1990a; pp 345-361.

Sanders, C.J. Responses of Male Spruce Budworm Moths to Sex Pheromone Released from Filter Paper and Rubber Septa. *Can. Entomol.,* **1990b**, 122, 263-269.

Sartwell, C.; Daterman, G.E.; Overhulser, D.L.; Sower, L.L. Mating Disruption of Western Pine Shoot Borer (Lepidoptera: Tortricidae) with Widely Spaced Releases of Synthetic Pheromone. *J. Econ. Entomol.,* **1983**, 76, 1148-1151.

Schoonhoven, L.M. Host-Marking Pheromones in Lepidoptera, with Special Reference to Two Pieris spp. *J. Chem. Ecol.,* **1990**, 16, 3043-3052.

Shea, P.J.; McGregor, M.D.; Daterman, G.E. Aerial Application of Verbenone Reduces Attack of Lodgepole Pine by Mountain Pine Beetle. *Can. J. For. Res.,* **1992**, 22, 436-441.

Smith, R.H. Trapping Western Pine Beetles with Baited Toxic Trees. U.S. Dept. Agric., For. Serv. Res. Note PSW-382, 1986.

Sower, L.L.; Overhulser, D.L.; Daterman, G.E.; Sartwell, C.; Laws, D.E.; Koerber, T.W. Control of Eucosma sonomana by Mating Disruption with Synthetic Sex Attractant. *J. Econ. Entomol.,* **1982**, 75, 315-318.

Sower, L.L.; Wenz, J.M.; Dahlsten, D.L.; Daterman, G.E. Field Testing of Pheromone Disruption on Preoutbreak Populations of Douglas-fir Tussock Moth (Lepidoptera: Lymantriidae). *J. Econ. Entomol.,* **1990**, 83, 1487-1491.

Städler, E. Host Plant Stimuli Affecting Oviposition Behavior of the Eastern Spruce Budworm. *Entomol. Exp. Appl.,* **1974**, 17, 176-188.

Stock, A.J. Use of Pheromone Baited Lindgren Funnel Traps for Monitoring Mountain Pine Beetle Flights. B.C. For. Serv. Int. Rep. PM-PR-2, 1984.

Stock, A.J. The Western Balsam Bark Beetle, Dryocoetes confusus Swaine: Impact and Semiochemical-Based Management. Ph.D. Thesis, Simon Fraser Univ., Burnaby, B.C., 1991.

Thiéry, D.; LeQuére, J. Identification of an Oviposition-Deterring Pheromone in the Eggs of the European Corn Borer. *Naturwiss.,* **1991**, 78, 132-133.

Thiéry, D.; Gabel, B.; Farkas, P.; Pronier, V. Identification of an Oviposition-Regulating Pheromone in the European Grapevine Moth, Lobesia botrana (Lepidoptera: Tortricidae). *Experientia,* **1992**, 48, 698-699.

U.S.D.A. Pest Risk Assessment of the Importation of Larch from Siberia and the Soviet Far East. U.S. Dept. Agric., For. Serv. Misc. Pub., 1495, 1991.

Watterson, G.P.; Payne, T.L.; Richerson, J.V. The Effects of Verbenone and Brevicomin on the Within-Tree Populations of Dendroctonus frontalis. *J. Georgia Entomol. Soc.,* **1991**, 17, 118-126.

Webber, J.E. A Natural Biological Control of Dutch Elm Disease. *Nature,* **1981**, 292, 449-451.

Applications of Pheromones for Monitoring and Mating Disruption of Orchard Pests

R.E. Rice, University of California, Davis, CA 95616

Insect semiochemicals in orchard crop systems are used primarily for pest detection and monitoring in integrated pest management (IPM) programs, and more recently for control by mating disruption of several key lepidoptera pests. Pheromone trap data coupled with pest phenology models have enabled growers to reduce pesticide applications significantly in some crops such as peaches, nectarines, and almonds. Mating disruption holds great promise as a biorational pest control technique, but requires more monitoring and attention to application details than conventional pesticide use.

Isolation and identification of the sex pheromone of codling moth, *Cydia pomonella*, (Roelofs, 1971) opened the door in orchard crops to a rapidly expanding field of research with insect semiochemicals, particularly toward their proposed application in insect pest management. Initially, however, direct use of insect sex pheromones concentrated on improved studies of insect behavior (e.g. Baker and Carde, 1979; Carde, 1981). The programs and directions of many of the early researchers in this area have recently been reviewed by Silverstein (1990).

Aside from the more basic research and work on various aspects of insect behavior in both laboratory and field environments, the applied uses of pheromones have been concentrated in the areas of insect monitoring and detection, and in direct control of pest populations and manipulation of beneficial insects within the context of "soft" IPM programs.

Insect Monitoring

A recent compilation of commercially available pheromones (Inscoe et al., 1990) identified a total of 230 pheromones for monitoring, and 18 pheromones for mating disruption. Of these 248 pheromones, 79 were for arthropod species found in orchard crops. Detection and monitoring of pest species is a key component of any successful IPM program, whether it be in agricultural commodities such as orchard crops, or in urban, forest, or stored product environments. With the rapid identification of pheromones for orchard pests, primarily Lepidoptera species, use of synthetic pheromones in monitoring traps became a recommended procedure in most orchard pest management programs. However, it was soon realized that a number of factors could influence trapping results and data derived from the use of these traps. The positioning of traps in the trees, number of traps per unit area, trap design, frequency of trap servicing, and pheromone release rates are only some of the considerations that must be included in optimizing use of monitoring traps. With experience and continuing use, however, most trap systems became standardized within a given crop system and for specific pests.

With the development of standardized trapping techniques, the potential uses of the traps and pheromones became more apparent. These included the ability to detect pest infestations in previously uninfested areas, finding "hot spots" or high concentrations of pests in orchards that could be managed with localized or spot pesticide treatments, detecting prime pest sources outside the crop, evaluating the effect of various control measures, and, possible applications in decision-making for pest control.

After pheromone monitoring traps were first introduced, many pesticide sprays were simply timed as closely as possible to peak populations of a given pest in each

2726–8/93/0280$06.00/0 © 1993 American Chemical Society

generation within a season. It became apparent, however, that this approach could be severely misused by applicators and pesticide advisors, often leading to excessive use of pesticides. Such an approach was contrary to development of good IPM practices. Consequently, the use of monitoring traps in orchards stagnated for a period of time because of the potential for misuse of trap information and the inability to apply the trap data in a logical, scientific manner.

In the early 1980s, the development of insect phenology models that could be used in conjunction with pest monitoring traps led to a renewed interest in use of pheromone traps in orchard crops. These efforts eventually enabled researchers, pest control advisors, and growers to time pesticide sprays to optimum periods in a given pest's life cycle. This ultimately led to a reduction in pesticide applications during a growing season. For example, a phenology model for the oriental fruit moth (Croft et al. 1980; Rice et al. 1984) enabled growers to time one or perhaps two organophosphate sprays at precise periods to the second and third (or fourth) moth flights, resulting in a reduction of approximately 50% of the insecticide sprays previously used for this pest. Similar reductions in pesticide applications in apples, pears, and walnuts have been achieved by combining use of monitoring traps and phenology models for the codling moth (Riedl et al. 1976, Pitcairn et al. 1992). The phenology model developed for the peach twig borer, *Anarsia lineatella*, (Brunner and Rice 1984; Rice and Jones 1988) enabled growers to apply a single, precisely timed spray to achieve season-long control of this pest in stone fruits and almonds, whereas previously two or sometimes three sprays had been routinely applied.

Development of the phenology models has also enabled practitioners of biological controls to anticipate the development of various life stages of pest species. This allows augmentative or inundative releases of beneficials, such as various parasites or predators, when the most susceptible life stages of the pest species are present. These applications of monitoring traps and phenology models will undoubtedly see increased use as non-pesticidal control of orchard pests increases in IPM programs.

A question that is constantly asked of researchers and pest control advisors is, "How many moths in a trap indicate a need to treat or not treat?" In other words, can pheromone monitoring traps be used to establish treatment or economic injury thresholds for a pest species in an orchard. Many people familiar with the manner in which these monitoring traps are used are reluctant to develop information to answer this question because of the great variety of factors that can effect the number of insects trapped. For example, the trap type, load and release rate of the pheromone dispenser, trap height, number of traps used, servicing interval, and population trends in given generations can all effect the numbers of insects that are collected in a given period of time. The consequences of an advisor presenting a grower with the "wrong number" could in many cases be extremely expensive, both in crop loss and litigation. Therefore, this potential use of pheromone monitoring traps has not been exploited or developed extensively.

In spite of these shortcomings, the use of pheromone monitoring traps continues to expand and gain favor in almost all successful orchard IPM programs. One reason for this is the higher level of understanding by younger pest control advisors and growers. These individuals have been trained and exposed to the concepts and theory of pheromone uses, and are therefore willing to try new technology in their own IPM programs.

The value of pheromone traps in spray timing, evaluating the results of sprays, determining the sources of pest populations, distribution of pests within the orchard, seasonal population trends, etc. has been well documented.

Mating Disruption as a Control Strategy for Orchard Pests

One of the primary applications of semiochemicals in orchard ecosystems has been for mating disruption (or "male confusion") as a direct control for several major insect pests in the order Lepidoptera. Initial research efforts in this area concentrated on the codling moth in the U.S., Europe, and South America. The oriental fruit moth, a major pest of peaches and nectarines, has also recently been the target of intensive mating disruption programs in Australia, the U.S., South Africa, South America and southern Europe. Commercial product registrations for both codling moth and oriental fruit moth mating disruption have been obtained in most fruit production areas of the world. First product registrations for an oriental fruit moth mating disruption system was obtained in the U.S. in 1987 followed by codling moth in pome fruits in 1991. Other successful programs for mating disruption in stone fruits have included the peachtree borer, *Synanthedon exitiosa* and the lesser peachtree borer, *S. pictipes* in the southern U.S. However, registrations and commercial use of these latter two pheromones have lapsed due to low sales and application.

Additional mating disruption programs are under development in orchard crops and several are approaching registration. These include systems designed to control leafroller species such as tufted apple budmoth, *Platynota idaeusalis*; redbanded leafroller, *Argyrotaenia velutinana*; orange tortrix, *A. citrina*; omnivorous leafroller, *Platynota stultana*; *Pandemis pyrusana*, and the peach twig borer, *Anarsia lineatella*. Other research programs are currently investigating the potential of mating disruption as a control for leafrollers in the genera *Choristoneura* and *Archips*.

Some of the benefits of mating disruption for control of pests include fewer regulatory restrictions on application of these materials in contrast to requirements for Class I pesticides. For example, mating disruption pheromones do not require notices of application intent to local regulatory officials, fields do not require hazard posting, and re-entry periods are not required following application of the pheromone dispensers. There also are no fruit residues when pheromones are hand applied using "single

point source" dispensers. This is an extremely important factor in processed fruit products such as baby food. The pheromones are totally compatible with scheduling of other cultural operations, and do not interfere with irrigation, cultivation, fruit thinning, tree propping, or harvest at any time following application. The pheromones are also considered "soft" materials because they preserve predators, parasites, and pollinating insects and do not result in disruption of phytophagous mite populations, a problem associated with the use of many standard insecticides. Growers and applicators using mating disruption are also learning that there are no container or rinse water disposal problems such as occur with the use of standard insecticides.

The objective of mating disruption is long-term population suppression over a period of several generations or years. Mating disruption cannot be used to solve an immediate or short-term problem with a particular pest, particularly just prior to harvest. Consequently, growers must plan in advance for the application and continued use of mating disruption to achieve maximum benefits. Experiences in California with oriental fruit moth in peaches and nectarines have shown that the control with mating disruption is equivalent to conventional spray programs and that the cost is generally competitive with two to three insecticide sprays, which may include the cost of a miticide application following organophosphate or pyrethroid insecticide applications. The pheromone dispensers are easy and safe to apply, and require little or no worker training in contrast to pesticide applicator training, protective clothing and equipment, and use of expensive spray equipment as is required for standard pesticide applications.

Along with the obvious advantages of mating disruption for orchard pests, however, there are also some limitations or drawbacks to their use. A major problem is migration of mated females from sources outside of a pheromone-treated orchard. For example, in California, almond orchards can serve as a major source of oriental fruit moths since this crop is seldom, if ever, treated for control of this insect. Consequently, large populations of mated females can migrate from almonds to adjacent peaches or nectarines, particularly as these crops begin to ripen and become more attractive to egg-laying females. This single factor alone requires a greatly improved monitoring and surveillance program in mating disruption orchards, particularly along the edges adjacent to untreated hosts. The experience of most pest control advisors has shown an increase in monitoring costs associated with mating disrupted orchards because of the additional time required for observation and monitoring.

Mating disruption has rightfully been called a "high-tech" approach to pest control, and as such requires a higher level of understanding and attention to detail than many previous methods based primarily on chemical controls. For example, timing of dispenser applications based on an accurate pest biofix, coupled with knowledge of dispenser release rates and field longevity are essential to obtain maximum benefits from mating disruption. Correct dispenser numbers per unit area, placement, and coverage are also critical factors that must be understood if mating disruption is to be successful.

Another problem observed in mating disruption orchards is an increase of secondary pests. In the past these have been suppressed or controlled with insecticides used for control of the primary pest. Examples include the resurgence of the omnivorous leafroller, *P. sultana*, in peaches and nectarines treated with mating disruption for oriental fruit moth, and increased populations of leafrollers (*P. pyrusana* and *Choristoneura rosaceana*) in apple or pear orchards using mating disruption for codling moth. Standard insecticide sprays normally applied in late May for oriental fruit moth usually suppress omnivorous leafroller to subeconomic levels. However, when pheromones for mating disruption are used, omnivorous leafroller can become a significant pest with damage levels approaching two to five percent in some cases, levels which are unacceptable for fresh market stone fruit sales.

Growers planning to use mating disruption are often advised that they should use a standard insecticide program to spray out high populations of the target pest concurrently or prior to placement of the pheromone dispensers. The reason for this is to reduce the pest population to low, or at least moderate, levels that would then enable the pheromones to outcompete the virgin pest "calling" females and reduce the possibility that they could become mated even in the presence of disruption pheromones. In this context, a standard practice for control of oriental fruit moth and codling moth is evolving that includes early season application of mating disruption pheromones followed by a single organophosphate insecticide application in the generation of moths just prior to harvest of a particular cultivar. This integrated approach to control achieves several objectives including reducing the season-long cost by using only one application of pheromone, but also reducing the number of insecticide applications from as many as four, down to only one in a given season. This strategy also helps slow the development of resistance to the organophosphate insecticides that has recently been observed in some production areas. Semiochemicals used for mating disruption are in many cases not going to be the sole method of control, but they will become an additional tool for use in IPM programs along with biological controls, cultural controls, and standard pesticides.

Another problem that has been observed with mating disruption in some locations is that sloped or uneven terrain will result in poor control. While it is presently difficult to measure the actual pheromone concentration in tree canopies, it is believed the problem with uneven terrain is due to the sinking and drifting of the pheromones out of the tree canopy down to the orchard floor and ultimately draining away from high ground. This results in little or no control in the higher areas of the treated orchard but good disruption and control in the lower areas. This phenomenon has been observed with several pests in different areas of the world, including Germany, British Columbia, and California, particularly for codling moth in pome fruit orchards planted on relatively steep hillsides.

In summary, the use of pheromones for

monitoring and control of orchard pests has been the subject of intensive and increasing research throughout the world over the past 20 years. This developing technology has been addressed at numerous conferences, symposia, and informal meetings, particularly within the framework of the West Palearctic IOBC Working Group on Pheromones and Other Semiochemicals in Europe. Other conferences that have convened on this subject include the recent Symposium at the Entomological Society of America National Conference in 1987 in Boston.

Based on several early successes with semiochemical approaches to pest control, it is apparent that this emphasis on developing soft or biorational approaches to pest control will continue to receive major attention in research and IPM programs. Increased use of these materials has shown, however, that in many cases they will not be a stand alone control strategy, but will become part of an overall IPM approach within a cropping system. One inhibiting factor that should be recognized is that even though pheromones are applied at extremely low concentrations in orchards, the cost of these materials is still relatively high compared to standard insecticides. It is hoped that with increasing development and use of pheromones, costs will decline and mating disruption (and monitoring) of orchard pests will continue to increase, to the benefit of growers, applicators, consumers, and the environment.

References

Baker, T.C.; Cardé, R.T. "Endogenous and exogenous factors affecting periodicities of female calling and male sex pheromone response in *Grapholitha molesta* (Busck)." *J. Insect Physiology*, **1979**, 25, 943-950.

Brunner, J.F.; Rice, R.E. "Peach twig borer, *Anarsia lineatella* Zeller (Lepidoptera: Gelechiidae), development in Washington and California." *Environ. Entomol.*, **1984**, 13, 607-610.

Cardé, R.T. In *Management of Insect Pests with Semiochemicals*; Mitchell, E.R., Ed.; Disruption of long distance pheromone communication in the oriental fruit moth: camouflaging the natural aerial trails from females?; Plenum Press: New York, 1981; pp. 385-398.

Croft, B.A.; Michels, M.F.; Rice, R.E. "Validation of a PETE timing model for the oriental fruit moth in Michigan and central California (Lepidoptera: Olethreulidae)." *Great Lakes Entomol.*, **1980**, 13, 211-217.

Inscoe, M.N.; Leonhardt, B.A.; Ridgeway, R.L. In *Behavior Modifying Chemicals for Insect Management*; Ridgeway, R.L.; Silverstein, R.M.; Inscoe, M.N., Eds.; Commercial availability of insect pheromones and other attractants; Marcel Dekker, Inc.: New York, 1990; pp. 631-715.

Pitcairn, M.J.; Zalom, F.G.; Rice, R.E. "Degree-day forecasting of generation time of *Cydia pomonella* (Lepidoptera: Tortricidae) populations in California." *Environ. Entomol.*, **1992**, 21, 441-446.

Rice, R.E.; Jones, R.A. "Timing post-bloom sprays for peach twig borer (Lepidoptera: Gelechiidae) and San Jose scale (Homoptera: Diaspididae)." *J. Econ. Entomol.*, **1988**, 81, 293-299.

Rice, R.E.; Weakley, C.V.; Jones, R.A. "Using day-degrees to determine optimum spray timing for the oriental fruit moth." *J. Econ. Enotmol.*, **1984**, 77, 698-700.

Riedl, H.; Croft, B.A.; Howitt, A.J. "Forecasting codling moth phenology based on pheromone trap catches and physiological-time models." *Can. Entomol.*, **1976**, 108, 449-460.

Roelofs, W.; Comeau, A.; Hill, A.; Milicevic, C. "Sex attractant of the codling moth: characterization with electroantennogram technique." *Science*, **1971**, 174, 297-299.

Silverstein, R.M. In *Behavior Modifying Chemicals for Insect Management*; Ridgeway, R.L.; Silverstein, R.M.; Inscoe, M.N., Eds.; Practical uses of pheromones and other behavior-modifying compounds: Overview; Marcel Dekker, Inc.: New York, 1990; pp. 1-8.

Identification of Sucrose Esters from *Nicotiana* Active Against the Greenhouse Whitefly

J. George Buta, William R. Lusby, John W. Neal, Jr., Rolland M. Waters, and George W. Pittarelli, Beltsville Agricultural Research Center, Agricultural Research Service, U.S. Department of Agriculture, Beltsville, MD 20705

N. gossei was found to be highly effective among several species of *Nicotiana* showing resistance to the greenhouse whitefly. Mixtures of 2,3-di-*O*-acyl-6'-*O*-acetylsucrose and 2,3-di-*O*-acyl-1',6'-*O*-diacetylsucrose esters where the acyl groups were 5-methylhexanoyl and 5-methylheptanoyl were isolated from the leaf surface and identified as compounds active in causing mortality of early instar greenhouse whitefly nymphs when applied topically.

Several species of *Nicotiana* grown in greenhouses at Beltsville were observed to be resistant to infestation by the Greenhouse Whitefly, *Trialeuroides vaporariorum*. Investigation based on these observations resulted in the finding that several species of *Nicotiana, N. africana, N. fragrans, N. gossei, N. repanda* caused nymphal mortality and a high degree of ovipositional non-preference by the greenhouse whitefly. These effects were in contrast to the large infestation of the insects on the susceptible commercial species, *N. tabacum* MD 609 (Neal, 1987).

The cuticular chemistry of *Nicotiana*, in particular *N. tabacum*, has been found to be very complex (Severson, 1991) with insecticidal activity often being attributed to nicotine alkaloids. However, a differential susceptibility to insect attack not explained by alkaloids suggested the possibility of other cuticular compounds having important insecticidal activity (Thurston, 1962).

Isolation of Active Compounds

After determining the relative activities of the surface extracts of 13 species among the subgroups of *Nicotiana* (65 species) by use of the nymph bioassay, the highly active cuticular extract of *N. gossei* was selected for systematic fractionation. The polar lipid fraction obtained by solvent partition was separated further by gel permeation chromatography yielding an active fraction having an apparent molecular weight range (500-600). This fraction was chromatographed on silica gel yielding a Fraction I eluted with EtOAc and Fraction II with MeOH-EtOAc (5:95), both causing nymph mortality.

Fraction II was then examined further by mass spectrometry. Molecular weights of 626'A', 636'B', 650, 664'C', 678'D' were found for 5 compounds in this fraction only by use of NH_3-CIMS. CIMS and EIMS indicated that the compounds were a series of carbohydrate esters. The individual compounds comprising Fraction II were isolated by HPLC using a polymeric semi-preparative column and a CH_3CN-H_2O solvent system. The four major components of the active fraction was isolated in the proportions (A=2%, B=26%, C=35%, D=37%) and their purity confirmed by NH_3-CIMS.

The MS study indicated the presence of two 5-methylheptanoate esters on one hexose portion of the disaccharide and the likelihood of acetate substituents on the other hexose moiety. The G2,G3 location of the 5-methylheptanoyl moieties were assigned by extensive NMR decoupling studies. The 2 acetate substituents were established on F1 and F6 by NMR study of the perdeuteroacetylated D. The structure of D was identified as 2,3-di-*O*-5-methylheptanoyl-1',6'-di-*O*-acetylsucrose. The structures of A, B, C were then elucidated as follows: A is

2, 3-di-*O*-acyl-6'-*O*-acetylsucrose where 2, 3 acyl groups are 5-methylhexanoyl or 5-methylheptanoyl (1:1). B is 2, 3-di-*O*-5-methylheptanoyl-6'-*O*-acetylsucrose. C is 2, 3-di-*O*-acyl-1', 6'-*O*-diacetylsucrose with the 2, 3-acyl groups being 5-methylhexanoyl or 5-methylheptanoyl (1:1). Each of these compounds was tested individually and was found to have significant insecticidal activity against the whitefly (Buta, 1993).

Structure-activity Relationship

The insecticidal sugar esters in *N. gossei* are exuded by the trichomes located mainly on the adaxial leaf surface (Cutler, 1986). When adult greenhouse whiteflies were observed on the leaves of *N. gossei*, they appeared transparent and dehydrated, particularly those in contact with the trichomes. In contrast, they were unaffected by contact with the trichomes of *N. tabacum* and other whitefly-susceptible Nicotiana species. Approximately two-thirds of the *Nicotiana* species produce sucrose or glucose esters in varying quantities and of a highly variable chemistry both in substitution pattern and acyl substituent size. Some preliminary experiments concerning the effectiveness of various sugar ester fractions as contact-type insecticides have been done using adult greenhouse whiteflies and green peach aphids (*Myzus persicae*) in mortality bioassays. The greatest mortality to these two insects was achieved with 0.1% solutions of the sucrose ester fractions of *N. gossei, N. benthamiana, N. cavicola > N. glutinosa, N. trigonophylla >> N. tabacum, N. sylvestris*. Very little mortality was caused by sucrose esters from the latter two species. In addition, low mortality was obtained with comparable solutions of glucose esters such the *N. gossei* glucose esters (Fraction I). The degree of insect mortality obtained in these assays is in good agreement with the resistance to the greenhouse whitefly observed with the plants. The concentrations of sugar esters [glucose (GE) and sucrose (SE)] were found to vary considerably on leaf surfaces of various Nicotiana species where the SE chemistry is known. For example, the quantities of SE of a few species that cause

differing degrees of whitefly mortality are *N. gossei*, 8 $\mu g/cm^2$ of leaf surface; *N. glutinosa* 38 $\mu g/cm^2$; *N. tabacum*, 57 $\mu g/cm^2$ (Severson, 1991). However, the insecticidal activity does not correlate with the relative quantities of SE on the leaf surfaces. If structural differences are examined, then different patterns of acyl substitution and composition are found for these three species. The *N. gossei* SE have 2,3-acyl-1', 6' acetyl substitution with C_7 or C_8 acyl groups while the *N. glutinosa* SE are 2, 3, 4-acyl-3'-acetyl and 2, 3, 4-acyl substituted with C_7 acyl groups and *N. tabacum* SE are 2, 3, 4-acyl-6-acetyl substituted with C_5 acyl groups.

These differences in acyl substitution and composition appear to be an explanation for the differences in insecticidal activity observed where the SE with both longer acyl chain lengths and a more dispersed pattern of acylation appear to be more active. The chemistry of several other active *Nicotiana* species remain yet to be investigated before this interesting structure-activity relationship of the sugar esters and insecticidal activity of *Nicotiana* is well understood.

References

Neal, J. W.; Pittarelli, G. W.; Gott, K. M. "*Nicotiana* Species with High Resistance to Greenhouse Whitefly." *Tob. Sci.*, **1987**, *31*, 61-62.

Severson, R. F.; Jackson, D. M.; Johnson, A. W.; Sisson, V. A.; Stephenson, M. G. *Naturally Occurring Pest Bioregulators*, American Chemical Society: Washington, DC, 1991, pp. 264-277.

Thurston, R.; Webster, J. A. "Toxicity of *Nicotiana gossei* to *Myzus persicae*." *Ent. Exp. & Appl.*, **1962**, *5*, 233-238.

Buta, J. G.; Lusby, W. R.; Neal, J. W., Jr.; Waters, R. M.; Pittarelli, G. W. "Sucrose Esters from *Nicotiana gossei* Active against the Greenhouse Whitefly." *Phytochemistry*, **1993**, *32*, 859-864.

Cutler, H. G.; Severson, R. F.; Cole, P. D.; Jackson, D. M.; Johnson, A. W. "Natural Resistance of Plants to Pests." American Chemical Society: Washington, DC, 1986, pp. 178-196.

Efficacy of Clarified Neem Seed Oil Against Foliar Fungal Pathogens and Greenhouse Whiteflies

James C. Locke, USDA, Florist and Nursery Crops Lab., BARC-West, Beltsville, MD 20705
Hiram G. Larew, USAID, Office of Strategic Planning, Policy Dir., Washington, DC 20523
James F. Walter, W.R. Grace & Co., Washington Research Center, Columbia, MD 21044

Hydrophobic neem seed oil, solvent extracted from ground, mature seeds of the neem tree (*Azadirachta indica*), has been clarified and formulated to produce an emulsifiable spray product. Rate studies showed that 0.5 to 1% oil can give protection equal to synthetic fungicides and 1 to 3% oil significantly reduces whitefly oviposition. The mode of action is primarily protective against the pathogens and repellent to whitefly oviposition.

Botanical insecticides (Margosan-O[R], Azatin[TM], Bioneem[TM]), containing the active ingredient azadirachtin which is extracted from neem tree seeds, have recently entered the marketplace as alternatives to synthetic insecticides. The literature also refers to fungicidal activity of various components of the neem tree, including the seed oil, but no commercial products have yet been developed. Northover and Schneider (1993) have reported on the activity of various botanical oils against powdery mildew and scab of apple and white rust of spinach. The efficacy of clarified neem seed oil (NSO) has been demonstrated against various rusts (Locke and Stavely,1991) and powdery mildew (Locke, 1992). This research reports the dual potential of a natural product to control both fungal pathogens and foliar insects.

Materials and Methods

The clarified NSO used in these studies is a unique vegetable oil produced by organic solvent extraction of the hydrophobic oils and fatty acids from ground neem seeds followed by removal of phytotoxic components through clarification. The clarified oil is formulated with a surfactant to allow emulsification in water. The NSO conists of glycerides (70-95%), free fatty acids (4-20%), and a small amount of inorganic and organic salts.

A) Bean rust. Pinto bean seedlings (*Phaseolus vulgaris* cv. Pinto 111) were grown by planting 4 seeds per 10 cm pot at a depth of 1.5 cm and germinating them in a greenhouse at approx. 25C for 7-8 days. When the primary leaves were half expanded, the seedlings were thinned to the most uniform three plants per pot, treated, and/or inoculated following the method of Stavely (1983). Treatments consisted of water, aqueous emulsions of NSO (0.25, 0.5, 1%) or Sunspray[R] 6E Plus Horticultural Oil (1, 2%) which were applied to both the upper and lower leaf surfaces. When dry, urediniospores of *Uromyces appendiculatus* (2×10^4 spores/ml in tap water containing 0.1 ml/L Tween 20) were atomized onto the upper and lower leaf surfaces. The inoculated plants were held in a dew chamber over night at 20C then placed in the greenhouse for incubation. Disease assessment was made 7-10 days later, when the uredinia were sporulating, by counting the number of uredinia/unit area of leaf.

B) Snapdragon rust. Snapdragon seedlings (*Antirrhinum majus* cv. Panama) in the 5-6 leaf pair stage, planted individually in 10 cm pots, were treated as described above. Urediniospores of *Puccinia antirrhini* were atomized onto the lower leaf surface. Inoculated seedlings were held in a dew chamber over night at 20C before placing them in a cool (15C night) greenhouse. Disease assessment was made after 14 days when the uredinia were beginning to erupt from the leaf. The number of uredinia/unit area of leaf was recorded for each of two leaf pairs/plant.

C) Hydrangea powdery mildew. Individual, rooted tip cuttings of hydrangea (*H. macrophylla*) in 10 cm pots were sprayed with NSO (1%), Sunspray[R] (1%), or Benlate[R] 50WP fungicide, then exposed to natural powdery mildew infection in a cool (15C night) greenhouse. Repeat treatments (4-5) were made at 2-wk intervals. Disease assessment was made by counting the number of leaves with mildew and estimating the percent leaf area affected.

D) Chrysanthemum whitefly. Commercially rooted cuttings of chrysanthemum (*Dendranthema morifolium* cv. Iceberg) were grown for 3 wk in 10 cm square plastic pots. All leaves were removed except for 3 upper, fuly expanded leaves which were sprayed to runoff with treatment emulsions. Choice and no choice experiments were conducted. For choice experiments, treatments were water, 1% or 3% NSO. After drying, plants were placed for 24 h in a colony of greenhouse whiteflies (*Trialeurodes vaporariorum*) maintained on tobacco and tomato plants. For no-choice experiments, plants were sprayed with water or 2% NSO. All plants from both treatments were placed into separate cages with approx. 100 whiteflies in each. The no-choice experiments were conducted for 24 h or 48 h. Following exposure, plants in both experiments were water misted to remove adult flies. The number of eggs laid was counted on 2 leaves per plant (choice) or 3 leaves per plant (no-choice), and leaf area determined. Egg counts were adjusted to number/100 cm^2.

Results and Discussion

A) Bean rust. Rust pustule development was reduced by all NSO and Sunspray[R] treatments. The lowest rate of NSO (0.25%) reduced pustule counts by at least 88%. Equivalent control required 2% Sunspray[R]. Both oils are primarily protectants and are much less effective if applied after infection.

B) Snapdragon rust. Results with snapdragon rust followed the same pattern as with bean. Protectant activity was obtained with as low as

1% NSO. NSO was more effective than the petroleum-based oil applied at an equivalent rate but not as effective as commercially available fungicides.

C) Hydrangea powdery mildew. NSO (1%) and Sunspray[R] (1%) provided greater protection against powdery mildew than the standard Benlate[R] treatment. Although Benlate[R] did not give complete protection, it significantly reduced the severity on infected leaves. This assay also demonstrated that if a light infection does begin, the use of either NSO or Sunspray[R] has some eradication activity. It appears that the exposed mycelium of the powdery mildew is sensitive to the oil treatments whereas the internal development of the rust mycelium serves to protect it.

D) Chrysanthemum whitefly. Under choice conditions, 1% and 3% NSO treatments significantly reduced the number of eggs laid compared to the untreated, but there was no significant difference between the two rates. Under no-choice conditions, oil-treated (2%) plants had significantly fewer eggs laid than the untreated indicating repellency as has been reported with other plant oils (Butler et al, 1989). Repellency has been reported with the sweet potato whitefly, *Bemisia tabaci*, as well as nymphicidal and ovicidal activity with several insects (Larew, 1990). These results give promise for the development and commercialization of a natural product which could serve as a general protectant, providing an alternative to a multi-pesticide program.

References

Butler, G.D.; Jr., Coudriet, D.L.; Henneberry, T.J. "Sweet potato whitefly: Host plant preference and repellent effect of plant-derived oils on cotton, squash, lettuce, and cantaloupe." *Southwest Entomol.* **1989**, 14, 9-16.

Larew, H.G. "Activity of neem seed oil against greenhouse pests." In *Neem's Potential in Pest Management Programs, Proceedings of the USDA Neem Workshop*; Locke, J.C. and Lawson, R.H., Eds.; USDA, ARS-86, **1990**, pp 128-131.

Locke, J.C. "Comparison of neem oil, Sunspray[R] 6E Plus Horticultural Oil and

Benlate[R] for control of powdery mildew on greenhouse hydrangea." *Phytopathology* **1992**, 82, 1121.

Locke, J.C.; Stavely, J.R. "Use of oil and wax compounds extracted from neem seeds to control bean and snapdragon rusts." *Phytopathology* **1991**, 81, 703.

Northover, J.; Schneider, K.E. "Activity of plant oils on diseases caused by *Podosphaera leucotricha, Venturia inaequalis,* and *Albugo occidentalis.*" *Plant Dis.* **1993**, 77, 152-157.

Stavely, J.R. "A rapid technique for inoculation of *Phaseolus vulgaris* with multiple pathotypes of *Uromyces phaseoli.*" *Phytopathology* **1983**, 73, 676-679.

Allelopathy for Weed Suppression

Horace G. Cutler, USDA, Agricultural Research Service, Richard B. Russell Agricultural Research Center, P. O. Box 5677, Athens, GA 30613

A number of biologically active natural products have been isolated from fungi and higher plants that exhibit either selective or broad spectrum herbicidal activity. These secondary metabolites are diverse in structure and select compounds obtained from fungi control growth and development of lambsquarters, johnsongrass, evening primrose, spotted knapweed, and purple nutsedge. Plant derived phytotoxins, which tend to be more broad spectrum in activity, have been isolated from radish seedlings, Podocarpus, and Magnolia. Additionally, some broad spectrum phytotoxins have been isolated from Pythium, Cochliobolus, and Scopulariopsis. A selection of natural products, their structures and their biological activities are described. In many cases these structures are ideal templates for making derivatized, either by synthesis or biotransformation, useful biodegradable herbicides.

While allelopathy has been described as, "The reputed baneful influence of one living plant upon another due to secretion of toxic substances" (Websters, 1971) perhaps a broader interpretation should be applied to include all organisms so that any biologically active natural product derived from a living source that detrimentally affects another may be termed an allelochemical that has allelopathic properties. Insofar as natural products that affect weed control are concerned there are two major classes that attract attention. Within each of these classes are metabolites that have been isolated from higher plants, fungi, and bacteria, and all are diverse in structure. The first major class consists of those compounds that are highly target specific against singular weed species and in rare instances some of those metabolites are only active against selected weed races.

The second class consists of compounds that have both pre- and post-emergence activity against model test plants, have a broader spectrum of activity and, according to the literature, have not been tested against weeds. Both classes have environmentally attractive features in that they tend not to be persistent and generally have high specific activity. Furthermore, they may be subjected to biotransformation to produce metabolites that possess different target specificity than their parent, or they may be altered by chemical synthesis. At present, the limiting factors to successfully producing natural product herbicides are that natural products are generally isolated in small amounts and, in this regard, their practical evolution somewhat parallels that of penicillin. In that case, enough penicillin was produced by fermentation to keep an Oxford policeman alive for three days, but only after his urine had been extracted for the excreted unmetabolized drug which was then readministered to the patient. The cascade effect was that eventually no penicillin was available from the urine and the patient died. Subsequent work at the Northern Regional Research Laboratory, USDA, ARS, Peoria, Illinois, led to the isolation of a high yielding strain of *Penicillium*, and titres were beefed up using corn steep liquor.

Relative to the isolation of natural products for weed control this means that because of the small yields of secondary metabolites from a singular source only a limited number of selected weed species can be challenged in the initial screening stage and, furthermore, only a limited number of associated economic crop species can be tested for resistance. Ideally, one seeks a high yielding strain of the producer microorganism or plant. Alternatively, synthesis of the natural product may provide larger quantities of a needed metabolite. In any event, the task is not easy and yet the approach continually reveals novel compounds some of which are structurally complicated but produced by nature with comparative ease. And it is precisely these compounds that are useful chemical templates for making practical compounds of economic importance.

In earlier reviews we have discussed natural products from plants and microorganisms that have phytotoxic or allelochemic properties (Cutler, 1987; Cutler, 1988; Cutler, 1988; Edwards et al. 1988; Cutler, 1988; Cutler, 1992; Cutler, 1992) and we now turn our attention to certain secondary metabolites and their derivatives that may have herbicidal application.

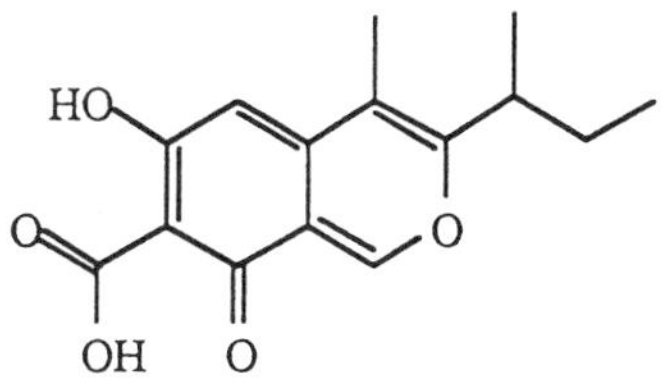

Specific Weed Control Natural Products

Lambsquarters. *Ascochyta hyalospora* (= *Diplodia hyalospora* = *Pleospora chenopodii*) is a pathogen of lambsquarters (*Chenopodium album*) and all organs of the plant are susceptible to attack, including the seed, making the microorganism a potential biocontrol agent. Three phytotoxic compounds were isolated from the culture filtrate of *A. hyalospora* and included ascochytine (Fig. 1), pyrenolide A (Fig. 2) and hyalopyrone (Fig. 3). Of these, both ascochytine and pyrenolide A have been previously isolated, but hyalopyrone is a novel metabolite (Venkatasubbaiah and Chilton, 1992). Subsequent testing of all the metabolites from 25-100μg/ detached leaf took place on lambsquarters (*C. album*), prickly sida (*Sida spinosa*), sicklepod (*Cassia obtusifolia*), morning glory (*Ipomoea* sp.), johnsongrass (*Sorghum halepense*), sorghum (*S. bicolor*), bentgrass (*Agrostis alba*), ragweed (*Ambrosia artemisiifolia*), watercress (*Naturtium officinale*), and jimsonweed (*Datura stramonium*) and the results were compared with unfractionated culture filtrate of the microorganism. Surprisingly, the culture filtrate was more toxic to johnsongrass and sorghum than to the original host plant, lambsquarters, and ascochytine was concomitantly more toxic to the former plants. Conversely, pyrenolide A and ascochytine were almost equally as active against lambsquarters while hyalopyrone was only approximately one-third as active (Venhatasubbaiah and Chilton, 1992).

Fig. 1. Structure of ascochytine.

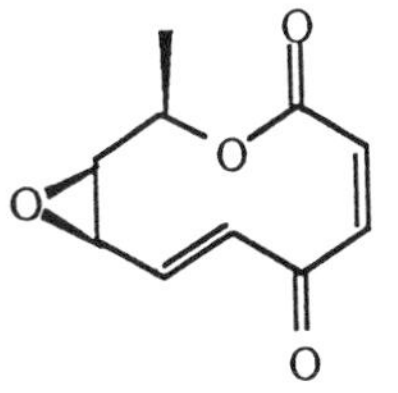

Fig. 2. Sturcture of pyrenolide A.

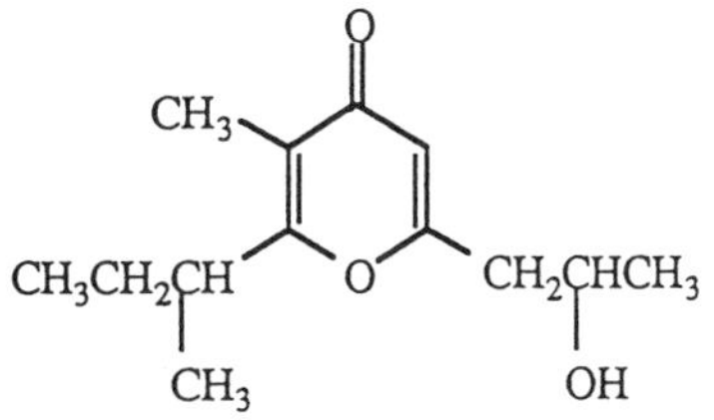

Fig. 3. Structure of hyalopyrone.

Unfortunately, no experiments were conducted using either mixtures of all three phytotoxins in the ratios at which they were isolated from *A. hyalospora*, or in any other ratios, but it must be emphasized that the experiments were conducted on detached leaves and, had intact plants been used, a longer term evaluation of whole plant treatments may have given a different response. Since only 12 mg of hyalopyrone were isolated and other assays including leaf electrolyte leakage, root growth inhibition of sorghum, and antibiotic disk assays against three genera of bacteria and three genera of fungi were conducted, a good deal of experimental territory was covered. Again, the problem alluded to earlier of isolating sufficient quantities of a metabolite for intensive biological investigation recurs.

Johnsongrass. Of all the weeds tested with natural products that exhibit herbicidal properties, johnsongrass appears to have received the most attention, with good reason.

291

It is perhaps one of the most pernicious weeds in the world and affects some 30 major economic crops in 53 countries (Holm et al., 1977). Among those biologically active microbial metabolites that have been tested, the following show promise for development as a herbicide. *Exserohilum turcicum* (= *Drechslera turcica*) and its many strains are prolific producers of secondary metabolites and among these is a strain that was isolated from diseased johnsongrass in the Ashad area of Israel. Upon fermentation in liquid shake culture and subsequent fractionation, monocerin, a known metabolite with broad spectrum biological activity, was isolated and unequivocally identified (Fig. 4). There followed extensive bioassays on johnsongrass (Robeson and Strobel, 1982). Excised leaves were sprayed at concentrations that ranged from 1.0 to 0.1 mg/ml and, at six days following treatment, the leaves became chlorotic and at eight days there was complete necrosis. These effects were seen at all the rates tested but no end point was pursued to determine the lowest possible concentration to induce necrosis. In germination assays, 1 mg/ml of monocerin was effective in almost fully inhibiting root elongation of pre-germinated seed, though the effect was not total. A dose of 33 mg/l induced a 50% reduction in root elongation within 96 hours relative to controls and, in turn, shoot growth was equally inhibited. Monocerin was also phytotoxic to Canada thistle (*Cirsium arvense*) and, when cuttings were placed in solutions containing 300 μg/ml, there was induction of a necrotic spot ~ 7 mm in diameter in the leaf lamina accompanied by necrotic flecks along the mid-rib, 16 hours following treatment. By 40 hours, necrosis had occurred in 50% of the apical leaf area while control plants appeared normal. However, monocerin was non-specific in its activity against plants and when tomato plant cuttings (*Lycopersicon esculentum* cv. Pixie hybrid) were treated with 330 μg/ml wilting occurred in 16 hours without necrosis or chlorosis (Robeson and Strobel, 1982).

The selective phytotoxic effects of prehelminthosporol (Fig. 5) and the synthetic derivative prehelminthosporol acetate (Fig. 5) have been described using the model plant corn (*Zea mays* cv. Norfolk Market White) (Cutler et al. 1982). Prehelminthosporol, isolated from

Drechslera sorokiana, induced stem collapse and general necrosis in twelve-day-old plants 24 hours following treatment with 10^{-2} M solutions at the rate of 100 μl per leaf whorl. Prehelminthosporol acetate took longer to induce a response and stem collapse and, marginal leaf necrosis only appeared 48 hours after treatment with 10^{-2} M solutions. Six-week-old tobacco plants (*Nicotiana tabacum* cv. Hick's) were not affected by either compound. Eleven-day-old bean plants (*Phaseolus vulgaris* cv. Black Valentine) exhibited mild plant growth regulatory responses wherein the first true leaves were either vertical or completely turned over because of bending of the pulvinus. Thus, selective herbicidal activity was seen with both metabolites.

Later, prehelminthosporol and dihydroprehelminthosporol (Pena-Rodriguez et al., 1988) (Fig. 6) were isolated from a *Bipolaris* sp. strain 36 leading to the isolation of victoxinine and prehelminthosporolactone (Fig. 5) (Pen-Rodriguez an Chilton, 1989). Prehelminthosporol, dihydroprehelminthosporol and prehelminthosporol lactone produced lesions on johnsongrass that resembled those obtained

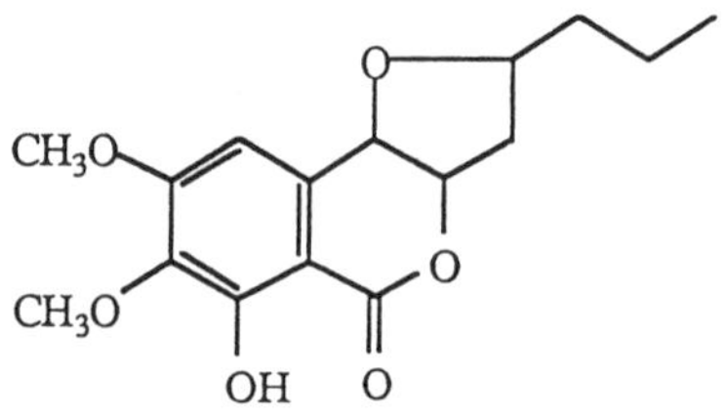

Fig. 4. Structure of monocerin.

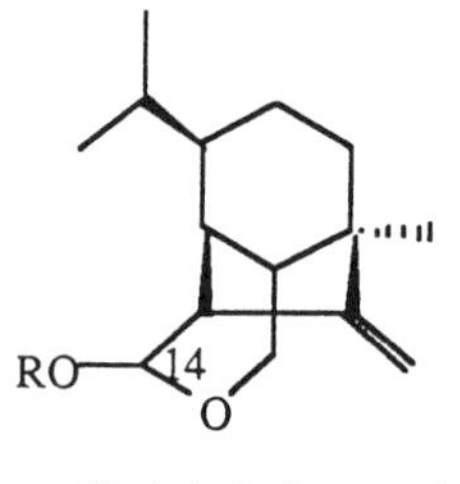

R = H   Prehelminthosporol
R = CH$_3$C=O Prehelminthosporol acetate
C14 = O =   Prehelminthosporol lactone

Fig. 5. Structure of prehelminthosporol, prehelminthosporol acetete, and prehelminthosporol lactone.

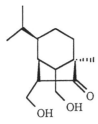

Fig. 6. Structure of dihydroprehelminthosporol.

with the phytopathogen when evaluated in the leaf spot assay. The lesions have been described as black circles that contain maroon colored centers with an outer chlorotic ring. However, all four metabolites were active against sorghum (*S. bicolor*), sicklepod (*C. obtusfolia*) and bentgrass (*A. alba*) at concentrations of 10 μg of metabolite per 5 μl droplet.

The ophiobolins have also been evaluated for plant growth regulatory and herbicidal properties and their history has been described in detail. *Drechslera maydis*, the phytopathogen that induces Southern corn leaf blight, and *D. sorghicola* that produces leaf spot in johnsongrass, were fermented in liquid M-1-D medium shake cultures to produce an array of ophiobolins. Two races of *D. maydis*, race T and race O, were examined (Sugawara et al., 1987). Both races produced ophiobolin I, ophiobolin A, ophiobolin C, 6-epianhydroophiobolin A, and 6-epiophiobolin in approximately equal amounts but, race T produced five times more 25-hydroxyophiobolin I (Fig. 7) than race O. On the other hand, *D. sorghicola* produced approximately the same amounts of ophiobolin I, ophiobolin A, 25-hydroxyophiobolin I, and 6-epiophiobolin as race T and O, but both 6-eipanhydroophiobolin A and ophiobolin C could not be detected. When all these fungal metabolites were tested against johnsongrass, the most active phytotoxically was 6-epiophiobolin which was common to both species of *Drechslera* tested. That activity, which included the appearance of lesions, was manifest at 10^{-2} to 10^{-5} M. The next most active metabolite against johnsongrass was ophiobolin A, again common to both species; it was moderately active at 10^{-3} and 10^{-4} M. Similarly, ophiobolin I was only slightly active at concentrations ranging from 10^{-3} to 10^{-5}

M. Ophiobolin C and 6-epianhydroophiobolin A, which were not detected in *D. sorghicola*, were only marginally active at 10^{-3}M against johnsongrass. Additionaly, 25-hydroxyophiobolin I, found in *D. sorghicola*, was slightly active against johnsongrass at 10^{-3} M and slightly active at both 10^{-3} and 10^{-4} M against sorghum (*S. bicolor*) (Sugawara et al., 1987).

One of the more recent structurally interesting natural products to be discovered that possesses potent herbicidal activity is cornexistin, (Fig. 8) Paecilomyces isolated from the culture filtrate of a newly isolated fungus, *Paecilomyces variotii* SANK 21086 (Nakajima et al. 1991). This purely Japanese contribution originated from the fungus of a deer dung sample being collected at Nojak, Alberta, Canada and, the genesis of how the fungus arrived in Japan must be an interesting tale. Be that as it may, some 920 liters of culture filtrate were extracted with ~ 2,760 liters of ethyl acetate, which was further worked up to eventually yield 115 grams of pure cornexistin. Obviously, the intention from the outset was to produce sufficient quantities for fairly large scale initial tests. Thus, these figures are important because they indicate the sort of commitment essential to any study involving natural products for practical use.

The metabolite was exceptionally active against johnsongrass and foxtail (*Setaria viridis*) and was fairly active against crabgrass (*Digtaria* sp.) and barnyardgrass (*Echinochloa crusgalli*) at 100 mg/liter, though, at 500 and 1000 mg/liter all species were killed. Several dicotyledonous weeds were also killed by cornexistin but corn (*Z. mays*) was unaffected by the compound. The authors claim that cornexistin would be a useful post emergence weed control herbicide in corn.

Cornexistin was further tested against ~ 30 species of microorganisms in disk assays but did not exhibit activity at rates up to 1 mg/ml. From a practical viewpoint, the metabolite had an $LD_{50} > 1$ g/kg orally in mice while the intraperitoneal LD_{50} was > 100 mg/kg. The structure, which is an anhydride, is closely related to rubratoxin B, a double anhydride (Nakajima et al., 1991), which has been reported to have herbicidal activity. Of singular

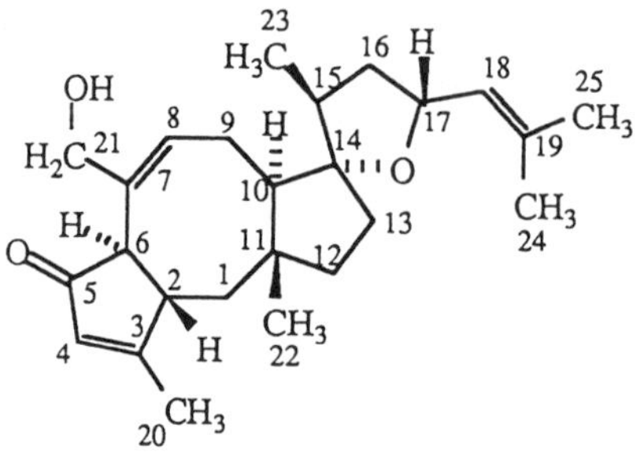

Fig. 7. Structures of ophiobolins.

Ophiobolin I

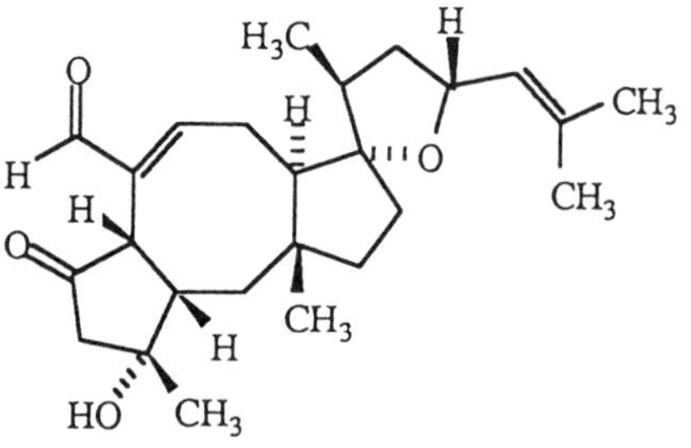

Ophiobolin A

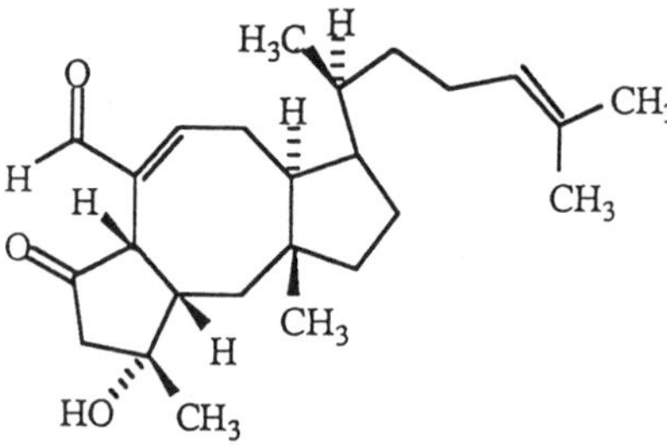

Ophiobolin C

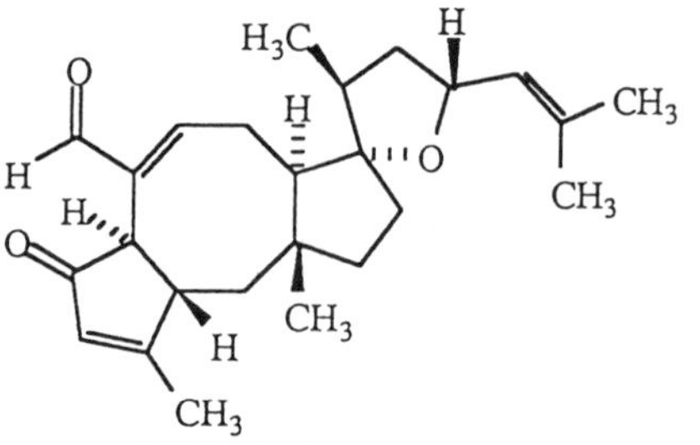

6-Epianhydroophiobolin A

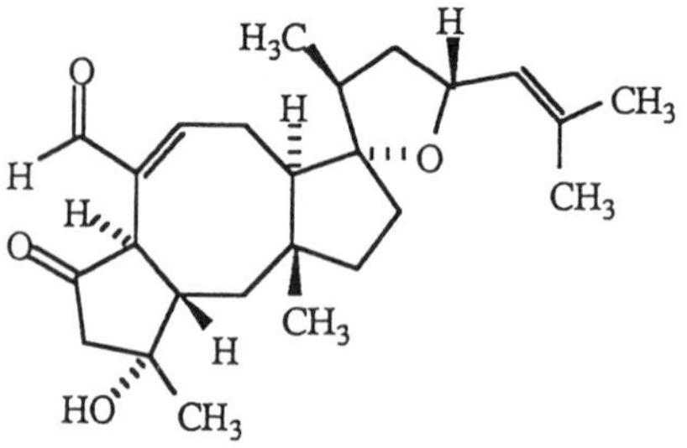

6-Epiophiobolin

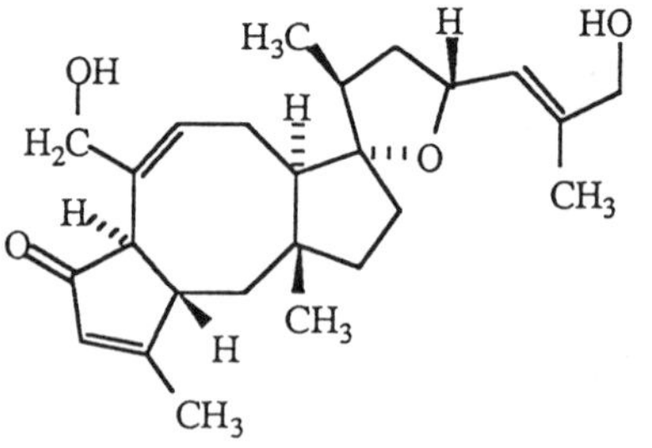

25-Hydroxyophiobolin I

importance is that cornexistin was discovered not from a phytopathogenic fungus but, from a saprophyte.

Evening Primrose. *Pestalotiopsis oenotherae* (IMI No. 334043) is a recently discovered pathogen of evening primrose (*Oenothera laciniata*). Upon liquid fermentation in potato-dextrose-broth shake cultures, extraction against ethyl acetate and chromatographic separation, four metabolites were isolated. These were oxysporone, pestalopyrone, pestalotin, and hydroxypostalotin (Ventkatasubbaiah et al., 1991). Of these, the two most active were

oxysporone (Adesogan and Alo, 1979) and pestalopyrone (Fig. 9, 10). In leaf disk assays against evening primrose and johnsongrass, oxysporone was the more active of the two. Concentrations of 1 mg/liter of oxysporone were almost equally as active against evening primrose and johnsongrass. Pestalopyrone is related to another secondary metabolite, nectriapyrone, which has an additional methyl group at C3 and has been isolated from *Gyrostoma missouriensis* (Nair and Carey, 1975) and *Phomopsis oblonga*. It is also related to fusalanipyrone, which lacks an O methyl group at C4, that was isolated from *Fusarium*

294

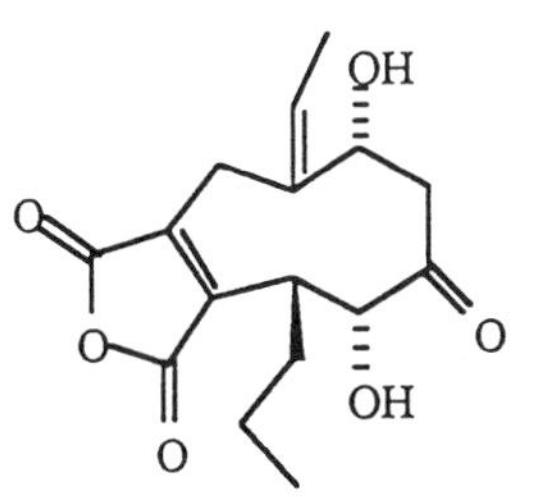

Fig. 8. Structure of cornexistin.

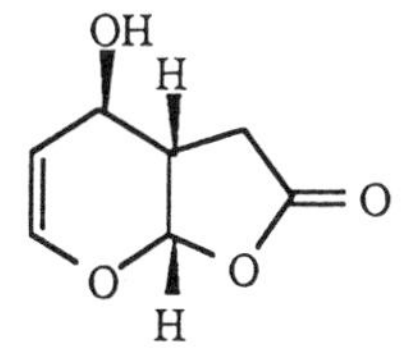

Fig. 9. Structure of oxysporone.

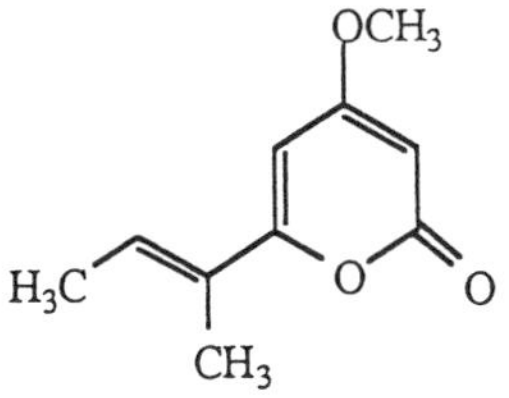

Fig. 10. Structure of pestalopyrone.

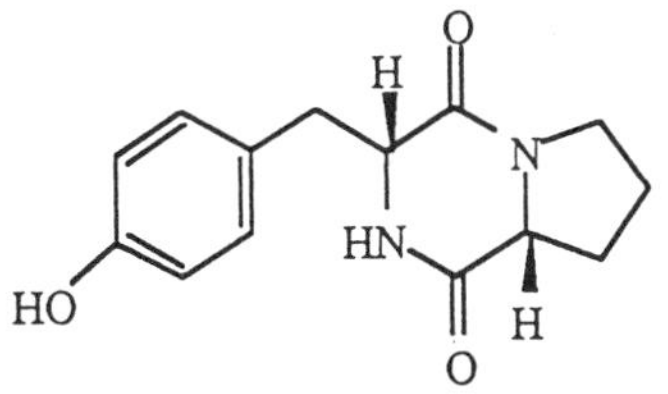

Fig. 11. Structure of maculosin.

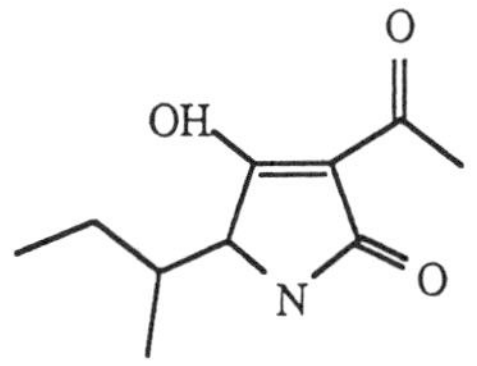

Fig. 12. Structure of tenuazonic acid.

solani (Abraham and Arfmann, 1988). Oxysporone has an interesting history in that it was found in a strain of *Fusarium oxysporum* that had been cultured from earthworm faeces (Adesogan and Alo, 1979).

Spotted Knapweed. One of the more noxious weeds of the western region of the North American continent is spotted knapweed (*Centaurea maculosa*). It is particularly concentrated in British Columbia, Washington, Idaho, and Montana. Over two million acres of rangeland have been infected with spotted knapweed resulting in an estimated loss of 70% in forage production (Harris and Cranston, 1979). A phytopathogen that selectively attacks spotted knapweed is *Alternaria alternata* which produces maculosin (Stierle et al. 1988) (Fig. 11). During the isolation of maculosin, tenuazonic acid (Fig. 12), a secondary metabolite known to have cytotoxic properties,

was also isolated. Subsequently, the herbicidal properties of both natural products were tested individually and in combination (Stierle and Cardellina II, 1989). Evaluations were conducted on 13 genera of dicotyledonous and six genera of monocotyledonous plants. Both maculosin and tenuazonic acid exhibited remarkable phytotoxic selectivity toward spotted knapweed. Tenuazonic acid was also active to a similar or lesser degree to 12 of the dicotyledonous and all of the monocotyledonous plants tested. At 10^{-3}M, both produced weeping necrotic lesions at application sites with moderate toxicity at 10^{-4}M, and slight activity at 10^{-5} M. When solutions of maculosin and tenuazonic acid were mixed in equimolar proportions the activity of the phytotoxins was logarithmically increased indicating that more than one phytotoxin may play a role in the etiology of certain plant diseases (Stierle and Cardellina II, 1989).

Purple Nutsedge. Purple nutsedge (*Cyperus rotundus*) occurs throughout most of the climatically warmer regions and is considered to be one of the world's most pernicious weeds. It is difficult to eradicate and is the bane of many farmers in cultivated crops (King, 1966). The recent discovery of a new pathogen of *C. rotundus* was made in India and was determined to be *Ascochyta cypericola*. Pure cultures of the organism were grown for three weeks on liquid M-1-D medium that contained 12 ml/liter of coconut milk. Following extraction with ethyl acetate, the organic layer was chromatographed and, eventually, pure cyperine (Fig. 13), a novel biphenyl ether, was isolated (Stierle et al., 1991). The metabolite was tested against various species of *Cyperus* and, while no hard data were given, it was stated that *C. rotundus* exhibited the greatest sensitivity to cyperine.

All the organisms examined so far are potential biological control agents and, furthermore, their chemical mode of action has been adequately elucidated to put a defined product on the market.

Phytotoxins of Plant Origin

A recent review of herbicidal compounds from higher plants includes a number of different structures that have plant growth inhibiting properties. Again, the major problem concerning these secondary metabolites, potent though they may be, is the isolation of sufficient quantities for practical use. There are notable exceptions; An example is the lichens, of which some eighteen thousand species have been classified (Kimball, 1978) were at one time considered abundant enough to supply cottage industries with enough material to produce dyestuff. The Harris tweed industry relied solely on lichen dyes to produce the Umbrian tones for which the cloth was famous. And certainly, enough lichen could be sensibly grown in the tundra to supply specialty chemicals, if needed, without affecting dependent wildlife. The enigma still remains, to some extent, as to whether the green or bluegreen alga is the producer organism of important secondary metabolites, or whether the associated fungus, generally an ascomycete but sometimes a basidiomycetes, is the major player. At the other end of the producer scale

are trees like *Magnolia* which biosynthesize metabolites that are relatively potent in small amounts. This will be discussed later.

Inhibitors from Radish Seedlings. Most of the phytotoxins of plant origin have only been tested against model plants. Some detailed and clever work has been executed with growth inhibitors from light-grown radish plants (*Raphanus sativus* var. *hortensis* f. *gigantissimus*) (Hasegawa and Miyamota, 1978). Four compounds, all having plant growth inhibitory properties, have been identified and some have been given trivial names. They include raphanusol A (Hase and Hasegawa, 1982), raphanusol B (Hasegawa and Hase, 1981), raphanusanin and 2-thioxothiazolidine-4-carboxylic acid (Hase et al., 1983). Another less polar metabolite, 3-(E)-(methylthio)methylene-2-pyrrolidinethione (Sakoda et al., 1990) (Fig. 14) has also been isolated from light grown radish and has been classified as a new plant growth inhibitor. The compound significantly inhibited the hypocotyl growth of etiolated cress seedlings (Sakoda et al., 1990). Practically, the compound may be useful in selectively controlling the height of cruciferous weed seedlings in economic crops.

Inhibitors from *Podocarpus*. The genus *Podocarpus* has wide distribution in tropical and subtropical areas, especially in Asia and the Southern Hemisphere where some 80 species are found. The species *P. nagi* has been intensively examined over the past 25 years and compounds have been isolated that have antitumor, plant growth regulatory, and insecticidal properties (Galbraith et al., 1970; Russell et al., 1973; Hayoshi and Matsumoto, 1982). Among the plant growth regulators that exhibit inhibitory properties are a subset of the nagilactones. The nagilactones are a most interesting group of natural products and can be

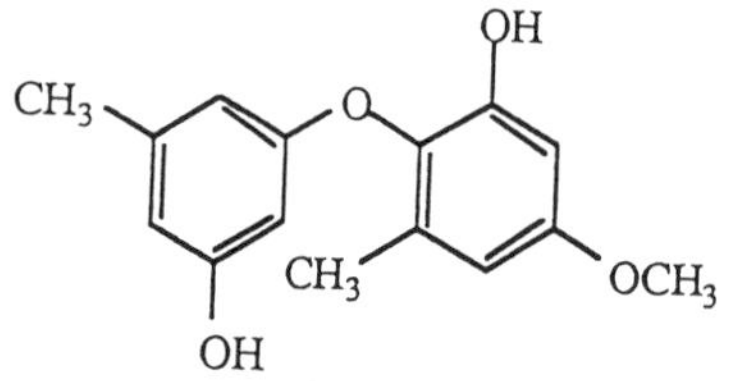

Fig. 13. Structure of cyperine.

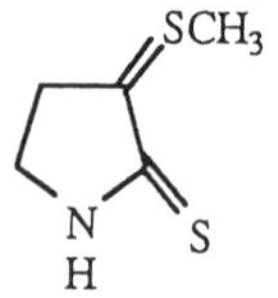

Fig. 14. Structure of 3-(E)-(methylthio)methylene-2-pyrrolidinethione.

categorized into two groups based on their biological activity. Group I consists of six members that promote growth of lettuce radicles when applied at concentrations of 1 to 10 μg/ml. They differ only by merit of the functional groups H, -0-, OH, or CH_3 at the R positions (Fig. 15). The Group II candidates, of which there are presently two natural products, also differ in the first case by having different functional groups at the R positions (Fig. 16a) and in the other case by a structural modification consisting of an epoxide (Fig. 16b) (Kubo, et al., 1991). At 5-10 mg/liter, both compounds significantly inhibited lettuce radicles, though the epoxide inhibited radicles 100% at 10 mg/liter. Because of the number of functional groups on the nagilactone structure, any one of the parent molecules becomes a viable candidate for derivatization and, consequent, evaluation of biological activity.

Phytotoxin from *Magnolia grandiflora*. There has been a common observation that virtually no plants will grow beneath a magnolia tree;

those that do have a short life and appear spindly, suggesting that an allelopathic effect may well be responsible for the phenomenon. In a bioassay - directed fractionation of the leaves from *M. grandiflora*, a sesquiterpene ketone of the aromadendrane class, cyclocolorenone (Fig. 17) was isolated and its structure proved by UV, IR, NMR, and MS (Jacyno et al., 1991). The metabolite significantly inhibited the growth of etiolated wheat coleoptiles (*Triticum aestivum*, cv. Wakeland) 100 and 58%, respectively, at 10^{-3} and 10^{-4} M, relative to controls. Nine-day-old bean plants (*Phaseolus vulgaris*, cv. Black Valentine) exhibited two types of response within 48 hours of treatment. Leaf blades had irregular, necrotic lesions and, leaves were bent at the pulvinus so that the surfaces were inverted. A similar response has been noted with prehelminthosporol. Nine-day-old corn plants (Z. *mays* cv. Norfolk Market White) treated with cyclocolorenone at 10^{-2} M showed severe leaf necrosis and stem collapse within 48 hours. Even after two weeks, plants appeared stunted, relative to controls. Six-week-old tobacco plants had slight necrosis at 10^{-2} M treatments, but recovered within a week and plant vigor was unaffected. Thus, cyclocolorenone appeared to be somewhat target specific in phytotoxic activity. In addition, this metabolite has antibacterial and antifungal properties. Unfortunately, only small amounts of cyclocolorenone were isolated and, furthermore, the titer of the compound appeared to vary seasonally so that synthesis may be a more suitable source for quantities of material.

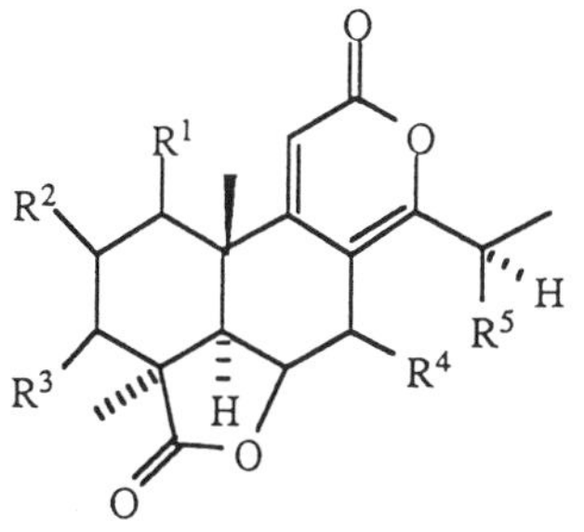

Fig. 15. Parent structure of the nagilactones. R positions have different substitutions.

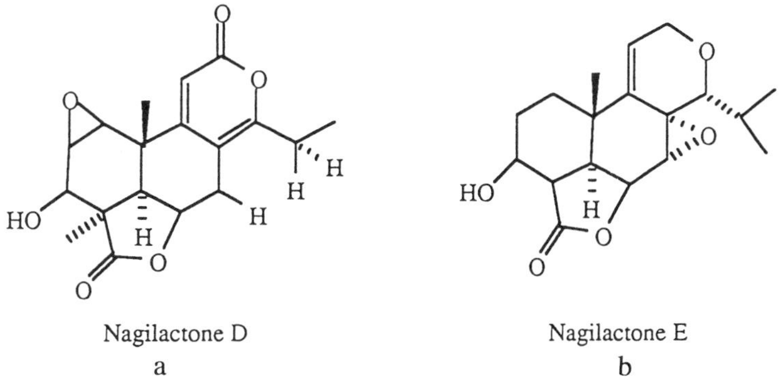

Nagilactone D
a

Nagilactone E
b

Fig. 16a,b. Structure of nagilactone D and E.

Phytotoxins of Fungal Origin

Fig. 18. Structure of (3R,5Z)-(-)-3-hydroxy-5-dodecanoic acid.

The most promising sources of useful allelochemicals are the microorganisms. They are readily available and even those that are phenotypically identical and therefore of the same genus and species, are different in their biochemical constitution: Thus, they give rise to several different types of secondary metabolites. Many are saprophytes, making their culturing relatively easy and they may be lyophilized so that a pure source of culture can be safely stored. If larger quantities of a selected metabolite are required, a suitably larger fermentation can be prepared. Ideally, a group working with fungal metabolites should consist of one team isolating novel compounds and a second that re-isolates natural products that have economic potential.

Pythium ultimum. Sixteen milligrams of (3R,5Z)-(-)-3-hydroxy-5-dodecenoic acid (Fig. 18) were obtained from 15 liters of potato-sucrose broth which served as the liquid medium for fermentation of *P. ultimum* (Ichihara et al. 1985). At high concentrations of 250 mg/liter, the metabolite darkened roots and shoots of sugar beet (*Beta vulgaris* L., cv Monhope), but at 50 mg/liter, the growth of shoots and roots were promoted. In contrast, the growth of rice (*Oryza sativa* L.) roots and shoots was inhibited 52 and 31%, respectively, at 500 mg/liter. At 25 mg/liter, roots were inhibited 8.8% and shoots 14%. This metabolite is of interest because of its bioactive diversity and the possibility that a grass herbicide could be developed for use in sugar beet that would have a dual role as a plant growth regulator. The parent molecule has two functional groups, one at C1, the other at C3, which allows for derivatization and,

additionally, the unsaturation at C5-C6 allows some opportunities for manipulating the molecule.

Cochliobolus spicifer. A number of phytotoxins that possess varying degrees of activity have been isolated from *Cochliobolus spicifer,* the causal organism of leaf spot in grasses. Among those of interest are spiciferones A (Nakajima et al., 1989), B (Nakajima et al., 1991), and C (Nakajima et al, 1991) (Fig. 19), all γ-pyrones, and the azaphilone, spiciferinone (Nakajima, et al, 1992) (Fig. 20). In bioassays, using 7-day-old wheat plants (*T. aestivum* cv. Ushio-Komugi) wherein the cotyledons were detached and cut to an 8 cm length followed by placing the basal ends in vials with 3 ml of test solution (3 cuttings per vial), it was observed that spiciferone A and B were approximately equally as active at 10^{-3} and 3×10^{-4} M. At 10^{-3} M, all the cotyledons were killed while, at 3×10^{-4} M, there were necrotic spots on the leaves, compared to controls. Spiciferone B which was inactive, differs from spiciferone C in that, while both have exactly the same CH_2OH functional group, they have been inverted. That is, in spiciferone B the CH_2OH is adjacent to the ketone of the γ-pyrone. In later assays, using the same wheat variety but not the same technique, wheat coleoptile protoplasts indicated

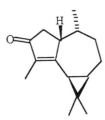

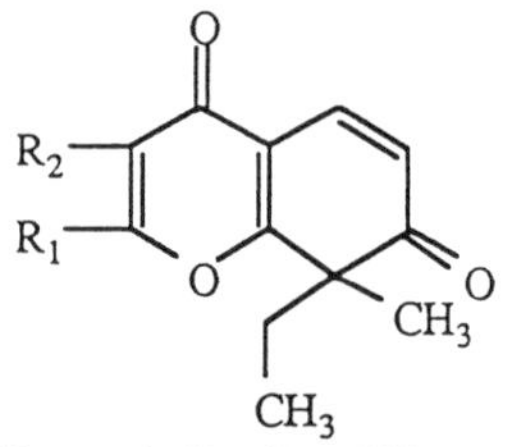

Fig. 17. Structure of cyclocolorenone.

Spiciferone A: $R_1=R_2=CH_3$
Spiciferone B: $R_1=CH_2OH, R_2=CH_3$
Spiciferone C: $R_1=CH_3, R_2=CH_2OH$

Fig. 19. Structure of spiciferones.

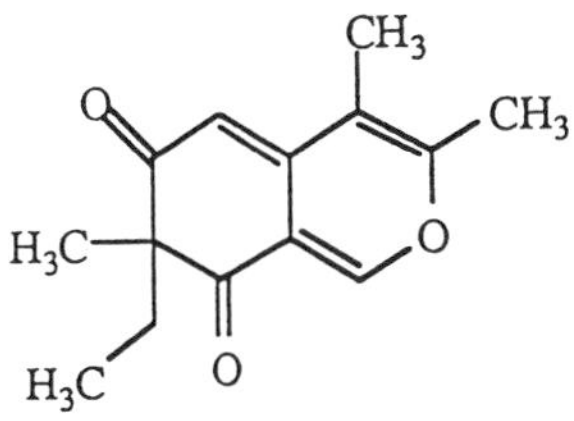

Fig. 20. Structure of spiciferinone.

that spiciferinone and spiciferone A were active. At 10^{-3}M spiciferinone (Fig.20) completely killed wheat protoplasts while spiciferone A had ~ 8.7% viability. At 3×10^{-4} M spiciferinone was slightly more active than spiciferone A, while at 1×10^{-4} M the reverse was true.

Scopulariopsis candidus. A number of potentially useful pharmaceutical compounds accidentally find their way into the agrochemical field, usually as the result of independent re-isolation from a novel fungal source. Furthermore, the literature refers only to the pharmaceutical properties with none to the agrochemical properties. Such is the case with the ß-lactone (E,E)-11-[3'-(hydroxymethyl)-4'-oxo-2'-oxetanyl]-3,5,7-trimethyl-2,4-undecanoic acid, known also trivially as antibiotic 1233A (Fig. 21). Compound 1233A was originally discovered in a *Cephalosporium* sp. It was later rediscovered in both a *Scopulariopsis* sp. and a *Fusarium* sp. This compound had potent inhibitory action on the enzyme 3-hydroxy-3-methylglutaryl coenzyme A synthase, making it an excellent candidate for reducing blood serum cholesterol. The compound was again isolated from *Scopulariopsis candidus* during the examination of microorganisms for biologically active natural products for use as agricultural chemicals. Both 1233A and the tetrahydro derivative, made by perhydrogenating the molecule using Pt_2O as a catalyst, were active in the etiolated wheat coleoptile (*T. aestivum* cv. Wakeland) (Jacyno

et al. 1991). 1233A significantly inhibited coleoptiles at 10^{-3}, 10^{-4}, and 10^{-5} M while tetrahydro 1233A inhibited at 10^{-3} and 10^{-4} M. Both compounds demonstrated selective phytotoxic action on greenhouse-grown plants. Six-week-old tobacco plants were unaffected by either metabolite. On the other hand, the first true leaves of bean (*P. vulgaris* cv. Black Valentine) showed necrotic lesions after 48 hours following treatment with tetrahydro 1233A at 10^{-2} M, but the trifoliates were not affected, while 1233A at 10^{-2} M produced only slight veinal necrosis during the same time. The 10^{-3} M treatments produced no visible effects in bean. In corn plants (*Z. mays* cv. Norfolk Market White), both 1233A and tetrahydro 1233A caused necrosis at the point of application, at 10^{-2} M, within 48 hours. Tetrahydro 1233A induced stem collapse in all treated corn plants, whereas, complete stem collapse was only produced in 20% of the corn plants treated with 1233A. Two weeks following application, corn plants treated with tetrahydro 1233A had recovered, but the overall height of the plants was only ~ 70% that of the controls. The molecule has functional sites that may be altered to develop a grass herbicide.

Conclusion

A number of plants and microorganisms produce secondary metabolite that may be developed as herbicides. Some of these microorganisms may be used as biocontrol agents to control specific weeds. If this does occur then the chemistry of the organism relative to a mode of action must be known. It is possible in these very discreet cases that the microorganism is the best delivery system for the biologically active metabolite. But all biologically active secondary metabolites are potential sources of novel templates which may be altered either by biotransformation to

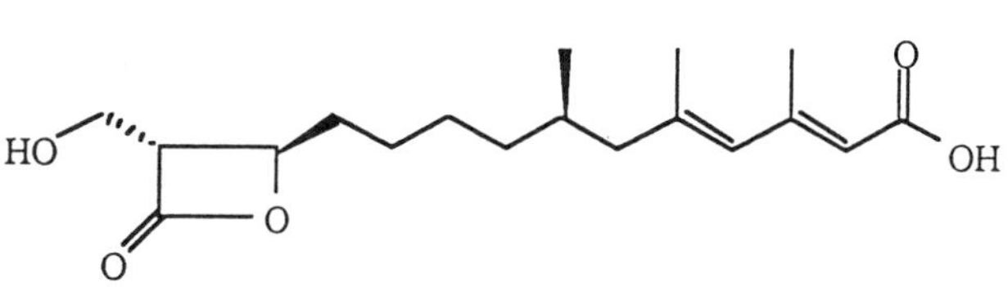

Fig. 21. Structure of 1233A.

produce further natural products, or, synthetically changed in the hope of retaining the biodegradable properties. Perhaps the most successfully derived natural product herbicide for weed control has been glyphosate, N-(phosphonomethyl) glycine. There are yet a number of "glyphosates" waiting to be discovered.

References

Abraham, W.-F.; Arfmann, H.-A. "Fusalanipyrone, A Monoterpenoid from *Fusarium solani*." *Phytochemistry* **1988**, *27*, 3310-3311.

Adesogan, E.K.; Alo, B.I. "Oxysporone, A New Metabolite from *Fusarium oxysporum*." *Phytochemistry* **1979**, *18*, 1886-1887.

Claydon, N.; Grove, J.F.; Pople, M. "Elm Bark Bettle and Feeding Deterrents from *Phomopsis oblonga*." *Phytochemistry* **1985**, *24*, 937-943.

Cutler, H.G. In *Allelochemicals: Role in Agriculture and forestry*; Waller, G.R., Ed.; ACS Symposium Series No. 330; American Chemical Society: Washington DC **1987**; pp 23-38.

Cutler, H.G. "Perspectives on Discovery of Microbial Phytotoxins with Herbicidal Activity." *Weed Technology.* **1988**, *2*, 525-532.

Cutler, H.G. "Unusual Plant-Growth Regulators from Microorganisms." *CRC Critical Reviews in Plant Sciences.* **1988**, *6*, 323-343.

Cutler, H.G. In *Biologically Active natural Products: Potential Use in Agriculture*; Cutler, H.G., Ed.; ACS Symposium Series No. 380: Washington Dc, **1988:** pp 1-22.

Cutler, H.G. In *Handbook of Natural Toxins*; Keeler, R.F.; Tu, A.T., Eds.; Marcel Dekker, Inc, NY, **1991**; pp 411-438.

Cutler, H.G. In *Plant Biochemical Regulators*; Gausman, H.W., Ed.; Marcel Dekker, Inc., NY, **1992**; pp 113-139.

Cutler, H.G. In *Phytochemical Resources for Medicine and Agriculture*; Nigg, H.N.; Seigler, Eds.; Plenum Press, NY, **1992**; pp 205-226.

Cutler, H.G.; Crumley, F.G.; Cox, R.H.; Davis, E.E.; Harper, J.L.; Cole, R.J.; Sumner, D.R. "Prehelminthosporol and Prehelminthosporol Acetate: Plant Growth Regulating Properties." *J. Agric. Food Chem.* **1982**, *30*, 658-662.

Edwards, J.V.; Dailey, O.D.; Bland, J.M.; Cutler, H.G. In *Biologically Active Natural Products: Potential Use in Agriculture*; Cutler, H.G., Ed.; ACS Symposium Series No. 380: Washington DC, **1988**; pp 35-56.

Galbraith, M.N.; Horn, D.H.S.; Sasse, J.M.; Adamson, D. "The Structures of Podolactones A and B., Inhibitors of Expansion and Division of Plant Cells." *Chem. Comm.* **1970**, 170-171.

Harris, P.; Cranston, R. "An Economic Evaluation of Control Methods for Diffuse and Spotted Knapweed in Western Canada." *Can. J. Plant Sci.* **1979**, *59*, 375-382.

Hase, T.; Hasegawa, K. "Raphanusol A, A New Growth Inhibitor From Sakurajima Radish Seedlings." *Phytochemistry* **1982**, *21*, 1021-1022.

Hase, T.; Koreeda, M.; Hasegawa, K. "A Growth Inhibitor, 2-thioxothiazolidine-4-carboxylic acid From Sakurajima Radish Seedlings." *Phytochemistry*, **1983**, *22*, 1275-1276.

Hasegawa, K.; Miyamota, K. "Light Growth Inhibition and Growth Inhibitors in Sakurajima Radish Seedings." *Plant & Cell Physiol.* **1978**, *19*, 1077-1083.

Hasegawa, K.; Hase, T. "Raphanusol B: A Growth Inhibitor of Light-grown Radish Seedlings." *Plant & Cell Physiol.* **1981**, *22*, 303-306.

Hayoshi, Y.; Matsumoto, T. "Reaction and Interconversion of Norditerpenoid Dialactones, Biologically Active Principles isolated from *Podocarpus* Plants." *J. Org. Chem.* **1982** *47*, 3421-3428.

Holm, L.G.; Plucknett, D.L.; Pancho, J.V.; Herberger, J.P. In *The World's Worst Weeds—Distribution and Biology*; The University Press of Hawaii, Honolulu, HI, **1977.**

Jacyno, J.M.; Cutler, H.G.; Roberts, R.G.; Waters, R.M. "Effects on Plant Growth of the HMG-CoA Synthase Inhibitor, 1233A/F-244/L-659,699, Isolated from *Scopulariopsis candidus." Agric. Biol. Chem.* **1991**, *55*, 3129-3131.

Jacyno, J.M.; Montemurro, N.; Bates, A.D.; Cutler, H.G. "Phytotoxic and Antimicrobial Properties of Cyclocolorenone from *Magnolia grandiflora* L." *J. Agric. Food Chem.* **1991**, *39*, 1166-1168.

Kimball, J.W. In *Biology*, Addison-Wesley Publishing Co., Reading, MA, 1978.

King, L.J. In *Weeds of the World, Biology and Control*. Leonard Hill Books, London, **1966**.

Kubo, I.; Sutisna, M.; Tan, K.-S. "Effects of Nagilactones on the Growth of Lettuce Seedlings." *Phytochemistry* **1991**, *30*, 455-456.

Ichihara, A.; Hashimoto, M.; Sakamura, S. "(3R,5Z)-(-)-3-hydroxy-5-dodecenoic Acid, A Phytotoxic Metabolite of *Pythium ultimum." Agric. Biol. Chem.* **1985**, *49*, 2207-2209.

Nair, M.S.R.; Carey, S.T. "Metabolites of Pyrenomycetes II: Nectriapyrone, An Antibiotic Monoterpenoid." *Tetrahedron Letters* **1975**, 1655-1658.

Nakajima, H.; Hamasaki, T.; Kimura, Y. "Structure of Spiciferone A, A Novel γ-pyrone Plant Growth Inhibitor Produced by the Fungus *Cochliobolus spicifer* Nelson." *Agric. Biol. Chem.* **1989**, *53*, 2297-2299.

Nakajima, H.; Itoi, K.; Takamatsu, Y.; Sato, S.; Furkawa, Y.; Furuya, K.; Honma, T.; Kadotani, J.; Kozasa, M.; Haneishi, T. "Cornexistin: A New Fungal Metabolite with Herbicidal Activity." *J. Antibiot.* **1991**, *44*, 1065-1072.

Nakajima, H.; Hamasaki, T.; Kohao, M.-A.; Kimura, Y. "Spiciferones B and C, Minor Phytotoxins from the Fungus *Cochliobolus spicifer." Phytochemistry*, **1991**, *30*, 2563-2565.

Nakajima, H.; Kimura, Y.; Hamasaki, T. "Spiciferinone, An Azaphilone Phytotoxin Produced by the Fungus *Cochliobolus spicifer." Phytochemistry* **1992**, *31*, 105-107.

Pena-Rodriquez, L.M.; Armingeon, N.A.; Chilton, W.S. "Toxins from Weed Pathogens, I. Phytotoxins from a *Bipolaris* Pathogen of Johnsongrass." *J. Nat. Prod.* **1988**, *51*, 821-828.

Pena-Rodriguez, L.M.; Chilton, W.S. "Victoxinine and Prehelminthosporollactone, Two minor Phytotoxic Metabolites Produced by *Bipolaris* sp., A Pathogen of Johnson grass." *J. Nat. Prod.* **1989**, *52*, 899-901.

Robeson, D.J.; Strobel, G.A. "Monocerin, A Phytotoxin from *Exserohilum turcicum* (≡ *Drechslera turcica*)." *Agric. Biol. Chem.* **1982**, *46*, 2681-2683.

Russell, G.B.; Fenemore, P.G.; Singh, P. "Structures of Hallactones A and B., Insect Toxins from *Podocarpus halii." J.C.S. Chem. Com.* **1973**, 166-167.

Sakoda, M.; Hase, T.; Hasegawa, K. "A Growth Inhibitor, 3-(E)-(methylthio)-methylene-2-pyrrolidinethione from Light-grown Radish Seedlings." *Phytochemistry* **1990**, *29*, 1031-1032.

Stierle, A.C.; Cardellina II, J.G.; Strobel, G.A. "Macolusin, A Host-specific Phytotoxin for Spotted Knapweed from *Alternaria alternata." Proc. Natl. Acad. Sci. USA* **1988**, *85*, 8008-8011.

Stierle, A.C.; Cardellina II, J.H. "Phytotoxins from *Alternaria alternata*, A Pathogen of Spotted Knapweed." *J. Nat. Prod.* **1989**, *52*, 42-47.

Stierle, A.; Upadhyay, R.; Strobel, G. "Cyperine, A Phytotoxin Produced by *Ascochyta cypericola*, a Fungal Pathogen of *Cyperus rotundus." Phytochemistry* **1991**, *30*, 2191-2192.

Sugawara, F.; Strobel; G.; Strange, R.N.; Siedow, J.N.; VanDuyne, G.D.; Clardy, J. "Phytotoxins from the Pythogenic Fungi *Drechslera maydia* and *Drechslera sorghicola." Proc. Natl. Acad. Sci. USA* **1987**, *84*, 3081-3085.

Ventkatasubbaiah, P.; VanDyke, C.G.; Chilton, W.S. "Phytotoxins Produced by *Pestalotiopsis oenotherae*, A Pathogen of Evening Primrose." *Phytochemistry* **1991**, *30*, 1471-1474.

Venkatasubbaiah, P.; Chilton, W.S.

"Phytotoxins of *Ascochyta hyalospora*, Causal Agent of Lambsquarters Leaf Spot." *J. Nat. Prod.* **1992**, *55*, 461-467.

In *Websters 3rd New International Dictionary*, G&C Merriam Co., Springfield, MA, **1971.**

Genetic Manipulation
of Biocontrol Agents

Expression of Viral Genes and Viral and Antiviral Proteins in Transgenic Plants to Confer Virus Resistance

Ramon L. Jordan and John Hammond, U.S. Department of Agriculture, Agricultural Research Service, Florist and Nursery Crops Laboratory, Plant Sciences Institute, Beltsville Agricultural Research Center, 10300 Baltimore Ave, Beltsville, Maryland 20705.

Recent attempts to confer resistance to plant viruses by transformation and genetic engineering with viral and antiviral genes have provided promising new strategies for virus control. The major approaches that have been developed to generate transgenic resistance include expression of viral coat protein coding sequences, transformation of plants with other viral gene products, such as viral replicase, and expression of untranslatable sense or anti-sense viral transcripts. Other novel approaches include expression of catalytic ribozyme RNAs and mouse antiviral antibody genes. Descriptions of these strategies and how the gene sequences may impart virus resistance are presented.

The control of virus diseases of plants is of considerable importance in modern agriculture. Although natural genes conferring virus resistance have been incorporated into the genomes of some crop plants by breeding, control of plant virus diseases has historically involved numerous, often combined, strategies to provide durable effective resistance.

Recent attempts to confer resistance against plant viruses using genetic engineering and transformation with viral and antiviral genes ("transgenes") has provided promising additional strategies for virus control. The major approaches that have been developed to generate transgenic resistance include: (i) expression of viral coat protein (full length or truncated) coding sequences to confer "coat protein-mediated resistance"; (ii) transformation of plants with other viral gene products, such as truncated or full length replicase components; and (iii) expression of untranslatable sense or antisense viral transcripts. Other, more novel, approaches include: (i) the use of synthetic antisense oligonucleotides expressed in plants; (ii) transgenic expression of RNAs with catalytic ribozyme activity; and (iii) the expression of mouse antiviral (anti-coat protein or non-structural protein) antibody genes in transformed plants.

Plant viruses are complex and their multiplication in plants involves many different steps, including transmission, uncoating, replication, gene expression, movement, assembly, and symptom expression. Each of these steps in the virus infection cycle are potential targets for interference. Table 1 lists the potential targets for interference and the possible active transgene(s) that could cause such interference. Some of these transgenes and their potential activities have been shown to be successful in providing virus resistance, while the others are speculative. Descriptions of the various strategies listed above, as well as several considerations as to how these gene sequences may impart virus resistance, are presented in this report.

Coat Protein-Mediated Resistance

In 1986 a new method to introduce induced resistance to a plant virus was reported (Powell-Abel et al., 1986). This method, based on genetic modification of the host plant, mimics the natural phenomenon of

Table 1. Potential targets in virus infection cycle for transgenic interference.

Stage	Interfering event	Potential transgene
Uncoating	Compete for RNA	Coat Protein (CP)
Replication	Inhibit enzymes Compete for enzymes Block RNA	Plant-expressed antibodies Untranslatable RNAs Antisense CP
Translation	Block RNA Cut MRNA	Antisense Ribozyme
Transcription	Inhibit enzymes	Plant-expressed antibodies
Assembly	Compete for capsids	Untranslatable RNAs Plant-expressed antibodies
Dissemination	Interfere with vector transmission	Non-transmissible CP Defective helper component Plant-expressed antibodies

Adapted and expanded from Mayo, 1992.

'cross protection' by which a plant can be protected from a severe strain of a virus by pre-inoculation with a mild strain of the same virus (McKinney, 1929). Powell-Abel et al. (1986) were the first to demonstrate that transgenic tobacco expressing the coat protein (CP) gene of tobacco mosaic virus (TMV) was resistant to TMV challenge infection. This phenomenon is now commonly referred to as coat protein-mediated protection which is resistance caused by the expression of a virus CP gene in transgenic plants. Accumulation of the CP confers resistance to infection and/or disease development by the virus from which the CP gene was derived and by related viruses. CP-mediated protection has since been used to develop resistance to viruses in at least 12 different virus groups, including the tobamo-, potex-, cucumo-, ilar-, tobra-, poty-, carla-, nepo-, furo-, luteo-, tospo-, and alfalfa mosaic virus groups (see references; and, for reviews see Beachy et al., 1990; Gadani et al., 1990; Mayo, 1992).

Initial experiments were performed with tobacco generally because it is easily transformed by *Agrobacterium tumefaciens* and it is susceptible to numerous plant viruses. More recently, however, coat protein-mediated protection has been expanded to other commercially important crops such as tomato (Nelson et al., 1988), alfalfa (Hill et al., 1991), cucumber (Gonzalves et al., 1992), sugarbeet (Kallerhof et al., 1990), melons (Fang and Grumet, 1993), papaya (Fitch et al., 1992) and potato (Kawchuk et al., 1991; Lawson et al., 1990; MacKenzie et al., 1991).

Even though CP-mediated protection has been effective in protecting plants against various viruses, the mechanism(s) of this protection remain imperfectly understood. The degree of protection ranges from delay in symptom expression to absence of disease symptoms and virus accumulation. Generally, the effects of CP-mediated protection on virus multiplication is that there are fewer sites where infection occurs on inoculated leaves, that the systemic rate of spread is reduced, and that less virus accumulates in tissues that do become infected (Beachy et al, 1990; Mayo, 1990).

However, the question of how CP confers resistance remains unanswered. The nature of virus infection and disease development make it likely that resistance is effected at several stages (Beachy et al., 1990). Several lines of evidence suggest that CP-gene expression interferes with early events in infection such as virus uncoating (Register and Beachy, 1988), or even replication and expression of viral RNA (Osbourn et al., 1989). Additional evidence also suggests interference with cell-to-cell and/or systemic spread of virus as a possible mode of action (Beachy et al., 1990; Lindbo and Dougherty, 1992a), leading in some cases to total recovery (Hammond and Kamo, 1993a,b).

In many cases the level of CP-mediated protection is proportional to the accumulation of the transgenic CP in the plant, and transgenic plants that accumulate CP gene transcripts but not the CP itself are not resistant to infection. However, this is not always the case. Indeed, transgenic plants transformed with CP genes have been obtained that are resistant to challenge infection even though very little to no transgenic CP could be detected (Lawson et al., 1990; Kawchuck et al., 1991; Pang et al., 1992; van der Vlugt et al., 1992). Van der Vlugt et al. (1992) reported that a translatable construct of the potato virus Y (PVY) CP gene and part of the 3' untranslated region conferred strong protection in some lines, although expression of the protein could not be detected. Elimination of the CP initiation codon to yield an untranslated transcript resulted in similar levels of protection, suggesting that RNA rather than CP was the protective entity. This 'RNA-mediated' resistance could be due to positive sense RNA transcripts hybridizing to viral negative sense RNA replication intermediates, thereby blocking further virus replication, or, conversely, the transcripts may be competing with viral RNA for host or viral factors

involved in RNA replication (see discussion below).

Pang et al. (1992) demonstrated resistance to tomato spotted wilt (TSWV) and impatiens necrotic spot (INSV) tospoviruses in plants transformed with the nucleocapsid protein (NP) gene of TSWV. Resistance to heterologous isolates of TSWV was mainly found in plants accumulating very low, if any, levels of NP, whereas transgenic plants accumulating high levels of NP were resistant to the INSV isolate. Apparently different mechanisms exist that mediate these different resistance modes.

Lindbo and Dougherty (1992a) demonstrated that transgenic plants expressing truncated forms of the tobacco etch potyvirus (TEV) CP were more effective in CP-mediated resistance to TEV than transgenic plants that expressed full-length CP. They speculated that the truncated Cps are in some way dysfunctional and are more effective in disrupting the normal virus-host relationship than full-length CP.

Other Viral Gene Products

Inhibition of the synthesis or function of viral replicase genes is another potential strategy to control virus disease. This 'replicase-mediated resistance' strategy has been applied to at least four different plant viruses. Golemboski et al. (1990) and Braun and Hemenway (1992) observed strong resistance to TMV and potato virus X (PVX), respectively, in plants transformed with part or all of translatable viral replicase genes. However, in neither case could the protein product be detected, despite accumulation of the RNA transcript, suggesting that in this case an RNA-mediated mechanism was effective. MacFarlane and Davies (1992) were also unable to detect the protein product of a putative 54K protein from the 3' region of the pea early browning virus replicase in transformed plants, but did observe strong resistance to infection. However, when they mutated the construct to cause premature translation the protection was abolished, suggesting that the protection was conferred by the protein rather than RNA, even though the level of protein was below detection limits.

Anderson et al. (1992) transformed tobacco plants with a modified and truncated replicase gene from CMV. The in vitro translated truncated gene produced a translation product ≈75% as large as the full-length protein. The presence of this defective protein in transgenic plants was not able to be determined due to a lack of antisera to the protein. However, plants transformed with the truncated gene were resistant to CMV when the inoculum consisted of either virions or RNA. And as was the case for the TMV and PVX 54K replicase domains above, the resistance was absolute, as neither symptoms nor virus could be detected in uninoculated leaves, even after prolonged incubation. Also, these plants were resistant to inocula containing very high doses of virus and viral RNA, inocula that would have been sufficient to rapidly overcome resistance mediated by CP expression. These findings define a promising new approach for controlling plant viral infection.

Untranslatable and Antisense Viral genes

Untranslatable copies of genes can be created from translatable protein-expressing genes by introducing premature stop codons close to the viral initiation codon, by introducing frame-shifts leading to termination at a pre-existing stop codon, or by eliminating ribosome binding sites and initiation codons. As no viral-related protein product will result from expression of RNA transcripts of such constructs in plants, resistance presumably results from RNA:RNA interactions or by competing with or disrupting the activity of the functional gene. Antisense (AS) constructs (complementary viral RNA sequences) are also presumed not to direct protein synthesis, but to regulate viral replication through RNA:RNA interactions, perhaps at different steps in the replication cycle. In the case of negative-strand RNA viruses, an untranslatable construct may also be effectively AS, as it is complementary to the genomic RNA (de Haan et al., 1992).

A special case of RNAs that are not necessarily translated, but have a well documented ability to interfere in replication and modulate symptoms, is found with defective interfering (DI) RNAs. Kollár et al. (1993) introduced a cloned copy of a naturally occurring DI RNA of cymbidium ringspot tombusvirus into transgenic plants in both plus and minus orientations. The DI RNA was replicated and the only plants protected from severe symptoms were those expressing the DI RNA in the plus orientation. Satellite RNAs (which typically do not yield translation products) are discussed in a separate paper in these proceedings and are not considered here.

CP-mediated resistance has been shown to be effective with TMV, whether or not the construct has the TMV 3' untranslated region (UTR), but an untranslatable TMV CP gene did not confer any resistance (Powell et al., 1990). An AS construct of the TMV CP gene and 3' UTR conferred lower levels of resistance than observed in CP-expressing plants, and removal of the 3' UTR from the construct completely abolished resistance (Powell et al., 1989). In plants with AS constructs of cucumber mosaic cucumovirus (CMV; Cuozzo et al., 1988) and potato virus X potexvirus (Hemenway et al., 1988), CP + 3' UTRs were each found to confer levels of resistance lower than their respective CP-expressors. Rezaian et al. (1988) found that only one transformant of one out of three CMV AS constructs conferred any resistance to infection. This one was derived from the 5' region of RNA 1, whereas constructs derived from the 5' end of RNA 3, or the 3' end of RNA 2 (and including over 100nt common to each of the genomic RNAs) conferred no protection. However, each of these constructs lacked 30-70nt of the 5' or 3' UTR.

Kawchuk et al. (1991) showed that both CP-expressing and AS potato leafroll luteovirus constructs conferred similar levels of resistance to Russet Burbank potato, although the CP itself was not detectable over a 50-fold range of CP transcript expression. The levels of CP transcript were comparable to levels of AS transcript. They proposed that both positive and negative sense transcripts interfere with virus replication by interacting with the opposite sense strand of viral RNA, and that the resistance occurs early in infection when both strands are present at low levels.

De Haan et al. (1992) compared two translatable constructs with one untranslatable construct of the TSWV NP gene and found similar levels of resistance in each case; many individual plants escaped infection with the homologous TSWV isolate. No resistance was observed in NP-expressing lines to either of two TSWV isolates with about 80% NP-gene nucleotide sequence homology, further suggesting that the resistance is primarily at the RNA level. It may also be significant that TSWV has an ambisense genome, with the NP-gene transcribed from the virion-complementary strand.

Lindbo and Dougherty (1992a,b) reported that a tobacco etch potyvirus (TEV) CP gene AS construct was much more effective than the CP-expressing version, and an untranslatable derivative of the CP-expressor construct was even more effective, plants being apparently immune to TEV infection (these constructs all omit the TEV 3' UTR). They also showed that there was no heterologous protection, as expected if an RNA:RNA interaction is responsible, and suggested that RNA regions involved in RNA:protein binding are less likely to form stable RNA:RNA interactions, and thus may be less effective for resistance. By comparison, Hammond and Kamo (1993a,b) have observed highly effective protection with a bean yellow mosaic potyvirus (BYMV) AS construct that includes the viral 3' UTR, and with a similar CP-expressing construct. One AS line was apparently immune to BYMV, and others recovered from BYMV infection (Hammond and Kamo, 1993a,b).

Fang and Grumet (1993) showed that melon plants transformed with zucchini yellow mosaic potyvirus (ZYMV) CP AS RNA, including the 3' UTR, had a 3-9 day delay in symptom expression and milder symptoms than controls, but that plants expressing CP were more effectively protected, with either a much longer delay (>30 days) or no symptoms. Tobacco plants with ZYMV AS RNA showed a delay in symptom development with two heterologous

potyviruses, TEV and PVY (Fang and Grumet, 1993). The AS resistance to TEV might conceivably be due to the use of the TEV 5' UTR, but the 5' UTR of PVY differs from that of TEV at multiple positions except for repeats of two heptanucleotide sequences common to several potyviruses; whether these repeats are sufficient to form a strong RNA:RNA interaction between heterologous potyvirus RNAs is not clear.

Although here are some obvious differences in the potential interactions between the different types of constructs used by different groups, and differences in the replication strategies of different viruses, the basic resistance effect is interference in normal viral replication through altered regulation at the RNA level. Viral RNA transcripts may compete for either host or viral factors necessary for translation or replication - but competition for translation seems rather unlikely at least in those cases where undetectable levels of translated protein are present. AS RNA constructs like that of Lindbo and Dougherty (1992a,b) have the potential to interfere with CP translation as well as to disrupt replication by interfering with replicase procession in forming the minus strand template. Constructs including 3' UTRs have the additional potential for blocking initial interaction of the replicase complex with the plus strand. Lindbo and Dougherty (1992b) suggested that the 3' and 5' UTRs will not be effective due to the ability of protein:RNA interactions to destabilize RNA:RNA hybrids. This may not always be the case, as Hammond and Kamo (1993a,b) and Fang and Grumet (1993) have shown significant resistance to the homologous potyvirus with AS RNAs containing the 3' UTR. Fang and Grumet (1993) have also shown resistance with heterologous potyviruses, in this case the most significant homology is in two conserved 5' UTR heptanucleotide repeats that may be protein binding motifs. Powell et al. (1989) also showed that removing the TRNA-like 3' UTR from a TMV AS construct abolished any resistance, while Rezaian et al. (1988) did not observe effective resistance with CMV AS RNAs that lacked parts of the 5' or 3' UTRs.

An extension of this AS technology could

include the expression in transgenic plants of short AS oligonucleotides (Agrawal, 1992) specific to the 5', or 3' UTRs and/or to internal promoter binding sites.

Kawchuk et al. (1991) and others have suggested that RNA-mediated (antisense or untranslated RNA) protection will be most effective early in an infection when levels of both viral RNA strands are low. Hammond and Kamo (1993a,b), however, have observed full recovery in upper leaves of some plants expressing BYMV AS RNA that initially showed symptoms indistinguishable from controls. AS RNA may thus be capable of mediating resistance to replication or transport as well as to establishment of initial infection.

Untranslatable constructs may not have universal application, for while effective against TEV and PVY (potyviruses) and TSWV (ambisense tospovirus), no resistance was observed for TMV.

One confounding factor in the interpretation of differences between constructs is the extent of variability between transformants of a single construct. This is probably due in large part to position effects, the chromosomal context affecting the regulation of the promoter used; variability can range from complete susceptibility to apparent immunity with transformants from the same experiment. This does make some of the interpretations discussed above suspect - mainly in regard to negative data, although singular instances of resistance could be attributed to somaclonal variation. At present too few comparisons have been made with multiple transformants of different constructs in a single host/virus system to allow certainty in mechanistic interpretations.

Other Novel Approaches

Other potential, yet more novel, approaches in developing virus resistance in plants through genetic engineering using antiviral sequences and proteins include the transgenic expression of RNAs with catalytic antiviral ribozyme activity and of mouse antiviral antibodies in plants.

Ribozymes are RNA molecules that can act as antiviral enzymes and catalyze cleavage reactions in the absence of proteins (Haseloff and Gerlach, 1988). These catalytic RNAs have the potential to inhibit the expression of specific genes by cleavage of the corresponding (homologous) mRNA. This technology should be an additional extension of the AS strategies. Antiviral ribozymes have been described for at least three plant viruses, potato leafroll luteovirus CP and replicase (Lamb and Hay, 1990), TMV RNA (Edington and Nelson, 1992), and PVY replicase (van der Vlugt et al., 1993). The antiviral application of ribozymes in transgenic plants may have promise in that the TMV ribozyme has been shown to apparently inhibit virus replication both in protoplasts and transgenic plants. However, the reported protection levels have not yet exceeded those obtained with AS RNAs, which in fact were very low themselves (Edington and Nelson, 1992).

Hiatt et al (1989) demonstrated that plants transformed with antibody genes were able to produce serologically active antibodies and even suggested that this technology would be applicable to studying plant-pathogen interactions. At least five reports of 'plantibody' or 'phytoantibody' production in plants have since been published. In most cases the plant-produced antibody, when purified, was found to possess the antigen binding properties of the original monoclonal antibody (During et al., 1990; Hein et al., 1991; Hiatt and Ma, 1992); however, a plant-produced variable heavy domain antibody did not (Benvenuto et al., 1991). None of the plant-expressed antibodies, however, were specific to a plant protein, much less a plant pathogen. More recently however, a functional (in planta) anti-phytochrome single-chain antibody was successfully produced in transgenic tobacco (Owen et al., 1992).

The expression in plants of antibodies specific to viral gene products might be another effective way to inhibit or interfere with specific steps involved in virus replication. Anti-CP antibodies could interfere with virion disassembly or assembly, systemic movement within the plant, symptom expression, and vector dissemination of virions. Antibodies specific to viral proteins involved in replication or polyprotein processing (i.e., replicase and protease) could interfere or block those essential steps. One of the advantages of this type of 'anti-viral antibody-mediated protection' would be that no viral genes need be introduced into the environment as part of the transgenic plant.

Concluding remarks

The expression of viral genes in transgenic plants has provided new strategies for genetically engineering virus-resistant plants. Many viral CP genes, either as translatable sense, untranslatable sense or as antisense constructs, have been expressed in transgenic plants and shown to confer resistance. As a consequence, CP-mediated cross-protection has become an important technology directed at the production of virus-resistant crop plants. The mechanisms of CP-mediated cross-protection are complex and not yet understood, and it is apparent that even with a given virus, the CP may be involved in more than one role in inducing resistance. New exciting resistance strategies, involving expression of viral replicase genes, have provided resistance and are likely to be very important in the development of virus-resistant crops in the near future. Other exciting possibilities for transgenic resistance have been suggested and/or are in the early stages of development. The expression of antiviral ribozymes and functional antiviral antibodies in transgenic plants to potentially inactivate viral genes or gene products, respectively, are the given examples. Future studies on all of these approaches, and their mechanisms of action, should lead to more refined approaches for engineering virus resistance.

References

Agrawal, S. "Antisense oligonucleotides as antiviral agents." *Trends in BioTechnology*, **1992**, 10,152-158.

Anderson, J.M., Palukaitis, P., and Zaitlin, M. "A defective replicase gene induces resistance to cucumber mosaic virus in transgenic tobacco plants." *Proc. Nat'l. Acad. Sci.*, **1992**, 89,8759-8763.

Beachy, R.N., Loesch-Fries, S., and Tumer, N.E. "Coat protein-mediated resistance against virus infection." *Annu. Rev. Phytopath.*, **1990**, 28,451-74.

Benvenuto, E., Ordas, R.J., Tavazza, R., Ancora, G., Biocca, S., Cattaneo, A., and Galeffi, P. "'Phytoantibodies': a general vector for the expression of immunoglobulin domains in transgenic plants." *Plant Mol. Biol.*, **1991**, 17,865-874.

Braun, C.J., and Hemenway, C.L. "Expression of amino-terminal portions or full-length viral replicase genes in transgenic plants confers resistance to potato virus X infection." *The Plant Cell*, **1992**, 4,735-744.

Cuozzo, M., O'Connell, K.M., Kaniewski, W., Fang, R.X., Chua, N-H., and Tumer, N.E. "Viral protection in transgenic tobacco plants expressing the cucumber mosaic virus coat protein or its antisense RNA." *Bio/Technology,* **1988**, 6,549-557.

De Haan, P., Gielen, J.J.L., Prins, M., Wijkamp, I.G., van Schepen, A., Peters, D., van Grinsven, M.Q.J.M., and Goldbach, R. "Characterization of RNA-mediated resistance to tomato spotted wilt virus in transgenic tobacco plants." *Bio/Technology*, **1992**, 10,1133-1137.

During, K., Hippe, S., Kreuzaler, F., and Schell, J. "Synthesis and self-assembly of a functional monoclonal antibody in transgenic *Nicotiana tabacum*." *Plant Mol. Biol.*, **1990**, 15,281-293.

Edington, B.V. and Nelson, R.S. "Utilization of ribozymes in plants. Plant viral resistance." In *Gene Regulation: Biology of Antisense RNA and DNA*. Ericksen, R.P. and Izant, J.G., Eds; Raven Press Ltd., New York, 1992; pp209-221.

Fang, G., and Grumet, R. "Genetic engineering of potyvirus resistance using constructs derived from the zucchini yellow mosaic virus coat protein gene." *Mol. Plant-Microbe Interact.*, **1993**, (in press).

Fitch, M.M.M., Manshardt, R.M., Gonsalves, D., Slightom, J.L., and Sanford, J.C. "Virus resistant papaya plants derived from tissues bombarded with the coat protein gene of papaya ringspot virus." *Bio/Technology,* **1992,** 10,1466-1472.

Gadani, F., Mansky, L.M., Medici, R., Miller, W.A., and Hill, J.H. "Genetic engineering plants for virus resistance." *Arch. Virology,* **1990,** 115,1-21.

Golemboski, D.B., Lomonossoff, G.P., and Zaitlin, M. "Plants transformed with a tobacco mosaic virus non-structural gene sequence are resistant to the virus." *Proc. Natl. Acad. Sci. USA,* **1990,** 87,6311-6315.

Gonzalves, D., Chee, P., Provvidenti, R., Seem, R., and Slightom, J.L. "Comparison of coat protein-mediated and genetically-derived resistance in cucumbers to infection by cucumber mosaic virus under field conditions with natural challenge inoculations by vectors." *Bio/Technology,* **1992,** 10,1562-1570.

Hammond, J., and Kamo, K.K. "Transgenic coat protein and antisense RNA resistance to bean yellow mosaic potyvirus." *Acta Hort.,* **1993a,** (in press).

Hammond, J., and Kamo, K.K. "Resistance to bean yellow mosaic virus (BYMV) and other potyviruses in transgenic plants expressing BYMV antisense RNA, coat protein, or chimeric coat proteins." *Proceedings of the Fifth International Symposium on Biotechnology and Plant Protection,* Oct. 19-21, College Park, 1993b, (in press).

Haseloff, J. and Gerlach, WL. "Simple RNA enzymes with new and highly specific endoribonuclease activities." *Nature,* **1988,** 344,585-591.

Hein, M.B., Tang, Y., McLeod, D.A., Janda, K.D., and Hiatt, A. "Evaluation of immunoglobulins from plant cells." *Biotechnol. Progress,* **1991,** 7,455-461.

Hemenway, C., Fang, R-X., Kaniewski, W., Chua, N-H., and Tumer, N.E. "Analysis of the mechanism of protection in transgenic plants expressing the potato virus X coat protein or its antisense RNA." *EMBO J.,* **1988,** 7,1273-1280.

Hiatt, A., Cafferty, R., and Bowdish, K. "Production of antibodies in transgenic plants." *Nature,* **1989,** 342,76-78.

Hiatt, A. and Ma, K.-C. "Monoclonal antibody engineering in plants." *FEBS Letters,* **1992,** 307,71-75.

Hill. K.K., Jarvis-Egan, N., Halk, E.L., Krahn, K.J., Liao, L.W., Mathewson, R.S., Merlo, D.J., Nelson, S.E., Rashks, K.E., and Loesch-Fries, L.S. "The development of virus-resistant alfalfa, *Medicago sativa* L." *Bio/Technology,* **1991,** 9,373-377.

Kallerhof, J., Perez, P., Bouxoubaa, S., Ben Tahar, S., and Perre, T.J. "Beet necrotic yellow vein virus coat protein-mediated protection in sugarbeet (*Beta vulgaris* L.) protoplasts." *Plant Cell Reports,* **1990,** 9,224-228.

Kawchuk, L.M., Martin, R.R., and McPherson, J. "Sense and antisense RNA-mediated resistance to potato leafroll virus in Russet Burbank potato plants." *Mol. Plant-Microbe Interact.,* **1991,** 4,247-253.

Kollår, Å., Dalmay, T., and Burgyån, J. "Defective interfering RNA-mediated resistance against cymbidium ringspot tombusvirus in transgenic plants, *Virology,* **1993,** 193,313-318.

Lamb, J.W., and Hay, R.T. "Ribozymes that cleave potato leafroll virus RNA within the coat protein and polymerase genes." *J. Gen. Virol.,* **1990,** 71,2257-2264.

Lawson, C., Kaniewski, W., Haley, L., Rozman, R., Newell, C., Sanders, P., and Tumer, N.E. "Engineering resistance to mixed virus infection in a commercial potato cultivar: resistance to potato virus X and potato virus Y in transgenic russet burbank." *Bio/Technology,* **1990,** 8,127-134.

Lindbo, J.A., and Dougherty, W.G. "Pathogen-derived resistance to a potyvirus: Immune and resistant phenotypes in transgenic tobacco expressing altered forms of a potyvirus coat protein nucleotide sequence." *Mol. Plant-Microbe Interact.,* **1992a,** 5,144-153.

Lindbo, J.A., and Dougherty, W.G. "Untranslatable transcripts of the tobacco etch virus coat protein gene sequence can interfere with tobacco etch virus replication in transgenic plants and protoplasts." *Virology,* **1992b,** 189,725-733.

MacFarlane, S.A., and Davies, J.W. "Plants transformed with a region of the 201-kilodalton replicase gene from pea early browning virus RNA 1 are resistant to virus infection." *Proc. Natl. Acad. Sci. USA*, **1992**, 89,5829-5833.

MacKenzie, D.J., Tremaine, J.H., and McPherson, J. "Genetically engineered resistance to potato virus S in potato cultivar russet burbank." *Mol. Plant Microbe Interact.*, **1991**, 4,95-102.

Mayo, M.A. "Organization of viral genomes: the potential of virus genes in the production of transgenic virus-resistant plants." In *Biotechnology and crop improvement in Asia*; Moss, J.P., Ed.; ICRISAT: Patancheru, India, 1992; pp251-263.

McKinney, H.H. "Mosaic diseases in the Canary Islands, West Africa and Gibraltar." *J. Agric. Res.*, **1929**, 39,557-578.

Nelson, R.S., McCormick, S.M., Delannay, X., Dube, P., Layton, J., Anderson, E.J., Kaniewska, M., Prokch, R.K., Horsch, R.B., Rogers, S.G., Fraley, R.T., and Beachy, R.N. "Virus tolerance, plant growth, and field performance of transgenic tomato plants expressing coat protein from tobacco mosaic virus." *Bio/Technology*, **1988**, 6,403-409.

Osbourn, J.K., Watts, J.W., Beachy, R.N., and Wilson, M.A. "Evidence that nucleocapsid disassembly and a later step in virus replication are inhibited in transgenic tobacco protoplasts expressing TMV coat protein." *Virology*, **1989**, 172,370-373.

Owen, M., Gandecha, A., Cockburn, B., and Whitelam, G. "Synthesis of a functional anti-phytochrome single-chain F_v protein in transgenic tobacco." *Bio/Technology*, **1992**, 10,790-794.

Pang, S.-Z., Nagpala, P., Wang, M., Slightom, J.L., and Gonsalves, D. "Resistance to heterologous isolates of tomato spotted wilt virus in transgenic tobacco expressing its nucleocapsid protein gene." *Phytopathology*, **1992**, 82,1223-1229.

Powell-Abel, P.A., Nelson, R.S., De, B., Hoffman, N., Rogers, S.G., Fraley, R.T., and Beachy, R.N. "Delay of disease development in transgenic plants that express the tobacco mosaic coat protein gene." *Science,* **1986**, 232,738-743.

Powell, P.A., Stark, D.M., Sanders, P.R., and Beachy, R.N. "Protection against tobacco mosaic virus in transgenic plants that express tobacco mosaic virus antisense RNA." *Proc. Natl. Acad. Sci. USA*, **1989**, 86,6949-6952.

Powell, P.A., Sanders, P.R., Tumer, N., Fraley, R.T., and Beachy, R.N. "Protection against tobacco mosaic virus infection in transgenic plants requires accumulation of coat protein rather than coat protein RNA sequences." *Virology*, **1990**, 175,124-130.

Register, J.C. III and Beachy, R.N. "Resistance to TMV in transgenic plants results from interference with an early event in infection." *Virology*, **1988**, 166,524-532.

Rezaian, M.A., Skene, K.G.M., and Ellis, J.G. "Anti-sense RNAs of cucumber mosaic virus in transgenic plants assessed for control of the virus." *Plant Mol. Biol.*, **1988**, 11,463-471.

Van der Vlugt, R.A.A., Ruiter, R.K., and Goldbach, R. "Evidence for sense RNA-mediated protection to PVY[N] in tobacco plants transformed with the viral coat protein cistron." *Plant Mol. Biol.*, **1992**, 20,631-639.

Van der Vlugt, R.A.A., Prins, M, and Goldbach, R. "Complex formation determines the activity of ribozymes directed against potato virus Y[N] genomic RNA sequences." *Virus Research*, **199?** 27,185-200.

Recombinant Baculoviruses as Biological Insecticides

Bruce D. Hammock, Departments of Entomology & Environmental Toxicology, University of California, Davis, CA 95616

Baculoviruses are potent biological control agents which act with high specificity on many serious arthropod pests. However, extensive crop damage often occurs before the infected insect dies. This manuscript describes successful approaches to engineer viruses for rapid control of insect pests. One group of recombinant viruses speed the death of pest insects and reduce crop damage by expression of toxic peptides or proteins while another group of recombinant viruses act by expressing one of the pest insect's own proteins at an inappropriate time.

For the development of effective strategies of integrated pest management (IPM), it is essential that we have a repertoire of highly active and specific pest control agents. However, the current market and regulatory climate usually preclude the development of classical chemicals to meet these needs (Hammock and Soderlund 1986). Insect viruses and specifically the baculoviruses are important components of the background of natural control. They have been and are being used successfully in augmentative biological control (Entwistle and Evans 1985), however they have failed to compete with classical insecticides in most cases. There are several reasons why the baculoviruses are not yet competitive with the high standard of cost effectiveness set by classical insecticides. As discussed below, four of these limitations are being overcome, while the major limitation, speed of kill, is the topic of this chapter.

The increasing commercial interest in the baculoviruses is leading to research to over come their limitations. For instance, formulation and production of baculoviruses have been a problem in the past, but great strides are being made in both areas. The high specificity of the baculoviruses is very attractive in IPM, but their high specificity would preclude development of many of the viruses if they are viewed as classical chemicals. Some baculoviruses are of sufficient activity on key pest species to warrant even high registration costs. Hopefully regulatory agencies will consider the mechanism of action and safety record of these agents when designing safety tests

so that regulation will not be a barrier to the development of even the highly selective baculovirus species. The major limitation to the use of these viruses as biological insecticides is slow kill. Such slow kill is likely of benefit to the virus in a natural situation in that it leads to large larvae which produce large amounts of virus for subsequent infection. However, under most of the conditions when a biological pesticide would be used, the resulting crop damage would be unacceptable. Several laboratories have engineered viruses to overcome this problem, and we now have recombinant viruses whose speeds of kill approach those of classical chemical insecticides.

These engineered viruses seem to retain the light stability of wild type viruses and are orally active in several of the most serious pest species in the world. The engineered viruses show a 30% reduction in time to kill compared to the wild type virus and over a 75% reduction in food consumption (Fig 1). The proper use of synergists can lead to a further reduction in time to kill making the virus competitive with many classical insecticides. These engineered viruses show no deleterious effects on the nontarget organisms tested and greater selectivity for pest insects than either classical insecticides or the toxin of *Bacillus thuringiensis*.

Of the 7 families of insect viruses, the baculoviruses (family Baculoviridae) show the most short term promise as engineered bioinsecticides. There are 3 subgroups of baculoviruses. The non-occluded viruses

Fig. 1. Reduction in feeding damage by the cotton bollworm following treatment with a recombinant baculovirus. Each of the two tomato plants shown above were infected with 3 third stadium larvae of the major U.S. insect pest, *Heliothis virescens*. Larvae on the left plant were treated with a virus containing a gene for the toxin AahIT. Larvae on the right plant were treated with the wildtype virus.

(NOVs) include a virus successfully used for the control of the coconut rhinoceros beetle, but improved formulation will be critical if NOVs are to be used more generally. Many granulosis viruses (GVs) are exceptionally active and are being developed as biological control agents. However, to date GVs have proven difficult to handle in cell culture and to engineer. The GVs are likely to be the target of much more intense research in the future. The nuclear polyhedrosis viruses (NPVs) are receiving the most attention as both wild type and engineered biological insecticides. These double stranded DNA viruses are known only from the arthropods and are usually selective to a single genus of family of insects (Entwistle and Evans 1985). These viruses produce several proteins, notably the polyhedron protein which form a microencapsulation system for the viral particles containing the nucleocapsids. The resulting polyhedra are of greatly increased stability to light and other adverse conditions and are orally active in permissive insects. As discussed below the early baculovirus cloning systems replaced the polyhedron gene by a foreign gene which was transcribed in large amounts under the strong polyhedron protein promoter. However, the resulting recombinant viruses were not stable to adverse conditions and were of reduced oral activity.

Baculovirus Cloning Systems

Several baculoviruses are easily maintained in cell culture and are the basis of efficient systems for the production of foreign proteins (Vlak 1993; Kitts et al. 1990; Maeda 1989b; Miller 1988). Two excellent laboratory manuals have just appeared on this topic (King and Possee 1992; O'Reilly et al. 1992). Some eukaryotic proteins can be efficiently produced in prokaryotes, however many proteins are biologically inactive due to faulty folding, glycosylation, and other post translational modifications. Since baculoviruses infect eukaryotic cells, most plant and animal proteins expressed with the system have correct post translational modifications and are biologically active. Several promoter systems in baculoviruses are among the most powerful known. For these reasons the baculovirus expression systems may be the dominant expression systems for the production of eukaryotic proteins in the pharmaceutical industry for years to come. Recombinant proteins can be produced inexpensively at levels exceeding a gram per liter both *in vivo* and *in vitro*. This clearly represents a situation where the agricultural industry can benefit from research carried out in the pharmaceutical field.

The purpose of this chapter is not to

provide a guide to baculovirus cloning techniques, so only an overview will be presented. Simplistically, one uses the powerful promoter machinery of the virus coupled with either inexpensive insect cell culture or insects such as the silkworm, *Bombyx mori*, as production factories. Thus the initial goal is to insert the foreign gene of choice into the viral genome. For this work one uses one of many available transfer vectors. Most of these vectors are derivatives of the pUC family of plasmids. These plasmids will have a unique cloning or multicloning site allowing simple cloning into a region surrounded by sequences in common with the viral genome. Once the foreign gene is in the transfer vector it can be grown easily in *E. coli* like any other plasmid. In common procedures the resulting plasmid DNA is mixed with viral DNA and transfected on permissive insect cells. In the magical transfection process the DNA of the transfer vector containing foreign DNA is integrated into the viral genome. This transfection does not result in random incorporation of the foreign gene into the viral genome, but rather results in an integration where one knows exactly where the gene is incorporated. One then selects for the recombinant viruses based on one of several strategies.

Baculovirus expression is a field which is developing rapidly. Even a few years ago the selection of recombinants produced under the nonessential polyhedron promoter required patience and skill to search for the polyhedron negative viruses producing the desired recombinant protein. Now there are numerous systems of selection which lead to almost every group of virus infected cells (plaque) being the desired recombinant. Also a few years ago most expression work was done with the polyhedron promoter with the disadvantages discussed above. There are now systems which allow the recombinant virus to be enclosed in the polyhedron resulting in viruses that can be stored simply in a vial on the bench for years and which are biologically active under field conditions.

There are a variety of techniques for baculovirus cloning which are very efficient and simple. Several of these techniques are outlined in the two manuals introduced above. Several companies are selling kits, some of which are of very high quality. There are a variety of expression vectors available using different or multiple promoters, with or without blue-white selection, with or without M13 centers of replication, and other attributes (Bishop 1992). There are likely to be many new advances further simplifying baculovirus expression systems such as the treatment of the viral DNA as a large plasmid (Luckow, personal communication), but one series of advances will be outlined below as an example.

Rather than use the polyhedron protein for expression, two laboratories developed expression systems based on the promoter for the p10 protein. Vlak et al. (1990) developed a nice selection system based upon replacing the p10 protein gene with a recombinant gene while work from the laboratory of Possee led to vectors producing both the p10 and polyhedron proteins with the recombinant protein being produced under a p10 promoter (Weyer et al. 1990; Weyer and Possee 1991). By using a transfer vector which inserted the polyhedron protein and the recombinant protein back into a polyhedron negative virus, one could select for an easily visualized, stable virus against a background of an unstable virus. Kitts et al (1990) further advanced this technology by use of a unique restriction site to linearlize the host virus which then could be repaired when the recombinant transfer vector was added. This advance led to a transfection system so efficient that most plaques observed were recombinant plaques. Thus, the work of weeks was reduced to days.

An additional cloning step beyond many *E. coli* based systems is required for the use of most baculovirus expression systems, but this single cloning step is offset by the very high production of recombinant protein usually obtained. Baculovirus expression represents a simple system for producing proteins and peptides with possible biological activity on pest insects. In addition, the recombinant virus may be valuable itself in insect control as illustrated below.

Recombinant Baculoviruses with Improved Speed of Kill

Several approaches have been suggested for the use of recombinant DNA technology to improve crop protection (Hammock 1985; Kirschbaum 1985; Miller et al. 1983), and examples of most of these approaches have been published. The first success was reported by Maeda (1989a) who expressed a putative diuretic hormone gene from *M. sexta* in the baculovirus of the silkworm *Bombyx mori* (BmNPV). The infected insects were observed to die somewhat faster than those not expressing the neurohormone. In general this avenue of research has been disappointing. For example expression of eclosion hormone had little effect (Eldridge et al. 1991) and several other attempts remain unpublished. However the approach of expressing peptide neurohormones or other peptide chemical mediators holds great promise in the future if several problems can be overcome. First, expression of some small peptides in the baculovirus system is not as high as one could hope for based on expression of proteins. Thus, better expression of small molecules is needed and possibly improvements in post translational modification of some of these peptides. Second, the majority of neurohormones cloned to date should not give a physiological lesion expected to kill the insect. Thus additional structures are needed and a more intimate understanding of chemical mediation in insects is critical. Finally, expressed neurohormones will have to overcome natural homeostatic mechanisms and the insect's systems for the metabolism of these hormones. In retrospect we were short sighted to not anticipate that important peptide chemical mediators would probably be associated with highly active systems for the inactivation of these peptides. All of these problems can be overcome, but they all require additional research in practical aspects of insect neuroendocrinology.

The peptide field offers many potential leads which have not yet been explored. In addition to future neurohormones, there undoubtedly are numerous other peptide chemical mediators under various names such as local hormones or neurotransmitters. There may, in fact, be sufficient homology between vertebrate and insect peptides that vertebrate peptides expressed in the baculovirus system will have effects on insects. The expression of receptors or receptor fragments also could lead to biological effects. For instance, expression of a binding domain of a receptor could act as an antihormone

by effectively reducing the titer of the peptide. Finally, the juvenile hormone esterase example, discussed below, illustrates that there may be promise in the over-expressing of enzymes involved in the processing or degradative metabolism of peptide hormones.

A second approach is illustrated by elegant work by O'Reilly and Miller (1989) who noted that the NPV genome of the *Autographa californica* virus (AcNPV) coded for an ecdysteroid UDP-glucosyl transferase. This enzyme is expressed late in virus infection and in some insects will reduce ecdysone titers and keep the insect from molting. One can argue that such an event is to the benefit of the virus since it keeps the insect in the feeding stage rather than wasting energy on a molt just before death. By removing the EGT gene, insect feeding damage was reduced. Although the effects are not dramatic the EGT deletion has several practical implications. First, it illustrates enhanced speed of kill by gene deletion rather than addition. Second, EGT minus viruses can be isolated from natural populations so this effect can be bred into natural virus populations without recombinant DNA technology. Third, in a stepwise process of registration of recombinant viruses, the EGT minus viruses appear environmentally benign and could be an important tool in a stepwise program to gain acceptance of recombinant baculoviruses (Bishop et al. 1988; Wood et al. 1990). Finally, the EGT minus viruses like the p10 minus viruses illustrate that we probably can remove a variety of virus genes which may be needed for recycling of the virus in the field and obtain tailored viruses for such things as improved production of recombinant proteins in expression systems, improved virus production for field release, or viruses which will be less likely to recycle in the field and thus more attractive to regulatory agencies.

A third approach involves the insertion of a gene for an insect toxin into the recombinant baculovirus. Carbonell et al. (1988) failed to observe improved activity following the insertion of a scorpion neurotoxin into the AcNPV genome, illustrating that not every toxin gene will be successful. Several laboratories expressed the δ-endotoxin of *B. thuringiensis* in AcNPV with disappointing results (Merryweather et al. 1990). As candidate toxins are selected one should consider insect selectivity, but also the mechanism of toxicity. Toxins which are active on the gut but not internal tissues or toxins that kill the cells infected by the virus are not attractive targets for the system.

Several approaches involving the expression of toxin genes have been successful. A gene (tox34) for a protein toxin of the straw itch mite, *Pyemotes tritici,* termed TxP-I led to a dramatic improvement in the speed of kill in AcNPV (Tomalski and Miller 1991; Tomalski and Miller 1992). Two laboratories inserted genes for the highly selective insect toxin from the Algerian scorpion *Androctonous australis* Hector (AahIT) into BmNPV and AcNPV (Maeda et al. 1991; McCutchen et al. 1991; Stewart et al. 1991), and in both cases there was a dramatic increase in speed of kill. These studies are important in proving the concept and these viruses are of sufficient activity to be useful in the field. It is very exciting that such high activity could be obtained with a tiny research investment compared to that needed for a similar development with classical insecticide chemicals. However, these viruses should be viewed as leads which can be improved in a variety of ways (McCutchen and Hammock 1993). Some more general ways to improve recombinant viruses are outlined below, but an obvious approach is to use different toxins. TxP-I is not a particularly potent insect toxin. AahIT is very potent on some insects, but is of surprisingly poor activity on the lepidopterous larvae which were the targets of the above study (Herrmann et al. 1990). It is a credit to the concept that both the TxP-I and the AahIT viruses showed such dramatic effects, and exciting to observe that there are leads for many toxins which are far more potent. Zlotkin provides an overview of arthropod toxins affecting insects (1991; 1985) and there are a wealth of studies of toxins of arachnids in general (Jackson and Usherwood 1988) and scorpions specifically (Loret and Hammock 1993; Zlotkin et al. 1978). There are, of course many other potential sources of materials with greater biological activity, but many of the most promising substances are from members of the Hymenoptera (Piek 1991; Piek 1986). Certainly the use of a toxin of 10 fold greater potency will not translate into a virus with a 10 fold increase in speed of kill. However, there clearly are

benefits to be gained by the use of more potent toxins, and the pharmacokinetic parameters to be considered in their use have been discussed by Maeda and Hammock (1993).

A fourth approach involves the insertion of a gene for an enzyme into the baculovirus. Since the resulting enzyme is catalytic, this approach has the theoretical advantage of an amplification step. So far only one enzyme has shown a clear biological effect. The juvenile hormone esterase (JHE) of *H. virescens* when expressed in a polyhedron minus strain of AcNPV led to reduced feeding in early instars of the cabbage looper *Trichoplusia ni.* These effects were the ones anticipated from the biology of the JHE as discussed below, but the effects were not dramatic (Hammock et al. 1990). Eldridge et al (1992) tested the hypothesis that EGT expression could interfere with JHE activity, but failed to find positive results. Under many conditions, JHE is an exceptionally stable enzyme. However, Ichinose et al. (1992a) noted that it was rapidly removed following injection into the hemolymph. The enzyme was stable in hemolymph held *in vitro* in addition to numerous other tissues either homogenized or held *in vitro.* It was observed that natural and recombinant JHE whether injected or produced *in situ* by the virus accumulated in the pericardial cells (Booth et al. 1992; Ichinose et al. 1992b). The pericardial cells comprise a small thread like tissue that lies along the insect heart and is thought to act as a filter system for the blood. The data of Ichinose et al (1992a) strongly suggest that there is a specific uptake system for JHE based on active transport and that the enzyme is rapidly degraded once it enters the pericardial cells. Minor changes in the primary sequence of JHE led to viruses which would kill *H. virescens* and *T. ni* as rapidly as the viruses with the AahIT (Bonning and Hammock 1992; Bonning et al. 1993; Hammock et al. 1993). These data clearly demonstrate the potential of using enzymes or even enzyme systems to enhance the effectiveness of baculoviruses. They also indicate that investigation of the pharmacokinetics of the recombinant protein can result in materials of greatly improved efficacy.

There are other obvious approaches which remain to be tested. Possibly the most frustrating, but potentially exciting, possibility is serendipity. As more proteins are expressed in the baculovirus system, we may find proteins or peptides that are biologically active. We can hope that large pharmaceutical companies will develop a systematic process of screening viruses on insects. If the more modern high level expression systems are used which yield polyhedron positive viruses, such bioassays are easy.

Techniques to Improve Existing Recombinant Baculoviruses

At this time the recombinant AcNPVs expressing TxP-I, AahIT, and JHE are of sufficiently enhanced activity that it is attractive to use them in the field. However, if we put these discoveries in the context of classical chemical insecticides, we would be at the stage of a primary lead. One of course does not rush ahead with this primary lead in classical chemistry, rather one makes changes in the material to improve its efficacy and physical properties and reduce its cost. There clearly are differences between classical insecticides and the recombinant viruses with a major difference being that the viruses were developed with a fraction of the resources that normally go into classical insecticides. We really do not know how much more these viruses can be improved, however, it seems worth some investment in improving their activity to see where we reach a point of diminishing returns. Approaches to the improvement of recombinant viruses have been outlined in several articles (Bonning and Hammock 1993; Hammock et al. 1993; Maeda and Hammock 1993; McCutchen and Hammock 1993) and an overview is presented below.

The first obvious improvement involves the use of the recombinant gene in other viruses. AcNPV has numerous advantages as a model virus and even as a pest control agent. However, one of the advantages of the recombinant virus approach is that there are many viruses available of varying specificity. Thus, one could make recombinant viruses to target individual pest species such as the gypsy moth, spruce budworm, diamondback moth, or pink bollworm. One also could use viruses which attack a range of pests common to a crop situation or even prepare a cocktail of recombinant viruses designed to control those insects on a given crop in a single region such as for lepidopterous pests of cotton. Advances in this area will depend upon the development of systems which allow one to transfer engineered genes tested in AcNPV rapidly into other viruses as well as the regulatory climate. If extensive testing is required for each new recombinant virus even when the recombinant protein has been exhaustively tested previously, we will destroy the economic incentive to develop selective recombinant viruses just as we destroyed the economic incentive for the development of selective synthetic insecticides in a series (Hammock and Soderlund 1986).

There are numerous approaches for improvement which can apply generally to a variety of recombinant viruses and in some cases wild type viruses. For instance improved formulation systems to bring the virus in better contact with the pest, to stabilize the virus against UV light and other environmental factors, or even synergize its activity all are attractive. Both chemical and biological factors to enhance infectivity are very attractive. If modern formulations can be developed which will allow the use of budded virus (polyhedron minus virus) in the field rather than the virus enclosed in the polyhedron, one can produce orders of magnitude more infective particles.

As mentioned earlier improved *in vivo* and *in vitro* production methods will make both wild type and recombinant viruses more attractive. With dramatic improvements in the techniques of plant molecular biology, it may also be possible to engineer the crop plants to be more compatible with the use of the recombinant virus. Improved production is especially important with the viruses leading to rapid mortality since *in vivo* production will be decreased. More efficient production can be accomplished by *in vitro* production, the use of antidotes to the active agent, the use of insensitive stages or species for virus production, or molecular switches to turn off the recombinant protein during the production cycle.

There are other approaches which would have to be engineered for each virus, but where the concept is general. Clearly techniques to improve virus replication and spread among tissues is advantageous. It is unlikely that we will

318

have recombinant proteins of sufficient activity to kill an insect without at least some replication of the virus in the pest insect. Possibly the removal of genes which are not essential for the action of a viral insecticide will speed replication in the host, improve production, as well as reduce the competition of the engineered virus in the field (Hammock 1992). Enhanced transcription already has been demonstrated by the use of alternate promoters (Bonning et al. 1993) and hybrid promoters (Tomalski and Miller 1992). Enhanced translation can be accomplished in many ways varying from improved codon usage to message stabilization.

There are systematic ways to improve the recombinant protein and peptide as well, however most of these methods will be unique to the peptide in question. One always can hope that one can improve the activity of the peptide with the target site. At two extremes techniques to carry out such improvements could be development of systems for the rapid screening of large numbers of mutants of the target peptide on insects or receptor systems to select materials of improved properties. With the advances in design of numerous random or systematic mutants and improved panning systems, such large scale screening may be attractive. At the other end of the spectrum one could employ a rational approach of structure optimization relying heavily on the modeling of the target receptor and the recombinant protein.

Major improvements in classical chemicals often are based on changes not in the binding to the target molecule, but changes which influence distribution and stability. The success with the JHE attests to the value of such an approach. Such approaches will justify the study of the pharmacokinetics and dynamics of proteins and peptides in the future just as analogous studies on small molecules have proved advantageous in the past. Classical insecticide synergists such as tri-o-cresol phosphate and piperonyl butoxide are successful because they block the metabolism of the target molecule. Inhibitors of the degradation of peptides and proteins may similarly prove successful. Such inhibitors of metabolism could be either small molecules or a protein or peptide produced in the host plant or by multiple expression in a baculovirus.

One can envision other synergists based either upon interaction at the target site or a physiological approach. Certainly with neurotoxins, we often observe a cocktail of toxins in a single venom. One can envision either a series of toxins acting synergistically at a single receptor or the effects of a receptor agonist or antagonist being amplified at a physiological level.

In several insect species the reduction of juvenile hormone leading to cessation of feeding behavior and an initiation of a sequence of events leading to metamorphosis is regulated in part by the production of JHE. On a physiological level, there are a variety of avenues where one could anticipate synergistic interaction among JHE and either other peptides and proteins or small molecules. If commercially attractive anti-juvenile hormone molecules are developed in the future, they clearly should be synergistic with the virus (Staal 1986). Regulation of the JH titer undoubtedly involves regulation of receptor density, hormone biosynthesis and other factors. If peptides are found that regulate these processes, they should be very active synergists of JHE. Juvenile hormone metabolism is known to occur by several pathways including both ester cleavage and epoxide hydration. In theory epoxide hydrolases also should synergize the activity of JHE (Bonning et al. 1993; Grant et al. 1993; Beetham et al. 1993).

The above arguments illustrate that like with classical insecticides, an understanding of the mechanism of action of the agent can lead to systematic ways to improve its efficacy. Of course there is an added complexity or opportunity in that one can improve the recombinant gene, the virus, or even the combination of the two.

Role of Recombinant Baculoviruses in Agriculture

The effectiveness of even the first generation of recombinant baculoviruses in reducing feeding damage of some of our most serious crop pests is beyond question. These viruses offer the promise of reducing deleterious effects of synthetic pesticides on nontarget organisms and the ecosystem, reducing food residues on crop plants, improved insect pest management and resistance management, and improved profitability in agriculture. Whether these agents will have a role in agriculture will depend upon many factors not the least of which is that the opportunities for insect pest control constantly are changing. Perceived acceptance by the public, by regulatory agencies, and by food processors will be a major factor in the development of the field. The art of risk assessment commonly is used to evaluate the relative benefits and limitations of classical insecticides, and papers relevant to this topic recently have been reviewed (McCutchen and Hammock 1993). However, even the use of the term risk assessment is damning since it implies that significant risk is associated with the use of recombinant baculoviruses.

All information to date supports the observations that the natural and recombinant viruses are safe to man and to the ecosystem and promise to circumvent many problems associated with classical pesticides (Cory and Entwistle 1990; Summers et al. 1975). The technology encompasses some words that often are viewed as alarmist, so some attention must be given to interaction with regulatory agencies and the public so that critical evaluations of the safety of the viruses are carried out without alarming society. Possibly as more viruses and recombinant techniques are used in human medicine their application in agriculture will be perceived with less alarm. Clearly it will be important to stress the advantages this technology offers human and environmental health in addition to improved agricultural profitability.

It is critical to view the limitations of these viruses as well. First, classical insecticides have set a very high standard of efficacy, profitability, ease of use and safety. It is likely that the recombinant viruses will have to meet or exceed these standards if they are to be adopted. Although farming methods are evolving, one cannot expect farmers to change their basic practices immediately to accommodate the recombinant viruses. Instead the viruses will have to be integrated into existing cropping practice. A scenario where classical insecticides are banned by public referendum following accidental spillage or bad publicity is easy to imagine. Under such circumstances, the recombinant viruses could represent one of our few alternatives for control of our major pests.

Possibly the most serious limitation will be that of the very narrow spectrum of activity of the viruses. Commonly there are a variety of pests that the farmer will need to control. If classical insecticides must be applied to control aphids and beetles, for instance, there is no need to use the virus if the same insecticides will control lepidopterous pests as well. There are several ways to counter this argument. One way is to develop viruses of increased host range. Although there is some promising research on the molecular basis of specificity and improved host range will be important, we probably will see the improved viruses infecting only a few additional species. There are several reasons for this limitation. One reason is that as a virus becomes more of a generalist in its acceptance of hosts, its potency is likely to drop. A second reason is that for early releases regulatory agencies probably will look favorably on narrow host ranges. Finally, the host range needed in some agricultural situations is too wide for any known virus and even if a broad spectrum virus could be made, we would then lose the primary attraction of the recombinant viruses in their selectivity.

The specificity of the virus is one of its major attributes. AcNPV for instance has a sufficiently wide host range that there are many cropping situations where every serious pest can be controlled by this one virus. Many of our pest problems in fact come from the use of broad spectrum materials which release other pests from control by their natural enemies (McCutchen and Hammock 1993). The greatest value of the recombinant viruses will be as precision instruments to control specific pest species without entering the so called pesticide tread mill. However, it should be possible to extend the host range by proper selection of the host virus or a combination of viruses. Alternatively, these biological insecticides probably can be tank mixed with many classical compounds. This process will facilitate their use, possibly lead to synergism, and may allow environmentally softer classical compounds to be used if the lepidopterous pests can be controlled with the recombinant viruses.

Anytime that a new pesticide is found,

there are claims that it will be resistance proof. The development of resistance against a strong selection pressure is so certain with rapidly reproducing species like many insects, that resistance can be considered a badge of success for an insect control product. It certainly is conceivable that pests will be able to evolve resistance to recombinant viruses. The rapid evolution of resistance to pyrethroid insecticides can be attributed, in part, to previous selection with DDT which affects the same site. Insect pests were exposed to pathogenic viruses long before humans had evolved and coined the term pest. As biologists we know there must be effective traits of resistance, or we would find neither the insect nor its virus. However, there are many arguments that can be made that the recombinant viruses will be recalcitrant if not immune to resistance.

First, their selectivity and instability in the environment suggest that vast populations of insects will not be inadvertently exposed to the virus. Second, the wild type virus has been in an evolutionary struggle with its host. As discussed above, this struggle cautions the possibility for resistance to the wild type virus. However, since both virus and host are dynamic, one can anticipate that this struggle will continue. There are without a doubt genes in pest populations capable of leading to some resistance to the wild type viruses which can be selected for if the use of the viruses as biological insecticides increases. A variety of resistance mechanisms can be envisioned varying from behavioral changes to resistance to infection. However, all infected insects will die and the recombinant genes at this point only hasten death. Thus one would not expect selection pressure against the recombinant virus at a higher rate than the wild type virus. With a favorable regulatory climate, one could over come resistance by placing the recombinant gene in a new viral isolate. Depending upon the recombinant gene used, one could anticipate finding a totally naive pest population, one with some degree of cross resistance due to prior treatment with pesticides or exposure to plant natural products, or even negatively correlated cross resistance.

The resistance phenomenon with viruses is likely to be the subject of great discussion but offer little if any limitations to the use of either wild type or recombinant viruses. However, resistance existing now in the field offers a great opportunity with the recombinant viruses. Since most mechanisms of cross resistance from classical pesticides are not relevant, the viruses are likely to be as active on resistant as susceptible insects. Thus, the viruses could be used in a variety of scenarios to over come existing resistance problems and preclude future problems (Glass 1984).

Enhanced speed of kill clearly is not critical for all situations. For instance enhanced speed of kill of the gypsy moth probably is not important in many forest management situations if the outbreak can be caught early. However, at late stages of defoliation or in urban situations, rapid kill of the same pest would be very valuable. The existing recombinant viruses which lead to roughly a 30% increase in speed of kill and up to 80% reduction in food consumption would be very valuable for the control of many foliage feeders. However, faster speed of kill is likely to be important for insects like the cotton bollworm or pink bollworm on cotton since the damage to the boll could occur before the larva is killed. By integration of several techniques we have increased speed of kill to under 30 hours with recombinant viruses which is comparable with many commercial insecticides (Betana and McCutchen, unpublished). It is hard to predict if we have reached a point of diminishing returns with regard to virus improvement, but farmers and IPM specialists will likely favor viruses with even faster kill.

Possibly the advantages of the recombinant viruses will be best realized in developing countries. In developed countries we can afford sophisticated IPM systems to control insects with a minimum of pesticides. We also can afford very expensive insecticides and alternative control methods. We even can afford no pest control with the resulting increase in food prices adversely affecting only the poorest members of our society. Many developing countries cannot afford modern pesticides, cannot afford the investment needed for complex IPM programs, and can ill afford crop loss. However, the disaster at Bhopal in the production of carbaryl illustrates some of the dangers involved with the production of synthetic pesticides on site (Lepkowski 1992). In contrast, baculoviruses can be produced *in vivo* in developing countries better than in developed countries due to reduced labor costs. This can be accomplished by a variety of ways beginning with simply infecting pest insects grown in a local area to a cottage industry. Innocula for the recombinant virus can easily be mailed to local industries.

How the recombinant viruses will compete with or complement the rapidly evolving technology surrounding the toxins of *B. thuringiensis* remains to be seen. The cost effectiveness of BT certainly will be a standard against which the viruses must be compared. In some cases the spectrum of activity of BT is complementary to that of the viruses. There are situations such as a forest ecosystem, where the high specificity of some baculoviruses will present a distinct advantage over BT's which will kill both target and endangered lepidopterous pests. The viruses can also be considered as resistance management tools in an IPM program incorporating BT produced either by fermentation or by plants.

How the recombinant viruses will interact with plant systems expressing BT and other insect toxins remains to be seen. At one extreme we can envision every crop plant expressing such high levels of BT that no heterozygotes survive and no resistance develops. At the other extreme one can envision such rapid resistance that the recombinant plants are rendered useless in a few seasons. Reality probably is somewhere between where the recombinant viruses are useful in the control of insects on crops where cloning systems do not exist or where the value of the crop is not sufficient to warrant development of recombinant lines. Certainly the driving force in seed selection will be for traits other than resistance to insect damage. It may not be possible to develop varieties with other desirable properties and insect resistance fast enough to meet competition. In these cases the recombinant viruses also have a clear role. As with classical insecticides, developers of the recombinant viruses must keep in mind the changing pest problems, changing agronomic practices, changing regulations, and changing public perception.

In government, industrial and academic laboratories serious thought is being given to the best way to introduce recombinant baculoviruses into agriculture. We must not only demonstrate efficacy and profitability with current farming practice but gain public and regulatory acceptance. Certainly the most important component for acceptance will be a series of rigorous tests of the safety of these agents. However, we are at a decisive point. If regulations are too strict the technology effectively will be killed in developed nations. We can prohibit investigators from providing recombinant organisms to developing nations. However, developed nations do not have exclusive access to the techniques of DNA manipulation. Baculovirus expression is a very simple technique which can be accomplished with no industrial base and very little equipment. Apparently, large scale field studies in biotechnology which would be considered premature in most Western nations are proceeding with few restrictions in China (Coghlan 1993). With many recombinant materials we may be faced with the option of either testing them under carefully controlled conditions in developed countries or finding that recombinant materials of uncertain origin appear in the field with out proper testing in developing countries. Since there are fewer potential barriers to recombinant organisms than to classical chemical residues, recalcitrance to evaluate recombinant technology in developed nations may lead to its use with out testing in developing countries.

Acknowledgements

This work was supported in part by the National Science Foundation DCB-91-19332, the U.S. Department of Agriculture 91-37302-6185, the Binational Agricultural Research and Development Fund IS-2139-92, and the U.S. Forest Service 23-696.

References

Beetham, J.K.; Tian, T.; Hammock, B.D. "cDNA cloning and expression of a soluble epoxide hydrolase from human liver." *Arch. Biochem. Biophys.* **1993,** (in press).

Bishop, D.H.L. "Baculovirus expression vectors." *Seminars in Virology* **1992,** *3,* 253-264.

Bishop, D.H.L.; Entwistle, P.F.; Cameron, I.R.; Allen, I.R.; Possee, R.D. In *The Release of*

Genetically-Engineered Microorganisms; Sussman, M.; Collins, C.H.; Skinner, F.A.; Stewart-Tull, D.E., Eds.; Academic Press: New York, NY, 1988, pp 143-179.

Bonning, B.C.; Hammock, B.D. "Development and potential of genetically engineered viral insecticides." *Biotechnology and Genetic Engineering Reviews* **1992**, *10*, 453-487.

Bonning, B.C.; Hammock, B.D. In *Natural and Derived Pest Management Agents*; Hedin, P.; Menn, J.J.; Hollingworth, R., Eds.; American Chemical Society Symposium Series: Washington, D.C., 1993, (in press).

Bonning, B.C.; Roelvink, P.W.; Vlak, J.M.; Possee, R.D.; Hammock, B.D. "Comparison of expression characteristics of various promoters in the *Autographa californica* nuclear polyhedrosis virus using juvenile hormone esterase as a reporter enzyme." *J. Gen. Virol.* **1993**, (in press).

Booth, T.F.; Bonning, B.C.; Hammock, B.D. "Localization of juvenile hormone esterase during development in normal and in recombinant baculovirus-infected larvae of the moth *Trichoplusia ni*." *Tissue and Cell* **1992**, *24*, 267-282.

Carbonell, L.F.; Hodge, M.R.; Tomalski, M.D.; Miller, L.K. "Synthesis of a gene coding for an insect specific scorpion neurotoxin and attempts to express it using baculovirus vectors." *Gene* **1988**, *73*, 409-418.

Coghlan, A. "China's New Cultural Revolution." *New Scientist* **1993**, *137*, 3-4.

Cory, J.S.; Entwistle, P.F. "Assessing the risk of releasing genetically manipulated baculoviruses." *Aspects of Applied Biology* **1990**, *24*, 187-194.

Eldridge, R.; Horodyski, F.M.; Morton, D.B.; O'Reilly, D.R.; Truman, J.W.; Riddiford, L.M.; Miller, L.K. "Expression of an eclosion hormone gene in insect cells using baculovirus vectors." *Insect Biochem.* **1991**, *21*, 341-351.

Eldridge, R.; O'Reilly, D.R.; Hammock, B.D.; Miller, L.K. "Insecticidal properties of genetically engineered baculoviruses expressing an insect juvenile hormone esterase gene." *Appl. Environ. Microbiol.* **1992**, *58*, 1583-1591.

Entwistle, P.L.; Evans, H.F. In *Comprehensive Insect Physiology, Biochemistry and Pharmacology*; Gilbert, L.I.; Kerkut, G.A., Eds.; Pergamon Press: Oxford, UK, 1985, Vol. 12; pp 347-412.

Glass, E.H. *Pesticide Resistance Strategies and Tactics for Management*; National Academy Press: Washington D. C., 1986.

Grant, D.F.; Storms, D.H.; Hammock, B.D. "Molecular cloning and expression of murine liver soluble epoxide hydrolase." *J. Biol. Chem.* **1993**, (in press).

Hammock, B.D. In *Comprehensive Insect Physiology, Biochemistry, and Pharmacology*; Kerkut, G.A.; Gilbert, L.I., Eds.; Pergamon Press: New York, NY, 1985; pp 431-472.

Hammock, B.D. "Virus release evaluation." *Nature* **1992**, *355*, 119.

Hammock, B.D.; Bonning, B.; Possee, R.D.; Hanzlik, T.N.; Maeda, S. "Expression and effects of the juvenile hormone esterase in a baculovirus vector." *Nature* **1990**, *344*, 458-461.

Hammock, B.D.; McCutchen, B.F.; Beetham, J.; Choudary, P.; Fowler, E.; Ichinose, R.; Ward, V.K.; Vickers, J.; Bonning, B.C.; Harshman, L.G.; Grant, D.; Uematsu, T.; Maeda, S. "Development of recombinant viral insecticides by expression of an insect specific toxin and insect specific enzyme in nuclear polyhedrosis viruses." *Arch. Biochem. Physiol.* **1993**, *22*, 315-344.

Hammock, B.D.; Soderlund, D.M. In *Pesticide Resistance: Strategies and Tactics for Management*; Glass, E., Ed.; National Academy Press: Washington D.C., 1986, pp 111-129.

Herrmann, R.; Fishman, L.; Zlotkin, E. "The tolerance of lepidopterous larvae to an insect-selective neurotoxin." *Insect Biochem.* **1990**, *20*, 625-637.

Ichinose, R.; Kamita, S.G.; Maeda, S.; Hammock, B.D. "Pharmacokinetic studies of the recombinant juvenile hormone esterase in *Manduca sexta*." *Pesticide Biochem. Physiol.* **1992a**, *42*, 13-23.

Ichinose, R.; Nakamura, A.; Yamoto, T.; Booth, T.F.; Maeda, S.; Hammock, B.D. "Uptake of juvenile hormone esterase by pericardial cells of *Manduca sexta*." *Insect Biochem. Molec. Biol.* **1992b**, *22*, 893-904.

Jackson, H.; Usherwood, P.N.R. "Spider toxins

as tools for dissecting elements of excitatory amino acid transmission." *Trends in Neuro Sciences* **1988,** *11,* 278-283.

King, L.A.; Possee, R.D. *The Baculovirus Expression System;* Chapman and Hall: London, UK, 1992; pp 1-229.

Kirschbaum, J.B. "Potential implication of genetic engineering and other biotechnologies to insect control." *Ann. Rev. Entomol.* **1985,** *30,* 51-70.

Kitts, P.A.; Ayres, M.D.; Possee, R.D. "Linearization of baculovirus DNA enhances the recovery of recombinant virus expression vectors." *Nucleic Acids Res.* **1990,** *18,* 5667-5672.

Lepkowski, W. "Union Carbide-Bhopal saga continues as criminal proceedings begin in India." *Chem. Eng. News* **1992,** *70(11),* 7-14.

Loret, E.P.; Hammock, B.D. In *Scorpion Biology and Research*; Brownell, P.H.; Polis, G., Eds.; Oxford University Press: Oxford, U.K., 1993, (in press).

Maeda, S. "Increased insecticidal effect by a recombinant baculovirus carrying a synthetic diuretic hormone gene." *Biochem. Biophys. Res. Commun.* **1989a,** *165,* 1177-1183.

Maeda, S. "Expression of foreign genes in insects using baculovirus vectors." *Ann. Rev. Entomol.* **1989b,** *34,* p 351-372.

Maeda, S.; Hammock, B.D. In *Pest Control with Enhanced Environmental Safety*; Duke, S.O.; Menn, J.J.; Plimmer, J.R., Eds.; American Chemical Society Symposium Series 524: Washington, D.C., 1993, pp 281-297.

Maeda, S.; Volrath, S.L.; Hanzlik, T.N.; Harper, S.A.; Maddox, D.W.; Hammock, B.D.; Fowler, E. "Insecticidal effects of an insect-specific neurotoxin expressed by a recombinant baculovirus." *Virology* **1991,** *184,* 777-780.

McCutchen, B.F.; Choudary, P.V.; Crenshaw, R.; Maddox, D.; Kamita, S.G.; Palekar, N.; Volrath, S.; Fowler, E.; Hammock, B.D.; Maeda, S. "Development of a recombinant baculovirus expressing an insect-selective neurotoxin: Potential for pest control." *Bio/Technology* **1991,** *9,* 848-852.

McCutchen, B.F.; Hammock, B.D. In *Natural and Derived Pest Management Agents*;

Hedin, P.; Menn, J.J.; Hollingworth, R., Eds.; American Chemical Society Symposium Series. Washington, D. C., 1993, (in press).

Merryweather, A.T.; Weyer, U.; Harris, M.P.G.; Hirst, M.; Booth, T.; Possee, R.D. "Construction of genetically engineered baculovirus insecticides containing the *Bacillus thuringiensis* ssp. *kurstaki* HD-73 delta endotoxin." *J. Gen. Virol.* **1990,** *71,* 1535-1544.

Miller, L.K. "Baculoviruses as gene expression vectors." *Ann. Rev. Microbiol.* **1988,** *42,* 177-199.

Miller, L.K.; Lingg, A.J.; Bulla, L.A. "Bacterial, viral, and fungal insecticides." *Science* **1983,** *219,* 715-721.

O'Reilly, D.R.; Miller, L.K. "A baculovirus blocks insect molting by producing ecdysteroid UDP-Glucosyl transferase." *Science* **1989,** *245,* 1110-1112.

O'Reilly, D.R.; Miller, L.K.; Luckow, V.A. *Baculovirus Expression Vectors: A Laboratory Manual,* W. H. Freeman and Company: New York, 1992; pp 1-347

Piek, T. In *Pesticide Chemistry*; Frehse, H., Ed.; VCH: Weinheim, 1991, pp 75-85.

Piek, T. *Venoms of the Hymenoptera: Biochemical, Pharmacological and Behavioural Aspects,* Academic Press: Orlando, FL, 1986; pp 1-570.

Staal, G.B. "Anti juvenile hormone agents." *Ann. Rev. Entomol.* **1986,** *31,* 391-429.

Stewart, L.M.D.; Hirst, M.; Ferber, M.L.; Merryweather, A.T.; Cayley, P.J.; Possee, R.D. "Construction of an improved baculovirus insecticide containing an insect-specific toxin gene." *Nature* **1991,** *352,* 85-88.

Summers, M.; Engler, R.; Falcon, L.A.; Vail, P., Eds.; *Baculoviruses for Insect Pest Control: Safety Considerations,* American Society for Microbiology: Washington, DC,1975.

Tomalski, M.D.; Miller, L.K. "Expression of a paralytic neurotoxin gene to improve insect baculoviruses as biopesticides." *Bio/Technology* **1992,** *10,* 545-549.

Tomalski, M.D.; Miller, L.K. "Insect paralysis by baculovirus-mediated expression of a mite neurotoxin gene." *Nature* **1991,** *352,* 82-85.

Vlak, J.M. In *Molecular Approaches to*

Fundamental and Applied Entomology; Oakeshott, J.; Whitten, M.J., Eds.; Springer-Verlag: New York, NY, 1993, pp 90-127.

Vlak, J.M.; Schouten, A.; Usmany, M.; Belsham, G.J.; Klinge-Roode, E.C.; Maule, A.; Lent, J.W.M.; Zuidema, D. "Expression of a cauliflower mosaic virus gene I using a baculovirus vector based on the p10 gene and a novel selection method." *Virology* **1990**, *178*, 312-320.

Weyer, U.; Knight, S.; Possee, R.D. "Analysis of very late gene expression by *Autographa californica* nuclear polyhedrosis virus and the further development of multiple expression vectors." *J. Gen. Virol.* **1990**, *71*, 1525-1534.

Weyer, U.; Possee, R.D. "A baculovirus dual expression vector derived from the *Autographa californica* nuclear polyhedrosis virus polyhedrin and p10 promoters: co-expression of two influenza virus genes in insect cells." *J. Gen. Virol.* **1991**, *72*, 2967-2974.

Wood, H.A.; Hughes, P.R.; van Beek, N.; Hamblin, M. In *Insect Neurochemistry and Neurophysiology*; Borkovec, A.B.; Masler, E.P., Ed.; Humana: Clifton, NJ, 1990, pp 285-288.

Zlotkin, E. "Venon neurotoxins - models for selective insecticides." *Phytoparasitica* **1991**, *19*, 177-182.

Zlotkin, E. In *Comprehensive Insect Physiology*; Kerkut, G.A.; Gilbert, L.I., Eds.; Pergamon Press: Oxford, UK, 1985, pp 499-546.

Zlotkin, E.; Miranda, F.; Rochat, H. In *Arthropod Venoms*; Bettini, S., Ed.; Springer-Verlag: New York/Berlin, 1978, pp 317-369.

Genetics of BT Insecticidal Crystal Proteins and Strategies for the Construction of Improved Strains

Bruce C. Carlton, Ecogen Inc., Langhorne, PA 19047

Bacillus thuringiensis (BT) is a common soil bacterium that has insecticidal properties due to its propensity to harbor extrachromosomal plasmids that carry genes termed *cry* genes that encode a diverse array of proteins. These proteins form a variety of crystalline inclusion bodies in sporulating BT cells. In recent years many such *cry* genes have been cloned and sequenced, and expressed in a variety of bacterial hosts as well as in transgenic plants. This paper reviews the diversity of BT *cry* genes and their insecticidal crystal proteins (ICP's), and describes various genetic approaches for constructing improved strains of BT, using both natural plasmid-mediated conjugation and recombinant techniques.

Over the past ten years there has been a major renewal of interest in the development of safer and environmentally-compatible pest control products. The principle thrust has been with biologically-based materials including microbials, plant-derived compounds, and beneficial organisms such as parasites, predators, and other natural enemies of pests.

This discussion will focus on the most commercially successful of these agents, namely the insecticidal bacterium *Bacillus thuringiensis* (or BT as it is commonly referred to). BT-derived products currently account for over 90 per cent of all biological pest control products, but still only about one per cent of all annual pesticide sales.

Thus, while the most successful of the alternatives to synthetic chemical pesticides, BT products have not made major incursions into the insect control markets since their initial introduction over 25 years ago. There are several historic reasons for this slow acceptance, including inconsistent performance, non-competitive price, misconceptions about the speed of insecticidal activity, and last but not least, a fundamental lack of knowledge of the genetic and physiological basis for how BT activities are expressed and how they act on their target insects. The major positive aspect of the commercial history of BT products is their recognized safety to non-target organisms, including humans, and their environmentally-benign nature. This paper will summarize how one company, Ecogen Inc., has developed ways to improve upon the BT strains that originally launched this bioinsecticide business, and provide some insights into future prospects for new products based on this natural bacterial system.

Why Are Improved Products Needed

For reasons cited above, there has traditionally been little reason to select a BT product over a chemical insecticide. The situation is different in 1993, however. The public, and even the grower/user, is demanding safer pest control products, but both of these factions still want the highest quality food and fiber products at the lowest prices. This is difficult to achieve, not only with chemical pesticides but with biologically-based products as well. What, therefore, can be done to achieve these goals? At Ecogen we have been working for several years to develop BT-based products that overcome the shortcomings of earlier products. Basic research studies on the genetics and mode of action of BT have provided a much better understanding of the diversity and genetic complexity of the insecticidal crystal proteins (ICP's) that are responsible for the bioactivity of BT. New analytical methods have provided more accurate ways to assess the levels of these proteins in both experimental materials and in products for sale. Our scientists have isolated thousands of strains of BT and new ICP genes that have higher levels of potency against important target insect pests than strains and genes contained in earlier commercial products. Through molecular biology we are learning how to control the expression of ICP genes to improve productivity of the insecticidal proteins.

Bacillus thuringiensis as a Bioinsecticide

BT has been recognized as an insect pathogen since the early part of the 20th century, although the differentiation between the roles of spores and crystalline inclusions in insecticidal activity is fairly recent (reviewed by Fast, 1981). The natural diversity of BT strains has been well-documented with respect to their flagellar antigen patterns (deBarjac and Frachon, 1990) and their ICP's (Krywienczyk and Angus, 1967; Yamamoto, 1983). These studies have led to the realization that many BT strains in nature possess a complexity of ICP activities, which undoubtedly explains the diversity of insecticidal activities observed, even between strains of the same flagellar serotype (Dulmage and Cooperators, 1981).

It is now recognized that several events are involved in the BT insecticidal response. These include the ingestion of the crystalline inclusion by a susceptible insect, solubilization of the crystalline proteins in the insect mid-gut, and, at least for many ICP's, a proteolytic processing to generate the active "toxin" (Luthy and Ebersold, 1981).

A key step in the mode of insecticidal action is the binding of the active toxin to receptor sites on the gut epithelium. Much interest has recently been generated in understanding this processs, in large part due to its implication in the possible development of insect resistance to BT proteins. Although a detailed discussion of this topic is beyond the scope of this review, a few points are worth summarizing. First, it has been shown that there exist both specific, strong binding sites for certain ICP's on the insect brush border membranes, as well as relatively non-specific, weak binding sites. Second, binding studies with two different ICP's for two Lepidoptera species showed a good correlation between binding affinity and *in vivo* insect toxicity. Third, competition experiments with different ICP's have shown that highly-related ICP's generally compete well (i.e, bind to the same sites) while less-related ICP's compete poorly if at all (Hofmann *et al*, 1988a; Hofmann *et al*, 1988b). This situation is reflected in the binding data obtained using preparations derived from *Plodia interpunctella* selected in the laboratory to be resistant to a certain BT strain. There was a reduction in binding of the ICP associated with the resistance, and an actual increase in binding affinity for a different ICP (Van Rie *et al*, 1990). In a *Plutella xylostella* field-derived population presumed to have developed resistance to a BT product, a major reduction in binding capability to one ICP (CryIAb) was observed, with no change in binding for two other ICP's (CryIB and CryIC) (Ferre *et al*, 1991). A laboratory-derived resistant colony of *Heliothis virescens* showed what appeared to be altered binding specificities for two different ICP's accompanied by compensating alterations in binding affinity and number of binding sites, suggesting that a complexity of mechanisms can be involved in the development of resistance (MacIntosh *et al*, 1991). Clearly, there is much more to be learned regarding the nature and function(s) of receptors and their mode of action in ICP binding.

The final stage in the BT toxin mode of action is the actual disintegration of the gut epithelium. This presumably results from the physical insertion of the toxin into the brush border membrane structure, creating "pores" in the integrity of the cell membrane (Knowles and Ellar, 1987) that allow leakage of ions and water into the epithelial cells and their ultimate lysis.

In summary, the mode of action of the BT toxin proteins is multifaceted. There are several aspects of this action that can be perturbed by events controlled by either the BT bacterium or the target insect. Due to the concern regarding the potential for insects to develop resistance to the BT toxins, it is therefore very important to fully understand the interactions between the toxin proteins and the insect midgut.

Strategies for Improvements to BT-based Bioinsecticides

There have been a number of different approaches to the development of "improved" BT-based bioinsecticides. These approaches have ranged from those directed at BT itself, such as the isolation of new native strains having improved or novel activities, to the expression of BT insecticidal crystal protein genes in an array of alternative delivery systems, such as plant endophytes, plant epiphytes, and the plant genome itself.

Although each of these approaches has its own merits and disadvantages, this presentation will focus on the improvement of the BT organism itself as the host strain for genetic improvement. There are several reasons for this rationale. First, BT is already a very efficient producer of ICP's. It has been shown that up to 20 to 30 per cent of the total protein complement of sporulating BT cultures is composed of insecticidal crystal protein (Lecadet and Dedonder, 1971). Second, BT is already well-known as being able to stably maintain several ICP genes. In view of the recent concerns about insects developing resistance to BT-based products, it would be an obvious advantage to utilize a host expression system that is capable of supporting the expression of multiple genes having different insecticidal activities. Third, from a registration point of view we believe that a recombinant construct that ultimately contains only BT DNA, with no foreign gene sequences, should evoke considerably less regulatory agency concern than heterologous constructs that incorporate genes from one bacterium into a different bacterial species that normally would not be able to receive the gene, or a transgenic plant that contains inserted bacterial genes.

Based on these rationales, Ecogen embarked upon a program of BT strain improvement that incorporated these principles. In addition, we built into our strain improvemnt strategy a two-tiered regulatory approach that utilized less-controversial, non-recombinant genetic manipulations in the construction of "second-generation" improved strains, while at the same time developing the cloning vectors and expression systems that would facilitate the construction and evaluation of strains generated using gene-splicing or recombinant DNA approaches as required to make the ultimate desired improvements.

In order to utilize these genetic improvement approaches it was necessary to understand three major features of the BT system. First was the genetic localization and inheritance patterns for the ICP genes. Second was the number and types of genes that might be present in any given strain of BT. Studies in several laboratories provided evidence that the ICP genes were localized, for the most part, on extrachromosomal "plasmid" elements (see Carlton and Gonzalez, 1985a and 1985b, for reviews). Subsequently, the use of cloned ICP genes or gene fragments in DNA hybridization experiments with these plasmids allowed the direct demonstration that ICP genes were located on specific plasmids, and that

many strains contained not one but two or three genes of a particular type (Kronstad, Schnepf, and Whiteley, 1983). Third, was the discovery of a a genetic transfer system that could promote the exchange of these plasmids between strains, as had been shown for other bacterial systems. In 1982, results reported by Gonzalez et al revealed the existence of a conjugation-like transfer system in BT that promoted the high-frequency exchange of certain plasmids between strains (Gonzalez, Brown, and Carlton, 1982). Subsequent studies showed that this exchange could be prevented by subjecting the parental strains to conditions that did not allow direct cell-to-cell contact, and that the process is bi-directional in the sense that both parental strains can act as either donor or recipient of plasmids in the exchange process (Chapman and Carlton, 1985). From these three approaches a general picture has emerged for the genetic organization of ICP's that can be summarized as shown in Figure 1, namely that BT strains typically harbor multiple extrachromosomal plasmids, some of which contain ICP genes (sometimes more than one) and some of which are capable of being transferred by conjugation.

The plasmid localization of ICP functions has important implications with respect to the opportunities for genetic improvement of BT, as well as to the genetic stability of these functions from a production viewpoint. First, it is quite clear that the genetic information contained on the plasmids, including the ICP functions, is non-essential for survival of BT, since variants can be isolated that have been cured of (i.e., lost) any or all of the plasmids. This non-essentiality of plasmids facilitates various types of genetic manipulations with BT. Second, the fact that certain of the plasmids in a given strain of BT can be spontaneously cured provides a means to isolate variants that are altered with respect to their ICP properties. For example if a certain strain undergoes curing of an ICP-encoding plasmid that specifies an activity that is relatively poor against a particular target insect, then the strain can express higher levels of the remaining, more active ICP's thus leading to a greater level of potency. Further, by losing a less-active ICP-plasmid the opportunity exists to replace it with a more-active ICP-plasmid, by the natural conjugal transfer system described above. Thus, by utilizing a combination of these natural genetic processes it is possible to generate strains that contain more potent ICP combinations

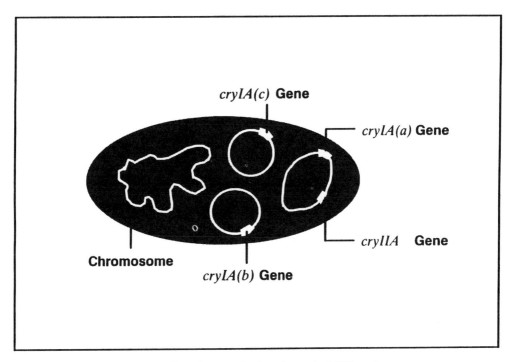

Fig. 1. Genetic complexity of a typical BT strain

or those having broadened insecticidal activity spectra.

The cloning and isolation of the first ICP gene was reported in 1981 (Schnepf and Whiteley, 1981). Over an 8-year period the list of characterized ICP genes had grown to 14, and in 1989 Hofte and Whiteley proposed a gene/protein nomenclature system that had become sorely needed to bring about a semblance of order in the naming of an ever-increasing array of ICP genes. This system groups the ICP genes and proteins into families based on two major criteria; insecticidal activity spectra and ICP size and sequence homology relationships. Although this system has definite limitations that do not fully take into account subtle differences in structure of nearly identical genes and the problems associated with access to information of what designations have already been assigned to new genes, we nevertheless have a reasonably workable system to which new ICP genes and proteins can be assigned (Table I). Currently there are five different ICP groups that have been established for approximately 24 ICP genes and proteins that differ significantly in either insecticidal activity spectra or sequence. CryI represents the largest array, with ten representatives, including the CryIA subgroup of three members that share greater than 80 per cent sequence homology. CryII has three representatives, including one (CryIIA) that is bifunctional in activity on lepidopteran targets as well as at least one dipteran species. CryIII currently has six members, two of which (CryIIIB2 and CryIIIB3) are 99% identical, although having quite different activities on target coleopteran insects (Donovan *et al*, 1992b). CryIV has four members, two of which are highly homologous with one another, and quite distinct from CryIVC and CryIVD. In addition, this group contains a fifth element referred to as CytA. This gene encodes a 28kd protein which, by itself, has little or no insecticidal activity, but which in combination with the complement of CryIV genes appears to contribute to a synergistic effect on dipteran insecticidal activity (Federici, 1993). Finally, three reports have appeared that describe Cry genes and proteins that fit none of the previously described categories. These represent genes encoding protein products of 81kd and that have low homologies with the other families (closest resemblance to CryIB, with 62% identity). One of them has been shown to be bifunctionally active against both lepidopteran insects targets and coleopteran targets. Based upon the

329

TABLE 1

CLASSIFICATION OF BT INSECTICIDAL CRYSTAL PROTEINS

ICP Gene	ICP Host-Range	ICP Size (kd)	Reference
cryIA(a)	L	133	Schnepf, Wong, and Whiteley (1985)
cryIA(b)	L	131	Wabiko, Raymond, and Bulla (1986)
cryIA(c)	L	133	Adang et al. (1985)
cryIB	L/C*	138	Brizzard and Whiteley (1988)
cryIC	L	135	Honee, van der Salm, and Visser (1988)
cryIC(b)	L	134	Bosse et al. (1991)
cryID	L	133	Hofte and Whiteley (1989)
cryIE	L	132	Visser et al. (1990)
cryIF	L	133	Chambers et al. (1991)
cryIG	L	130	Smulevitch et al. (1991)
cryIIA	L/D	71	Donovan et al. (1988a)
cryIIB	L	71	Widner and Whiteley (1989)
cryIIC	L	70	Wu et al. (1991)
cryIIIA	C	73	Herrnstadt et al. (1987)
cryIIIB	C	74	Sick et al. (1989)
cryIIIB2	C	74	Donovan et al. (1992a)
cryIIIB3	C	70	Donovan et al. (1992b)
cryIIIC	C*	129	Lambert et al. (1992a)
cryIIID	C	73	Lambert et al. (1992b)
cryIVA	D	134	Ward and Ellar (1987)
cryIVB	D	128	Chungjatupornchai et al. (1988)
cryIVC	D	78	Thorne et al. (1986)
cryIVD	D	72	Donovan et al. (1988b)
cryV	L/C	81	Tailor et al. (1992)
cytA	D/cytd.**	28	Waalwijck et al. (1985)

*   Activity detected only following proteolytic activation.
**   Marginal insecticidal activity.

difference in ICP size, the bifunctional activity, and the low level of sequence homology with other ICP genes, these genes have been assigned the designation cryV (Tailor et al, 1992; Gleave et al, 1993). The relationships between these genes and proteins to a truncated gene of very similar structure isolated from a strain of subsp. aizawai (Chambers et al, 1991) remains to be established at this writing.

Does this array of ICP's represent the full scope of BT activities? Undoubtedly not! Based upon the recent novel activities that have been discovered, it seems reasonable to expect that the coming months and years will reveal additional novel activities directed against new insect orders. In fact, companies such as Mycogen have reportedly claimed new isolates with activities against ants (Hymenoptera), (Payne *et al,* 1992) and non-insect targets such as spider mites and plant and animal-parasitic nematodes (Feitelson *et al,* 1992). The properties and full characterization of these strains remain to be described.

With this brief description of the genetic diversity of BT ICP activities, let us now move on to a consideration of various approaches to exploit these activities in new strain constructs. What is the liklihood that major increases in ICP potency will be possible, either by discovery of novel, more potent genes or by mutational alterations of existing ICP genes? The answer to this question has to be very positive, based on several experimental observations. First, are data from a number of laboratories showing that different ICP's have widely-varying activities against different target insects. These data have been obtained by separating the crystalline ICP's from spores by physical means. This is an important point because it has been demonstrated that certain insects (beet armyworm, as a prime example) are much more sensitive to ICP's in the presence of the BT spores than in their absence (Moar *et al,* 1989). Bioassays conducted with a number of different purified ICP's, either in their native state (MacIntosh *et al,* 1990) or in their activated state (van Frankenhuyzen *et al,* 1991) have shown that different insect targets can vary by a factor of more than 1,000-fold in their inherent susceptibility to different ICP's. Scientists at Ecogen have obtained similar findings. In a survey of the bioassay activity spectra of seven different lepidopteran-active ICP's and four different coleopteran-active ICP's (Jany *et al,* unpublished) it was observed that insecticidal potency differences of well over 1000-fold were observed between the most active and the least active ICP's on a broad array of target insects. These results lead to the conclusion that, for any given strain of BT, the genetic composition of its ICP activities plays a major role in the strain's potency against any pest insect target. There are two additional complicating factors in predicting the potencies of strains that contain several genes for ICP activities. First is that one needs to know what the relative expression efficiencies of each gene is (i.e., how many molecules of each ICP are produced in the cell). This is difficult to measure, even with

today's highly sophisticated techniques, particularly for those ICP's that are of the same size such that they cannot be separated by the SDS-PAGE quantitative assay described by Brussock and Currier (1990). A second complicating factor in predicting the potencies of multi-genic strains is the possibility for synergistic interactions between different ICP's, either postive or negative, with regard to insecticidal activity. In summary, there are a number of factors that need to be considered when developing a strategy for improving the potency of BT strains. Although the available evidence with respect to diversity of ICP genes and their bioactivities would suggest that the construction of new strains containing novel ICP gene combinations should generate significant improvements in potency, this approach is not necessarily straightforward and highly predictive, as discussed above.

The approach that was chosen at Ecogen Inc. in 1984 was to exploit the capabilities of the BT bacterium itself, by identifying new and novel ICP activities, by developing techniques for moving these activities from one strain of BT to the other, to optimize the BT organism for the expression of these activities, and to develop improved formulations and delivery methodologies so as to maximize the translation of increased laboratory potency to the field. A development flow scheme was devised that would integrate all phases of the work, including discovery, strain construction, laboratory, greenhouse and field evaluation, and process development and formulations scale-up. It is important to note that, particularly for the field evaluation and process development stages, these activities can be conducted in parallel so as to compress the development program as much as possible.

Since the program's inception, Ecogen has developed, registered with the Environmental Protection Agency, and introduced commercially three new BT-based bioinsecticide products, namely Cutlass®, Condor®, and Foil®. Each of these products is based on a unique, patented strain of BT that has resulted from a genetic improvement program that involved the use of the non-recombinant plasmid manipulation methods to construct new combinations of ICP's for specific target insect applications. Condor is directed at lepidopteran insects of row crops, particularly the budworm\bollworm complex on cotton and soybean loopers. Recently a granular formulation of Condor was registered for the control of European corn borer on corn. Cutlass was developed to have superior performance on insects such as beet armyworms and other caterpillar pests of vegetable crops as well as a broad array of lepidopteran pests on grapes and tree crops (e.g., almonds, walnuts, peaches, apples, etc.). Foil is the most unique of the three products in that it is the only BT-based product on the market that was constructed to have bifunctional activity against both lepidopteran insects and coleopteran insects. It is of interest to note that, according to flagellar serotype analysis, all three of these strains belong to serovar H3a,3b; subsp. *kurstaki* according to the classification system of deBarjac, despite their widely different insecticidal activity spectra. This points out an important feature of BT strains that has been widely misunderstood in the past, and that is that **there is no direct relationship between the subspecies of a strain, as measured by flagellar serotyping, and its insecticidal activity as determined by specific ICP genes.** Thus, to state that work has been conducted on "subspecies kurstaki", for example, is a meaningless statement with respect to a defined strain or insecticidal activity.

Although the approaches described above have been important in the development of new BT-based products (and, in fact, Condor, Cutlass, and Foil represent the first genetically-modified BT products registered in the U.S.) there are certain limitations to the ICP gene combinations that can be generated by plasmid curing and conjugal transfer. In addition, some ICP genes, such as cryIIB (Dankocsik *et al*, 1990) and cryV (Tailor *et al*, 1992) are apparently cryptic (non-expressed) in their native host strains and can only be expressed by fusing them to functional promoters. For these reasons we have devised a strategy for expanding the combinations of ICP genes that can be produced in BT through the use of recombinant DNA techniques. This strategy utilizes BT as the host expression system, for the same reasons that the BT bacterium itself was chosen as the host system for non-recombinant improvement approaches. In choosing the source of cloning vector materials we selected native BT plasmids, modified to utilize only those segments necessary for their maintainance in BT during cell growth and division. Several such *ori* elements were initially derived (Baum *et al*, 1990), from which three unique representatives were chosen for further development. These elements were shown to be widely distributed among many different subspecies of BT strains in nature. The *ori* elements were then used to build a series of cloning vectors which incorporated "cassettes" of additional elements, including an *E. coli* replicon and reporter gene (*lac z*), a gram-positive antibiotic determinant (chloramphenicol resistance), and a synthetic DNA segment containing a number of different restriction endonuclease cleavage sites into which genes to be cloned can be inserted. The cassette design of the vectors is such that either the *E. coli* elements, or the antibiotic resistance gene, or both, can be readily removed by digestion with an appropriate restriction enzyme (Figure 2). This design allows the cloning of a *cry* gene into the multi-cloning site, propagation of the recombinant vector in either *E. coli* or BT, and excision of all the foreign DNA sequences to generate a plasmid containing only BT gene sequences. Thus, this system provides for flexibility in the early-stage characterization of new constructs and for the assessment of stability of the chimeric plasmid in the presence or absence of selection, while ultimately allowing the removal of all foreign sequences so that the strain to be commercialized is technically no different than any naturally-derived strain of BT. Furthermore, it has been shown that the vector plasmids containing *ori* segments from normally highly-transferable plasmids have much reduced transfer capability (10^{-6} to 10^{-7} of the parental strain).

Using this vector system scientists at Ecogen have constructed a number of strains that have either increased potencies of the ICP complement, are more productive in fermentation, or have other improved properties. One of the earliest of these constructs was directed at the improvement of the bifunctional Foil bioinsecticide. As shown in Table II, this strain (EG2424) contains three ICP genes, two cryIA(c) genes and one cryIIIA gene. All three of these genes are located on native plasmids, originally derived from three different parental strains. Since the primary target for this product is Colorado potato beetle, the objective of the improvement program was to increase the amount of the beetle-active protein relative to the caterpillar-active protein. A new strain was constructed in which the original EG2424 strain was transformed with a chimeric plasmid containing the cryIIIB2 gene referred to previously (Donovan *et al*,

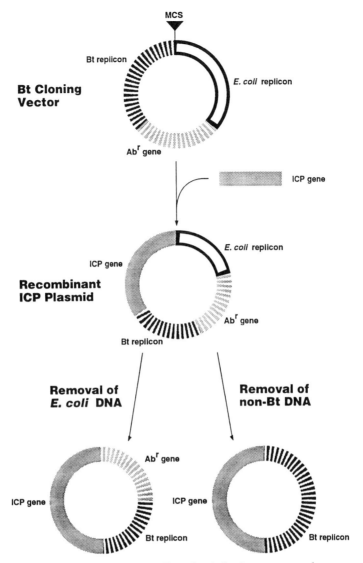

Fig. 2. Structure of a BT-derived cloning vector and removal of non BT sequences from ICP constructs

1992b). As a consequence of introducing this plasmid into the EG2424 background, the native 60 md plasmid having the same incompatibility functions (as well as a cryIA(c) gene) was deleted. The result was a strain (EG7618) that now had a single cryIA(c) gene and two cryIII genes; the original cryIIIA gene on its native 88 md plasmid and a cryIIIB2 gene on the recombinant vector plasmid. This construct proved to be not only extremely stable during fermentation, but also produced significantly more CryIII protein than its parental strain EG2424. The recombinant

strain retains the chloramphenicol-resistance marker of the original vector, which was used to conduct an environmental fate study in 1991 that showed that EG7618 had no greater persistence in the soil than its EG2424 parent (Baum and Rutkowski, unpublished data).

Field trials were conducted with this strain in both 1991 (two sites) and 1992 (four sites). EG7618 provided superior performance in both larval reduction and decreased defoliation, as related to the amount of ICP protein applied per acre, in

TABLE 2

CONSTRUCTION OF BT EG7618

Strain	ICP Genes/Location	ICPs	Insecticidal Activity
EG2424	cryIA(c)/44 Md	CryIA(c)	ECB
(Foil)	cryIA(c)/60 Md	CryIA(c)	ECB
	cryIIIA/88 Md	CryIIIA	CPB
EG7618	cryIA(c)/44 Md	CryIA(c)	ECB
	cryIIIA/88 Md	CryIIIA	CPB
	cryIIIB2/pEG894	CryIIIB2	CPB/SCRW

comparison to original Foil or competitive products such as Novodor and M-Trak. Subsequently a derivative strain has been produced that has had the chloramphenicol-resistance marker removed, and which is now an all-BT strain of commercial potential. This construct is currently being field-tested in seven states in the U.S.

This same BT-based vector technology is also being utilized to construct new strains for other applications, including vegetable crops, tree fruits, vineyards, and row crops. While we have not yet brought to commercial status a new product using this technology, we believe that the approach used in this improvement program will be acceptable to the Environmental Protection Agency and other regulatory bodies, and that the advantages to be gained in commercializing these products will far outweigh any perceived negatives that may focus on the trivial fact that recombinant techniques have been used in their construction.

If our expectations are met it will be another clear demonstration of the value of biotechnology to the world we all live in and share.

Acknowledgements

I acknowledge the work of the research and development staff at Ecogen Inc., whose contributions have helped to make this presentation possible. This paper is dedicated to the memory of the late Howard T. Dulmage, whose contributions to the field of BT are immeasurable. In particular I wish to acknowledge the early discussions and collaborations with Dr. Dulmage that provided many of the bases for my own work in this field.

References

Adang, M. J.; Staver, M.J.; Rocheleau, T. A.; Leighton, J.; Barker, R. F.; Thompson, D. V. "Characterized Full-length and Truncated Plasmid Clones of the Crystal Protein of *Bacillus thuringiensis* subsp. *kurstaki* HD-73 and Their Toxicity to *Manduca sexta*." *Gene*, **1985**, *36*, pp. 289-300.

Baum, J. A.; Coyle, D. M.; Gilbert, M. P.; Jany, C. S.; Gawron-Burke, C. "Novel Cloning Vectors for *Bacillus thuringiensis*."

Appl. Environ. Microbiol., **1990**, *56*, pp. 3420-3428.

Bossé, M.; Masson L.; Brousseau, R. "Nucleotide Sequence of a Novel Crystal Protein Gene Isolated from *Bacillus thuringiensis* subspecies *kenyae*." *Nucleic Acids Res.*, **1990**, *18*, p. 7443.

Brizzard, B. L.; Whiteley, H. R. " Nucleotide Sequence of an Additional Crystal Protein Gene Cloned from *Bacillus thuringiensis* subsp. *thuringiensis*." *Nucleic Acids Res.*, **1988**, *16*, pp. 2723-2724.

Brussock, S. M.; Currier, T. C. In *Analytical Chemistry of Bacillus thuringiensis*; Hickle, L. A. and Fitch, W. L., Eds.; ACS Symposium Series 432; **1990**, pp. 78-87.

Carlton, B. C.; González, J. M., Jr. In *The Molecular Biology of the Bacilli*; Dubnau D. A., Ed.; Academic Press: New York, NY, **1985a**; pp. 211-249.

Carlton, B. C.; González, J. M., Jr. In *Molecular Biology of Microbial Differentiation*; Hoch, J. A. and Setlow P., Eds.; American Society for Microbiology: Washington, DC, **1985b**; pp. 246-252.

Chambers, J. A.; Jelen, A.; Gilbert, M. P.; Jany, C. S.; Johnson, T. B.; Gawron-Burke, C. "Isolation and Characterization of a Novel Insecticidal Crystal Protein Gene from *Bacillus thuringiensis* subsp. *aizawai*." *Journal of Bacteriol.*, **1991**, *173*, pp. 3966-3976.

Chapman, J. S.; Carlton, B. C. In *Plasmids in Bacteria*; Helinski, D. R.; Cohen, S. N.; Clewell, D. B.; Jackson, D. A.; Hollaender, A., Eds.; Plenum Publishing Corp., **1985**, pp. 453-467.

Chungjatupornchai, W.; Höfte, H.; Seurinck, J.; Angsuthanasombat, C.; Vaeck, M. "Common Features of *Bacillus thuringiensis* Toxins Specific for Diptera and Lepidoptera." *Eur. J. Biochem*, **1988**, *173*, pp. 9-16.

Dankocsik, C.; Donovan, W. P.; Jany, C. S. "Activation of a Cryptic Crystal Protein Gene of *Bacillus thuringiensis* subspecies *kurstaki* by Gene Fusion and Determination of the Crystal Protein Insecticidal Specificity." *Mol. Microbiol.*, **1990**, *4*, pp. 2087-2094.

de Barjac H.; Frachon E. "Classification of *Bacillus thuringiensis* strains." *Entomophaga*, **1990**, *35*, p. 233-240.

Donovan, W. P.; Rupar, M. J.; Slaney, A. S.; Malvar, T.; Gawron-Burke, M. C.; Johnson, T. B. "Characterization of Two Genes Encoding *Bacillus thuringiensis* Insecticidal Crystal Proteins Toxic to *Coleoptera* Species." *Appl. Environ. Microbiol.*, **1992a**, *58*, pp. 3921-3927.

Donovan, W. P.; Dankocsik, C. C.; Gilbert, M. P.; Gawron-Burke, M. C.; Groat, R. G.; Carlton, B. C. "Amino Acid Sequence and Entomocidal Activity of the P2 Crystal Protein." *J. Biol. Chem.*, **1988a**, *263*, pp. 561-567.

Donovan, W. P.; Dankocsik; Gilbert, M. P. "Molecular Characterization of a Gene Encoding a 72-Kilodalton Mosquito- Toxic Crystal Protein from *Bacillus thuringiensis* subsp. *israelensis*." *J. Bacteriol.*, **1988b**, *170*, pp. 4732-4738.

Donovan, W. P.; Rupar, M. J.; Slaney, A. S. "*Bacillus thuringiensis cryIIIC(b)* Toxin Gene and Protein Toxic to Coleopteran Insects." International Patent Application #WO 9213954, August 20, **1992b**.

Dulmage, H. T.; Cooperators In *Microbial Control of Pests and Plant Diseases 1970-1980*; Burges, H. D., Ed.; Academic Press, London, **1981**, pp. 193-222.

Fast, P. G. In *Microbial Control of Pests and Plant Diseases 1970-1980*; Burges, H. D., Ed.; Academic Press, Inc.: London, **1981**, pp. 223-248.

Federici, B. A. "Insecticidal Bacterial Proteins Identify the Midgut Epithelium as a Source of Novel Target Sites for Insect Control." *Arch. of Insect Biochem. and Physiol.*, **1993**, *22*, pp. 357-371.

Feitelson, J. S.; Payne, J.; Kim, L. "*Bacillus thuringiensis:* Insects and Beyond." *Biotechnology*, **1992**, *10*, pp. 271-275.

Ferré, J.; Real, M. D.; Van Rie, J.; Jansens, S.; Peferoen, M. "Resistance to the *Bacillus thuringiensis* Bioinsecticide in a Field Population of *Plutella xylostella* is Due to a Change in a Midgut Membrane Receptor." *Proc. Natl. Acad. Sci. USA*, **1991**, *88*, pp. 5119-5123.

Gleave, A. P.; Williams, R.; Hedges, R. J. "Screening by Polymerase Chain Reaction of *Bacillus thuringiensis* Serotypes for the Presence of *cryV*-Like Insecticidal Protein Genes and Characterization of a *cryV* Gene Cloned from *B. thuringiensis* subsp. *kurstaki*." *Appl. Environ. Microbiol.*, **1993**, *59*, pp. 1683-1687.

González, J. M., Jr.; Brown, B. S., Carlton, B. C. "Transfer of *Bacillus thuringiensis* Plasmids Coding for delta-Endotoxin Among Strains of *B. thuringiensis* and *B. cereus*." *Proc. Natl. Acad. Sci. USA*, **1982**, *79*, pp. 6951-6955.

Herrnstadt, C.; Gilroy, T. E.; Sobieski, D. A.; Bennett, B. D.; Gaertner, F. H. "Nucleotide Sequence and Deduced Amino Acid Sequence of a Coleopteran-active delta-Endotoxin Gene from *Bacillus* thuringiensis subsp. *san diego*." **1987**, *57*, pp. 37-46.

Hoffman, C.; Lüthy, P.; Hutter, Pliska, V. "Binding of the delta-Endotoxin from *Bacillus thuringiensis* to Brush Border Membrane Vesicles of the Cabbage Butterfly (*Pieris brassicae*)." *Eur. J. Biochem.*, **1988**, *173*, pp. 85-91.

Hoffman, C.; Vanderbruggen, H.; Höfte, H.; Van Rie, J.; Jansens, S. "Specificity of *Bacillus thuringiensis* delta-Endotoxins is Correlated with the Presence of High-affinity Binding Sites in the Brush Border Membrane of Target Insect Midguts." *Proc. Natl. Acad. Sci. USA*, **1988**, *85*, pp. 7844-7848.

Höfte, H.; Whiteley, H. R. "Insecticidal Crystal Proteins of *Bacillus thuringiensis*." *Microbiol. Rev.*, **1989**, *53*, pp. 242-255.

Honée, G.; van der Salm, T.; Visser, B. "Nucleotide Sequence of Crystal Protein Gene Isolated from *B. thuringiensis* subspecies *entomocidus* 60.5 Coding for a Toxin Highly Active Against *Spodoptera* species." *Nucleic Acids Res.*, **1988**, *16*, p. 6240.

Knowles, B. H.; Ellar, D. J. "Colloid-Osmotic Lysis is a General Feature of the Mechanism of Action of *Bacillus thuringiensis* delta-Endotoxins with Different Insect Specificities." *Biochem Biophys. Acta.*, **1987**, *924*, p. 509-518.

Kronstad, J. W.; Schnepf, H. E.; Whiteley, H. R. "Diversity of Locations for the *Bacillus thuringiensis* Crystal Protein Genes." *J. Bacteriol.*, **1983**, *154*, pp. 419-428.

Krywienczyk, J.; Angus, T. A. "Serological Comparison of Several Crystalliferous Insect Pathogens." *J. Invert. Path.*, **1967**, *9*, pp. 126-128.

Lambert, B.; Höfte, H.; Annys, K.; Jansens, S.; Soetaert, P.; Peferoen, M. "Novel *Bacillus thuringiensis* Insecticidal Crystal Protein with a Silent Activity against Coleopteran Larvae." *Appl. Environ. Microbiol.*, **1992a**, *58*, pp. 2536-2542.

Lambert, B.; Theunis, W.; Aguda, R.; Van Audenhove, K.; Decock, C.; Jansens, S.; Seurinck, J.; Peferoen, M. "Nucleotide Sequence of Gene *crylIID* Encoding a Novel Coleopteran-active Crystal Protein from Strain BTI109P of *Bacillus thuringiensis* subsp. *kurstaki*." *Gene*, **1992b**, *110*, pp. 131-132.

Lecadet, M.-M.; Dedonder R. "Biogenesis of the Crystalline Inclusion of *Bacillus thuringiensis* During Sporulation." *Eur. J. Biochem.*, **1971**, *23*, pp. 282-294.

Lüthy, P.; Ebersold, H. R. In *Pathogenesis of Invertebrate Microbial Diseases*; Davidson, E. W., Ed.; Allenheld, Osmun & Co.: Totowa, NJ, **1981**, pp. 235-267.

MacIntosh, S. C.; Stone, T. B.; Jokerst, R. S.; Fuchs, R. L. "Binding of *Bacillus thuringiensis* Proteins to a Laboratory-Selected Line of *Heliothis virescens*." *Proc. Natl. Acad. Sci. USA*, **1991**, *88*, pp. 8930-8933.

MacIntosh, S. C.; Stone, T. B.; Sims, S. R.; Hunst, P. L.; Greenplate, J. T.; Marrone, P. G.; Perlak, F. J. Fischhoff, D. A.; Fuchs, R. L. "Specificity and Efficacy of Purified *Bacillus thuringiensis* Proteins Against Agronomically Important Insects." *J. Inv. Path.*, **1990**, *56*, pp. 258-266.

Moar, W. J.; Trumble, J. T.; Federici, B. A. "Comparative Toxicity of Spores and Crystals from the NRD-12 and HD-1 Strains of *Bacillus thuringiensis* subsp. *kurstaki* to Neonate Beet Armyworm (Lepidoptera: Noctuidae)." *J. Econ. Ent.*, **1989**, *82*, pp. 1593-1603.

Payne, J.M.; Kennedy, M.K.; Randall, J.B.; Meier, H.; Uick, H.J. "Novel *Bacillus thuringiensis* Isolates Active Against Hymenopteran Pests and Gene(s) Encoding Hymenopteran-active Toxins." *European Patent Application.* #92304228.7, December 2, **1992**.

Schnepf, H. E.; Whiteley, H. R. "Cloning and Expression of the *Bacillus thuringiensis* Crystal Protein Gene in *Escherichia coli*." *Proc. Natl. Acad. Sci. USA*, **1981**, *78*, p. 2893- 2897.

Sick, A.; Gaertner, F.; Wong, A. "Nucleotide Sequence of a Coleopteran-active Toxin Gene from a New Isolate of *Bacillus thuringiensis* subsp. *tolworthi*." *Nucleic Acids Res.*, **1989**, *18*, p. 1305.

Smulevitch, S. V.; Osterman, A. L.; Shevelev, A. B.; Kaluger, S. V.; Karasin, A. I.; Kadyrov, R. M.; Zagnitko, O. P.; Chestukhina, G. G.; Stepanov, V. M. "Nucleotide Sequence of a Novel delta-endotoxin Gene *crylg* of *Bacillus thuringiensis* ssp. *galleriae*." *FEBS*, **1991**, *293*, pp. 25-28.

Tailor, R.; Tippett, J.; Gibb, G.; Pells, S.; Pike, D.; Jordan, L.; Ely, S. "Identification and Characterization of a Novel *Bacillus thuringiensis* delta-Endotoxin Entomocidal to Coleopteran and Lepidopteran Larvae." *Mol. Microbiol.*, **1992**, *6*, pp. 1211-1217.

Thorne, L.; Garduno, F.; Thompson, T.; Decker, D.; Zounes, M. Wild, M.; Walfield, A. M.; Pollock, T. "Structural Similarity Between the Lepidoptera-and Diptera-specific Insecticidal Endotoxin Genes of *Bacillus thuringiensis* subsp. *kurstaki* and *israelensis*." *J. Bacteriol.*, **1986**, *166*, pp. 801-811.

336

van Frankenhuyzen, K.; Gringorten, J. L.; Milne, R. E.; Gauthier, D.; Pusztai, M.; Brousseau, R.; Masson, L. "Specificity of Activated CryIA Proteins from *Bacillus thuringiensis* subsp. *kurstaki* HD-1 for Defoliating Forest Lepidoptera." *Appl. Environ. Microbiol.*, **1991**, *57*, pp. 1650-1655.

Van Rie, J.; McGaughey, W. H.; Johnson, D. E.; Barnett, B. D.; Van Mellaert, H. "Mechanism of Insect Resistance to the Microbial Insecticide *Bacillus thuringiensis.*" *Science*, **1990**, *247*, pp. 72-74.

Visser, B.; Munsterman, E.; Stoker, A.; Dirkse, W. G. "A Novel *Bacillus thuringiensis* Gene Encoding a *Spodoptera exigua*-Specific Crystal Protein." *J. Bacteriol.*, **1990**, *172*, pp. 6783-6788.

Waalwijck, C.; Dullemans, A. M.; van Workum, M. E. S.; Visser, B. "Molecular Cloning and the Nucleotide Sequence of the Mr 28 000 Crystal Protein Gene of *Bacillus thuringiensis* subsp. *israelensis.*" *Nucleic Acids Res.* **1985**, *13*, pp. 8207-8217.

Ward, E. S.; Ellar, D. J. "Nucleotide Sequence of a *Bacillus thuringiensis* var. *israelensis* Gene Encoding a 130 kDa delta-Endotoxin, *Nucleic Acids Res.* **1987**, *15*, p. 7195.

Widner, W. R.; Whiteley, H. R. "Two Highly Related Insecticidal Crystal Proteins of *Bacillus thuringiensis* subsp. *kurstaki* Possess Different Host Range Specificities." *J. Bacteriol.*, **1989**, *171*, pp. 965-974.

Genetically Modified Bacteria for Biocontrol of Soilborne Plant Pathogens

D. P. Roberts, Biocontrol of Plant Diseases Laboratory, USDA, ARS, Beltsville, MD 20705

Bacterial biocontrol agents often provide inconsistent biocontrol performance in the field. The performance of bacterial biocontrol agents must become more consistent and be enhanced before there is increased commercialization of these bacteria. Molecular techniques are being employed to identify the traits important to the biocontrol interaction and to determine how the soil environment affects the expression of these traits. Information regarding biocontrol traits has been used to construct bacterial biocontrol agents with enhanced performance capabilities. In addition, genetic containment strategies are being developed for bacteria that are introduced into the environment.

Disease control strategies involving the use of introduced bacterial biocontrol agents may eventually reduce or eliminate the use of chemical pesticides for control of certain soilborne plant pathogenic bacteria and fungi. Unfortunately, bacterial biocontrol agents, as with biocontrol agents in general, often provide inconsistent biocontrol performance in the field (Weller, 1988). Consequently, the tools of molecular genetics have been employed in an effort to understand these performance problems and to enhance the biocontrol capabilities of bacterial biocontrol agents. This chapter is a brief overview of current research applying molecular genetics towards the identification and characterization of traits possessed by biocontrol bacteria that are important to the biocontrol of soilborne plant pathogens and towards the development of genetically modified bacterial biocontrol agents. In addition, genetic environmental containment strategies for bacteria will be discussed.

Genetic Analysis of Bacterial Biocontrol Agents

Fourteen bacterial biocontrol agents were registered with the U.S. Environmental Protection Agency for use as microbial pesticides as of November, 1992. Of these biocontrol agents, only *Agrobacterium radiobacter, Pseudomonas fluorescens*, and *Bacillus subtilus* were registered for use in controlling plant diseases caused by soilborne plant pathogens (D. Bays, U.S. EPA; personal

communication). One of the primary problems limiting the widespread commercialization of bacterial biocontrol agents is their inconsistent biocontrol performance (O'Sullivan and O'Gara, 1992; Weller, 1988). The biocontrol phenotype is the result of the concurrent or sequential expression of traits (Nelson and Maloney, 1992). To understand these inconsistencies in biocontrol performance it is necessary to identify the traits that contribute to the biocontrol phenotype and to understand how various soil environmental conditions affect the expression of these traits (Gutterson, 1990; Nelson and Maloney, 1992).

Identification of Biocontrol Traits. Genetic techniques allow construction of bacterial strains that can be used as tools for evaluation of the importance of genes or traits to biocontrol *in situ*. Construction of near-isogenic strains deficient in particular genes or traits allows analysis of the relevance of these genes or traits to biocontrol in comparison tests with wild-type strains. For example, Thomashow and Weller (1988) determined that the production of the antibiotic phenazine-1-carboxylate by the biocontrol bacterium *Pseudomonas fluorescens* 2-79 was important to its control of the fungal plant pathogen *Gaeumannomyces graminis* var. *tritici*. Six near-isogenic strains of *P. fluorescens* were constructed, each containing a single transposon insertion, by random transposon mutagenesis with Tn5. Each of these six mutants was deficient in the production of phenazine-1-carboxylate and had reduced biocontrol properties when compared with the parental strain. These data suggest that production of the phenazine antibiotic is important in the biocontrol interaction (Thomashow and Weller, 1988). Near-isogenic strains deficient in a particular gene or trait have been constructed by insertional inactivation to test the importance of other genes and traits in biocontrol. Some examples are listed in Table 1.

In an alternate approach, genes were isolated by molecular cloning and their relevance to biocontrol determined in a second strain containing a different genetic background. A chitinase gene from *Serratia marscesens* was cloned into a cosmid and mobilized by conjugation into the biocontrol bacterium, *Pseudomonas fluorescens* strain NRRL B-15135. The resultant strain, containing the cosmid and the *S. marscesens* chitinase gene, had increased biocontrol capabilities against *Fusarium oxysporum* f.sp. *redolens* on radish (Sundheim et al., 1988). A strain of *Escherichia coli* which reduced the incidence of disease caused by *Sclerotium rolfsii* on bean was constructed by introducing the *chi*A gene from *S. marscesens* into *E. coli* (Shapira et al., 1989). Enhanced biocontrol properties of the test strain suggested a role for the *chi*A gene in biocontrol. It should be noted that both of the genetic approaches outlined above are usually preceded by a significant quantity of research identifying potential biocontrol genes.

Environmental Effects on Expression of Biocontrol Traits. Performance of plant-beneficial biocontrol agents can vary greatly with the soil environment (Burr and Caesar,

1984). For example, *Enterobacter cloacae* is an effective biocontrol agent for control of *Pythium* seed and seedling diseases with certain plants (Hadar et al., 1983; Nelson, 1988; Nelson et al., 1986; Nelson and Maloney, 1992). However, biocontrol performance of this bacterium varies greatly with plant spermosphere (Nelson and Maloney, 1992). *Enterobacter cloacae* is antagonistic to *Pythium ultimum* when applied to seeds that exude low levels of carbohydrate during the onset of seed germination. In contrast, this bacterium is ineffective as a biocontrol agent when coated onto seeds that exude large quantities of sugar during germination (Nelson et al., 1986). This variation in biocontrol performance by *E. cloacae* may be due, at least in part, to effects of the spermosphere environment on the expression of traits important to biocontrol by this bacterium.

Attachment of *E. cloacae* to hyphae of *P. ultimum* was correlated with the ability of E. cloacae to control seed and seedling diseases caused by *P. ultimum* (Nelson et al., 1986). Adding certain sugars that have been detected in plant spermosphere and rhizosphere blocked attachment of *E. cloacae* to hyphae of *P. ultimum in vitro*. Addition of these sugars to *E. cloacae*-treated cucumber seed inhibited biocontrol of *P. ultimum*. *Enterobacter cloacae* is an effective biocontrol agent of this fungus on cucumber in the absence of these sugars (Nelson et al., 1986). Ammonia production by *E. cloacae*, a second trait correlated with biocontrol of *P. ultimum* by this bacterium, was also suppressed in the presence of certain sugars *in vitro* (Howell et al., 1988).

Expression of other traits that have been

Table 1. Demonstration of importance of traits to biocontrol using insertional inactivation

Biocontrol agent	Trait	Reference
Pseudomonas fluorescens HV37A	oomycin A	(Howie and Suslow, 1991)
Pseudomonas fluorescens 3551	siderophore	(Loper, 1988)
Pseudomonas fluorescens CHA0	2,4 diacetylphloro -glucinol	(Keel et al., 1992)
Pseudomonas fluorescens CHA0	hydrogen cyanide	(Voisard et al., 1989)
Serratia marscesens QMB1466	chitinase	(Jones et al. 1986)

implicated in biocontrol varied with environmental conditions *in vitro*. For example, the production of oomycin A by *P. fluorescens* HV37A required glucose for induction (Gutterson et al., 1988). Production of the *gac* gene product, a global regulatory molecule necessary for the expression of 2,4-diacetylphloroglucinol, hydrogen cyanide, and pyoluterin by *P. fluorescens* CHAO, was responsive to environmental conditions. The *gac* gene product was expressed preferentially *in vitro* when there was nutrient limitation or other conditions that restricted growth (Laville et al., 1992). In addition, the production of siderophores has been shown to be influenced by iron, phosphate, pH, and trace elements (Barbhaiya and Rao, 1985; Kloepper et al., 1980; reviewed in O'Sullivan and O'Gara, 1992).

The use of genetically modified biocontrol bacteria containing reporter genes should be valuable in determining the environmental parameters that affect the expression of biocontrol traits *in situ*. Reporter genes are genes whose expression are easily assayed. Therefore, placing a reporter gene under the direction of a biocontrol promoter through gene fusion allows the detection of expression of biocontrol genes that may not otherwise have been possible. In many cases the use of reporter genes greatly simplifies the measurement of gene expression. The *gus* (Jefferson et al., 1986), *lacZ* (Kroos and Kaiser, 1984), *lux* (Shaw and Kado, 1986), and *xylE* (Zukowski et al., 1983) reporter genes are commonly used. Gene fusions with the *LacZ* reporter gene allowed analysis of expression of antibiotic production by *P. fluorescens* HV37A *in vitro* (Gutterson et al., 1988) and *in situ* on cotton seed (Howie and Suslow, 1991).

Genetically Modified Bacterial Biocontrol Agents

Little work has been performed toward strain improvement due to the general lack of knowledge of traits important to biocontrol (Gutterson, 1990) and to the genetic complexity of the biocontrol phenotype. In experiments where major biocontrol determinants have been inactivated by genetic insertion, the resultant strain was reduced in its biocontrol properties.

The inactivation of individual biocontrol genes did not result in a complete loss of the biocontrol phenotype (Howie and Suslow, 1991; Thomashow and Weller, 1988; Loper, 1988; Keel et al., 1992; Voisard et al., 1989; Jones et al., 1986) indicating that more than one gene is involved in biocontrol in these systems. The identity of the other genes functioning in biocontrol in these systems are unknown.

However, genetic analysis of biocontrol bacteria indicates that it may be possible to improve the performance of these bacteria through genetic engineering (O'Sullivan and O'Gara, 1992). Attempts at enhancing biocontrol performance include: 1) adding biocontrol traits to bacterial biocontrol agents, as with the mobilization of genes encoding chitinase enzyme synthesis by conjugation described above (Sundheim et al., 1988), 2) modifying the regulation of expression of traits important to biocontrol, and 3) enhancing the stability of the biocontrol interaction.

Minor genetic alterations in the regulation of traits important to a particular biocontrol interaction have led to potential gains in biocontrol efficacy. For example, it may be possible to enhance biocontrol performance by altering the regulation of antifungal metabolites (Gutterson, 1990). The antibiotic oomycin A, produced by *Pseudomonas fluorescens* strain HV37A, plays a major role in the biocontrol of *Pythium ultimum* by this bacterium on cotton (Howie and Suslow, 1991). However, this antibiotic is catabolite-induced by glucose (Gutterson et al., 1988). It is likely that glucose concentrations in the rhizosphere are inadequate for expression of oomycin A or are highly variable, leading to spatially inconsistent production of oomycin A in the rhizosphere and therefore to inconsistent biocontrol performance. In an effort to improve biocontrol performance by *P. fluorescens* strain HV37A, the regulation of oomycin A was altered. The *afuE* locus, a locus involved in the *Escherichia coli* regulation of oomycin A production (Gutterson et al., 1988), was fused to the *tac* promoter *in vitro* and returned to *P. fluorescens* strain HV37A (Gutterson, 1990). The recombinant construct produced oomycin A in the presence and absence of glucose at levels equivalent to the parental strain when the parental strain was grown under optimal conditions (Gutterson,

1990). For *in situ* analysis, the *tac-afu*E fusion was integrated into the *P. fluorescens* chromosome. The altered regulation of oomycin A production due to the fusion of the *afu*E gene to the *tac* promoter led to increased production of this antibiotic in the rhizosphere and to increased biocontrol performance (Gutterson, 1990).

Siderophores are important to biocontrol in certain antagonist-pathogen interactions (reviewed in Loper and Buyer, 1991). However, siderophore production is repressed under high iron concentrations (Kloepper et al., 1980). Therefore, it may be possible to increase biocontrol performance in soil environments containing inhibitory iron concentrations (O'Sullivan and O'Gara, 1992). A mutant of *Pseudomonas* sp. strain M114 was isolated that produces siderophore in the presence of iron. This mutant has *in vitro* biocontrol qualities and shows inhibition under iron conditions where the parental strain is non-inhibitory (reviewed in O'Sullivan and O'Gara, 1992).

It may be possible to increase the stability of the biocontrol interaction through the genetic modification of the biocontrol agent. The biocontrol agent *Agrobacterium radiobacter* strain K84 inhibits the plant pathogen *Agrobacterium tumefaciens* largely through the production of the antibiotic Agrocin 84 (Kerr and Htay, 1974; Kerr and Tate, 1984). The genes for the production of Agrocin 84 and for immunity to this antibiotic are carried on the conjugative plasmid pAgK84 (Farrand et al., 1985; Ryder et al., 1987). Transfer of plasmid pAgK84 to *A. tumefaciens* by conjugation may lead to the breakdown in control of this pathogen since *A. tumefaciens* strains harboring this plasmid are resistant to Agrocin 84 (Ellis and Kerr, 1979; Panagopoulos et al., 1979). To prevent plasmid pAgK84 from being transferred to *A. tumefaciens* by conjugation, a transfer-deficient derivative of this plasmid was constructed by deleting the genes required for conjugation (Jones et al., 1988).

With an alternate approach, it may be possible to effect biocontrol through competitive niche exclusion using avirulent-derivatives of bacterial plant pathogens (Lindow et al., 1989). A biocontrol agent can be constructed by identifying the traits possessed by a plant pathogen that are necessary for plant pathogenesis and subsequently inactivating these traits through genetic manipulation. Since the plant pathogen and the near-isogenic, avirulent derivative possess essentially the same genotype they are capable of occupying the same ecological niche. The result is the competitive exclusion of the plant pathogen from its niche by the avirulent biocontrol agent (Lindow et al., 1989).

Lindow and coworkers constructed a biocontrol agent for the foliar plant pathogen *Pseudomonas syringae* in this fashion. This pathogen is common on the leaves of many plants that are susceptible to frost injury. Ice nucleation activity by this plant pathogen causes warm temperature frost damage. A non-pathogenic, near-isogenic strain of *P. syringae* was constructed by inactivating the genetic determinant responsible for ice nucleation activity through gene replacement (Green and Warren, 1985; Orcer et al., 1985). Application of the avirulent mutant results in biocontrol through pre-emptive competitive exclusion of the pathogen from potentially colonizable niches on the leaf. Preliminary results indicate that this strategy of using an avirulent pathogen for biocontrol may be successful (reviewed in Lindow et al., 1989).

It may be possible to construct avirulent derivatives of soilborne plant pathogenic bacteria for the biocontrol of soilborne plant pathogens using the strategy of Lindow and coworkers. A number of genetic loci have been demonstrated to have a role in pathogenesis by *Pseudomonas solanacearum* (Boucher et al., 1985; Roberts et al., 1988; Schell et al., 1988; Trigalet et al., 1986; Xu et al., 1988). Avirulent derivatives of *Pseudomonas solanacearum*, containing inactivated virulence genes or combinations of these inactivated genes, may be useful in biocontrol strategies relying on competitive niche exclusion.

Genetically Modified Bacteria for Environmental Containment

One obstacle to the use of genetically engineered microorganisms in agriculture is their potential risk to man, animals, plants and the ecosystem (Smit et al., 1992). Areas of potential risk posed by the introduction of genetically engineered microbes include: 1) potential pathogenicity to man, other animals, and plants, 2) disturbance of the ecological balance and community structure, and 3) dissemination of heterologous genes from the genetically engineered microbe to other organisms (Smit et al., 1992). Approaches involving genetically modified bacteria are being developed to address these concerns. Genetically modified bacteria have been developed to aid in monitoring the longevity and spread of genetically engineered bacteria in the environment. In addition, genetic strategies are being developed for the environmental containment of introduced bacteria.

Genetically Modified Bacteria for Environmental Monitoring. Biocontrol bacteria can be tagged genetically to aid in the analysis of populations of these bacterial biocontrol agents in the environment. For culturable bacteria, a number of reporter genes or unique genetic sequences can be introduced to aid in the detection and quantification of these bacteria during plating procedures. This allows for analysis of populations of specific strains of bacteria when selective media are unavailable or inadequate.

Reporter genes provide the bacterium with a unique property such as a unique coloration or bioluminescence that can be used alone or in conjunction with other properties to specifically identify the bacterium. For example, introduction of the *xyl*E gene into a bacterium allows detection of the bacterium carrying the *xyl*E gene on a non-selective medium. The *xyl*E gene encodes the enzyme 2,3-catechol dioxygenase. This enzyme catalyzes the conversion of catechol into 2-hydroxymuconic semialdehyde, a compound which is bright yellow in color. Therefore, colonies containing the *xyl*E gene are bright yellow after the application of catechol to the culture medium (Winstanley et al., 1989). The *lux* gene cassette can also be used in this fashion (Fravel et al., 1990; Shaw and Kado, 1986). Hybridization procedures, where the probe is specific for a unique genetic sequence can be used to detect and enumerate populations of bacteria during dilution plating procedures (VanElsas and Waalwijk, 1991). Hybridization procedures can also be used to monitor the persistence of a particular gene in the environment independent of the host bacterial strain.

Genetic Approaches for Environmental Containment of Bacterial Biocontrol Agents. Two fundamentally different approaches to biological containment are being developed to reduce the risk of adverse effects on the ecosystem due to the introduction of bacteria. With the first approach a genetically debilitated strain that can not effectively compete in the environment is used. The second approach involves the construction of conditional suicide systems that kill the genetically engineered bacterium under certain conditions (Knudsen and Karlstrom, 1991).

Certain agricultural applications may allow the use of genetically debilitated plant-beneficial bacteria that can not persist in the environment. For example, strains of *E. cloacae* are effective in suppressing the early stages of seed infection caused by *Pythium ultimum* when *E. cloacae* is introduced into the soil as a seed treatment. Seeds of a number of different crops, including cucumber, appear to be especially vulnerable to *Pythium* infection during the first 12 hr of seed germination (Nelson and Maloney, 1992). After this time seedlings become less susceptible to infection. Therefore, the traits important to biocontrol of *P. ultimum* by *E. cloacae* must be expressed during the first 12 hr of seed germination (Nelson and Maloney, 1992).

We are currently testing a strategy for environmental containment of *E. cloacae* which takes advantage of the time constraints of this particular disease interaction. Application of a non-persistent strain of *E. cloacae* should provide protection of the seed since the cucumber plant is only vulnerable to infection by *P. ultimum* for a short period of time. We have chosen to screen *E. cloacae* auxotrophs for mutants that do not persist for extended periods of time in plant spermosphere. Auxotrophs seem advantageous since: 1) they are easily isolated from a transposon mutant library by screening on minimal media. 2) They have the potential for instability in the environment. And 3) their biochemical requirements for growth can be applied to the spermosphere along with the bacterium in the seed treatment (Roberts and Marty, unpublished). Therefore, it is potentially possible to regulate the persistence of the bacterium in the environment through application of varying amounts of the biochemical growth requirement.

The approach appears promising since a number of auxotrophs affected in spermosphere proliferation were isolated. One *E. cloacae* auxotroph, strain A-46, is especially promising for this environmental containment strategy. In some experiments, this strain could not be detected by dilution plating seven days after placing in corn, cucumber, pea, and radish spermospheres. In contrast, the prototrophic parental strain, strain 501R3, increased in number several-fold in all spermospheres (Table 2). It may be possible to introduce further biochemical requirements into the *E. cloacae* strain through a second round of transposon mutagenesis involving an auxotroph such as strain A-46 that shows particular promise. This may enhance the level and consistency of environmental containment.

A second approach employs the use of so-called suicide containment systems (Bej et al., 1988; Contreras et al., 1991; Knudsen and Karlstrom 1991). With these suicide systems a killing function is directly or indirectly regulated by an environmental signal. Change in quantity of this chemical signal leads to expression of the killing function and ultimately, bacterial cell death. The *gef*, *hok*, and *rel*F genes, which encode killing functions, have been employed in

these suicide containment systems (Bej et al., 1988; Contreras et al., 1991, Knudsen and Karlstrom 1991). The *hok* gene has been fused to the *lac* promoter (Bej et al., 1988). Induction of the *hok* gene through addition of isopropyl-*beta*-thiopyranogalactoside, an inducer of the *lac* promoter, led to expression of the *hok* gene. The *hok* gene encodes a lethal polypeptide that, when overexpressed, collapses bacterial membrane potential (Bej et al., 1988).

In a second suicide containment system, a more complex strategy was employed. The promoter for the *Pseudomonas putida meta* cleavage pathway, Pm, was fused to the *lac*I and *xyl*S genes. The *lac*I gene encodes the Lac repressor while the *xyl*S gene encodes for a positive regulator of Pm. A second gene fusion between the *tac* promoter and the *gef* gene was also employed. The *gef* gene encodes the killing function while the *tac* promoter is repressed by the Lac repressor. Therefore, in the presence of *xyl*S effectors, the Lac repressor protein is produced and the production of *gef* gene product is repressed. In the absence of *xyl*S effectors, the *gef* gene product is expressed resulting in bacterial cell death (Contreras et al., 1991). There are two drawbacks to this system. The concentration of the *xyl*S effector required to maintain repression of the *gef* gene product is high and may not be present in the environment at sufficient levels. In addition, clones resistant

Table 2. Proliferation of *Enterobacter cloacae* strains 501R3 and A-46 in corn, cucumber, pea, and radish rhizosphere

| Seed | strain | Mean log10 colony-forming units | |
		0 h	168 h
corn	A-46	3.52	< 2.00
	501R3	3.41	5.59
cucumber	A-46	4.57	< 2.00
	501R3	4.23	5.62
pea	A-46	3.81	< 2.00
	501R3	3.45	5.09
radish	A-46	2.93	< 2.00
	501R3	3.11	5.76

to the *gef* gene product appear. The authors suggest that a suicide system employing two killing functions may overcome this problem with resistance (Contreras et al., 1991).

Conclusions

The consistency of biocontrol performance must be improved before there can be an increase in the use of bacterial biocontrol agents for the control of soilborne plant pathogens. Current research efforts are focussed on the identification of traits possessed by the biocontrol agent that function in the biocontrol interaction and the way that environmental conditions impact on the expression of these traits. This information is necessary before strains can be genetically modified for enhanced biocontrol performance. Genetic strategies for environmental containment of introduced bacterial biocontrol agents must be developed to allay both the real and the perceived fears concerning the detrimental effects on the ecosystem due to the introduction of genetically altered biocontrol agents.

References

Barbhaiya, H.B., Rao, K.K. "Production of pyoverdine, the fluorescent pigment of *Pseudomonas aeruginosa PA01.*" *FEMS Microbiol. Lett.* **1985**, 27:233-235.

Bej, A.K., Perlin, M.H., Atlas, R.M. "Model suicide vector for containment of genetically engineered microorganisms." *Appl. Environ. Microbiol.* **1988**, 54:2472-2477.

Boucher, C.A., Barberis, P.A., Trigalet, A.P., Demery, D.A. "Transposon mutagenesis of *Pseudomonas solanacearum*: isolation of Tn5-induced avirulent mutants." *J. Gen. Microbiol.* **1985**, 131:2449-2457.

Burr, T.J., Caesar, A. "Beneficial plant bacteria." *Crit. Rev. Plant Science* **1984**, 2:1-20.

Contreras, A., Molin, S., Ramos, J. -L. "Conditional -suicide containment system for bacteria which mineralize aromatics." *Appl. Environ. Microbiol.* **1991**, 57:1504-1508.

Ellis, J.G., Kerr, A. "*Agrobacterium*: genetic studies on agrocin 84 production and the biological control of crown gall." *Physiol. Plant Pathol.* **1979**, 15:311-319.

Farrand, S.K., Slota, J.E., Shim, J.-S., Kerr, A. "Tn5 insertions in the agrocin 84 plasmid: the conjugal nature of pAgK84 and the locations of determinants for transfer and agrocin 84 production." *Plasmid* **1985**, 13:106-117.

Fravel, D.R., Lumsden, R.D., Roberts, D.P. "*In situ* visualization of the biocontrol rhizobacterium *Enterobacter cloacae* with bioluminescence." *Plant Soil* **1990**, 125:233-238.

Green, R.L., Warren, G.J. "Physical and functional repetition in a bacterial ice nucleation gene." *Nature* **1985**, 317:645-648.

Gutterson, N. "Microbial fungicides: recent approaches to elucidating mechanisms." *Crit. Rev. Biotechnol.* **1990**, 10:69-91.

Gutterson, N., Ziegle, J.S., Warren, G. J., Layton, T.J. "Genetic determinants for catabolite induction of antibiotic biosynthesis in *Pseudomonas fluorescens* HV37A." *J. Bacteriol.* **1988**, 170:380-385.

Hadar, Y., Harman, G.E., Taylor, A.G., Norton, J.M. "Effects of pregermination of pea and cucumber seeds and of seed treatment with *Enterobacter cloacae* on rots caused by *Pythium* spp." *Phytopathology* **1983**, 73:1322-1325.

Howie, W.J., Suslow, T.V. "Role of antibiotic biosynthesis in the inhibition of *Pythium ultimum* in the cotton spermosphere and rhizosphere by *Pseudomonas fluorescens.*" *Molec. Plant-Microbe Interact.* **1991**, 4:393-399.

Howell, C.R., Beier, R.C., Stipanovic, R.D. "Production of ammonia by *Enterobacter cloacae* and its possible role in the biological control of Pythium preemergence damping-off by the bacterium." *Phytopathology* **1988**, 78:1075-1078.

Jefferson, R.A., Burgess, S.M., Hirish, D. "Beta-glucuronidase from *Escherichia coli* as a gene-fusion marker." *Proc. Nat'l. Sci. U.S.A.* **1986**, 83:8447-8451.

Jones, D.A., Ryder, M.H., Clare, B.G., Farrand, S.K., Kerr, A. "Construction of a Tra- deletion mutant of pAgK84 to safeguard the biological control of crown gall." *Molec. Gen. Genet.* **1988**, 212:207-214.

Jones, J.D.G., Grady, K.L., Suslow, T.V., Bedbrook, J.R. "Isolation and characterization of genes encoding two chitinase enzymes from *Serratia marcescens*." *EMBO J.* **1986**, 5:467-473.

Keel, C., Schnider, U., Maurhofer, M., Voisard, C., Laville, J., Burger, U., Wirthner, P., Haas, D., Defago, G. "Suppression of root diseases by *Pseudomonas fluorescens* CHA0: importance of the bacterial secondary metabolite2,4-diacetylphloroglucinol."*Molec. Plant-Microbe Interact.* **1992**, 54-13.

Kerr, A., Htay, K. "Biological control of crown gall through bacteriocin production." *Physiol. Plant Pathol.* **1974**, 4:37-44.

Kerr, A., Tate, M.E. "Agrocins and the biological control of crown gall." *Microbiol. Sci.* **1984**, 1:1-4.

Kloepper, J.W., Leong, J., Teintze, M., Schroth, M.N. "Enhanced growth promotion by siderophores produced by plant growth-promoting rhizobacteria." *Nature* **1980**, 286:885-886.

Knudsen, S.M., Karlstrom, O.H. "Development of efficient suicide mechanisms for biological containment of bacteria." *Appl. Environ. Microbiol.* **1991**, 57: 85-92.

Kroos, L., Kaiser, D. "Construction of Tn*5 lac*, a transposon that fuses *lac*Z expression to exogenous promoters, and its introduction into *Myxococcus xanthus*." *Proc. Nat'l. Acad. Sci.* U.S.A. **1984**, 81:5816-5820.

Laville, J., Voisard, C., Keel, C., Maurhofer, M., Defago, G., Haas, D. "Global control in *Pseudomonas fluorescens* mediating antibiotic synthesis and suppression of black root rot of tobacco." *Proc. Nat'l. Acad. Sci.* U.S.A. **1992**, 89:1562-1566.

Lindow, S.E., Panopoulos, N.J., McFarland, B.L. "Genetic engineering of bacteria from managed and natural habitats." *Science* **1989**, 244:1300-1307.

Loper, J.E. "Role of fluorescent siderophore production in biological control of *Pythium ultimum* by a *Pseudomonas fluorescens* strain." *Phytopathology* **1988**, 78:166-172.

Loper, J.E., Buyer, J.S. "Siderophores in microbial interactions on plant surfaces." *Molec. Plant-Microbe Interact.* **1991**, 4:5-13.

Nelson, E.B. "Biological control of *Pythium* seed rot and preemergence damping-off of cotton with *Enterobacter cloacae* and *Erwinia herbicola.*" *Plant Disease* **1988**, 72:140-142.

Nelson, E.B., Maloney, A.P. "Molecular

approaches for understanding biological control mechanisms in bacteria: studies of the interaction of *Enterobacter cloacae* with *Pythium ultimum*." *Can. J. Microbiol.* **1992**, 14:106-114.

Nelson, E.B., Chao, W.L., Norton, G.T., Harman, G.E. "Attachment of *Enterobacter cloacae* to hyphae of *Pythium ultimum*: possible role in the biological control of Pythium pre-emergence damping-off." *Phytopathology* **1986**, 76:327-335.

Orcer, C., Staskawicz, B.J., Panopoulos, N.J., Dahlbeck, D., Lindow, S.E. "Cloning and expression of bacterial ice nucleation genes in *Escherichia coli*." *J. Bacteriol.* **1985**, 164:359-366.

O'Sullivan, D.J., O'Gara, F. "Traits of fluorescent *Pseudomonas* spp. involved in suppression of plant root pathogens." *Microbiology Reviews* **1992**, 56:662-676.

Panagopoulos, C.G., Psallidas, P.G., Allivizatos, A.S. In *Soil-borne plant pathogens*; Schippers, B., Gams, W. Ed.; *Academic Press*: London, 1979, pp 569-578.

Roberts, D.P., Denny, T.P., Schell, M.A. "Cloning of the *egl* gene of *Pseudomonas solanacearum* and analysis of its role in phytopathogenicity." *J. Bacteriol.* **1988**, 170:1445-1451.

Ryder, M.H., Slota, J.E., Scarim, A., Farrand, S.K. "Genetic analysis of agrocin 84 production and immunity in *Agrobacterium* Ti plasmids." *J. Bacteriol.* **1986**, 169:4184-4189.

Schell, M.A., Roberts, D.P., Denny, T.P. "Analysis of the *Pseudomonas solanacearum* polygalacturonase encoded by *pgl*A and its involvement in phytopathogenicity." *J. Bacteriol.* **1988**, 170:4501-4508.

Shapira, R., Ordentlich, A., Chet, I., Oppenheim, A.B. "Control of plant diseases by chitinase expressed from cloned DNA in *Escherichia coli*." *Phytopathology* **1989**, 79:1246-1249.

Shaw, J.J., Kado, C.I. "Development of a Vibrio bioluminescence gene-set to monitor phytopathogenic bacteria during the ongoing disease process in a non-disruptive manner." *Bio/Technol.* **1986**, 4:560-564.

Smit, E., van Elsas, J.D., van Veen, J.A. "Risks associated with the application of genetically modified microorganisms in terrestrial ecosystems." *FEMS Microbiol. Rev.* **1992**, 88:263-278.

Sundheim, L., Poplawsky, A.R., Ellingboe, A.H. "Molecular cloning of two chitinase genes from *Serratia marcescens* and their expression in *Pseudomonas* species." *Physiol. Molec. Plant Pathol.* **1988**, 33:483-491.

Thomashow, L.S., Weller, D.M. "Role of a phenazine antibiotic from *Pseudomonas fluorescens* in biological control of *Gaeumannomyces graminis* var. *tritici.*" *J. Bacteriol.* **1988**, 170:3499-3508.

Trigalet, A., Demery, D. "Invasiveness in tomato plants of Tn5-induced avirulent mutants of *Pseudomonas solanacearum.*" *Physiol. Molec. Plant Pathol.* **1986**, 28: 423-430.

VanElsas, J.D., Waalwijk, C. "Methods for the detection of specific bacteria and their genes in soil." *Agric. Ecosys. Environ.* **1991**, 34:97-105.

Voisard, C., Keel, C., Haas, D., Defago, G. "Cyanide production by *Pseudomonas fluorescens* helps suppress black root rot of tobacco under gnotobiotic conditions." *EMBO J.* **1989**, 8:351-358.

Weller, D.M. "Biological control of soilborne plant pathogens in the rhizosphere with bacteria." *Annu. Rev. of Phytopathol.* **1988**, 26:379-407.

Winstanley, C., Morgan, A.W., Pickup, R.W., Jones, J.G., Saunders, J.R. "Differential regulation of Lambda *p*l and *p*r promoters by a *c*I repressor in a broad-host range thermoregulated plasmid marker system." *Appl. Environ. Microbiol.* **1989**, 55:771-777.

Xu, P., Leong, S., Sequeira, L. "Molecular cloning of genes that specify virulence in *Pseudomonas solanacearum.*" *J. Bacteriol.* **1988**, 170: 617-622.

Zukowski, M.M., Gaffney, D.F., Speck, D., Kaufmann, M., Findeli, A., Wisecup, A., Lecocq, J.P. "Chromogenic identification of genetic regulatory signals in *Bacillus subtilis* based on expression of a cloned *Pseudomonas* gene." *Proc. Nat'l. Acad. Sci. U.S.A.* **1983**,80:1101-1105.

The Genome of Biocontrol Fungi: Modification and Genetic Components for Plant Disease Management Strategies

G. E. Harman, C. K. Hayes, and M. Lorito
Departments of Horticultural Sciences and Plant Pathology, Cornell University, Geneva, NY 14456

A number of techniques are available for modification of fungal genomes of biocontrol fungi, including mutation, protoplast fusion, and transformation. The genome of *Trichoderma* and *Gliocladium* has unusual organization and mechanisms of variation. A process of interstrain gene transfer appears to be a major method of asexual genetic exchange and the classical parasexual cycle seems to occur rarely, if ever. The genomes of *Trichoderma* and *Gliocladium* contain a number of genes coding for different chitinolytic and glucanolytic enzymes that degrade cell walls of target fungi. These enzymes act synergistically with each other, with synthetic and natural fungicides, and with the biocontrol bacterium *Enterobacter cloacae* to control plant pathogenic fungi.

Successful use of biological control of plant diseases or other pests requires (a) effective biocontrol agents, (b) production and formulation methods that give rise to high yields of biomass consisting of appropriate efficacious propagules of high viability and stability, and (c) delivery systems that provide a conducive milieu and minimize growth of competitive microflora (Harman, 1991; Jin, et al., 1992). Highly effective biocontrol agents can either be selected from nature or produced using some method of genetic manipulation. Further, if biocontrol organisms are effective, they must contain genes coding for useful biocontrol products. Such genes and gene products may provide effective biological control, and in addition, the genes can be used to determine mechanisms of biocontrol. The purpose of this paper is to summarize methods of genetic manipulation to produce superior biocontrol fungi, and to describe the unusual genetic events that occur in *Trichoderma* spp. The paper will also summarize our discovery of synergistic, efficacious genes and gene products from these fungi.

Methods for genetic manipulation

General procedures for genetic manipulation in *Trichoderma* and *Gliocladium* were recently reviewed (Hayes, 1992) and so will be dealt with only briefly. Physical (e.g. UV irradiation) or chemical mutagenesis were probably the first methods to alter the genotype of biocontrol fungi to enhance biocontrol. An initial approach was to produce fungicide-resistant strains for use in integrated biological-chemical approaches to plant disease control (Ahmad and Baker, 1987; Papavizas et al., 1982). Such approaches provided the expected fungicide-resistant strains, but also some mutants were more effective biocontrol agents than the wild type (Papavizas et al., 1982). Further, other mutants were more strongly able to colonize root surfaces (i.e. were more strongly rhizosphere competent) and also produced higher levels of cellulase than the parental types (Ahmad and Baker, 1988). However, while mutation has been used to produce useful strains, other methods provide opportunities for producing greater or more precise changes.

Protoplast fusion is another process that has been used to produce superior strains (Harman, et al., 1989, Stasz et al., 1988). In this process, parental strains are usually modified by mutation to contain complementary selectable markers. Protoplasts are produced by digestion of hyphal cells with appropriate enzymes. Protoplasts are fused, usually in the presence of polyethylene glycol and $CaCl_2$, and the resulting mixture plated on a medium that contains amendments permitting growth only of fusants (Harman and Stasz, 1991; Harman and Hayes, 1993).

Protoplast fusion gives rise to great diversity within progeny following inter- or intrastrain fusions in *Trichoderma* and *Gliocladium*. Progeny frequently are unstable and sector often, but if progeny strains and sectors are cultivated through a number of generations, many, but not all, strains become stable (Harman and Hayes, 1993; Stasz et al., 1989, Stasz and Harman, 1990). Only a very small percentage of fusion progeny possess improved biocontrol ability relative to the wild type, but a few strains are substantially improved. One of these has been extensively tested, was registered with the U.S. Environmental Protection Agency by a major corporation as a microbial pesticide, and controls a wide range of plant pathogenic fungi, including *Pythium ultimum., Rhizoctonia solani, Fusarium graminearum, Botrytis cinerea, Sclerotinia homeocarpa,* and *Guignardia bidwelli* but not *Phytophthora* spp. It is effective on a wide range of crops and can be used as a seed treatment, a granule for in-furrow or broadcast application to soil, or as a spray to control fruit or foliar pathogens.

Transformation is a third major method of genetic manipulation of fungal genomes. In this procedure, specific genetic sequences can be introduced into fungal cells, where they typically integrate into the fungal genome (Fincham, 1989; Herrera-Estrella et al., 1990; Ossanna and Mischke, 1990; Sivan et al., 1992; Thomas and Kenerley, 1989).

Typically, the process of transformation uses the techniques of protoplast fusion. Plasmids containing the gene of interest are mixed with protoplasts of the strain to be transformed and the cells are fused by addition of polyethylene glycol and $CaCl_2$. Other procedures are possible; for example, treatment with alkali metal ions (Ito et al., 1983) or glass beads (Costanzo and Fox, 1988) may permit plasmids to be integrated into cells without the requirement of protoplast formation.

We recently have used biolistic ("gene gun") protocols to transform *Trichoderma* and *Gliocladium* spp. The procedure was relatively simple; single conidia were plated on the surface of appropriate media in petri plates and inserted into the vessel of the apparatus. DNA (either plasmids, or in some cases, genomic DNA containing a hygromycin B gene from *Escherichia coli*) was coated onto tungsten beads and the coated beads placed on the launch surface membrane. The chamber was evacuated and a charge of helium explosively propelled the coated beads through the fungal walls into the cytoplasm. This method resulted in higher levels of transformation and higher percentages of transformed nuclei within cells than with protoplast-mediated transformation. Either method resulted in integration of heterologous sequences into the fungal genome. Moreover, use of genomic DNA resulted in recovery of transformants with biolistic but not with protoplast-mediated transformation. Both methods gave transformants when plasmid DNA was used.(Lorito et al., 1993b).

Either heterologous genes can be introduced to confer a new property to the strain, or else an homologous sequence can be used. Homologous sequences can be portions of genes, and if these are used, they may insert into homologous sites in the fungal genome. Consequently, the gene into which they are inserted may be disrupted and in this way mutants deficient in the production of particular products can be produced. Such insertional

mutants may be of great utility in mechanistic studies (Thomashow and Weller, 1988). Further, transformation can also be used to add genes back to strains, frequently under the control of different promoter sequences. Such procedures permit a molecular version of Koch's Postulates to be performed to provide quantitative measures of the role of specific genes in biocontrol. Moreover, change of the control of critical metabolites in biocontrol from inducible to constitutive promoters may (a) change the temporal relationships between biocontrol agents and plant pathogens to favor the biocontrol agent since the time of induction can be eliminated, and (b) increase production of critical materials through the use of more powerful promoters. Such techniques should produce more effective biocontrol agents and also be of use in mechanistic studies.

Organization and Variation within the *Trichoderma* and *Gliocladium* Genome

The genome of *Trichoderma* and *Gliocladium* is rather unconventionally organized and there are quite unusual mechanisms for genetic variation within these fungi. Other similar fungi probably have similar mechanisms. These aspects have important implications both for scientists wishing to alter or study these fungi, and for understanding their variation and adaptation to changing environmental conditions.

First, filamentous Deuteromycetes and Ascomycetes usually contain haploid nuclei and the genome is relatively small, frequently about 25 to 50 mbp (million base pairs) (Gilly and Sands, 1991; Rambosek and Leach, 1987). *Trichoderma* spp. are no exception to this; in our studies with isozyme analysis we rarely, if ever, detected any strains giving multiple bands for single loci. Recently, several studies have been conducted that measured the sizes and numbers of chromosomes in *Trichoderma* spp. using some variant of pulsed field electrophoresis. In one study with *T. reesei* Rut-C30, six chromosomes were detected and these totalled 38 mbp. These bands were estimated to be of 3.3, 3.8, 4.5, 6.0, 8.4, and 11.9 mbp (Gilly and Sands, 1991), while another group found six chromosomes of 3.4, 3.7, 4.6, 5.5, and 7.4 mbp (Herrera-Estrella et al., 1992). In a third study (Mäntylä et al., 1992), wild-type and mutant strains were compared. Chromosome numbers ranged from 5 to 7, and differed substantially in size. We (Hayes et al., 1993) examined three biocontrol strains of *T. harzianum*, i.e. T12, T95, and a protoplast fusion progeny strain 1295-22. Only the largest chromosome banded in a similar position among the three strains, at about 5.4 mbp. Strain T12 had a second chromosome band at about 4.2 mbp, and no other bands, 1295-22 had a second band at 4.5 mbp, and T95 gave four bands of about 5.4, 4.7, 4.0, and 2.2 mbp. In our estimations, more than one chromosome probably occurs in at least some bands, otherwise the size of the genome of these strains would be unusually small. Southern analysis of the blot of T95 using the genomic DNA of 1295-22 as a probe demonstrated that the smallest chromosome of T95 has no homology to the genome of 1295-22. Conversely, another strain of *T. harzianum* provided six chromosome bands of 2.2, 3.7, 5.6, 6.5, and 7 mbp, while a strain of *T. viride* contained five bands of 4.2, 5.3, 6.0, 7.0, and 7.2 mbp (Herrera-Estrella et al., 1992). All of these data indicate that chromosomes of *Trichoderma* vary in number and size. Perhaps variation in numbers and size of chromosomes is tolerated in imperfect fungi because meiosis does not occur and so chromosome pairing is unnecessary.

Further, nearly all cells within *Trichoderma* and *Gliocladium* are polynucleate. All hyphal cells, and protoplasts derived from them (Sivan et al., 1990), contain numerous nuclei. Nuclear staining of some hyphal cells reveals a very high number of nuclei which we estimate to be >30. Conidia in the strains of *T. harzianum* that we have examined are also polynucleate (Stasz et al., 1988), but *T. reesei* and *Gliocladium virens* may produce at least some uninucleate conidia (Toyama et al., 1984; Ossanna and Mischke, 1990). Any manipulation that alters the nuclear genotype of a multinucleate structure is likely to give rise to heterokaryons. Such heterokaryons can be reduced to homokaryons by isolating subprogeny derived from single conidia since each conidium receives a single nucleus from its phialid (Stasz et al., 1988; Fincham, 1989). Mature conidia may contain several nuclei; this apparently arises from nuclear division during conidial maturation.

Studies on protoplast fusion indicate that some unusual events may alter the genome of *Trichoderma* spp., and perhaps other similar fungi as well. As described earlier, protoplast fusion progeny are extremely variable and frequently are unstable. While some variability is introduced by the protoplasting process itself, the variability and instability among progeny rising from fusion between two dissimilar strains is much greater than that occurring in progeny derived in a fusion within a single strain.

The expected genetic process subsequent to protoplast fusion is parasexuality. In this process, heterokaryons are formed, occasionally some nuclei fuse to give rise to a diploid, and then during subsequent growth, the diploid nucleus segregates to give rise to haploid or aneuploid nuclei containing some chromosomes from one parental strain and some from the second. Occasionally, mitotic crossing over may occur (Alexopoulos, 1964). This process clearly indicates (a) that initial progeny should be heterokaryotic, (b) that progeny resulting from parasexuality should contain genes from both strains, and (c) that the genes from opposite parental strains should usually be contained in DNA segments that are of chromosomal size, or if crossing over occurs, that represents a substantial fraction of a chromosome. Further, since entire cells fuse, extranuclear genetic sequences (e.g mitochondria) should also be heterokaryotic.

We examined the progeny of protoplast fusion in a wide range of inter- and intraspecific fusions. We identified a series of isozyme markers that unequivocally distinguished among parental strains (Stasz et al., 1988). Initial studies were with progeny following fusion between strain T12 his- and T95 lys- of *T. harzianum*. Progeny were extremely variable but when they were examined for their isozyme phenotype, nearly all were of the T12 phenotype. The few which exhibited the T95 phenotype were those which sectored and resembled the T95 wild type (prototrophic) parent. We could find no progeny which exhibited any recombination among isozyme characters. The original progeny strains were apparently highly unbalanced heterokaryons, with nuclei giving rise to a T12 phenotype outnumbering ones giving rise to the T95 phenotype by several orders of magnitude. In this and other fusions, we identified these two types as prevalent and nonprevalent parents or phenotypes, respectively. We then examined a larger sample of inter- and intrastrain fusions (over 1000 separate progeny) using 17 separate isozyme assays and again could find no evidence for recombination of genomes (Stasz and Harman, 1990).

We further examined a single progeny (1295-22) from the fusion between T12 his- and T95 lys- that had greater biocontrol ability than either wild type parent strain using chromosome electrophoresis. The isozyme phenotype of 1295-22 was that of T12, and the chromosome banding pattern was also similar to T12, except that the smallest chromosome band was slightly larger than that of T12 his- (Hayes et al., 1993).

RFLP analysis of 1295-22, another group of progeny from the fusion between T12 his- and T95 lys- that all arose from sectoring or single sporing from a single thallus, and a third set representing progeny of a fusion between *T. hamatum* strain 52198 lys- and *T. viride* T105 leu- gave results consistent with isozyme analysis. The repetitive RFLP pattern of mitochondrial DNA was always the same as that obtained with isozyme and nuclear RFLP analysis, and no evidence of recombination was seen, which suggests that mitochondrial and nuclear genomes are not independently segregated in protoplast fusion progeny (unpublished). These data all provided additional evidence against the operation of the parasexual cycle, or in fact, against the recombination of any very large DNA sequences.

Nonetheless, great variability did occur. We hypothesized that a process we have designated as interstrain gene transfer operates instead of the parasexual cycle. This process consists of the following steps: (a) cell fusion occurs through protoplast fusion or anastomosis, and therefore heterokaryons are formed. (b) Most nuclei of the nonprevalent parent are degraded and give rise to imbalanced heterokaryons. (c) Small DNA segments are released from the degraded nuclei, and these are eventually inserted into the nucleus of the prevalent strain. Different nuclei may obtain different genetic segments from the nonprevalent nuclei, and so nuclei may each differ slightly from its neighboring nuclei in the cell or thallus.

We have evidence that this process does indeed occur. We produced strain T95 his-hyg^R by transformation with a plasmid containing the *E. coli* gene for hygromycin B resistance (*Hyg*). *Hyg* was integrated into the genome of the strain and no free plasmids were detected. This transformed strain, which contained a dominant selectable marker which could be detected by Southern analysis, was fused with strain T12 wild type. We selected for strains that were prototrophic, were hyg^R, and which resembled strain T12 in colony morphology. Such strains were readily detected, and Southern analysis indicated that they contained *Hyg*. However, RFLP and isozyme analysis indicated that the genome was otherwise that of strain T12. Clearly, *Hyg* was transferred to the genome of T12 in the absence of other detectable genetic recombination. Extensive analysis of other progeny indicate that other genes, e.g. those conferring prototrophy to an auxotrophic mutant, can also be transferred in the absence of changes detectable by RFLP or isozyme analysis. The only requirement for detecting such shifts is an appropriate selection system which will identify the particular change of interest. Probably many other genes are also transferred to different nuclei, but in the absence of selection pressure, are not evident.

If such a system does occur, this has substantial implications for variability and adaptation in these and similar fungi. If anastomosis occurs in nature (we have evidence of its occurrence *in vitro* (Harman and Hayes, 1993)), then heterokaryons would be produced that contain many nuclei that differ from one another by the insertion of rather small genetic sequences. In addition, the nonprevalent nuclear type may also be present, perhaps also in modified form. In protoplast fusion progeny, we can detect nonprevalent types by the occasional occurrence of sectors of the nonprevalent parental type and also in subprogeny isolated from single conidia. Strains containing such a nuclear mosaic could be extremely adaptable. When environmental changes occur, different phenotypes would be selected for, just as we select for specific phenotypes with various selection pressures. In addition, when conidiation occurs, each spore may contain a subtly different genome than its sister spores, and so variation occurs again, and environmental pressures will select for those which best fit the immediate temporal and spatial environment. For biocontrol strains, it may be possible to produce synthetic strains that contain nuclei with specific traits for adaptation to variable environments or changing pathogen mixes. Such variation may also explain some of the variability and instability of Deuteromycete isolates in culture.

Genes and Gene Products

Biocontrol agents must contain useful genes and gene products if they are able to control plant pathogens. There are several requirements for such genes and their products to be of maximum utility, i.e. (a) they should code for proteins so that the gene can directly produce an active product, (b) the gene product should be highly active against the target organism, and (c) they should not be toxic to higher plants or vertebrates.

Enzymes that degrade fungal cell walls, especially chitinolytic enzymes, are attractive candidates for useful products. (1) They attack chitin, which is a constituent of the cell walls of higher fungi but which is not found in higher plants or vertebrates. Therefore, these enzymes should not be toxic to crops or higher animals. (2) Proteins with similar activity are probably part of the natural defense mechanisms of plants against infection (Legrand et al., 1987; Roberts and Selitrennikoff, 1988; Roby and Esquerre-Tugaye, 1987; Schlumbaum et al., 1986). This suggests that future transgenic crops containing genes that encode for cell wall degrading enzymes will be healthy and have little or no detrimental effect on the environment. Either the gene(s) or the gene product(s) can be used in pest control strategies. The remainder of this review will consider the cell wall degrading enzymes produced by *T. harzianum* and *G. virens*, and why we believe that they will have an important role in future plant disease management strategies.

Identification and use of gene products useful in biocontrol

Strains of *Trichoderma* and *Gliocladium* have

been shown to possess biocontrol characteristics, with different mechanism(s) proposed (Chet, 1987). The mechanism of mycoparasitism includes degradation of the target fungus by the action of extracellular hydrolytic enzymes (including chitinolytic enzymes) since fungal cell walls of Basidiomycetes, Ascomycetes and Zygomycetes contain chitin microfibrils (Kuhn et al., 1990). Three chitinolytic enzymes; N-acetyl-ß-glucosaminidase (NAGase, EC 3.2.1.30), chitin 1,4-ß-chitobiosidase (chitobiosidase), endochitinase, and a glucan 1,3-ß-glucosidase from *Trichoderma harzianum* strain P1, as well as an endochitinase from *Gliocladium virens* strain 41 have been identified in our laboratory and purified to electrophoretic homogeneity. The enzymes were purified using gel filtration followed by chromatofocusing and if necessary, isoelectric focusing from culture filtrates of strains grown under induced conditions (Di Pietro et al., 1993; Harman et al., 1993; Lorito et al., 1993a). The endochitinase and chitobiosidase had a molecular mass of approximately 40 kDa, a pI around 3.9 and displayed enzyme activity in the pH range of 4.5 to 7. The NAGase was approximately 72 kDa, a pI about 4.6 and had enzyme activity in the pH range of 4 to 7. Glucosidase was the largest enzyme, with a molecular mass of approximately 78 kDa, a pI about 6.2 and an active pH range from 4 to 7. None of the four enzymes purified from *T. harzianum* strain P1 were serologically related (Hayes et al., unpublished). The endochitinase from *G. virens* strain 41 had a molecular mass of about 41 kDa, a pI of 7.8 and a pH optimum for activity of about 3 to 7. This enzyme was shown to be serologically related to the endochitinase from P1 since it partially reacted with antibodies prepared against the P1 endochitinase. Other chitinolytic and glucanolytic enzymes are also produced by these fungi and remain to be purified and characterized (Tronsmo and Harman, 1993).

The endochitinase and chitobiosidase from *T. harzianum* had significant antifungal activity when tested individually. Only 40 to 180 µg/ml of protein were required for 50% inhibition (ED$_{50}$) of cell replication or spore germination of a variety of chitinous plant pathogens (Lorito et al, 1993a). The glucosidase and NAGase demonstrated similar levels of antifungal activity at comparable concentrations (unpublished). Chitinases from higher plants and bacteria required >100 µg/ml to obtain similar inhibition values. The endochitinase from *G. virens* also demonstrated antifungal activity, but required about 4-fold greater concentrations than *Trichoderma* enzymes to obtain 50% inhibition (Di Pietro et al., 1993).

Interestingly, chitinolytic enzymes from *Trichoderma* can attack and degrade strains of this fungus. *T. harzianum* strain 1295-22 is about as susceptible to enzymes from *T. harzianum* strain

P1 as plant pathogenic fungi; strain P1 apparently protects itself from its own enzymes through production of an inhibitor (Harman et al., 1993). If *T. harzianum* strain 1295-22 is grown with ß-1,3 glucan as a sole carbon source, the enzymes it produces severely damages its own hyphae (unpublished).

Synergy was observed when combinations of chitinolytic enzymes were tested (Lorito et al., 1993a). Only 10 to 34 µg/ml of total protein were required to reach the ED$_{50}$ value with the combination of endochitinase and chitobiosidase. These values were at least 10-fold lower than that reported for chitinolytic enzymes from other sources. Synergy was observed not only within chitinolytic enzymes, but also with these cell wall degrading enzymes and synthetic fungicides. In combination with sterol synthesis inhibitors used in both animal and human therapy and in plant protection, the enzymes reduced the ED$_{50}$ of some toxicants by more than 100-fold (unpublished). With other toxicants, the increase in sensitivity was about 6-fold. The enzymes may enhance sensitivity of the target pathogen to the fungitoxic compounds by increasing the uptake of the fungicides via cell wall digestion.

Synergy also occurred between enzymes and a biocontrol bacterium (Lorito et al., 1993c). Previous work (Hadar et al., 1983) has demonstrated that *Enterobacter cloacae* was an effective biocontrol bacterium but that it produced only very low levels of chitinolytic enzymes (Lorito et al., 1993c). Bacterial cells were shown to bind to hyphal walls via lectin-mediated attachments (Nelson et al., 1986). We hypothesized that this binding was advantageous to the bacterium because this placed the cell in a favorable location to obtain nutrients when the fungal cell wall became leaky due to damage or aging. Thus when in the presence of active chitinolytic enzymes the bacteria were perfectly positioned for the assimilation of released nutrients. Experiments were performed by adding the bacterium or the enzyme at sufficiently low concentrations so that individually there was little effect on the target fungus, *Botrytis cinerea*. However, when the enzyme and the bacteria were combined, the target fungi were essentially destroyed (Lorito et al., 1993c). These data indicate that there was a strong synergy between *E. cloacae* and chitinolytic enzymes from *T. harzianum*. From these results, it appears that transgenic *E. cloacae* expressing and secreting chitinolytic enzymes from *T. harzianum* would provide a very potent organism for biological control.

Gene isolation and sequencing

We have isolated three genes that code for an

endochitinase, chitobiosidase and glucosidase from *T. harzianum* strain P1. A cDNA library was constructed, in the phage expression vector λgtll, using mRNA isolated from *T. harzianum* strain P1, grown with chitin as the sole carbon source. Unique polyclonal antibodies, i.e. with no cross reactivity to other *T. harzianum* extracellular proteins, were produced for each enzyme and used to screen the library. Several positive plaques were detected with each antibody and further purified. In the case of a endochitinase clone, a fusion protein was produced that released fluorescent methylumbelliferone from a substrate for endochitinase activity, i.e. 4-methylumbelliferyl ß-D-N,N',N"-triacetyl-chitotriose (Tronsmo and Harman, 1993). The gene coding for this protein was sequenced and found to be 1096 bases in length. Comparison of the nucleic acid sequence data with the N-terminal amino acid sequence indicated that a portion of the 5'-end was missing. Therefore we prepared a 20-mer internal primer 89 bp downstream from the 5'-end of the endochitinase gene and used this and a 24-mer forward or reverse λ primers in PCR to identify viral clones containing the missing sequence. This effort has been successful; the entire gene from *T. harzianum* is 1554 bp in length, and it has some similarity to to class III endochitinases from higher plants and to endochitinases from prokaryotic sources (unpublished).

A similar procedure for gene isolation has been performed by screening the cDNA library with antibodies for chitobiosidase and glucosidase. Clones have been identified that may contain the gene for each enzyme. These clones are now being sequenced.

Uses of the genes that code for chitinolytic enzymes

When genes are isolated for enzymes of interest, portions of sequences will be used for insertional mutation and specific inactivation of genes of interest to determine the role each enzyme assumes in mycoparasitism. Once deficient mutants are available, we will also add the gene coding for the missing enzyme back to the mutant and in this way be able to perform a molecular version of Koch's Postulate.

Endochitinase minus mutants have already been prepared and are currently being tested. The mutants were formed by inserting a segment of the endochitinase gene into a plasmid and transforming *T. harzianum* strain P1 using biolistic transformation (Lorito et al., 1993b). The inability of these strains to produce active endochitinase was probably due to disruption of the native gene by homologous insertion of the plasmid into the gene. The inability of the transformants to produce functional endochitinase has had no major affect on growth of the strains, which indicates that the enzyme does not have a major role in growth and differentiation of the fungus (unpublished).

Once intact genes are isolated, the genetic sequences coding for products that are inhibitory to plant pests could have great potential for use in producing transgenic microbes with enhanced biocontrol activity and in production of transgenic plants resistant to pests. Transgenic plants with enhanced ability to produce endochitinase have been shown to be either only partially resistant to (Broglie et al., 1991) or else to be susceptible (Neuhaus et al., 1991) to plant pathogenic fungi. Since these plants express only single enzymes, the enzymes would have to be produced in quite high levels for any substantial disease control to be exhibited, and so we would expect the results obtained in these studies. In addition, transgenic *E. coli* obtained biocontrol ability when engineered to express the endochitinase from *Serratia marcescens* (Shapira et al., 1989) These early studies with transgenic plants or microbes transformed to produce single proteins are encouraging, since even with these relatively inefficient systems positive results have been obtained. Once transgenic organisms capable of producing highly efficient synergistic combinations of enzymes are produced, excellent disease control should be obtained. Transgenic organisms, therefore, offer the potential of substantial reductions in the quantity of chemical pesticide applications required for producing disease free plants.

Acknowledgement: The gene for hygromycin resistance used in some portions or this work was provided by the Eli Lilly Co. This research was supported in part by BARD Grant US-1723-89.

References

Ahmad, J. S., Baker, R. "Rhizosphere competence of *Trichoderma harzianum.*" *Phytopathology* **1987**, *77*, 182-189.

Ahmad, J. S., Baker, R. "Growth of rhizosphere-competent mutants of *Trichoderma harzianum* on carbon substrates." *Can. J. Microbiol.* **1988**, *34*, 807-814.

Alexopoulos, C. J. *Introductory Mycology, 2nd ed.*; John Wiley & Sons: New York, 1964.

Broglie, K., Chet, I., Holliday, M., Cressman, R., Biddle, P., Knowlton, S., Mauvais, C. J., Broglie, R. "Transgenic plants with enhanced resistance to the fungal pathogen *Rhizoctonia solani.*" *Science* **1991**, *254*, 1194-1196.

Chet, I. "*Trichoderma*-Application, mode of action, and potential as a biocontrol agent of soilborne plant pathogenic fungi." In *Innovative Approaches to Plant Disease Control*; I. Chet, Ed., J. Wiley and Sons: New York, 1987; pp 137-160.

Costanzo, M. C., Fox, T. D. "Transformation of yeast by agitation with glass beads." *Genetics* **1988**, *120*, 667-670.

Di Pietro, A., Lorito, M., Hayes, C.K., and Harman, G.E. "Endochitinase from *Gliocladium virens:* isolation, characterization and synergistic antifungal activity in combination with gliotoxin." *Phytopathology* **1993**, *83*, 308-313.

Fincham, J. R. S. "Transformation In Fungi." *Microbiol.Rev.* **1989**; *53*; 148-170.

Gilly, J. A., Sands, J. A. "Electrophoretic karyotype of Trichoderma reesei." *Biotechnol Letters* **1991**, *13*, 477-482.

Hadar, Y., Harman, G. E., Taylor, A. G., Norton, J. M. "Effects of pregermination of pea and cucumber seeds and of seed treatment with *Enterobacter cloaca*e on rots caused by *Pythium* spp." *Phytopathology* **1983**, *73*, 1322-1325.

Harman, G. E. "Seed treatments for biological control of plant disease." *Crop Protect.* **1991**, *10*, 166-171.

Harman, G. E., Taylor, A. G., Stasz, T. E. "Combining effective strains of *Trichoderma harzianum* and solid matrix priming to improve biological seed treatments." *Plant Dis.* **1989**, *73*, 631-637.

Harman, G. E., Hayes, C. K. "The genetic nature and biocontrol ability of progeny from protoplast fusion in *Trichoderma.*" In *Biotechnology in Plant Disease Control*; I. Chet, Ed.; 1993, pp. 237-255.

Harman, G. E., Hayes, C. K., Lorito, M., Broadway, R. M., Di Pietro, A., Tronsmo, A. "Chitinolytic enzymes of *Trichoderma harzianum*: purification of chitobiase and endochitinase." *Phytopathology* **1993**, *83*, 313-318.

Harman, G. E., Stasz, T. E. "Protoplast fusion for the production of superior biocontrol fungi." In *Microbial Control of Weeds*; D. O. TeBeest, Ed.; Chapman and Hall: New York, 1991; pp 171-186.

Hayes, C. K. "Improvement of *Trichoderma* and *Gliocladium* by genetic manipulation." In *Biological Control of Plant Diseases. Progress and Challenges for the Future*; E. C. Tjamos Papavizas, G., and Cook, R. J., Ed.; Plenum Press: New York, 1992; pp 277-286.

Hayes, C. K., Harman, G. E., Woo, S. L., Gullino, M. L., Lorito, M. "Methods for electrophoretic karyotyping of filamentous fungi in the genus *Trichoderma.*" *Anal. Biochem.* **1993**, 176-182.

Herrera-Estrella, A., Goldman, G. H., Van Montagu, M., Geremia, R. A. "Electrophoretic karyotype and gene assignment to resolved chromosomes of Trichoderma spp." *Molec. Microbiol.* **1993**, *7*, 515-521.

Herrera-Estrella, A., Goldman, G. H., van Montagu, M. "High-efficiency transformation system for the biocontrol agents, *Trichoderma* spp." *Molec. Microbiol.* **1990**, *4*, 839-843.

Ito, H., Fukuda, Y., Murata, K., Kimura, A. "Transformation of intact yeast cells treated with alkali cations." *J. Bacteriol.* **1983**, *153*, 163-168.

Jin, X., Hayes, C. K., Harman, G. E. "Principles in the development of biological control systems employing *Trichoderma* species against soil-borne plant pathogenic fungi." In *Frontiers in Industrial Mycology;* G. C. Letham (Ed.); Am. Mycol. Soc., Chapman and Hall: 1992; pp. 174-195.

Kuhn, P. J., Trinci, A. P. J., Jung, M. J., Goosey, M. W., Copping, L. G. Eds.; *Biochemistry of Cell Walls and Membranes in Fungi;* Springer-Verlag: New York, NY 1990.

Legrand, M., Kauffmann, S., Geoffroy, P., Fritig, B. "Biological function of pathogenesis-related proteins: four tobacco pathogenesis-related proteins are chitinases." *Proc. Nat. Acad. Sci. USA* **1987**, *84*, 6750-6754.

Lorito, M., Harman, G., Hayes, C., Broadway, R., Tronsmo, A., Woo, S. L., Di Pietro, A. "Chitinolytic enzymes produced by *Trichoderma harzianum*: antifungal activity of purified endochitinase and chitobiosidase." *Phytopathology* **1993a**, *83*, 302-307.

Lorito, M., Hayes, C. K., Di Pietro, A., Harman, G. E. "Biolistic transformation of *Trichoderma* spp. and *Gliocladium* spp. using plasmid and genomic DNA." *Curr. Genet.* **1993b**, (in press).

Lorito, M., Hayes, C. K., Woo, S. L., Di Pietro, A., Harman, G. E. "Antifungal synergistic interaction between chitinolytic enzymes from *Trichoderma harzianum* and *Enterobacter cloacae.*" *Phytopathology* **1993c**, *(In press).*

Mäntylä, A. L., Rossi, K. H., Vanhanen, S. A., Penttilä, M. E., Suominen, P. L., Helena Nevalainen, K. M. "Electrophoretic karyotyping of wild-type and mutant *Trichoderma longibrachiatum (reesei)* strains." Curr. Genet. **1992**, *21*,471-477.

Nelson, E. B., Chao, W.-L., Norton, J. M., Nash, G. T., Harman, G. E. "Attachement of *Enterobacter cloacae* to hyphae of *Pythium ultimum*: possible role in the biological control of Pythium preemergence damping-off." *Phytopathology* **1986**, *76*, 327-335.

Neuhaus, J-M, Ahl-Goy, P., Hinz, U., Flores, S., Meins, F. "High-level expression of a tobacco chitinase gene in *Nicotiana sylverstris*. Suceptibility of transgenic plants to *Cercospora nicotianae* infection." *Plant Molec. Biol.* **1991**. *16*, 141-151.

353

Ossanna, N., Mischke, S. "Genetic transformation of the biocontrol fungus *Gliocladium virens* to benomyl resistance." *Appl Environ Microbiol* **1990**, *56*, 3052-3056.

Papavizas, G. C., Lewis, J. A., Abd-El Moity, T. H. "Evaluation of new biotypes of *Trichoderma harzianum* for tolerance to benomyl and enhanced biocontrol capabilities." *Phytopathology* **1982**, *72*, 126-132.

Rambosek, J. A. L., Leach, J. "Recombinant DNA in filamentous fungi: progress and prospects." *CRC Crit. Rev. Biotechnol.* **1987**, *6*, 357-393.

Roberts, W. K., Selitrennikoff, C. P. "Plant and bacterial chitinases differ in antifungal activity." *J. Gen. Microbiol.* **1988**, *134*, 169-176.

Roby, D., Esquerre-Tugaye, M.-T. "Induction of chitinases and of translatable mRNA for these enzymes in melon plants infected with *Colletotrichum lagenarium*." *Plant Sci* **1987**, *52*, 175-185.

Schlumbaum, A., Mauch, G., Voglie, U., Boller, T. "Plant chitinases are potent inhibitors of fungal growth." *Nature* **1986**, *234*:, 365-367.

Shapira, R., Ordentlich, A., Chet, I., Oppenheim, A. B. "Control of plant diseases by chitinase expressed from cloned DNA in *Escherichia coli*." *Phytopathology* **1989**, *79*, 1246-1249.

Sivan, A., Harman, G. E., Stasz, T. E. "Transfer of isolated nuclei into protoplasts of *Trichoderma harzianum*." *Appl. Environ. Microbiol.* **1990**, *56*, 2404-2409.

Sivan, A., Stasz, T. E., Hemmat, M., Hayes, C. K., Harman, G. E. "Transformation of *Trichoderma* spp. with plasmids conferring hygromycin B resistance." *Mycologia* **1992**, *84*, 687-694.

Stasz, T. E., Harman, G. E. "Nonparental progeny resulting from protoplast fusion in *Trichoderma* in the absence of parasexuality." *Exp. Mycol.* **1990**, *14*, 145-159.

Stasz, T. E., Harman, G. E., Gullino, M. L. "Limited vegetative compatibility following intra- and interspecifc protoplast fusion in *Trichoderma*." *Exp. Mycol.* **1989**, *13*, 364-371.

Stasz, T. E., Harman, G. E., Weeden, N. F. "Protoplast preparation and fusion in two biocontrol strains of *Trichoderma harzianum*." *Mycologia* **1988**, *80*, 141-150.

Stasz, T. E., Weeden, N. F., Harman, G. E. "Methods of isozyme electrophoresis for *Trichoderma* and *Gliocladium* species." *Mycologia* **1988**, *80*, 870-874.

Thomas, M. D., Kenerley, C. M. "Transformation of the mycoparasitie *Gliocladium*." *Curr. Genet.* **1989**, *15*, 415-420.

Thomashow, L. S., Weller, D. M. "Role of a phenazine antibiotic from *Pseudomonas fluorescens* in biological control of *Gaeumannomyces graminis* var. *tritici*." *J. Bacteriol.* **1988**, *170*, 3499-3508.

Toyama, H., Yamaguchi, K., Shinmyo, A., Okada, H. "Protoplast fusion of *Trichoderma reesei* using immature conidia." *Appl. Environ. Microbiol.* **1984**, *47*, 363-368.

Tronsmo, A., Harman, G. E. "Detection and quantification of N-acetyl-ß-D-glucosaminidase, chitobiosidase and endochitinase in solution and on gels." *Anal. Biochem.* **1993**, *208*, 74-79.

Improvement of the Biocontrol Fungus, *Gliocladium virens,* by Genetic Manipulation

Sue Mischke, Biocontrol of Plant Diseases Laboratory, USDA, ARS, Beltsville, MD 20705

Gliocladium virens is a biocontrol agent for several soilborne plant pathogens. Transgenic strains, created by the introduction of a mutant *Neurospora crassa* gene for β-tubulin, were screened for resistance to benomyl. Ability of transformants to control Rhizoctonia damping-off of cucumber was tested in growth chambers. Transgenic strains which performed at least as well as the isogenic wild-type progenitor are potential biocontrol agents suitable for use in integrated pest management (IPM) applications with benomyl.

Gliocladium virens, a biocontrol fungus related to *Trichoderma* spp. (Stasz et al., 1989), has biocontrol activity against several important fungal pathogens (Lumsden and Locke, 1989; Papavizas and Lewis, 1989). Because *G. virens* is sensitive to benzimidazole (MBC) fungicides, it cannot be used in combination with benomyl in an IPM program. Production of *G. virens* biotypes tolerant to benomyl by mutagenesis has been difficult (Papavizas et al., 1990) and the rare benomyl-tolerant mutants were less efficient as biocontrol agents than the respective wild-type strains (Papavizas, 1992). Transgenic benomyl-resistant biotypes of *G. virens* were created by introduction of foreign DNA into multiple sites in the genome (Ossanna and Mischke, 1990). It cannot be assumed that efficacy was not impaired by this "improvement." *In vitro* testing indicated that transformed strains were still antagonistic to *Rhizoctonia solani* (Mischke, unpublished). The objective of this study was to test *in vivo* biocontrol characteristics of the transgenic *G. virens.*

Materials and Methods

The wild-type (+) strain Gl-21 of *G. virens* was transformed to benomyl resistance (Ossanna and Mischke, 1990) with a mutant β-tubulin gene cloned from *Neurospora crassa* (Orbach et al., 1986). Fermentor biomass (FB) of each strain (Table 1) was grown in a molasses/yeast medium. Medium for transgenic

Table 1. Effect of Biotypes on Cucumber Stands After 2 Weeks.

Treatment	Percent Healthy Plants	
	Mean ± SE	Duncan Group[*]
Noninfested	82.9 ± 4.3	A
R. solani	22.1 ± 5.7	D
Gl-21 (+)	46.4 ± 6.6	C
Gl-21-T1	30.7 ± 8.0	C D
Gl-21-T2	65.0 ± 8.0	B

[*]Treatments followed by different letters are significantly different ($P=.05$).

biotypes was amended with 0.5 µg/ml benomyl to prevent growth of revertants. Biomass was air-dried, milled to pass through a 425-µm screen, and formulated as 6% (w/w) in vermiculite-bran (VBA-FB; Lewis et al., 1991). Two days after activation with 0.05N HCl, the VBA-FB was added at the rate of 5% (w/w) to autoclaved potting soil or to autoclaved potting soil infested with isolate Rs-23 of *Rhizoctonia solani* (AG-4). Controls (noninfested or *R. solani*) contained VBA (without biocontrol organisms). Six days later, pots containing 180 g amended potting soil were planted with 20 cucumber seeds (Pepino, Straight Eight). Pots were placed in a growth chamber with a 12-hr light cycle at 20.5±2 C. Distilled water was applied daily as needed to maintain pots at the same moisture level. Healthy plant stand was assessed after two weeks and expressed as a percent of seeds planted. The randomized complete block design was performed twice with at least three replicates. Data were analyzed by ANOVA (SAS Institute, Inc., Cary, NC) for significant differences using Duncan's multiple-range test.

Results and Discussion

The preservation of antagonism against the target pest by the transgenic biotypes was demonstrated in this study. Results from separate experiments were similar and data presented in Table 1 were pooled prior to analysis. Statistically, the transformed strains were at least as efficient as Gl-21. When data were classed by the time of disease appearance, preemergence losses were greater with Gl-21-T1 than with Gl-21-T2, suggesting that the transformants may be different in their respective abilities to control disease prior to germination (data not shown). These differences were not significant at the $P = .05$ level.

Demonstrating retention of biocontrol efficacy of transgenic organisms is crucial. Impairment of biological processes following transformation has been documented for *Pseudomonas putida* (Hong et al. 1991) and for *Cochliobolus heterostrophus* (Keller et al. 1990). However an insect biocontrol fungus, transformed to benomyl resistance with an *Aspergillus nidulans* gene, retained desirable pest management characteristics (Goettel et al., 1990). The demonstration that biocontrol ability was conserved in transgenic *G. virens* permits use of the improved biotypes in an IPM program employing benomyl.

References

Goettel, M.S; Leger, R.J.S; Bhairi, S.; Jung, M.K.; Oakley, B.R.; Roberts, D.W.; Staples, R.C. "Pathogenicity and growth of *Metarhizium anisopliae* stably transformed to benomyl resistance." *Curr. Genet.,* **1990,** *17,* 129-132.

Hong, Y.; Pasternak, J.J.; Glick, B.R. "Biological consequences of plasmid transformation of the plant growth promoting rhizobacterium *Pseudomonas putida* GR12-2." *Can. J. Microbiol.,* **1991,** *37,* 796-799.

Keller, N.P.; Bergstrom, G.C.; Yoder, O.C. "Effects of genetic transformation on fitness of *Cochliobolus heterostrophus.*" *Phytopathology,* **1990,** *80,* 1166-1173.

Lewis, J.A.; Papavizas, G.C.; Lumsden, R.D. "A new formulation system for the application of biocontrol fungi to soil." *Biocontrol Sci. Technol.,* **1991,** *1,* 59-69.

Lumsden, R.D.; Locke, J.C. "Biological control of damping-off caused by *Pythium ultimum* and *Rhizoctonia solani* with *Gliocladium virens* in soilless mix." *Phytopathology,* **1989,** *79,* 361-366.

Orbach, M.J.; Porro, E.B.; Yanofsky, C. "Cloning and characterization of the gene for β-tubulin from a benomyl-resistant mutant of *Neurospora crassa* and its use as a dominant selectable marker." *Mol. Cell. Biol.,* **1986,** *6,* 2452-2461.

Ossanna, N.; Mischke, S. "Genetic transformation of the biocontrol fungus *Gliocladium virens* to benomyl resistance." *Appl. Environ. Microbiol.,* **1990,** *56,* 3052-3056.

Papavizas, G.C. In *Biological Control of Plant Diseases;* Tjamos, E.C.; Papavizas, G.C.; Cook, R.J., Eds.; NATO ASI Series A: Life Sciences; Plenum Press: New York, NY, 1992, Vol. 230; pp 223-230.

Papavizas, G.C.; Lewis, J.A. "Effect of *Gliocladium* and *Trichoderma* on damping-off and blight of snapbean caused by *Sclerotium rolfsii* in the greenhouse." *Plant Pathol.,* **1989,** *38,* 277-286.

Papavizas, G.C.; Roberts, D.P.; Kim, K.K. "Development of mutants of *Gliocladium virens* tolerant to benomyl." *Can. J. Microbiol.,* **1990,** *36,* 484-489.

Stasz, T.E.; Nixon, K.; Harman, G.E.; Weeden, N.F.; Kuter, G.A. "Evaluation of phenetic species and phylogenetic relationships in the genus *Trichoderma* by cladistic analysis of isozyme polymorphism." *Mycologia,* **1989,** *81,* 391-403.

Transgenic Beneficial Arthropods for Pest Management Programs: An Assessment of their Practicality and Risks

Marjorie A. Hoy, Department of Entomology and Nematology, University of Florida, Gainesville, FL 32611-0620

Techniques to develop transgenic beneficial arthropods have advanced and are improving rapidly. Future use of transgenic arthropods in pest management programs will require both fundamental advances in techniques and resolution of issues surrounding the release of transgenic arthropods into the environment.

Thirty-two years ago, Sailer (1961) suggested that interspecific hybridization might provide useful genes for genetic improvement of beneficial arthropods. In so doing, he anticipated the possibilities offered by recombinant DNA (rDNA) techniques. It is likely that advances in molecular genetics now will provide the opportunities for employing genes from a wide array of species to modify beneficial arthropod species for use in pest management programs. The goal of molecular genetic manipulations of beneficial arthropods is to *enhance* the natural enemy so that improved biological pest control is achieved. Competitiveness and ability to function effectively under field conditions is required. A significant constraint on genetic manipulation is the anticipated difficulty of maintaining quality in mass-reared populations. One of the significant benefits of rDNA techniques may be that it will be easier to measure and maintain 'quality' in transgenic arthropods.

Genetic improvement of arthropod natural enemies has been achieved by artificial selection. Hybridization of different strains to achieve heterosis or the use of mutagens to obtain a specific trait have been proposed, but these approaches have been employed only rarely (Hoy, 1990a). Beckendorf & Hoy (1985) suggested that rDNA techniques could make genetic improvement of arthropod natural enemies more efficient and less expensive, because once a gene has been cloned it could be inserted into a number of beneficial species. If the species need not be reared for long periods of time, as is often the case with artificial selection projects, there is less likelihood that laboratory selection and inbreeding would occur.

The ability to manipulate genetic material and insert it into the genome of *Drosophila* has been used to develop a fundamental understanding of genetics, biochemistry, development and behavior (Lawrence, 1992). Genetic engineering of arthropods other than *Drosophila* has been attempted, with limited successful results (Handler & O'Brochta, 1991; Walker, 1989). However, the techniques available to achieve stable transformation of arthropods are changing rapidly.

Traditional genetic manipulation projects involving artificial selection or mutagenesis by irradiation typically have three phases: conceiving and identifying the problem, developing the genetically-manipulated strain, and evaluating and implementing (Hoy, 1990a). The use of rDNA techniques may alter this sequence because risks associated with releases into the environment need to be resolved. This paper reviews the state of the art of genetic manipulation of beneficial arthropods by rDNA techniques from the point of view of improving pest management in agriculture.

What Genetic Changes Might Be Useful?

A number of arthropod natural enemies have been selected for resistance to pesticides, lack of diapause, and temperature tolerance, although most projects have involved selection for resistance to pesticides (Hoy, 1990b). Pesticide-resistant strains have been evaluated in the field and are being implemented in several integrated pest management (IPM) programs (Hoy, 1990b). Genetic improvement has proven to be practical and cost effective when the trait(s) limiting efficacy can be identified, the improved strain retains its fitness, and methods for implementation have been developed (Headley & Hoy, 1987). Traits primarily determined by single major genes are most appropriate for manipulation at this time because methods for manipulating and stabilizing traits that are determined by complex genetic mechanisms are not yet available. Genetic improvement can be useful when the natural enemy is known to be a potentially effective biological control agent except for a limiting factor and the limiting trait primarily is influenced by a single major gene.

Genetic improvement requires that the gene be obtained by selection, mutagenesis, or cloning; the manipulated strain is fit and effective; and the released strain can be maintained by some form of reproductive isolation. Typically, applications of pesticides reduce populations of susceptible natural enemies, allowing the resistant population to establish and persist. Alternatively, genetically-manipulated strains can be released into greenhouses or in new geographic regions so that the reproductively-isolated population is maintained with the desired trait intact. Whether it is necessary to maintain reproductive isolation is unclear. We have no data to indicate whether a gene of interest can be introduced into a wild population by introgression to produce a well-adapted new 'hybrid' population that carries the new attribute (Hoy, 1990a).

Steps In Genetic Manipulation by rDNA Methods

Genetic improvement by rDNA techniques involves several steps (Fig.1). Success generally requires that we have a thorough knowledge of the biology, ecology and behavior of the target species. The first step is critically important together with identifying one or more specific traits that, if altered, potentially would achieve the goals of the project. Next, suitable genes must be identified and cloned. Appropriate regulatory sequences must be identified so that the inserted gene will be expressed at appropriate levels in the correct tissues and at a relevant time. Effective germ line transformation methods are needed so that stable transformation of organisms can be achieved in an efficient and predictable manner. Next, the relative fitness of the transgenic strain should be evaluated. These steps can take place in the laboratory. If the laboratory tests are positive, the transgenic strain(s) should be evaluated in small field plots to confirm their efficacy and fitness. Before any field tests can occur, however, regulatory issues relating to the safety of releasing transgenic beneficial arthropods must be resolved (Hoy, 1990a, 1992).

Potential Germ Line Transformation Methods

Inserting cloned DNA into pest or beneficial arthropods could be accomplished by several different techniques. The effects of the inserted DNA could be transient and short term, or stable and long term. Cloned DNA can be isolated from the same or other species; it is feasible to insert genes from microorganisms into arthropods and have the DNA transcribed and translated if

promoters (controlling elements) and other regulatory DNA sequences derived from a eukaryote are present. Most research on stable (germ line) transformation methods has been accomplished with *Drosophila melanogaster*. Initial efforts to genetically engineer *D. melanogaster* resulted in rare successes until a transposable element called the P element was genetically manipulated to serve as a vector to carry exogenous genes into the chromosomes of germ line cells (Rubin & Spradling, 1982; Spradling & Rubin, 1982). This pioneering work has created an explosion of research on fundamental genetic analyses of gene structure, function, and regulation in *Drosophila* and has given us a broad understanding of how flies develop (Lawrence, 1992). Many genes have been identified, isolated, and cloned from *Drosophila*. Unfortunately, only a few of these genes appear to be potentially useful in genetic manipulations of beneficial arthropods.

P-element vectors have been used effectively with other *Drosophila* species such as *D. similans* and *D. hawaiiensis*. A number of other insect species, including three mosquitoes and the Mediterranean fruit fly, *Ceratitis capitata*, have received microinjected DNA cloned into P-element vectors. *Aedes aegypti* (Morris et al., 1989), *Anopheles gambiae* (Miller et al., 1987), and *Aedes triseriatus* (McGrane et al., 1988) have been transformed in a stable manner. However, the rate of transformation was very low (less than 0.1% of the microinjected embryos), and there is no evidence that the process of transformation was P element-mediated. It appears that P element-mediated transposition may be limited to *Drosophila* species (Handler & O'Brochta, 1991). As a result, a variety of other methods for achieving transformation have been considered and evaluated.

Other transposable element vectors Transposable elements are commonly found in organisms, but they have been relatively unstudied in arthropods other than *Drosophila* (Berg & Howe, 1989). Species-specific transposable elements could be isolated and genetically modified for use as vectors, but this is time consuming and expensive and this approach may be limited to arthropods that are of major economic importance such as silk moths (Michaille et al., 1990). Furthermore, because transgenic arthropods being released into the environment should be stably transformed, transposable element vectors should be incapable of subsequent movements. Thus, issues of risk assessment should be considered in designing a genetic manipulation project involving 'native' transposable element vectors (Hoy, 1992).

Microinjection Microinjecting exogenous DNA carried in P-element vectors into

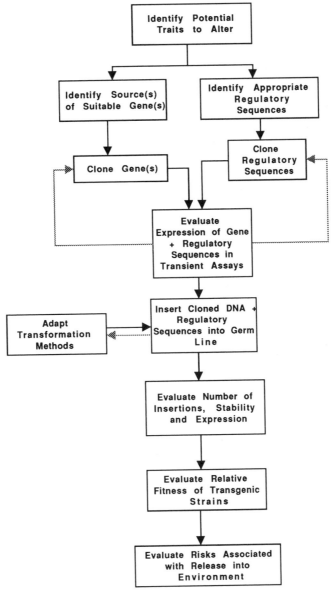

Fig. 1. Steps in genetic engineering of beneficial arthropods to the time when they are ready to be released into the environment.

Drosophila eggs is a well-developed technique (Santamaria, 1986).These microinjection methods had to be modified for mosquito eggs (McGrane et al., 1988; Miller et al., 1987; Morris et al., 1989). Milne et al. (1988) developed a method for microinjecting early honeybee, *Apis mellifera*, embryos. Presnail & Hoy (1992) found that eggs of the phytoseiid predator *Metaseiulus occidentalis* were extremely difficult to dechorionate and dehydrate and that the needle

tip had to be modified. It appears that the *Drosophila* microinjection methods will have to be adapted empirically to each insect species and will not be feasible with all. Variables to consider include whether to dechorionate or not, whether to dehydrate and for how long, what age/stage to inject, what holding conditions to implement after injection, and what size and shape of needle to use. It may be feasible to microinject DNA into insect embryos without using a transposable element vector. Transformation of a number of organisms, including *Drosophila*, can be achieved by such a method, although at a relatively low rate (Walker, 1989). Steller & Pirrota (1985) suggested that the frequency of genomic integration might be stimulated by the use of nicked plasmid DNA. Others have suggested that integration might be enhanced if single-stranded DNA were used.

Maternal microinjection Early preblastoderm eggs present within adult females of the predatory mite *M. occidentalis* were microinjected by inserting a needle through the cuticle of gravid females. This technique, called 'maternal microinjection', resulted in relatively high levels of survival and stable transformation without the aid of a transposase-producing helper plasmid (Presnail & Hoy, 1992). The *lacZ* reporter gene regulated by the *Drosophila hsp70* promoter was expressed in larvae developing from the injected eggs and in subsequent generations. Stable transformation was confirmed in the sixth generation by polymerase chain reaction (PCR) amplification of a region spanning the *Drosophila*/*E. coli* DNA sequences inserted into the mite.

Maternal microinjection of *M. occidentalis* is less laborious than microinjection of eggs that have been laid by females, because the eggs do not need to be dechorionated or dehydrated prior to injection. Survival rates of injected females were comparable to survival of microinjected *Drosophila* eggs. The transformation rate was approximately 1/10th the efficiency of P element-mediated transformation of *Drosophila*, but comparable to techniques employed for species in which transformation is achieved without a P-element vector. It is possible that maternal microinjection will provide a 'universal' DNA delivery system for arthropod species (Presnail & Hoy, 1992).

Other techniques Before the development of P-element vectors, exogenous DNA was introduced into *Drosophila* embryos by soaking them in DNA solutions after dechorionation (reviewed by Walker, 1989). Some of the adult flies produced exhibited somatic mosaicism, which appeared to be due to incorporation and expression of exogenous DNA. However, the method was not much used because of low uptake (<2%), variable phenotypes, and the difficulty of establishing stably-transformed lines of flies. Most experiments used total genomic DNA and Walker (1989) speculated that soaking embryos in specific cloned sequences rather than total DNA could yield higher rates of stable transformation.

The use of sperm as carriers of exogenous DNA has been evaluated for honey bees, *Apis mellifera*, and the Australian sheep blowfly, *Lucilia cuprina* (Atkinson et al., 1991). This approach to gene transfer would probably be limited to species, such as the honey bee, for which semen can be collected and used for artificial insemination. Stable transformation has not been achieved using this technique and considerable controversy surrounds this approach (Barinaga, 1989; Milne et al., 1989). The technique was originally applied to transfer foreign genes into mice, but the original success apparently has not been replicated in other laboratories (Barinaga, 1989). Atkinson et al. (1991) indicate that sperm from both *L. cuprina* and *A. mellifera* are capable of binding labeled DNA, but the DNA is associated entirely with the outer sperm membrane. There is no evidence that the DNA is taken into the sperm and these authors concluded that internalization of DNA is important if the sperm is to serve as a vector.

A novel method for delivering DNA into living plant cells involves coating microprojectiles with DNA or RNA, then shooting them into a plant cell with a gun. The so-called 'gene gun' has been used successfully to transform major crop plants, yeast and cultured cells. The number of attempts with arthropod eggs is, however, limited. Baldarelli & Lengyel (1990) obtained transient expression of DNA in *Drosophila* embryos after ballistic introduction with DNA-carrying tungsten particles. The authors suggest this method may, with some modification, be suitable for stable germ line transformation. Whether other arthropod eggs can be dechorionated and transformed has not been determined. This technique may be particularly useful with species that deposit large numbers of eggs. It would not be advantageous for species, such as parasitic Hymenoptera, that deposit their eggs into the body of their insect host because obtaining large numbers of eggs by dissection would be extremely tedious.

Electroporation of DNA into insect embryos allows many more embryos to be transformed at one time than does microinjection into individual dechorionated eggs. With short electric impulses above a certain strength, membranes temporarily are made more permeable, which allows material to cross the perturbed membrane. Non-dechorionated *Drosophila* embryos, incubated in a solution of DNA, can take up and transiently express this DNA (Kamdar et al., 1992).

Artificial chromosomes have been constructed in yeast that behave much like natural ones do. Apparently, the essential components needed for chromosomes include genes, centromeres, sequences that serve as origins of chromosome replication, and telomeres (the chromosome ends). It may be possible to develop artificial chromosomes for arthropods. Recently, transgenic mice were obtained by injecting a yeast artificial chromosome (YAC) into fertilized mouse oocytes (Schedl et al., 1992). The resulting mice carried the YAC DNA and expressed the YAC-encoded tyrosinase gene so that the albino mice were pigmented. The YAC integrated into the mouse genome and the presence of yeast telomeric sequences apparently did not reduce the efficiency of integration. Artificial chromosomes may be particularly useful for situations where it is desirable to insert a number of genes that should be linked.

Zalokar (1981) reported methods for injecting and transplanting nuclei and pole cells into eggs of *Drosophila*. Thus, it might be possible to genetically transform insect cells in cell culture, isolate the nuclei, and transplant them into the region in embryos where the germ line cells (pole cells) will develop.

Beard et al. (1992) demonstrated that genetic engineering of insect symbionts is feasible by transforming a bacterial symbiont, *Rhodococcus rhodnii*, of the Chagas' disease vector *Rhodnius prolixus*. The symbiont lives extracellularly in the insect gut lumen and is transmitted from adult to progeny by egg shell contamination or by contamination of food with infected feces. *R. rhodnii* was genetically engineered to be resistant to an antibiotic and the resistant symbionts were transmitted to insects lacking symbionts. The insects containing the resistant symbionts subsequently were treated with the antibiotic and survived, transmitting the transformed symbiont to successive generations of insects.

Cultured insect cells can be induced to take up exogenous DNA by several methods, and the transformed cells can be used to evaluate the expression of genes and promoters (Fallon, 1991). Calcium phosphate coprecipitation has been used successfully with *Drosophila* cells (Walker, 1989). Other methods to transfer exogenous DNA into insect cells have been tried, including electroporation, liposomes, laser micropuncture, and several types of microinjection (Walker, 1989; Fallon, 1991). At present, developing these other methods appears to be an empirical process. Stable transformation of other insect cell lines apparently has not yet been achieved.

Baculoviruses infect insects and several have been used as biological pesticides (Wood & Granados, 1991). *Autographa californica* nuclear polyhedrosis virus, silkworm *Bombyx mori* nuclear polyhedrosis virus, and gypsy moth *Lymantria dispar* NPV have been exploited as vectors to carry exogenous DNA into insect cells (Miller, 1988; Iatrou & Meidinger, 1990; Yu et al., 1992). However, because insect cells or larvae die from their infection, baculovirus vectors do not produce stably-transformed insects. If baculovirus vectors are developed that are nonlethal to their hosts, these vectors could be used for stable transformation.

A gene coding for a yeast recombinase, FLP, has been found on a plasmid isolated from the yeast *Saccharomyces cerevisiae*. This plasmid also carries two inverted recombination target sites (FRT) that are specifically recognized by the FLP recombinase. FLP recombinase will catalyze recombination of the DNA between the FRT sites in the plasmid, inverting the sequences between them. FLP will catalyze both intramolecular and intermolecular recombination, and the possibility thus exists that FLP-mediated recombination could be used to insert foreign DNA into a specific site in an arthropod chromosome after it had been engineered to have the FRT sites. This would allow insertion of foreign DNA into the same site in a strain each time and could eliminate some of the position-effect influences on gene expression associated with random insertion of exogenous DNA. Morris et al. (1991) showed that FLP-mediated, site-specific intermolecular recombination occurred in microinjected embryos of the mosquito *A. aegypti*. The experiments conducted did not allow them to determine whether the mosquitoes were stably transformed. This system could provide a rapid method of inserting different DNA sequences into a specific chromosomal site (at the FRT site). However, because a stable FRT site in the genome is necessary and is integrated into the arthropod genome through nonspecific recombination, different lines will have to be evaluated to determine which is best for allowing expression of the foreign genes. The FLP system may be best suited for those species undergoing intensive and long term genetic analysis and manipulation.

What Genes Are Available?

Genes theoretically can be isolated from either closely- or distantly-related organisms for insertion into arthropods. It may also be possible to isolate a gene from the species being manipulated, alter it, and reinsert it into the germ line. Assuming that a transformation method is available so that either transient or stable transformation can be achieved, the major issue then becomes whether the exogenous gene is expressed appropriately and effectively in its new host. Expression requires an appropriate promoter and other regulatory elements.

A number of genes have been cloned from *Drosophila* and other species and inserted into *Drosophila* by P element-mediated transformation. Most of these cloned genes may not be particularly useful for improving efficacy of beneficial arthropods. Microbial (neomycin or G418 resistance, chloramphenicol acetyltransferase and ß-galactosidase) and *Drosophila* eye color genes such as *rosy* or *white* have been used to identify transformants. Some

Drosophila genes could serve as probes for homologous sequences in other insect species. Cloned genes also could be modified by *in vitro* mutation to achieve a desired phenotype. For the foreseeable future, resistance genes will probably be the most available and useful (Table 1). Potentially-useful resistance genes that have been cloned include a parathion hydrolase gene (*opd*) from *Pseudomonas diminuta* (Serdar et al., 1989) and *Flavobacterium* (Mulbry & Karns, 1989), a cyclodiene resistance gene (GABA$_A$) from *Drosophila* (ffrench-Constant et al., 1993), ß-tubulin genes isolated from *Neurospora crassa* (Orbach et al., 1986) and *Septoria nodorum* (Cooley et al., 1991) conferring resistance to benomyl, an acetylcholinesterase gene (*Ace*) from *D. melanogaster* and the mosquito *Anopheles stephensi* (Fournier et al., 1989; Hall & Spierer, 1986; Hall & Malcolm, 1991; Hoffman et al., 1992), a glutathione transferase gene (GST1) from *Musca domestica* (Wang et al., 1991; Fournier et al., 1992), a cytochrome P450-B1 gene (*CYP6A2*) associated with DDT resistance in *Drosophila* (Waters et al., 1992),

and the amplification core and esterase B1 gene isolated from *Culex* mosquitoes that are responsible for organophosphorus insecticide resistance (Mouches et al., 1990).

Metallothionein genes cloned from *Drosophila* and other organisms appear to function in homeostasis of copper and cadmium and in their detoxification (Theodore et al., 1991). Perhaps these genes could provide resistance in arthropod natural enemies to fungicides containing copper. Multidrug resistance genes, *mdr* or *pgp*, in mammals become amplified and overexpressed in multidrug-resistant cell lines, resulting in cross-resistances to a broad spectrum of compounds, including those used in cancer chemotherapy. The multi-drug resistance genes code for a family of membrane glycoproteins that appear to function as an energy-dependent transport pump. Two members of this multigene family were isolated from *D. melanogaster* and these genes (*Mdr49* and *Mdr65*) could provide resistances to a number of exogenous chemicals (Wu et al., 1991). For example, *D. melanogaster* strains that were made deficient for

Table 1. Some Cloned Resistance Genes Possibly Useful For Genetic Manipulation Of Beneficial Arthropods

Gene (Resistance)	Source(s)	Reference(s)
acetylcholinesterase	*D. melanogaster*	Hall & Spierer 1986
	Anopheles stephensi	Hall & Malcolm 1991
		Hoffmann et al. 1992
B-tubulin	*Neurospora crassa*	Orbach et al. 1986
(benomyl)	*Septoria nodorum*	Cooley et al. 1991
catalase	*D. melanogaster*	Orr & Sohal 1992
(H$_2$O$_2$)		
cytochrome P450-B1	*D. melanogaster*	Waters et al. 1992
(DDT)		
esterase B1	*Culex* species	Mouches et al. 1986, 1990
amplification core		
(OPs)		
GABA$_A$ receptor	*D. melanogaster*	ffrench-Constant et al.1991
(dieldrin)		
glutathione S-transferase	*D. melanogaster*	Toung et al. 1990
(DDT)		
glutathione S-transferase	*Musca domestica*	Wang et al. 1991
(OPs)		Fournier et al. 1992
metallothionein genes	*D. melanogaster*	Theodore et al. 1991
(copper)		
multidrug resistance	*D. melanogaster*	Wu et al. 1991
(colchicine)		
neomycin	Transposon Tn5	Beck et al. 1982
phosphotransferase		
(kanamycin, neomycin, G418)		
parathion hydrolase	*Pseudomonas diminuta*	Serdar et al. 1989
(parathion, paraoxon)		Dumas et al. 1990
		Phillips et al. 1990
	Flavobacterium sp.	Mulbry & Karns 1989

Mdr49 were viable and fertile, but had an increased sensitivity to colchicine during development. Whether the insertion of multidrug resistance genes would provide a useful increase in tolerance to chemicals that arthropods might encounter in the environment remains to be determined.

Preliminary results suggest that microbial genes conferring resistance to pesticides can function in arthropods. The *opd* gene isolated from *Pseudomonas* and conferring resistance to organophosphorus insecticides has been inserted, using a baculovirus expression vector, into cultured fall armyworm, *Spodoptera frugiperda*, cells and larvae (Dumas et al., 1990). Phillips et al. (1990) also transferred the *opd* gene into *D. melanogaster*. The *opd* gene was put under control of the *Drosophila* heat shock promoter, *hsp70*, and stable active enzyme was produced which accumulated with repeated induction. It is likely that this gene could be used to confer resistance to organophosphorus pesticides in beneficial arthropod species.

Increased freeze resistance in frost-susceptible arthropods may be increased by gene transfer. Antifreeze protein genes cloned from the wolffish, *Anarhichas lupus*, have been expressed in transgenic *Drosophila* (Rancourt et al., 1990, 1992) using the *hsp70* promoter and yolk polypeptide promoters of *Drosophila*. Altering longevity of certain arthropods might be beneficial and research on mechanisms of aging may provide useful genes in the future. A cloned catalase gene inserted into *D. melanogaster* by P element-mediated transformation provided resistance to hydrogen peroxide, although it did not prolong the life span of flies (Orr & Sohal, 1992).

As basic research progresses other traits that might be useful to introduce into beneficial insects will become obvious. Shortening developmental time, enhancing progeny production, altering sex ratio, extending temperature and relative humidity tolerances, and altering host or habitat preferences could enhance biological control (Hoy, 1976). However, it is not simple to document that changes in one or more of these attributes would actually improve the performance of a biological control agent.

The Importance of Appropriate Regulatory Signals

Genes consist of coding segments that determine the amino acid sequences in the enzyme or structural proteins produced. However, whether a coding region is transcribed and translated in a specific tissue is determined by a number of regulatory sequences in the DNA, including promoters and enhancers. Some of these regulatory structures are in close proximity to the coding region, while others may be located farther away. The stability of messenger RNA is influenced by polyadenylation (polyA) signals at the 3' end of the RNA, which can influence the amount of protein produced. It is crucial to obtain expression of the inserted gene at appropriate times, levels, and tissues. Another factor that may be important in maintaining the inserted DNA in the transgenic line over time is the presence of origins of replication that regulate DNA replication of the chromosomes. If exogenous DNA is inserted into a region of the chromosome far from a site where an origin of replication occurs naturally, the exogenous DNA could be lost over time because it is not replicated.

Because regulatory sequences may vary from species to species, the source of regulatory sequences chosen for cloning may be as important, or even more important, than the source of the protein-coding sequences (Fig. 1). Furthermore, some regulatory sequences allow genes to be expressed only in particular tissues or in response to particular stimuli (such as heat shock), while other genes are expressed in most tissues most of the time. If it is important that the inserted gene function in a tissue- or stimulus-specific manner, it is essential to identify tissue- or stimulus-specific promoters.

Currently, the number of regulatory sequences available for genetic manipulation of arthropods is limited. The heat shock (*hsp70*) promoter from *Drosophila* is commonly used as an inducible promoter. It is the strongest promoter known in *Drosophila* and appears to function in all cells. However, induction of the *hsp70* promoter may be different in different species (Miller et al., 1987; Sakai & Miller, 1992; McInnis et al., 1990).

Other commonly-used regulatory sequences from *Drosophila* are the actin 5C promoter, the α1-tubulin promoter, and the metallothionein (*Mtn*) promoter. Angelichio et al. (1991) compared the ability of these four promoters in cultured *D. melanogaster* cells and found that the actin 5C and the metallothionein promoters generated comparable levels of RNA and protein.

The α1-tubulin promoter generated about 4-fold lower levels and the fibroin promoter had no detectable activity in these cells. However, the fibroin promoter was cloned from the silk worm *Bombyx mori* and may not function in the fruit fly. The effects of poly(A) signals also need to be evaluated to determine their impact on stability of the transcribed messenger RNA (Angelichio et al., 1991).

Chromosome replication in higher eukaryotes is not well understood, but it is known that origins of replication are located at intervals along each chromosome. Origins of replication involved in amplification of chorion genes in *D. melanogaster* have been identified (Carminati et al., 1992). The ACE3 chorion element has been cloned and shown to be sufficient to regulate amplification of the chorion gene cluster. During genetic manipulation of beneficial species, it may be useful to insert ACE3, or similar, elements along with the exogenous genes to ensure that replication of this region of the chromosome occurs in order to increase the stability of the exogenous DNA in the transgenic strain.

Identification, cloning, or genetic modification of promoters and other regulatory sequences may increase the precision with which desired proteins are transcribed and expressed in transgenic arthropods. Research to understand the structure and function of regulatory sequences for use in transgenic arthropods should be a high priority. Project goals will dictate what type of regulatory sequences are most useful. In some cases, low level constitutive production of transgenic proteins will be useful, while in other cases high levels of protein production will be required after inducement by a specific cue. Researchers will have to evaluate the trade offs between high levels of protein production and the subsequent impact on relative fitness of the transgenic arthropod strain based on the specific goals of each program.

Identify Transformed Arthropods

After inserting the desired genes, the next issue is how to detect whether the exogenous gene has in fact been incorporated into the germ line. A screening method is needed to identify transformed individuals. This process is relatively simple in *Drosophila*, where there is a wealth of genetic information, including visible markers that can identify transgenic individuals. Most beneficial arthropods lack such extensive genetic information. Identifying transformed individuals could be achieved by using a pesticide resistance gene, such as the *opd* gene, as the selectable marker. Other options are to use the neomycin (*neo*) antibiotic resistance gene or the ß-galactosidase gene (*lacZ*) isolated from *E. coli* and regulated by the *Drosophila hsp70* promoter. If an appropriate selectable marker is not available, identifying transformed lines can be accomplished with the polymerase chain reaction (PCR) and subsequent analysis by Southern blot hybridization or an immunological procedure.

Risks Associated With Releases of Transgenic Beneficial Arthropods

Until recently, most practitioners of biological control asserted that biological control of arthropod pests or weeds by arthropod natural enemies was environmentally-safe and risk free if carried out by trained scientists. However, questions about the safety of classical biological control have been raised, particularly where environmentalists are concerned about the preservation of native flora and fauna (Howarth, 1991) and the era of accepting classical biological control as environmentally risk free appears to have passed (Ehler, 1990; Harris, 1985; Hoy, 1992). Protocols for evaluating the risks associated with releasing parasitoids and predators that have been manipulated with rDNA techniques do not currently exist but will likely include, as a minimum, the questions or principles discussed by Hoy (1990a, 1992) and Tiedje et al. (1989).

The evidence presented by Raymond et al. (1991) indicates that there has been a worldwide migration of *Culex pipiens* mosquitoes carrying naturally-amplified organophosphorus resistance genes. This suggests that dispersal of arthropods can be rapid and extensive. DNA sequences or restriction fragment length polymorphisms of esterase B2 from mosquitoes collected in Africa, Asia and North America were identical, leading the authors to conclude that all amplified esterase B2 genes rose from a single initial mutational event; resistant mosquitoes subsequently migrated between continents via international airline flights or ships. This example will trigger intensive scrutiny of releases of transgenic pest arthropods into the environment.

Another risk issue involves the possibility that horizontal transfer of genes may occur between one arthropod species and another (Houck et al., 1991). The P element appears to have invaded *D. melanogaster* populations within the last 50 years. Houck et al. (1991) showed that P elements may have been transferred between *Drosophila* species by a semiparasitic mite, *Proctolaelaps regalis*. Horizontal transfer of P elements from *D. willistoni* to *D. melanogaster* must be a very rare event, requiring that two *Drosophila* females of different species lay their eggs in proximity so that a mite can feed on one and then on the other (in the correct order). The mite must carry the P element to the recipient egg, which must be in a very early stage of embryonic development, the recipient embryo must incorporate a complete copy of the P element into a chromosome before it is degraded by enzymes in the cytoplasm, the recipient embryo must survive the feeding by the mite, and the adult that develops from the embryo must transmit the P element to its progeny. If each event is rare, and the combined probability is multiplicative, then the probability that horizontal gene transfer between different arthropod species will occur must be low.

Interspecific transfer of another transposable element (*mariner*) has been suggested as an explanation of *mariner* elements in the drosophilid genera *Drosophila* and *Zaprionus* (Maruyama & Hartl, 1991). The *mariner* element occurs in five of eight species in the *D. melanogaster* species group, but is found only in the genus *Zaprionus* outside the *Drosophila* group even though *Zaprionus* is not closely related to *Drosophila*. DNA sequences indicate that the *mariner* elements in the two groups are 97% identical, although, by comparison, alcohol dehydrogenase(*Adh*) genes are not this close. A *mariner*-like sequence has been discovered in the genome of the lepidopteran *Hyalophora cecropia* (Lidholm et al., 1991). While the interspecific transfer of *mariner* is suspected only on the basis of DNA sequence similarities, and no specific vector has been identified, the data are consistent with the hypothesis that transposable elements can move between different species.

Evidence from sequencing DNA isolated from bacterial endosymbionts of mosquitoes, Coleoptera and *Drosophila* suggest that the symbionts may have been horizontally transferred between these species (O'Neill et al., 1992). Bacterial endosymbionts of insects are involved in many examples of cytoplasmic incompatibility, in which certain crosses between symbiont-infected individuals lead to death of embryos or distortion of the progeny sex ratio (Rousset et al., 1992; Stouthamer et al., 1993). An analysis of the 16S rRNA genes specific to prokaryotes from *Culex pipiens, Tribolium confusum, Hypera postica, Aedes albopictus*, two populations of *Drosophila simulans* and *Ephestia cautella* indicated that their symbionts are all closely related (O'Neill et al., 1992). O'Neill et al. (1992) speculated that cytoplasmic incompatibility is due to infection with a specialized bacterium that infects a wide range of different arthropods and that the symbiont has been acquired more than once by different insects. Preliminary surveys with a DNA probe indicated that additional insects, including *Corcyra cephalonica, Sitotroga cerealella, Diabrotica virgifera, Attagenus unicolor, Rhagoletis pomonella, Rhagoletis mendax* and *Anastrepha suspensa* carry the symbiont, although cytoplasmic incompatibility has not been demonstrated in these species.

If horizontal transmission of DNA (or microorganisms) between arthropods occurs, even if exceedingly rare, there is no guarantee that genes inserted into any species are completely stable. Naturally-occurring horizontal transmission of DNA between species may have provided some of the variability upon which evolution has acted, but the extent and nature of this kind of gene transfer are just being determined. Thus, releases of transgenic arthropods will have to be evaluated on the basis of their probable benefits and potential risks.

Experience indicates that the probability that a new organism will establish is small (Williamson, 1992). Discussions of risk include questions about survival, reproduction and dispersal of transgenic species and their effects on other species. Questions also are asked about the inserted DNA, its stability, and its possible effect on other species should the genetic material move (U. S. Dept. Agric., 1991; Hoy, 1992). Historical examples of biological invasions or classical biological control demonstrate the lack of predictability, the low level of successful establishment, the importance of scale, specificity, and the speed of evolution (Ehler, 1990). Transgenic organisms could pose risks because they will be released in large numbers. Williamson (1992) speculated that the greater the genetic novelty, the greater the possibility of surprising results, and recommended using molecular markers to begin to understand dispersal and the interactions between species.

Hadrys et al. (1992) point out that within the past two years several molecular genetic techniques have become available to analyze behavioral ecology and population biology. Thus, DNA fingerprinting techniques can be used to determine taxonomic identity, assess kinship, analyse mixed genome samples and create specific probes (Hadrys et al., 1992). The RAPD method of PCR is useful in situations in which limited amounts of DNA are available, for species with minimal genetic information, and because it is relatively efficient and inexpensive. The use of molecular techniques in ecological studies promises to provide powerful tools to help assess the risks of releasing transgenic arthropods. These techniques and others, such as population genetic models that incorporate information on dispersal rates and gene frequencies (Caprio et al., 1991), will provide methods for improving our knowledge of the ecology and behavior of beneficial arthropods in pest management programs, whether they have been genetically manipulated or not.

Conclusions

Genetic manipulation projects of beneficial arthropods for use in pest management programs require methods for efficient and stable transformation. Knowledge of appropriate promoters and other regulatory elements required to obtain an effective expression of the inserted gene in both space and time is critical. The number of genes that are cloned and of potential value for pest management programs remains limited at this time, primarily to single genes for resistance to pesticides or other toxins.

Eventually, we will need cloned genes that code for other desired traits. Ultimately, it may be useful to develop strains with either high or low tolerances to abiotic factors such as temperature and relative humidity, modified sex ratios, or altered developmental rates and fecundities. Before transgenic arthropods can be developed with these traits, we must understand the underlying mechanisms and identify the critical genes involved.

One factor hindering progress in the genetic manipulation of beneficial arthropods is the lack of a 'universal' transformation system. The availability of an efficient transformation method that would provide a rapid and general system for introducing exogenous DNA into species for which little genetic information is available would revolutionize the genetic engineering of arthropods (Presnail & Hoy, 1992).

Currently, we lack an example that demonstrates that rDNA technology can yield an effective beneficial arthropod in a pest management program. For many years, genetic manipulation of arthropod natural enemies was considered to be impractical for pest management programs (Hoy, 1976), and this limited the resources devoted to this tactic. The demonstration that a laboratory-selected strain of predatory mite could provide cost- effective control of spider mites in an agricultural crop (Headley & Hoy, 1987) provided an impetus to this tactic in biological control. It is important to demonstrate that a transgenic beneficial arthropod can regulate pest populations and have no negative impacts on the environment. Until this has been achieved, adequate resources and funding will be difficult to obtain because it is considered to be high risk research.

Because the potential risks of releasing transgenic arthropods into the environment have not been resolved, it may be proper to first consider releasing a relatively risk-free example. This might involve releasing a transgenic beneficial arthropod that is carrying either a noncoding segment of exogenous DNA or a gene such as β-galactosidase (Hoy, 1992). For example, a transgenic strain of the phytoseiid predator *M. occidentalis* carrying a *lacZ* construct would be relatively simple to evaluate. It is an obligatory predator, has a low dispersal rate, and is unlikely to become a pest (Hoy, 1992). Ideally, the transgenic *M. occidentalis* could be released into a site where it is unlikely to become permanently established. Risk assessment of transgenic arthropods, as it has with transgenic crops and microorganisms, adds a significant cost in both time and resources to the project. It has taken years for companies to get to the point where transgenic crops are commercially available, and it remains unclear how successful they will be in the market place. Thus, it will be important to conduct benefit-cost analyses when transgenic beneficial arthropods are used in pest management programs.

Significant, exciting, and unpredictable advances are being achieved in molecular biology and genetics. It is very difficult to anticipate the opportunities for genetic manipulation of pest and beneficial arthropods over even the next few years. Despite these anticipated advances, much research remains to be done if we are to gain an understanding of the attributes other than resistance to pesticides that we might manipulate. Getting a transgenic arthropod to the field may be an even greater challenge.

Acknowledgements

This is University of Florida paper No. R-02993. I thank L. Caprio for assistance with the manuscript and reviewers for helpful comments.

References

Angelichio, M.L.; Beck, J.A.; Johansen, H.; Ivey-Hoyle, M. "Comparison of several promoters and polyadenylation signals for use in heterologous gene expression in cultured *Drosophila* cells." *Nuc. Acids Res.* **1991**, *19*, 5037-5043.

Atkinson, P.W.; Hines, E.R.; Beaton, S; Matthaei, K.I.; Reed, K.C.; Bradley, M.P. "Association of exogenous DNA with cattle and insect spermatozoa in vitro." *Molec. Reprod. Develop.* **1991**, *23*, 1-5.

Baldarelli, R.M.; Lengyel, J.A. "Transient expression of DNA after ballistic introduction into *Drosophila* embryos." *Nuc. Acids Res.* **1990**, *18*, 5903-5904.

Barinaga, M. "Gene-transfer method fails test." *Science* **1989**, *246*, 446.

Beard, C.B.; Mason, P.W.; Aksoy, S.; Tesh, R.B.; Richards, F.F. "Transformation of an insect symbiont and expression of a foreign gene in the Chagas' disease vector *Rhodnius prolixus*." *Am. J. Trop. Med. Hyg.* **1992**, *46*,195-200.

Beck, E.; Ludwig, G.; Auerswald, E.A.; Reiss, B.; Schaller, H. "Nucleotide sequence and exact localization of the neomycin phosphotransferase gene from transposon tn5." *Gene* **1982**, *19*, 327-336.

Beckendorf, S.K.; Hoy, M.A."Genetic improvement of arthropod natural enemies through selection, hybridization or genetic engineering techniques," In *Biological Control In Agricultural IPM Systems,* M. A. Hoy, D. C. Herzog, Eds., Academic Press, Orlando, 1985, Pp. 167-187.

Berg, D.E.; Howe, M.M., Eds.; *Mobile DNA;*

Am. Soc. Microbiol.: Washington, DC, **1989**; Vol. 1.

Carminati, J.; Johnston, C.G.; Orr-Weaver, T. L. "The *Drosophila ACE3* chorion element autonomously induces amplification." *Molecular and Cellular Biology* **1992**, *12*, 2444-2453.

Caprio, M.A.; Hoy, M.A.; Tabashnik, B.E. "A model for implementing a genetically-improved strain of the parasitoid *Trioxys pallidus* Haliday (Hymenoptera: Aphidiidae)." *Am. Entomol.* **1991**, *34(4)*, 232-239.

Cooley, R.N.; Van Gorcom, R.F.M.; van den Hondel, C.A.M.J.J.; Caten, C.E. "Isolation of a benomyl-resistant allele of the β-tubulin gene from *Septoria nodorum* and its use as a dominant selectable marker." *J. Gen. Microbiol.* **1991**, *137*, 2085-91.

Dumas, D.P.; Wild, J.R.; Rauschel, F.M. "Expression of *Pseudomonas* phosphotriesterase activity in the fall armyworm confers resistance to insecticides." *Experientia* **1990**, *46*, 729-734.

Ehler, L. E. "Environmental impact of introduced biological-control agents: implications for agricultural biotechnology," In *Risk Assessment In Agricultural Biotechnology*. Proc. Intern. Conf., J. J. Marois, G. Bruyening, Eds., Univ. Calif., Div. Agric. Natur. Res. Publ. No. 1928, Pp. 85-96, 1990.

Fallon, A. M. "DNA-mediated gene transfer: applications to mosquitoes." *Nature* **1991**, *352*, 828-29.

ffrench-Constant, R.H.; Mortlock, D.P.; Shaffer, C.D.; Macintyre, R.J.; Roush, R.T. "Molecular cloning and transformation of cyclodiene resistance in *Drosophila*: an invertebrate γ-aminobutyric acid subtype A receptor locus." *Proc. Natl. Acad. Sci. USA* **1991**, *88*, 7209-7213.

ffrench-Constant, R.H.; Steichen, J.C.; Rocheleau, T.A.; Araonstein, K.; Roush, R.T. "A single-amino acid substitution in a γ-aminobutyric acid subtype A receptor locus is associated with cyclodiene insecticide resistance in *Drosophila* populations." *Proc. Natl. Acad. Sci. USA* **1993**, 90, 1957-1961.

Fournier, D.; Karch, F.; Bride, J.; Hall, L.M.C.; Berge, J.B.; Spierer, P. "*Drosophila melanogaster* acetylcholinesterase gene structure, evolution and mutations." *J. Mol. Biol.* **1989**, *210*, 15-22.

Fournier, D.; Bride, J.M.; Poirie, M.; Berge, J.; Plapp, F.W., Jr. "Insect glutathione s-transferases, biochemical characteristics of the major forms from houseflies susceptible and resistant to insecticides." *J. Biol. Chem.* **1992**, *267*, 1840-1845.

Hadrys, H.; Balick, M.; Schierwater, B. "Applications of random amplified polymorphic DNA (RAPD) in molecular ecology." *Molec. Ecol.* **1992**, *1*, 55-63.

Hall, L.M.C.; Malcolm, C.A. "The acetycholinesterase gene of *Anopheles stephensi*." *Cell. Molec. Neurobiol.* **1991**, *11*, 131-141.

Hall, L.M.C.; Spierer, P. "The *Ace* locus of *Drosophila melanogaster*: structural gene for acetylcholinesterase with an unusual 5' leader." *EMBO J.* **1986**, *5*, 2949-2954.

Handler, A.M.; O'Brochta, D.A. "Prospects for gene transformation in insects." *Annu. Rev. Entomol.* **1991**, *36*, 159-183.

Harris, P. "Biocontrol and the law." *Bull. Entomol. Soc. Can.* **1985**, *17(1)*, 1.

Headley, J.C.; Hoy, M.A. "Benefit/cost analysis of an integrated mite management program for almonds." *J. Econ. Entomol.* **1987**, *80*, 555-559.

Hoffman, F.; Fournier, D.; Spierer, P. "Minigene rescues acetylcholinesterase lethal mutations in *Drosophila melanogaster*." *J. Mol. Biol.* **1992**, *223*, 17-22.

Houck, M.A.; Clark, J.B.; Peterson, K.R.; Kidwell, M.G. "Possible horizontal transfer of *Drosophila* genes by the mite *Proctolaelaps regalis*." *Science* **1991**, *253*, 1125-1129.

Howarth, F.G. "Environmental impacts of classical biological control." *Annu. Rev. Entomol.* **1991**, *36*, 485-509.

Hoy, M.A. "Genetic improvement of insects: fact or fantasy." *Environ. Entomol.* **1976**, *5*, 833-839.

Hoy, M.A. "Genetic improvement of arthropod natural enemies: becoming a conventional tactic?" In *New Directions In Biological Control*, R. Baker and P. Dunn, [Eds.], UCLA Symp. Molec. Cell. Biol., New Series, Vol. 112, A. R. Liss, NY, pp. 405-417. 1990a.

Hoy, M.A. "Pesticide resistance in arthropod natural enemies: variability and selection responses." In *Pesticide Resistance In Arthropods*, R. T. Roush, B. E. Tabashnik, [Eds.], Chapman and Hall, NY, Pp. 203-236, 1990b.

Hoy, M.A. "Criteria for release of genetically-improved phytoseiids: an examination of the risks associated with release of biological control agents." *Exp. Appl. Acarol.* **1992**, *14*, 393-416.

Iatrou, K.; Meidinger, R.G. "Tissue-specific expression of silkmoth chorion genes *in vivo* using *Bombyx mori* nuclear polyhedrosis virus as a transducing vector." *Proc. Natl. Acad. Sci. USA,* **1990**, *87*, 3650-3654.

Kamdar, P.; von Allmen, G.; Finnerty, V. "Transient expression of DNA in *Drosophila* via electroporation." *Nuc. Acids Res.* **1992**, *20(13)*, 3526.

Lawrence, P.A. *The Making of a Fly. The Genetics of Animal Design*; Blackwell Scientific Publ.: London, 1992.

Lidholm, D.A.; Gudmundsson, G.H.; Boman, H.G. "A highly repetitive, *mariner*-like element in the genome of *Hyalophora cecropia*." *J. Biol. Chem.* **1991**, *266*, 11518-11521.

Maruyama, K.; Hartl, D.L. "Evidence for interspecific transfer of the transposable element mariner between *Drosophila* and *Zaprionus*." *J. Mol. Evol.* **1991**, *33*, 514-524.

McGrane, V.; Carlson, J.O.; Miller, B.R.; Beaty, B.J. "Microinjection of DNA into *Aedes triseriatus* ova and detection of integration." *Am. J. Trop. Med. Hyg.* **1988**, *39*, 502-510.

McInnis, D.O.; Haymer, D.S.; Tam, S.Y.T.; Thanaphum, S. "*Ceratitis capitata* (Diptera: Tephritidae): transient expression of a heterologous gene for resistance to the antibiotic geneticin." *Ann. Entomol. Soc. Am.* **1990**, *83*, 982-986.

Michaille; J.J., Mathavan, S.; Gaillard, J.; Garel, A. "The complete sequence of Mag, a new retrotransposon in *Bombyx mori*." *Nuc. Acids Res.* **1990**, *18*, 674.

Miller, L.H.; Sakai, R.K.; Romans, P.; Gwadz, W.; Kantoff, P.; Coon, H.G. "Stable integration and expression of a bacterial gene in the mosquito *Anopheles gambiae*." *Science* **1987**, *237*, 779-781.

Miller, L.K. "Baculoviruses as gene expression vectors." *Annu. Rev. Microbiol.* **1988**, *42*, 177-199.

Milne, C.P., Jr.; Phillips, J.P.; Krell, P.J. "Microinjection of early honeybee embryos." *J. Apicult. Res.* **1988**, *27*, 84-89.

Milne, C.P.; Eishen, F.A.; Collis, J.E.; Jensen, T.L. 1989. "Preliminary evidence for honey bee sperm-mediated DNA transfer." Int. Symp. Mol. Insect Science, Tucson, AZ, P. 71 (Abstract).

Morris, A.C.; Eggleston, P.; Crampton, J.M. "Genetic transformation of the mosquito *Aedes aegypti* by micro-injection of DNA." *Med. Vet. Entomol.* **1989**, *3*, 1-7.

Morris, A.C.; Schaub, T.L.; James, A.A. "FLP-mediated recombination in the vector mosquito, *Aedes aegypti*." *Nuc. Acids Res.* **1991**, *19(21)*, 5895-5900.

Mouches, C.; Pasteur, N.; Berge, J.B.; Hyrien, O.; Raymond, M.; de Saint Vincent, B.R.; de Silvestri, M.; Georghiou, G.P. "Amplification of an esterase gene is responsible for insecticide resistance in a California *Culex* mosquito." *Science* **1986**, *233*, 778-780

Mouches, C.; Pauplin, Y.; Agarwal, M.;

Lemieux, L.; Herzog, M.; Abadon, M.; Beyssat-Arnaouty, V.; Hyrien, O.; de Saint Vincent, B.R.; Georghiou, G.P.; Pasteur, N. "Characterization of amplification core and esterase B1 gene responsible for insecticide resistance in *Culex*." *Proc. Natl. Acad. Sci. USA*, **1990**, *98*, 2574-78.

Mulbry, W.W.; Karns, J.S. "Parathion hydrolase specified by the *Flavobacterium opd* gene: relationship between the gene and protein." *J. Bact.* **1989**, *171(12)*, 6740-6746.

O'Neill, S.L.; Giordano, R.; Colbert, A.M.E.; Karr, T.L.; Robertson, H.M. "16s rRNA phylogenetic analysis of the bacterial endosymbionts associated with cytoplasmic incompatibility in insects." *Proc. Natl. Acad. Sci. USA*, **1992**, *89*, 2699-2702.

Orbach, M.J.; Porro, E.B.; Yanofsky, C. "Cloning and characterization of the gene for β-tubulin from a benomyl-resistant mutant of *Neurospora crassa* and its use as a dominant selectable marker." *Mol. Cell. Biol.* **1986**, *6*, 2452-2461.

Orr, W.C.; Sohal, R.S. "The effects of catalase gene overexpression on life span and resistance to oxidative stress in transgenic *Drosophila melanogaster*." *Arch. Bioch. Biophy.* **1992**, *297*, 35-41.

Phillips, J.P.; Xin, J.H.; Kirby, K.; Milne, C.P., Jr.; Krell, P.; Wild, J.R. "Transfer and expression of an organophosphate insecticide-degrading gene from *Pseudomonas* in *Drosophila melanogaster*." *Proc. Natl. Acad. Sci. USA*, **1990**, *87*, 8155-8159.

Presnail, J.K.; Hoy, M.A. "Stable genetic transformation of a beneficial arthropod by microinjection." *Proc. Natl. Acad. Sci. USA*, **1992**, *89*, 7732-7736.

Rancourt, D.E.; Peters, I.D.; Walker, V.K.; Davies, P.L. "Wolffish antifreeze protein from transgenic *Drosophila*." *Bio/Technol.* **1990**, *8*, 453-457.

Rancourt, D.E.; Davies, P.L.; Walker, V.K. "Differential translatability of antifreeze protein mRNAs in a transgenic host." *Biochim. Biophys. Acta* **1992**, *1129*, 188-194.

Raymond, M.; Callaghan, A.; Fort, P.; Pasteur, N. "Worldwide migration of amplified insecticide resistance genes in mosquitoes." *Nature* **1991**, *350*, 151-153.

Rousset, F.; Bouchon, D.; Pintureau, B.; Juchault, P.; Solignac, M. "*Wolbachia* endosymbionts responsible for various alterations of sexuality in arthropods." *Proc. R. Soc. Lond.B.* **1992**, *250*, 91-98.

Rubin, G.M.; Spradling, A.C. "Genetic transformation of *Drosophila* with transposable element vectors." *Science* **1982**, *218*, 348-53.

Sailer, R. I. "Possibilities for genetic improvement of beneficial insects." In *Germ Plasm Resources*. Am. Assoc. Adv. Sci., Washington, DC., P. 295, 1961.

Sakai, R.K.; Miller, L.H. "Effects of heat shock on the survival of transgenic *Anopheles gambiae* (Diptera: Culicidae) under antibiotic selection." *J. Med. Entomol.* **1992**, *29(2)*, 374-375.

Santamaria, P. "Injecting Eggs," Pp. 159-73. In *Drosophila A Practical Approach*, D. B. Roberts, [Ed.], IRL Press, Oxford, 1986.

Schedl, A.; Beermann, F.; Thies, E.; Montoliu, L.; Kelsey, G.; Schutz, G. "Transgenic mice generated by pronuclear injection of a yeast artificial chromosome." *Nucl. Acids Res.* **1992**, *20*, 3073-3077.

Serdar, C.M.; Murdock, D.C.; Rohde, M.F. "Parathion hydrolase gene from *Pseudomonas diminuta* mg: subcloning, complete nucleotide sequence, and expression of the mature portion of the enzyme in *Escherichia coli*." *Bio/Technol.* **1989**, *7*, 1151-1155.

Spradling, A.C.; Rubin, G.M. "Transposition of cloned P elements into *Drosophila* germline chromosomes." *Science* **1982**, *218*, 341-347.

Steller, H.; Pirrota, V. "Fate of DNA injected into early *Drosophila* embryos." *Develop. Biol.* **1985**, *109*, 54-62.

Stouthamer, R.; Breeuwer, J.A.J.; Luck, R.F.; Werren, J.H. "Molecular identification of microorganisms associated with parthenogenesis." *Nature* **1993**, *361*,66-68.

Theodore, L.; Ho, A.; Maroni, G. "Recent evolutionary history of the metallothionein gene *Mtn* in *Drosophila*." *Genet. Res., Camb.* **1991**, *58*, 203-210.

Tiedje, J.M.; Colwell, R.K.; Grossman, Y.L.; Hodson, R.E.; Lenski, R.E.; Mack, R.M.; Regal, P.J. "The planned introduction of genetically engineered organisms: ecological considerations and recommendations." *Ecology* **1989**, *70*, 298-315.

Toung, Y.P.S.; Hsieh, T.S.; Tu, C.P.D. "*Drosophila* glutathione S-Transferase 1-1 shares a region of sequence homology with the maize glutathione S-Transferase III." *Proc. Natl. Acad. Sci. (USA)* **1990**, *87*, 31-35.

U. S. Dept. Agric. 1991. Part III. Proposed guidelines for research involving the planned introduction into the environment of organisms with deliberately modified hereditary traits; notice. Federal Register Vol. 56 (22), Friday, February 1, 1991, Pp. 4134-4151.

Walker, V.K. Gene transfer in insects. *Adv. Cell Culture* **1989**, *7*, 87-124.

Wang, J.Y.; Mccommas, S.; Syvanen, M. "Molecular cloning of a glutathione s-transferase overproduced in an insecticide-resistant strain of the housefly (*Musca domestica*)." *Mol. Genet.* **1991**, *227*, 260-266.

Waters, L.C., Zelhof, A.C.;Shaw, B.J.; Chang, L.Y. "Possible involvement of the long terminal repeat of transposable element 17.6 in regulating expression of an insecticide resistance-associated P450 gene in *Drosophila*." *Proc. Natl. Acad. Sci. USA*, **1992**, *89*, 4855-4859.

Williamson, M. "Environmental risks from the release of genetically modified organisms (GMOs)--the need for molecular ecology." *Molec. Ecol.* **1992**, *1*, 3-8.

Wood, H.A.; Granados, R.R. "Genetically engineered baculoviruses as agents for pest control." *Annu. Rev. Microbiol.* **1991**, *45*, 69-87.

Wu, C.T.; Budding, M.; Griffin, M.S.; Croop, J.M. "Isolation and characterization of *Drosophila* multidrug resistance gene homologs." *Molec. Cell. Biol.* **1991**, *11(8)*, 3940-3948.

Yu, Z.; Podgwaite, J.D.; Wood, H.A. "Genetic engineering of a *Lymantria dispar* nuclear polyhedrosis virus for expression of foreign genes." *J. Gen. Virol.* **1992**, *73*, 1509-1514.

Zalokar, M. "A method for injection and transplantation of nuclei and cells in *Drosophila* eggs." *Experientia* **1981**, *37*, 1354-1356.

Bt Maize for Control of European Corn Borer

Mary-Dell Chilton, Michael Koziel, Thomas C. Currier
Ciba-Geigy Corporation, Agricultural Biotechnology, PO Box 12257,
Research Triangle Park, NC 27709
Benjamin J. Miflin, Ciba-Geigy Ltd, Basel CH 4002, Switzerland

European corn borer (ECB), a major pest of maize, has been treated in the past by chemical or biocontrol insecticides or agents. Expression of a CryIA(b) endotoxin from *Bacillus thuringiensis* in transgenic maize affords excellent ECB control under field conditions. Some advantages of this approach over previous control measures are indicated. Issues of insect specificity, food safety, pricing, and breeding strategy are raised. Arguments for and against the possibility of selection of ECB resistant to CryIA(b) are discussed, as well as strategies for delaying and combatting such resistance if it does occur.

Characteristics of the Pest

European corn borer (*Ostrinia nubilalis*) is a major pest of maize, causing 4-7% yield loss in average years and up to 20% yield loss in years of heavy infestation. First generation larvae hatch early in the season from egg masses on the underside of leaves and migrate to the whorl. After feeding on leaf material for 7-10 days, third instar larvae bore into the stalk where they pupate and mature. Second generation eggs are laid in midsummer on leaves near the ear node. It is the second generation larvae that cause the greatest damage by eventually boring into the stalk in the ear region, causing stalk breakage and dropped ears.

Chemical and Biocontrol Products

Chemical insecticides are used against European corn borer (ECB) with limited success. They are problematical for several reasons. Timing of the application is critical because borers cannot be reached by chemical spray once they are inside the stalk; hence scouting the field for egg masses is crucial. Repeated applications are necessary because of the extended egg-laying period. Chemical control measures are not fully effective because inevitably some borers escape the treatment. For these reasons, many farmers elect to accept the yield penalty of ECB damage rather than attempt to combat it.

Various biocontrol measures are also used against ECB (Ferguson, 1992). *Trichogramma* is a parasitic wasp that deposits its egg in the ECB egg. *Beauvaria bassiana* is a fungus that attacks ECB. In addition, sprays containing the crystal proteins of certain strains of *Bacillus thuringiensis* are active against ECB (Hofte & Whiteley, 1989), although currently available products are not commonly used in field crop production because of cost and low efficacy.

Transgenic Maize Expressing CryIA(b)

The advent of gene transfer technology opened up the attractive possibility of incorporating various crystal protein genes from Bt directly into the plant genome. This turned out to be more difficult than anticipated, for a variety of reasons, some of which are still not fully understood. Chimeric gene constructs with a prolific plant-active gene promoter coupled to the coding region of the CryIA(b) protein were expressed poorly if at all in most plants. The discovery that only the N-terminal half of the protein is needed for insect toxicity led to the successful expression of a truncated form of the gene in dicot plants (Vaeck et al., 1987; Barton et al., 1987; Fischoff et al., 1987; Perlak et al., 1990; Carozzi et al., 1992). While the coding region from *Bacillus thuringiensis* formed active transcripts and insecticidal truncated protein in dicots, transfer of the same truncated gene constructs to maize produced no detectable gene product.

This impasse, while its sources are still unclear, has been surmounted (Koziel et al., 1993) by modifying the native truncated Bt gene sequence using maize-preferred codons rather than those of *Bacillus*. The protein sequence is unaltered, but the base composition (%GC = % guanine + cytosine) of the DNA is maize-like (65% GC) rather than Bacillus-like (38% GC). We have transformed an elite maize line with such a synthetic truncated *cry*IA(b) gene using the Biolistic device. The transforming DNA, a mixture of plasmids containing the *cry*IA(b) gene and a gene conferring resistance to phosphinothricin (PPT) (Thompson et al., 1987), was coated onto gold particles and fired at immature maize embryos (14-15 days after pollination) of an elite maize line. Transformed cells were selected on medium containing phosphinothricin, followed by regeneration of plants. Plants were screened for the presence of CryIA(b) protein by enzyme linked immunosorbant assay (ELISA), and for insecticidal activity by in vitro bioassay of leaf pieces using neonate ECB.

The two Bt maize lines tested in the field in 1992 differed in the *cry*IA(b) gene construct employed. One had the CaMV 35S gene promoter, which is expected to express throughout the plant. The other line contained two chimeric *cry*IA(b) genes, one with a maize pollen-specific promoter and one with a maize phosphoenolpyruvate carboxylase (PEPC) promoter, which is expected to express only in green tissue. The T_0 plants regenerated from culture were crossed to several elite lines, and the resulting segregating progeny were screened for activity against neonate ECB. Insecticidal plants, as well as some negative segregants as controls, were transplanted to the field and were rigorously tested for ECB tolerance by eight weekly infestations of 300 neonates each (total of 2400 larvae per plant). The first four infestations corresponded roughly to the timing of first brood ECB infestation; the remaining four corresponded to second brood infestation.

First brood ECB damage was measured using a visual rating scale (Fig. 1) that was modified at the low end of the scale to increase discrimination among transgenic plant lines. The 176-derived lines were marginally better protected than the 171-derived lines, but both were clearly superior to control plants (Fig. 1). Second brood damage was measured as length of tunneling observed in a 92 cm section of split stalks, taken 46 cm above and below the primary ear node. Both transgenic plant lines were clearly superior to control plants, as indicated by the average values in Fig. 2. The tunneling ranged from 28-114 cm for control plants and 1.7-5 cm for the transgenic line with *cry*IA(b) genes driven by tissue-specific promoters (Koziel et al., 1993).

Bt Expression in Plant Parts

By ELISA analysis, we determined the concentration of CryIA(b) in various parts of transgenic plants derived from 171 and 176 lines (Fig. 3). The 35S promoter in 171-derived plants caused expression in all parts of the plant except pollen/anther, with significant expression in seed

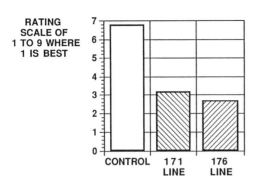

Fig. 1. Mean first generation ECB damage ratings for transgenic and control lines

The damage rating scale from 1-4 was made more stringent than usual in order to discriminate among the transgenic lines. Ratings were determined as follows: 1. No visible leaf injury. 2. Fine "window pane" damage only on the unfurled leaf where larvae were applied. No pinhole penetration. 3. Fine "window pane" damage on two unfurled leaves where larvae were applied. No pinhole penetration. 4. Pinhole or shot hole damage on two or more leaves that emerged from whorl. 5. Elongated lesions and/or midrib feeding on more than 3 leaves that emerged from whorl. Lesion < 1" in length. 6. Several leaves with elongated lesions (0.75-1.5") and/or no more than one leaf with broken midrib. 7. Long lesions (> 1") common on half of leaves and/or 2-3 leaves with broken midribs. 8. Long lesions (> 1") common on two thirds of leaves and/or >3 leaves with broken midribs. 9. Most leaves with long lesions. Several leaves with broken midribs.

(212 ng/mg soluble protein). In contrast the pollen- and PEPC-promoter driven genes of 176-derived plants expressed highly in pollen and leaf, parts consumed by ECB, with minimal expression in seed (16 ng/mg soluble protein).

Issues Raised by Bt Maize

The success of Bt maize at controlling even this artificially high rate of infestation by ECB clearly demonstrates the technical feasibility of the approach. The scientific success, however, is only the first of many challenges along the road to a labeled product that will be acceptable to regulatory agencies, growers and consumers, as well as profitable to a seed company.

Target vs. Non-Target Insects: Specificity of Bt Maize

Each Bt endotoxin exhibits differential activity on various insects. The CryIA type endotoxins in all lepidopteran-active Bt products are present as a crystal that is solubilized in the alkaline insect midgut and subsequently proteolytically activated (cleaved). For insecticidal activity, the activated endotoxin apparently must bind specifically to sites on the brush border membrane of the columnar cells in the insect midgut, and

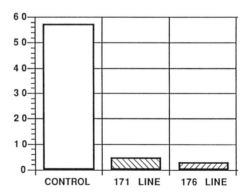

Fig. 2. Mean second generation ECB tunneling damage for transgenic and control lines

Selected stalks from transgenic and control lines were harvested. The extent of internal ECB tunneling damage in a 92 cm section of stalk 46 cm above and below the primary ear node was measured.

NG CRYIA(b) PROTEIN/MG SOLUBLE PROTEIN IN VARIOUS PARTS OF PLANTS

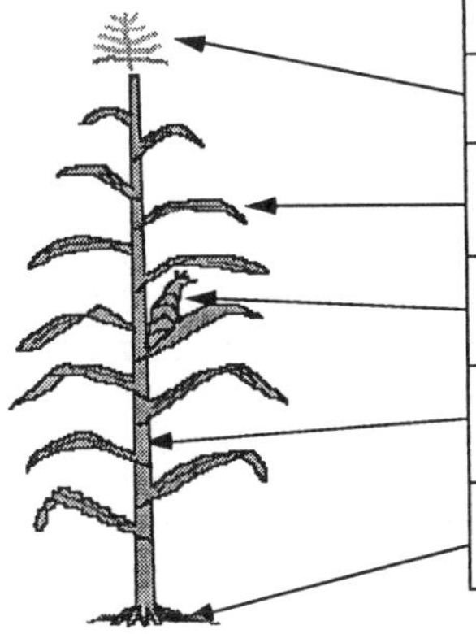

	171 LINE	176 LINE
POLLEN/ ANTHER	0	339
LEAF	711	1496
KERNEL	212	16
PITH	3410	87
ROOT	2278	53

Fig. 3. Concentration of CryIA(b) protein in various parts of transgenic plants.

Quantitative determination of CryIA(b) was monitored using enzyme-linked immunosorbant assays (Koziel et al., 1993).

integrate into the membrane to create a non-specific ion pore. The loss of the cell's ability to osmoregulate ultimately results in colloidal osmotic lysis (Knowles and Ellar, 1987).

The importance of binding sites in toxin specificity is clear from studies of resistant insects (Peferoen, 1992). Resistant populations have been obtained from both laboratory and field populations using a variety of selection schemes. In all but one case (Gould et al., 1992) where the mechanism has been analyzed, resistance has been shown to be due to loss or great reduction in ability of the toxin to bind to brush border microvillae. It is also clear from these studies that different endotoxins bind to the insect cell at different sites. While clearly important, binding sites are not the only determinants of specificity. In one study, proteolytic activation was found to play a role in specificity (Haider et al., 1986). It is empirically difficult to test whether specificity can be determined by events subsequent to binding. Because any measure of binding could be artifactual, biologically significant binding is difficult to prove unless toxicity to insects results.

The truncated form of CryIA(b) protein produced in our maize is expected to produce, on proteolytic cleavage in the insect gut, the same toxic peptide with the same receptor requirements as does full length CryIA(b) (Fig. 4). It has the same insect specificity and receptor requirements as the proteolytically activated endotoxin produced by the microbe. There is thus no reason to

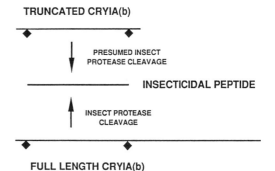

Fig. 4. Cleavage of CryIA(b) full length and truncated proteins. The truncated protein is expected to yield the same insecticidal peptide as full length CryIA(b) after proteolytic cleavage in the insect gut at amino acids 29 and 607 (diamonds in diagram).

373

anticipate any difference in specificity between truncated CryIA(b) and the full length product in microbial spray preparations.

There is one additional feature of transgenic maize that increases its specificity for the target pest: unlike sprayed insecticidal products, the CryIA(b) in the plant reaches only insects that ingest maize tissue (which, by definition, are target pests). An exception might be accidental ingestion of pollen expressing Bt by non-target insects. While to date there is no evidence of toxicity of CryIA(b) toward bees, toxicity testing on honeybees will be performed to assure that this poses no problem.

Food Safety Issues

Based on the arguments presented above, there is no reason to anticipate that the truncated Bt protein in maize will have any different animal toxicity than does the CryIA(b) crystal protein in Bt microbial sprays. There is a thirty year safe history of use of such spray preparations on vegetable crops; the products can be used right up to the day of harvest. Nevertheless, testing will be done as required by the EPA for registration of Bt maize seed (which is classified as a pesticidal product).

Will ECB become resistant to CryIA(b)?

One of the most difficult and potentially most important issues is the possibility that ECB may develop resistance to CryIA(b) if presented with enormous selection pressure by thousands of acres of Bt maize. This would be a serious business problem for a seed company that has invested heavily in development of Bt maize! Further, it would be, in the view of many, an ecological/ethical problem: Bt as a microbial spray is a favorite insecticide for many gardners as well as professional growers of vegetables. ECB is a significant backyard garden pest in several vegetable crops. Development of CryIA(b) resistance would deprive growers of use of a unique product that many see as a natural resource, not to be "squandered."

It is not easy to predict how rapidly, if at all, ECB might develop resistance to CryIA(b). This is a subject debated by insect population geneticists. As mentioned above, there is precedent for development of resistance in the field by other insect pests, but generally after many years of repeated spraying with traditional formulations of Bt as the only form of insecticide. The halflife of Bt spray activity in the field is short, affording opportunities for successive rounds of selection at sublethal toxicity levels.

All cases of resistance to Bt known thus far, whether from field or laboratory selection, are based on acute exposure regimens. Transgenic plants will provide chronic selection, a potentially important difference. Attempts at chronic selection in the laboratory result in such impairment of development and fecundity that the experiment is not feasible. Bt maize will produce high levels of CryIA(b) at all times during the growing season. Further, Bt maize exposes larvae immediately after hatching when they are most sensitive. Some larvae must survive in order for selection to take place. However arguments will not settle this question. We need more information.

Resistance management strategies can be developed to minimize the likelihood of resistance occurring and to deal with it if it does occur. Rotation to other means of pest control is a traditional part of integrated pest management. Development of maize with different insect control genes would present additional new approaches. An important parameter needed for population genetic models is the frequency with which any resistance allele exists in the population. It is also important to determine whether

374

that resistance allele is dominant, recessive or codominant. Resistance to the high CryIA(b) concentrations in our maize plants is unknown in ECB, and until it is identified and characterized, population genetic models remain highly speculative.

Maintaining refugia (non-Bt host plants--possibly non-Bt maize or even weeds adjoining the field) may help dilute any resistance genes selected for in the Bt maize field. If a resistance gene is recessive, such dilution can make a dramatic reduction in appearance of resistant individuals. It can be argued that use of Bt proteins that bind to more than one receptor in the insect gut should make it much less likely that the insect can become resistant through receptor mutations. The number of functional receptors in ECB for CryIA(b) is unknown. New information must be gathered before optimal resistance management plans can be made. Several ongoing research collaborations will bring needed facts on which to base models and develop strategies to combat resistance, if and when it becomes a problem.

Pricing Issues

Will Bt maize be priced out of reach of small farmers? If Bt maize gives higher yield with lower input by the grower, clearly the seed has added value and will be sold at a premium price. However the added value in the seed is clear and calculable, and seeds priced above what they are worth to the grower simply would never sell. It can be argued that Bt maize will be advantageous to small farmers because it will require less high tech care than ordinary maize. Bt maize seed will be available to all farmers on an equal basis; its efficacy and value will determine the extent to which it is used.

Breeding Issues

Seed companies will have to decide how many of their lines should receive the Bt gene, as well as future transgenes. This is a difficult question for several reasons. It could quickly become unmanagable to extend the number of seed lines offered for sale, by offering Bt and non-Bt versions of each hybrid. The numbers become even more unmanageable when introduction of additional kinds of transgenes is considered. The extra work of introgressing biotech's latest useful gene must not impair the traditional work of developing superior elite germplasm as the ideal vessel for such transgenes.

Public Perception

There has been significant opposition to biotechnology in the US, and highly active opposition in parts of Europe. Critics often couch their arguments in terms of safety concerns. Often a perception of risk is created where there is no scientific basis. It is difficult to allay fears with facts which are not understood by most of the public. Further, the public whose fears have been raised may see little benefit to counterbalance perceived risk. Products may also face significant opposition for reasons that have little to do with safety, as the case of bovine somatotropin (BST) has shown. Public acceptance will clearly be a key factor in marketing any transgenic seed product such as Bt maize.

It is important that the technology for introduction of transgenes into plants be demystified and made acceptable to the public, in order to diminish the layman's vulnerability to phantom risk arguments. This does not mean the public must understand all of the technical details. Few of us understand the workings of a color television set. It does mean that authoritative leadership is needed on the part of individuals perceived to be objective. That leadership can be most effective if it comes from a sector trusted by the public, such as university specialists, nutritionists, physicians--authorities with no financial stake in the sale of transgenic crops.

Bt maize is clearly improved from the grower's point of view. The advantage of this product to the public is indirect: savings of agricultural inputs (fuel, land, water, manpower) while maintaining production levels. Because of maize harvest and distribution practices, it is unlikely that the grain from Bt maize plants will be sold separately from non-Bt maize. The grain will not be priced above other grain. This is in contrast to the situation for Flavr Savr tomatoes developed by Calgene, which will be labeled as such and sold at a premium price based on improved flavor and texture. Bt maize will enhance a maize seed business; Flavr Savr tomatoes will enhance a consumer tomato business. The customer base for these products differs, but in the end, both will require public acceptance.

References

Barton, K. A.; Whiteley, H. R.; Yang, N.-S. "*Bacillus thuringiensis* δ -endotoxin expressed in transgenic *Nicotiana tabacum* provides resistance to lepidopteran insects." Plant Physiol. **1987**, *85*, 1103-1109.

Carozzi, N. B.; Warren, G. W.; Desai, N.; Jayne, S. M.; Lotstein, R.; Rice, D. A.; Evola, S.; Koziel, M. G. "Expression of a chimeric CaMV 35S *Bacillus thuringinesis* insecticidal protein gene in transgenic tobacco." Plant Molec. Biol. **1992**, *20*, 539-548.

Ferguson, J. S. "Biological Control: An Industrial Perspective." Florida Entomologist **1992**, *75*, 421-429.

Fischoff, D. A.; Bowdish, K. S.; Perlak, F. J.; Marrone, P. G.; McCormick, S. M.; Niedermeyer, J. G.; Dean, D. A.; Kusano-Kretzmer, K.; Mayer, E. J.; Rochester, D. E.; Rogers, S. G.; Fraley, R. T. "Insect tolerant transgenic tomato plants." Bio/Technology **1987**, *5*, 807-813.

Gould, F.; Martinez-Ramirez, A.; Anderson, A.; Ferre, J.; Silva, F. J.; Moar, W. J. "Broad-spectrum resistance to *Bacillus thuringiensis* toxins in *Heliothis virescens*." Proc. Natl. Acad. Sci. USA **1992**, *89*, 7986-7990.

Haider, M. Z.; Knowles, B; Ellar, D. J. "Specificity of *Bacillus thuringiensis* var. *colmeri* insecticidal delta-endotoxin by differential processing of the protoxin by larval gut proteases." Eur. J. Biochem. **1986**, *156*, 531-540.

Hofte, H.; Whiteley, H. R. "Insecticidal crystal proteins of *Bacillus thuringiensis*." Microbiol. Rev. **1989**, *53*, 242-255.

Knowles, B. H.; Ellar, D. J. "Colloid-osmotic lysis is a general feature of the mechanism of action of *Bacillus thuringiensis* δ−endotoxins with different insect specificity." Biochim. Biophys. Acta **1987**, *924*, 509-518.

Koziel, M. G.; Beland, G. L.; Bowman, C.; Carozzi, N. B.; Crenshaw, R.; Crossland, L.; Dawson, J.; Desai, N.; Hill, M.; Kadwell, S.; Launis, K.; Lewis, K.; Maddox, D.; McPherson, K.; Meghji, M.; Merlin, E.; Rhodes, R.; Warren, G. W.; Wright, M.; Evola, S. "Field performance of elite transgenic maize plants expressing an insecticidal protein derived from *Bacillus thuringiensis*." Bio/Technology **1993**, *11*, 194-200.

Peferoen, M. "Engineering of insect-resistant plants with *Bacillus thuringiensis* crystal protein genes." in: *Plant genetic manipulation for crop protection*, Gatehouse, A. M. R.; Hilder, V. A.; Boulter, D. Eds.; International: Tucson, AZ, **1992**, pp. 135-153.

Perlak, F. J.; Deaton, R. W.; Armstrong, T. A.; Fuchs, R. L.; Sims, S. R.; Greenplate, J. T.; Fischoff, D. A. "Insect resistant cotton plants." Bio/Technology **1990**, *8*, 939-943.

Thompson, C. J.; Movva, N. R.; Tizard, R.; Crameri, R.; Davies, J. E.; Lauwereys, M.; Botterman, J. "Characterization of the herbicide-resistance gene *bar* from

Streptomyces hygroscopicus." EMBO J., **1987**, *6*, 2519-2523.

Vaeck, M.; Reynaerts, A.; Hofte, H.; Jansens, S.; De Beukeleer, M.; Dean, C.; Zabeau, M.; Van Montagu, M.; Leemans, J. "Transgenic plants protected from insect attack." Nature (London) **1987**, *328*, 33-37.

Plant Hormone-Mediated Insect Resistance in Transgenic *Nicotiana*

Ann C. Smigocki, Plant Molecular Biology Laboratory, Agricultural Research Service, U.S. Department of Agriculture, Beltsville, MD 20705
John W. Neal, Jr., Florist and Nursery Crops Laboratory
Iris J. McCanna, Plant Molecular Biology Laboratory

Application of biotechnology to exploit natural plant defense mechanisms against pests and diseases holds much promise for reducing the usage of environmentally damaging synthetic pesticides. Cytokinins comprise a major group of plant hormones that regulate normal growth and development with a possible role in disease resistance. We modified the endogenous production of cytokinin in plants and studied its effect on insect resistance. Flowering tobacco plants stimulated by insect feeding to express a gene for an enzyme involved in cytokinin biosynthesis were more resistant to *Manduca sexta* and *Myzus persicae*.

Plant hormones are known to have pivotal roles in promoting normal growth and development of plants and may also contribute to the mechanisms of defense (Davis, 1987; Orr, 1992). Cytokinins as well as other plant hormones have commercial applications as bioregulators and in combination with endogenous hormones may protect plants from pests and pathogens by inducing physiological changes in the plants (Hallahan et al, 1992; Hedin et al, 1988; Thomas and Balkesley, 1987). Cytokinins have been shown to influence secondary metabolic pathways whose products exhibit insecticidal properties (Orr, 1992). Utilization of numerous secondary metabolites in crop protection, either by conventional plant breeding or by genetic engineering, is being evaluated (Hallahan et al, 1992). We evaluated the role cytokinin may play in insect resistance by genetically engineering *Nicotiana plumbaginifolia* with a wound-inducible cytokinin biosynthesis gene.

Chimeric cytokinin gene expression

A bacterial isopentenyl transferase (*ipt*) gene involved in cytokinin biosynthesis (Smigocki, 1991) was fused with a promoter from the proteinase inhibitor II (PI-IIK) gene (Thornburg, 1987) and introduced into tobacco (Fig. 1). Transcripts of the *ipt* gene were wound-inducible in leaves of transgenic PI-II-*ipt* plants and maximum increases of 25- to 35-fold were detected within 24 hours of wounding (Fig. 2). Concentrations of zeatin and zeatinriboside cytokinins, which are the major cytokinins produced in tissues transformed with the *ipt* gene (Smigocki and Owens, 1988), were determined using analytical kits (De Danske Sukkerfabrikker, Copenhagen; IDETEK, Inc., San Bruno, CA). Plant extracts were purified on columns packed with antizeatinriboside antibodies and quantified by ELISA. Cytokinin levels were elevated more than 70-fold in fully expanded leaves.

Insect Feeding Studies

Two independently transformed PI-II-*ipt* R2 plants (102 and 108) were used to test for defensive properties of cytokinins against insects. On the average, third-instar tobacco hornworm (*M. sexta*) larvae consumed 60% less when fed fully expanded leaves removed from flowering PI-II-*ipt* plants than larvae feeding on leaves from untransformed controls or transgenic controls without the cytokinin gene (Table 1). Mean larvae weight gain was reduced by approximately 20 to 60% in comparison to the controls (Table 1). When whole plants were infested with a single neonate hornworm larval, consumption of the PI-II-*ipt* plants was greatly reduced in comparison to the control plants in 3 experiments.

In similar studies with the green peach aphid

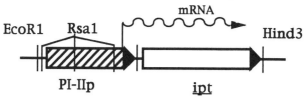

Fig. 1. Chimeric PI-II-*ipt* gene construct. The potato proteinase inhibitor IIK gene promoter (PI-II$_p$) was fused through its 5'-untranslated region to the coding region of the isopentenyl transferase gene *(ipt)* from pTiB6S3. A transcription initiation site is located 101 bp upstream of the *ipt* gene start codon.

nymphs, *M. persicae*, only 30 to 40% of the nymphs developed into adult females and of those 50 to 80% reproduced. On control transgenic and untransformed tissues, on the average, 74% of the nymphs reached adulthood and 93% of these adults reproduced within the 8 day time constraint of the test. Results represent averages of 4 independent tests done in replicates of 10 per treatment.

Cytokinin Mode of Action

Presently, the mode of action of the cytokinin gene product on enhanced resistance is not clear. We speculate that products of secondary metabolic pathways affected by elevated cytokinin levels are the likely candidates for the observed increase in insect resistance.

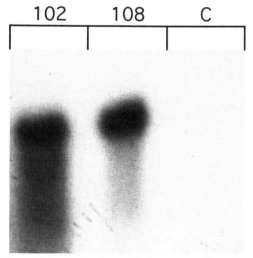

Fig. 2. Induction of *ipt* transcripts by wounding. Northern blot of total leaf RNA from wounded PI-II-*ipt* (102, 108) and transgenic control (C) tobacco plants.

Table 1. Enhanced resistance of PI-II-*ipt* plants (102, 108) to the tobacco hornworm larvae.

	Percent reduction as compared to controls in	
	mean leaf area consumed	mean larval weight gain
Experiment 1		
102	68	57
108	73	46
Experiment 2		
102	47	47
108	58	28
Experiment 3		
102	61	52
108	50	19

References

Davis, P. J., Ed.; *Plant hormones and their role in plant growth and development*; Kluwer Academic Publishers: Boston, MA, 1987.
Hallahan, D. L.; Pickett, J. A.; Wadhams, L. J.; Wallsgrove, R. M.; Woodcock, C. "Potential of secondary metabolites in genetic engineering of crops for resistance." In *Plant Genetic Manipulation for Crop Protection*; Gatehouse, A. M. R.; Hilder, V. A.; Boulter, D., Eds.; Redwood Press LTD, Medlsham, United Kingdom, 1992: pp 212-248.

Hedin, P. A.; Williams, W. P.; Davis, F. M.; Thompson, A. C. "Effects of plant bioregulators on nutrients, insect resistance, and yield of corn (*Zea mays* L.)." *J. Agri. Food Chem.*, **1988**, *36*, 746-748.

Orr, J. D.; Lynn, D. G. "Biosynthesis of dehydrodiconiferyl alcohol glucosides: implications for the control of tobacco cell growth." *Plant. Physiol.*, **1992**, *98*, 343-352.

Smigocki, A. C. "Cytokinin content and tissue distribution in plants transformed by a reconstructed isopentenyl transferase gene." *Plant Mol. Biol.* **1991**, *16*, 105-115.

Smigocki, A. C.; Owens, L. D. "Cytokinin gene fused with a strong promoter enhances shoot organogenesis and zeatin levels in transformed plant cells." *Proc. Natl. Acad. Sci. USA*, **1988**, *85*, 5131-5135.

Thomas, T. H.; Balkesley. "Cytokinins - plant hormones in search of a role - practical and potential uses of cytokinins in agriculture and horticulture." *Monograph British Growth Regulator Group*, **1987**, *14*, 69-83.

Thornburg, R. W.; An, G.; Cleveland, T. E.; Johnson, R.; Ryan C. A. "Wound-inducible expression of a potato inhibitor II-CAT gene fusion in transgenic tobacco plants." *Proc. Natl. Acad. Sci. USA* **1987**, *84*, 744-748.

IMPLEMENTATION:
NEEDS, ISSUES, AND CHALLENGES

The Role of Education in the Transfer of Biological Control Technologies

Michael S. Fitzner, USDA/Extension Service, 1400 Independence Avenue, SW, Washington, DC 20250-0900

A convergence of technical, environmental, and social forces is moving agriculture towards greater reliance on nonchemical pest management alternatives like biological control. However, technology transfer must be accelerated to ensure more widespread use of biological control technologies. The Cooperative Extension System can assist technology transfer by: 1) providing intensive biological control training to Extension agents and specialists; 2) allocating additional resources to developing educational materials for biological control; and 3) promoting greater collaboration among research and Extension personnel, producers, and the private sector. In many cases, however, additional applied research is needed before detailed biological control education programs can be provided to pest management practitioners.

Successful implementation of biological control for agricultural pests requires an adequate research base, a good understanding of biological control principles, and sound regulatory policies. Although research advances continue to provide agriculture with new and more effective biological control technologies and regulatory policies are continually debated, insufficient consideration is given to the transfer of biological control technologies to pest management practitioners. This paper describes the role of education in the transfer of biological control technologies and suggests the need for a renewed effort to blend biological control tactics into comprehensive integrated pest management (IPM) strategies.

The Evolution of Pest Management Strategies

The integrated control concept described by Stern et al. (1959) provides the philosophical basis for an approach to pest management that "combines and integrates biological and chemical control." Integrated control emphasizes biological control to maintain pest populations below the economic threshold, but also acknowledges that biological control sometimes must be supplemented with chemical control "as necessary and in a manner which is least disruptive to biological control." The value of biological control has been well established over the last century. Van Driesch and Fero (1987) report the successful use of biological control tactics in 253 projects worldwide since 1888, when the vedalia beetle was introduced to control cottony cushion scale in California. Each dollar invested in California biological control programs produces a $30 increase in net return due to reductions in crop damage and pesticide expenditures (DeBach, 1975). However, a variety of technical, economic, and social factors have restrained biological control implementation in commercial agriculture.

The integrated control concept provides the basis for IPM programs used throughout the world. IPM strategies save U.S. agricultural producers an estimated $500 million per year and significantly reduce pesticide use (Rajotte et al., 1987). The National Academy of Science (1989) estimates IPM strategies are used on 45 million acres (18 percent) of 12 major U.S. crops approximately 14 percent of total U.S. cropland. Mueller (1988) estimates 11 million acres (8 percent) of U.S. cropland are enrolled in the Cooperative Extension System's IPM programs. Although IPM has been successfully implemented throughout the United States, critics argue that many IPM strategies do not adhere to the principles of integrated control. This criticism may arise from a perception that IPM strategies stress improved pesticide management—based on field monitoring and economic thresholds—rather than practices that are least disruptive to biological control. While this criticism may be valid in some cases, there are many examples of comprehensive IPM strategies that successfully blend biological control with other tactics (e.g., Hoy and Herzog, 1985; Leslie and Cuperus, 1993; Rajotte et al., 1987). However, host plant resistance and cultural management, not biological control, are the biologically-based tactics most frequently employed in IPM strategies.

There are several factors limiting the integration of biological control tactics into comprehensive IPM strategies. Currently, a number of major agricultural pests cannot be maintained below economically damaging levels without chemical pesticides (National Academy of Science, 1989). In some cases, these pests are introduced exotics that are not amenable to biological control. In other cases, biological control has not been tried adequately, if at all. Participants at a national IPM workshop cited six factors limiting integration (Nelson, 1989): 1) reliance on chemically-based strategies; 2) lack of practical information about the use of natural enemies in pest suppression; 3) a better understanding of biological control in perennial, rather than annual, cropping systems; 4) limited participation of biological control researchers due to an emphasis on chemical control tactics in many early IPM programs; 5) absence of a strong biological control industry; and 6) the emergence of IPM as a response to pest crises (e.g., pest resistance and secondary pest outbreaks) rather than as a population management strategy based on natural control. Participants also emphasized the need to improve communication between biological control researchers and Extension personnel.

Various terms are used to describe IPM strategies that rely more heavily on biological control tactics. Frisbie and Smith (1989) use "biologically intensive" to describe comprehensive IPM strategies that "rely on biologically based IPM tactics rather than

agricultural chemicals." Like conventional IPM, biologically intensive IPM blends biological control, host resistance, and cultural management into a single comprehensive pest management strategy. However, it differs from conventional IPM in that agricultural chemicals are not a cornerstone tactic. Prokopy (1993) suggests the transition from chemically-based to biologically intensive IPM is part of a natural evolution towards higher-level pest, crop, and whole-farm management systems. Several other authors use "biologically intensive" to delineate IPM strategies that rely on biological control from those that do not (Edwards, 1991; Ferro, 1993; Pedigo and Higley, 1992; Zalom and Fry, 1992). Biologically intensive IPM is neither an attempt to redefine IPM, nor an effort to reinvent the integrated control concept. It is not a radical change in direction for IPM research and Extension programs since many current IPM strategies emphasize biologically-based tactics. Rather, biologically intensive IPM is an acknowledgement of a need to expand and accelerate efforts to make biologically-based tactics a cornerstone of IPM in theory and practice. Biologically intensive IPM strategies differ from conventional IPM strategies because they are constructed with an assumption that pesticides are not available and economic control can be achieved with augmentation of natural controls.

Transfer of Biologically Intensive IPM Strategies

During the past four decades, intensive education programs helped agricultural producers make efficient use of chemical pesticides and fertilizers. A comparable effort is now needed to teach producers and other pest management practitioners how to implement biologically intensive IPM strategies. The early stages of this educational program should emphasize the importance of employing all available biologically-based tactics disfavoring pest population development, and maximize natural enemy activity by delaying pesticide applications until the economic threshold is reached. This is not a new strategy—since the late 1950's, entomologists cautioned producers that pesticides and other production practices have destructive impacts on nature enemies. However, additional research progress and growing public pressure to reduce pesticide use increases opportunities for successful integration of biological control into comprehensive IPM strategies.

Agricultural producers are reluctant to adopt new technologies, particularly those that disrupt proven production practices, until they are provided convincing evidence the technology is economically viable, and the benefits of adoption justify real and perceived risks associated with their use. Educational programs must provide this evidence and must cultivate a level of producer confidence that is equal to the confidence they have in their current production system. Kennedy et al. (1992) provide a useful summary of the key factors contributing to a decision to adopt new technologies or practices: 1) field demonstrations and other evidence verifies that the new practice is cost effective compared to current practices; 2) the new practice is compatible with existing knowledge and resources; 3) implementation of the new practice is not too difficult; and 4) there is a willingness to change for economic, environmental, or social reasons. Good interdisciplinary and multi-organizational collaboration in the development and transfer of a new strategy is another component of many successful IPM implementation efforts (e.g., Ferro, 1993; Haney et al., 1992; and Zalom et al., 1987).

The basic elements of a successful agricultural technology transfer program are the same regardless of the technology being transferred. After the developmental research phase is completed, adaptive research must be conducted to: 1) fine-tune the new technology to accommodate local conditions and producer needs; 2) test the new technology on farms of innovative and well-respected producers; and 3) assess the profitability and commercial viability of the new technology. Finally, an educational program must be developed to provide pest management practitioners with practical instructions for implementation, including complete "how to" details and cost/benefit analyses.

Focus on Biological Control Education

Although biologically intensive IPM education programs stress the importance of a comprehensive approach to pest management, special emphasis is given to biological control due to its knowledge-intensive nature. As new biological control tactics are developed, validated, and blended into biologically intensive IPM strategies, educational programs must be developed concurrently to ensure successful technology transfer. Educational materials (e.g., manuals, fact sheets, videos, and decision support software) must provide detailed information about important biological control organisms and their use in comprehensive IPM strategies. Educational materials should include: 1) high quality photographs and drawings for identification of natural enemies; 2) detailed information about the life-cycles of biological control organisms (including development rates), their tolerances of environmental conditions, and their compatibility with various agricultural chemicals; 3) detailed instructions for monitoring pests and beneficials and for application of biological control organisms (including rate, timing, and sources), if augmentation is required; 4) information about biological and "soft" pesticides, including their efficacies against both pests and beneficials; and 5) step-by-step guidance on blending biologically-based tactics into comprehensive IPM strategies.

While the development of an education program is an essential element of the technology transfer process, there are other economic and social factors that have major impacts on the success or failure of the implementation effort. Reichelderfer (1981) lists six factors that strengthen the potential for successful implementation of biological control programs: 1) target pest is a major pest of a high-value crop; 2) biological control consistently provides good pest control; 3) biological control organisms are readily available at a competitive price; 4) low application costs; 5) profitability is as good or better than profitability using current pest management tactics; and 6) good coordination among institutions ensures region-wide implementation. It has been observed that biological control programs are easiest to implement when no alternatives are available, public awareness of pesticide use is high, and pesticide applications are particularly dangerous to human health or the environment (Marois and English, 1989). Although all of these factors should be considered before scarce resources are committed to a technology transfer effort, it is essential that biological control not be viewed as the tactic of last resort—the tactic that is tried when all others fail or are no longer available. Aggressive biological control education programs are needed to exploit the full potential of this pest management tactic.

Extension's Role in the Education Process

The Cooperative Extension System can assume a leadership role in the transfer of biological control tactics by further developing the capacity to provide comprehensive IPM education programs. Extension's IPM education programs can stimulate greater adoption of biological control technologies, but its ability to provide these programs continues to erode (Aylsworth, 1993). Before educational programs can be implemented on a large scale, Extension personnel must receive intensive training on the practical use of biological control and its integration into comprehensive IPM strategies.

Several states allocate significant Extension resources to biological control efforts, but most have not established formal biological control programs. In all, 27 states and 1 U.S. territory supported IPM education efforts in biological control during fiscal years 1991 and 1992. The speakers at a recent conference emphasized the need for Extension's leadership in biological control education in collaboration with researchers, private consultants, agribusiness companies, and cooperating producers. It was also observed that before Extension can provide biological control education programs, additional

applied research is needed to: 1) establish economic thresholds and practical sampling protocols that recognize biological control's value in pest suppression; 2) determine the distribution and importance of natural enemies; 3) develop management strategies that preserve and enhance natural enemy activity; and 4) incorporate pesticide impacts on natural enemy populations into economic thresholds. The insufficient detail characteristic of many existing Extension biological control education materials is a due, in large part, to the inadequacy of the applied research base.

A successful national biological control education program will require further development of Extension's human resources. To prepare for an increased demand for information, Extension personnel need intensive training in the use of biological control tactics, including specific information about their integration into comprehensive IPM strategies. Extension entomology specialists in 12 North Central states estimate the portion of their time devoted to biological control will increase from 10 percent to 20 percent in 5 years and to 30 percent in 10 years (Mahr, 1991). It is likely the increasing demands for information on biological control will cause this trend to occur nationally.

Future Needs, Future Challenges

Decades of research conducted by scientists in the United States and elsewhere has expanded our understanding of the agro-ecosystem. Biological control tactics have been implemented in many agricultural production systems, but the technology transfer process must be accelerated. To do this, the Cooperative Extension System must significantly increase its commitment to biological control education programs by implementing an organizational strategy to develop the human and information resources required. This strategy should: 1) provide biological control training for Extension agents and specialists; 2) provide regional and national coordination for the development of educational materials; and 3) promote greater collaboration among research and Extension personnel, producers, and the private sector. The first steps of this strategy are:

1) Hold a series of national and regional workshops on biological control to provide pest management practitioners with practical information about biological control and its integration into biologically intensive IPM strategies. The workshops should allow time for research, Extension, and private-sector personnel to plan coordinated regional and national biological control programs.

2) Increase biological control programming in on-going Extension pest management programs. Additional resources should be allocated for applied and adaptive research to fine-tune biological control tactics for specific regions or localities. The feasibility of biological control should be assessed in large-scale field demonstrations on key farms, and should include an analysis of their economic viability. Resources are also needed to develop and evaluate publications, slide sets, teaching collections, and videos describing biological control organisms and their use in comprehensive IPM strategies.

3) Enhance linkages among researchers, Extension personnel, and the private sector. Linkages can be strengthened by establishing competitive funding programs that require collaboration.

Successful biological control implementation requires coordinated interdisciplinary and multi-organizational efforts throughout the developmental research, adaptive research, and technology transfer phases. A key element of many successful programs is a well planned and executed transition between the research and technology transfer phases (e.g., Hoy, 1989). This transition is ensured by 1) including Extension and private-sector personnel in all phases of the research process and 2) including research and private-sector personnel in all phases of the education and technology transfer processes.

A convergence of technical, environmental, and social forces is moving agriculture towards greater reliance on nonchemical pest management alternatives like biological control. Extension must recognize this shift and enhance its capacity to integrate biological control tactics into comprehensive IPM education programs. However, Extension cannot accomplish this goal alone. The responsibility for the development and implementation of these strategies must be shared by researchers, Extension personnel, and the private sector. Biological control must be viewed as an essential component of comprehensive IPM strategies, not a competitor. If we fail to establish common goals and a collaborative spirit in the development and transfer of these technologies, the necessary resources will not materialize because our efforts will appear unfocused, uncoordinated, and wasteful. Our common goal is a productive, profitable, safe, and environmentally-friendly agricultural industry. We must strive to achieve this goal together in a well-structured and organized way.

Acknowledgments

The helpful comments and suggestions provided by Diane Alston, James Cate, Raymond Frisbie, Dennis Kopp, John Impson, Daniel Mahr, and Judy Rude are gratefully acknowledged.

References

Aylsworth, J.D. "Budget Cuts Squeeze Extension," *Am. Fruit Grower.*, **1993**, 113, 6-7.

DeBach, P. *Biological Control by Natural Enemies*. Cambridge University Press: Cambridge, England, 1975.

Edwards, C.R. "National Organization Promotes Integrated Pest Management." *Am. Entomol.*, **1991**, 37, 136-137.

Ferro, D.N. Integrated Pest Management in Vegetables in Massachusetts, In *Successful Implementation of Integrated Pest Management for Agricultural Crops*. Leslie, A.R; Cuperus, G.W., Eds., Lewis Publishers: Boca Raton, FL, 1993, 95-105.

Frisbie, R.E.; Smith, J.W. Biologically

Intensive Integrated Pest Management: The Future. In *Progress and Perspectives for the 21st Century*. Menn, J.J.; Steinhauer, A.L., Eds., Entomological Society of America Centennial National Symposium, Entomological Society of America: Lanham, MD, 1989, 151-164.

Haney, P.B.; Morse, J.G.; Luck, R.F.; Griffiths, H.; Grafton-Cardwell, E.E.; O'Connell, N.V. *Reducing Insecticide Use and Energy Costs In Citrus Pest Management*. IPM Education and Publications, UC IPM Publication 15: University of California, Davis, CA, 1992.

Hoy, M.A. Integrating Biological Control Into Agricultural IPM Systems: Reordering Priorities. In *Proceedings of the National Integrated Pest Management Symposium/Workshop*, Las Vegas, Nevada, April 25-28, 1989. New York State Agricultural Experiment Station, Cornell University, Communications Services: Geneva, NY, 1989, 41-57.

Hoy, M.A.; Herzog, D.C., Eds.; *Biological Control in Agricultural IPM Systems*; Academic Press: New York, 1985.

Kennedy, L; Klonsky,K.; Carter, H.O.; Auburn, J.; Barriers and Incentives to Grower Adoption of Biological Approaches to Pest Management. In *Beyond Pesticides: Biological Approaches to Pest Management in California*; Beall, T., Ed. ANR Publications: University of California, Publication No. 3354, Oakland, CA, 1992,; 166-178.

Leslie, A.R.; Cuperus, G.W.; Successful Implementation of Integrated Pest Management for Agricultural Crops; Lewis Publishers: Boca Raton, FL, 1993.

Mahr, D.L. Implementing Biological Control of Arthropods in the North Central States: An Extension Perspective. A report prepared for North Central Regional Administrative Committee (NCA-15), 1991. (*unpublished*)

Marois, J.J.; English, J.T. Integration of Biological Control of Plant Pathogens into IPM Programs. In *Proceedings of the National Integrated Pest Management Symposium/Workshop*. Nelson, M.R., Moderator; Las Vegas, NV, April 25-28, 1989. New York State Agricultural Experiment Station, Cornell University, Communications Services: Geneva, NY, 1989, 33-40.

Mueller, W. "IPM: A Wise Discipline. So Why Hasn't It Caught On?" *Agric. Chem. Age*, **1988**, 32, 6-7, 10, 22-23.

National Academy of Science. *Alternative Agriculture*. Board on Agriculture, National Research Council, National Academy Press: Washington, DC, 1989.

Nelson, M.R. Integration of biological control in IPM. In *Proceedings of the National Integrated Pest Management Symposium/Workshop*. Las Vegas, Nevada, April 25-28, 1989. Communications Services: New York State Agricultural Experiment Station, Cornell University, Geneva, NY, **1989**, pp 200-201.

Pedigo, L.P.; Higley, L.G. "The Economic Injury Level Concept and Environmental Quality: A New Perspective." *Am. Entomol.*, **1992**, 38, 12-21.

Prokopy, R.J. "Stepwise Progress Toward IPM and Sustainable Agriculture." *The IPM Practitioner*, **1993**, 15, 1-4.

Rajotte, E.G.; Norton, G.W; Kazmierczak, R.F; Lambur, M.T.; Allen, W.A. National Evaluation of Extension's Integrated Pest Management (IPM) Programs. Publication No. 491-010, Virginia Cooperative Extension Service: Blacksburg, VA, 1987.

Reichelderfer, K.H. Economic Feasibility of Biological Control of Crop Pests. In *Biological Control in Crop Production*; Papavizas, G.C.; Endo, B.Y.; Klingman, D.L.; Knutson, L.V.; Lumsden, R.D; Vaughn, J.L., Eds. Beltsville Symposia in Agricultural Research, No. 5, 1981, 403-417.

Stern, V.M., Smith, R.F.; van den Bosch, R.; Hagen, K.S. "The Integrated Control Concept," *Hilgardia*, **1959**, 29, 81-101.

Van Driesche, R.G.; Ferro, D.N. Will the Benefits of Classical Biological Control Be Lost in the "Biotechnology Stampede?" *Am. J. Alternative Agric.*, **1987**, 2, 50, 96.

Zalom, F.G.; Fry, W.E., Eds. *Food, Crop Pests, and the Environment*. APS Press: St. Paul, MN, 1992.

Zalom, F., K. Klonsky; Barnett, W. *Evaluation of California's Almond Program*. IPM Education and Publication No. 6, University of California, Davis, CA, 1987.

New Regulatory Strategies for Pheromones: EPA Policies, Interpretations, and Rules

Phillip O. Hutton, Insecticide/Rodenticide Branch, Registration Division H7505CU.S. Environmental Protection Agency, 401 M Street, S.W., Washington, DC 20460

The Environmental Protection Agency (EPA) is revising certain policies and establishing new rules and procedures regarding registrations and experimental use permits (EUP) for pheromones. These new policies and procedures are discussed along with a description of EPA's long term goals for the regulation of semiochemicals.

History

EPA has regulated pheromones for a long period of time. These products were originally handled under the Federal Insecticide, Fungicide, and Rodenticide Act (FIFRA) much as any other chemical, with a full set of toxicology and residue data required. It should be noted, however, that the data requirements for all pesticides were relatively simple at that time compared to today's standards. Over time, the data requirements for the registration of pesticides became more complex, and pheromones were still subjected to the full set of requirements. Through communications with other government agencies such as USDA and with the increasing popularity of IPM, the EPA began to consider these kinds of products not just as chemicals with a set of potential risks but as potential replacements for more hazardous traditional chemical pesticides.

As a result of this the Agency in 1975 published a policy statement which differentiated pheromones as "biorational pesticides" and proposed reduced data requirements as compared to conventional pesticides. This came to fruition in 1982 with the publication of the subpart M guidelines, which greatly reduced the data requirements for pheromones which were now referred to as biochemical pest control agents. In this regulatory sense, biochemicals are not simply organic compounds. Subpart M in 1982 provided a regulatory definition of biochemical pest control agent as including semiochemicals, plant regulators, hormones, or enzymes used as a pesticide. The reduction of data requirements as compared to traditional toxic compounds was based upon the natural occurrence of the compound in the environment, a low potential for exposure to non-target organisms, and maximum challenge testing as the initial tier of a tiered testing scheme. Subpart M eliminated most of the long-term and expensive chronic studies as a requirement for registration of most pheromone products.

As basic research identified more and more compounds, and field research discovered better delivery methods and use techniques, EPA saw an increase in the number of submissions for new pheromone products. However, EPA began receiving complaints that the registration requirements for pheromones, while greatly reduced as compared to conventional toxicant, were still greater than necessary to properly assess the potential hazard. Part of the reason for this complaint was that pheromones, though increasing in popularity, were basically niche products, and pheromone sales have difficulty in justifying the time and cost associated with the regulatory investment.

Much of this was brought out in a formal symposium at the Entomological Society of America (ESA) meeting at Boston, MA in 1987. Follow-up meetings were held at the 1988 and 1989 ESA meetings at Louisville, KY and San Antonio, TX, respectively. As a result of this meeting it was decided that a "white paper" assessing the current state of knowledge, and hopefully justifying a further reduction of data requirements, would be produced. Unfortu-nately, obtaining funding and support was difficult, and the White Paper has taken additional time to get out. Following the Louisville meeting, the EPA

received a number of petitions requesting various degrees of regulatory relief. The following requests were received:
1. Exempt all tie-on products from a tolerance.
2. Make all non-food use pheromones exempt from regulation.
3. All the U.S. Forest Service to test 200-500 acres without a permit.
4. Exempt all pheromones from the requirement of a tolerance.
5. Allow multi-year EUPs.
6. Form a formal USDA-EPA Committee on Pheromones.
7. Allow all pheromones to be tested on 200-500 acres without an EUP.

In the interim, many interested parties attended the Brighton Workshop in the UK in 1991. At this meeting, EPA representatives found that there was a great deal of variance in the regulatory approaches of various nations. Australia, for example, does not regulate naturally occurring pheromones, while most of Europe requires a full set of supporting data similar to traditional chemical compounds. After the Brighton meeting, the Agency received additional requests to:
1. Waive all ecological effects data for matrixes.
2. Waive all toxicological studies related to dietary exposure for matrixes.
3. Waive the acute oral toxicology study for non-food uses.
4. Further data reduction for ecological effects studies.
5. Allow batch testing of multiple actives.

A further significant event was the formation of the American Semiochemicals Association (ASA) in 1992. Prior to this companies tended to make requests related to their own self interest. Those companies which were comfortable submitting a full set of studies did not support relaxation of requirements for new companies getting started. In essence, they felt they had paid their dues and now wanted to make sure the competition had to jump the same set of hoops. The formation of the ASA has resulted in a compromise uniform position from industry, which is likely to be more effective than past individual conflicting requests.

As a result of this input from a multitude of diverse sources, EPA has been moving forward on several fronts to reduce not only the data burden but if possible the submission burden for industry, academia, and other state and federal agencies.

Policy Changes Currently in Effect

1. *Overall Acreages* - While EPA policy only allows 4000 acre permits without extensive and lengthy internal review, EPA allows much larger acreages for pheromone uses, as long as the request is justified by the applicant.

2. *Dropped Requirements from past years* - The avian dietary LC50 study is no longer required for pheromones in point source matrixes. Non-target plant and non-target insect studies have been eliminated. The subchronic toxicity series and teratogenicity studies are usually not required for low dosage applications involving matrix dispensers.

3. *Flexible Confidential Statement of Formula* - Applicants frequently complained that as study trials progress during the season, formulations need to be adjusted to provide optimum release rates from dispensers. Normally, this requires the submission of the revised confidential statement of formula, with a 120 day review time. This is an obvious burden to research. Therefore the EPA now permits the original formula to have a wide range of ingredients and percentages to accommodate foreseeable changes during the course of the testing.

4. *Multiyear EUPs* - EPA allows multi-year permits to reduce the paperwork burden on applicants.

5. *Batch Testing* - EPA allows new active ingredients that contain multiple compounds to be tested as a batch of the compounds contained rather than requiring a separate series of studies on each component in the formulation.

Additional Regulatory Options Presently Under Consideration

The current work on regulatory revisions involves the uses associated with lepidopteran pheromones in dispensers. There are several reasons for this focus. First and foremost, the most extensive data base regarding human toxicity and non-target effects are for this group of straight chain compounds. Also, there is less exposure for non-target organisms and greatly reduced potential for residues in food from pheromones used in dispenser-type applications. In general, dispensers are considered to be any device or matrix which can be easily removed from the site of application. Finally, because of the low-toxicity and low exposure scenario associated with these uses, this group of compounds appears to fit well into EPA's Reduced Risk Pesticides initiative.

1. *Relaxed 10 Acre Rule to 250 for EUPs* - Essentially, this would mean that an EUP would not be required until the application involved more than 250 acres total.

2. *Generic Tolerance for Lepidopteran Pheromones* - The 250 acre increase for testing without a permit would be critical but without a tolerance exemption, would not be helpful as products could not be applied to food or feed crops unless the use was crop-destruct. Few companies or agencies can afford to buy 250 acres of crop. EPA is evaluating whether the present data base will support a generic exemption.

3. *Relaxed Data Requirements for Inerts* - As with active ingredients, all inert ingredients must be support by a tolerance or exemption. EPA is considering whether a generic exemption, declaration as non-food use, or some other regulatory policy option would permit the inert components of such matrixes

to be used without formal application for tolerance or exemption.

A major hurdle in accomplishing these goals is the administrative burdens of changing already codified policies. If legal determinations find that these actions require rulemaking, then that success generally takes a minimum of two years to accomplish. Therefore we are concentrating on items that can be accomplished as simple notifications of policy change. Generic tolerance exemptions do not require the lengthy rulemaking process and are also considered to be a more expeditious method of accomplishing these goals.

Looking to the Future

The following additional projects are being looked at as long-term goals:

1. Reduced Requirements for Lepidopterous Straight Chain materials in broadcast applications, such as in microencapsulated formulations.

2. Inclusion of Other Pheromone Types as the data base becomes available. Investigation of other semiochemical classes may utilize structure/function analysis if reliable information can be submitted by interested parties.

3. Further reduction in acute toxicology and non-target data requirements for EUPs and registrations, as the existing data base is expanded.

4. Expansion of the current FIFRA Section 25(b) Exemption for pheromones used in traps (40CFR 152.25) to include other devices or matrixes.

Patents and Intellectual Property: an Overview

Barry U. Buchbinder, Life Technologies, Inc., P.O. Box 6009, Gaithersburg, MD 20884

The world of intellectual property as it concerns agriculture is changing. These changes come from new technology, relatively minor changes in patent law, and changes in how companies desire to do business. In this paper, these changes and their ramifications will be explored.

There are many types of intellectual property. Much intellectual property consists of **R&D results, know-how**, and **business information**. Its exclusivity can be secured *via* legal mechanisms such as **patents, plant variety protection (PVP) certificates**, and **trade secrets**. For some kinds of intellectual property only one mechanism is appropriate, for other kinds, more than one can be used. Other types of intellectual property include **trademarks**, which are much less important to public-sector developers of biologically based pest management technology, and **copyrights**, which are virtually irrelevant.

This article will start with a brief overview of all types of intellectual property, followed by an in depth discussion of patents and a brief section on trade secrets. In passing special mention will be made of what those in the public sector should know and will end with a section on possible effects to public sector research.

Please note that many nuances and subtleties are omitted, notably for some controversial points. Many things are simplified, resulting in some statements not being true in all circumstances. The purpose of this article is to give the uninitiated an introduction. If you need legal advice, do not depend on this; talk to a lawyer.

Overview of the Types of Intellectual Property

Utility patents are what the average citizen thinks of when the word "patent" is used. A U.S. patent may be granted for any *process, manufacture, machine, composition of matter*, or *improvement* that meets certain tests for being *new, useful*, and *unobvious*. A patent allows one to forbid others to *make, use*, or *sell* the claimed invention. A patent does *not* allow one to practice the invention; other patents may dominate or government regulations may apply. The courts have ruled that living organisms *per se* can be patented; methods involving living organisms have long been covered by patents.

Design patents cover the ornamental features of a useful object. They are largely inapplicable to pest management.

Plant patents protect asexually propagated varieties, except for those propagated by tubers. This exception was placed in the law for the potato farmers. Plant patents are used mostly for ornamentals. Patented plants may be freely used as parents for crosses to develop new varieties. The Agriculture Marketing Service (AMS) assists the U.S. Patent and Trademark Office in the examination of the merits of applications for plant patents.

Plant variety protection certificates protect sexually propagated crops that breed true. They are examined and issued by the AMS. Farmers are given explicit permission to use part of their production as seed. Breeders are explicitly permitted to use a PVP protected variety as a starting material.

Until recently, a problem with both plant patents and PVP certificates had been that it was difficult to detect and prove infringement. This was especially true in agronomy; flowers tend to be distinctive but the crop market selects for uniformity. The properties that distinguish varieties may be very sensitive to environmental factors, and thus may be difficult to reproduce every growing season or in all locations. Therefore, plant patents and PVP certificates tended to function as trademarks. An unauthorized party could not sell a protected variety using the real name. One sued if the infringer had already effectively admitted infringement by using the official name. Modern genetic identification techniques have made it easier to enforce plant patents and PVP certificates.

Trademarks protect names used in commerce for products or services. Trademarks protect more than the assets of a company. Indeed, their prime purpose is to protect consumers, who depend on them to be assured of what is being bought and from whom. Trademarks are governed by a mixture of federal, state, and common law.

The public is notified of the existence of a trademark by use of ™, or ˢᴹ for a service mark, or ® after registration. Grammatically, the mark is virtually always an adjective describing a generic noun, e.g., Tennis Ball™ bruise-resistant tomatoes. Two parties or products may use the same mark if they are not competitive and no public confusion would result. Trademarks are important in agriculture.

Trade secrets have often been used to protect plant intellectual property. The source of germplasm is not revealed and inbreds are kept secret. Though there is always the danger that a competitor will steal some inbreds to make hybrids ("road breeding"), trade secrets have been the primary means for protecting hybrid corn. Also, inbred seed may contaminate hybrid seed and may be recovered, propagated, and sold. Anything in a bought bag of seed may be legally used as breeding material.

Copyrights protect the form of expression of information, not the information itself. Copyrights exist inherently. When a document bearing a copyright notice of a certain form is made public (i.e., published), international copyrights are automatically created. In general, copyright protection is not useful for protecting agricultural technologies.

Among the tools that control technology, trade secrets are the fastest and least expensive to establish, while utility patents are the most costly, slowest, and least certain to obtain.

Intellectual Property as Property

An important point to understand about all these tools is that they are *property*. (That is why they are called intellectual property.) They can be bought, sold, and licensed (i.e., rented) just like any other property. The essence of property is the right to *exclude* others from using it, not the right to do with it as one wishes. For instance, a skyscraper may not be built in a residential neighborhood due to zoning restrictions but one can keep others from using one's patio. For each of these intellectual property tools, the law defines exactly what is the property right. For example, a stranger may have the right to walk on your sidewalk but not to park in your driveway. Similarly, the law says that a farmer can save PVP protected seed for planting on his or her own farm in subsequent growing seasons, but one has no right to propagate asexually a flower covered by a plant patent, even if the propagules are not sold. So a farmer has the right to sell a PVP protected crop to another farmer if the seed is to be fed to animals but not if it is to be planted; the crop is the property of the person producing it but remains the intellectual property of the intellectual property owner.

Trade Secrets

Most intellectual property which is not appropriate for patenting can only be maintained as an asset through secrecy. The laws of many states recognize that most kinds of business information or knowledge, e.g., sales figures, customer mailing lists, manufacturing or quality control procedures, are worthy of legal protection as property. To constitute a "trade secret", information must be **used**, must give one a **competitive advantage**, and must be **secret**, i.e., not known to the industry. A trade secret does not establish exclusivity; others may independently have the same information as a trade secret. A trade secret can exist until it is generally known in its industry; some have lasted centuries.

Trade secrets are relatively inexpensive to establish but can be difficult to enforce. ("*Three may keep a secret if two of them are*

392

dead." [Benjamin Franklin]) The same characteristics that allow keeping a secret make discovery of a violation difficult. Trade secrets are often "leaked" or sold by disgruntled employees.

Biological materials, *e.g.*, recombinant DNA molecules, strains, cell lines, sera, etc., can be considered a type of confidential information if there are any conditions or restrictions placed on them. As physical objects, they can be property in the common sense. Of course, they are more difficult to handle than other types of information because they are not documentary in nature.

Keeping trade secrets for a long time is impractical for a public sector institution. (Professors must publish!) In any case, since biological pest management involves public release of materials, trade secrets tend to disappear as soon as a product is marketed. However, trade secrets used in production may be practical in this area.

Patents

Some technology cannot, by its nature, both be used and kept secret. But its exclusivity can be secured through patents.

A patent is an incentive to invent.

"The patent system added the fuel of interest to the fire of genius." [Abraham Lincoln, inventor of "Manner of Buoying Vessels", U.S. Patent 6,469, issued May 22, 1849]

Historically, they go back to renaissance Italy. In the U.S., patents (and copyrights) are based on the Constitution [Article 1, Section 8].

"The Congress shall have the power ... To promote the progress of science and the useful arts, by securing for limited times to authors and inventors the exclusive right to their respective writings and discoveries."

A patent is a grant by a national government, to the patent owner, of the exclusive rights to make, use, and sell the invention described in the patent's claims within the particular nation **for a set period of time**, in the U.S., 17 years after grant. In the process of getting a patent, one surrenders secrecy of the invention (thus the word "patent", which means "open") and sues in federal court to enforce exclusivity. A patentee can prohibit others from **inducing infringement** by third parties or from **contributing to infringement** by selling an unpatented item that can only be used to infringe a patented invention.

The value of a patent derives from the exclusivity. The patent does not give the inventor a right to practice his or her own invention; only **the right to exclude others** from practicing it. Indeed, several patents, each owned independently, can cover a product or process. A patentee can stop an infringer from practicing the technology even if it was independently invented.

The Patent Application

A patent application has several parts. The largest part consists of text and is called the **specification**. The main purpose of the specification is to teach the public how to make and use the invention. Enough information to **enable** the public to practice the invention must be given; if an applicant tries to keep crucial information secret, he has not fulfilled the his side of the bargain and does not deserve a right to exclude. The **best mode** contemplated for practice of the invention at the time of filing must also be disclosed; the law will not grant exclusivity if the best is held back. The technical disclosure usually has two parts, a **detailed description** which tries to generalize the invention and specific **examples** of results. The specification will also contain a **summary**, **background** material, including a review of **prior art**. It may also present arguments for why the invention is **new, useful,** and **unobvious**, the three criteria for patentability, and often contains definitions of terms used in the claims. The structure lends itself to repetition that many scientists find annoying.

Optionally, the application may include one or more **figures**. Though not formally part of the application, **deposit** of microorganisms or cell lines may be required to enable others to reproduce results, especially the best mode, or if some materials are not widely available.

Though at the end, the **claims** are the heart of a patent. The claims define the invention's subject matter and scope. All questions such as patentability and infringement are in terms of **the invention** *as claimed*. There are usually several claims, varying in scope. Broad claims are preferred, because they control more technology, but narrower claims are usually present in case a broad claim is found invalid during litigation. (Indeed, a narrow claim that covers the commercially preferred embodiment of the invention may be all that is needed.) The claims are usually written using an independent-dependent format. An independent claim is the broadest of a group. Dependent claims are narrower and contain, by reference, all of the text of their parent, plus additional limitations.

To be complete, a U.S. patent application must also have a fee and a **declaration** or oath. In this, the inventor(s) state that they are the first and original inventors, meaning that they themselves conceived the invention. The declaration usually includes a **power of attorney**, which gives patent counsel the right to prosecute the application.

An **assignment** is a functional, though not formal, part of a patent application. The assignment informs the Patent Office that an inventor has given his or her patent rights to someone else. Even if one is contractually bound to give an employer patent rights, the Patent Office has no way of knowing that without the assignment. Similarly, sales of patents are registered with the Patent Office.

Prosecution

Prosecution of patent applications can be long, complicated, and expensive. A complete explanation is outside the scope of this article, but a simple description is appropriate. Patent prosecution is conducted *ex parte*, meaning that, in a formal sense, it is not adversarial (though it may often seem otherwise). The examiner and the patent counsel conduct a type of negotiation, basically concerned with the issue of how broad are the claims to which the applicant is entitled in light of the prior art and the disclosure of the specification. This negotiation is generally carried out by correspondence, though interviews by telephone or in person are not uncommon. The correspondence takes the form of rejections by the examiner followed by the applicant's response. The response must be within six months of the rejection or the application is dead, and within three months to avoid extra fees. The examiner tries to reject as much as possible for as many reasons as possible. While this may be aggravating, it actually is good because it strengthens any patent that may result by establishing that issues have been considered and disposed of. Sometimes a rejection is made "final"; this is not truly final but rather means that the possible responses are limited, at least temporarily. During prosecution, the application, especially the claims, may be amended, though no "new matter" may be added.

During prosecution, an application may give birth to other applications. A **divisional** application results if the examiner decides that several distinct inventions are claimed in the original. A **continuation** may be filed in response to a final rejection. A **continuation-in-part** (CIP) is a continuation that includes new matter; it is used to add new supporting data or to claim new improvements that might not be independently patentable.

Application Secrecy

With regard to patent disclosure, there are basically two systems. One is practiced by the U.S. and the other by the rest of the world. In the U.S., secrecy is relinquished by the patent applicant when the patent is issued and the legal strength and scope of the patent is known. Elsewhere, secrecy is lost by publication of the application eighteen months after the priority filing with no guarantee of ever getting a patent. Patenting in even one country means giving up the secret to the world.

This latter point is especially important if a deposit of a plasmid, strain, or hybridoma is required. As soon as a application becomes public, the deposit becomes publicly available world-wide. In this case, the inventor not only tells the world how to practice the invention, he makes available copies of the "factory" (and pays $500-3000 for the privilege of doing so).

Priority of Invention

To obtain a patent, certain procedural formalities must be followed, or the opportunity to obtain a patent may be lost. Further, to obtain a patent covering an invention one must be "first". In the **U.S.**, the **first to invent** is first. (How this is determined is complicated and very expensive.) **Abroad**, the **first to file** a patent application wins. The way in which the determination in the U.S. of who is first-to-invent places a burden on scientists with regard to documentation of their work. Signed and witnessed laboratory notebooks are important.

The law in the United States recognizes **conception** and **reduction to practice** as two stages in the inventive processes. The date on which the invention was first **conceived** may be recognized as the date of invention. Further "diligent" work must be maintained and documented from conception to reduction-to-practice. "**Actual** reduction-to-practice" takes place when the invention is actually made and tested if an apparatus or a method to actually work. The filing of a patent application constitutes what is known as "**constructive** reduction-to-practice". One need not achieve an actual reduction-to-practice to obtain a patent but it often results in a stronger application.

Inventorship

The question of inventorship is often troublesome, as the law and one's ego can be in conflict. **Inventorship is a legal decision**. It is made by attorneys based on information the scientists provide. If one deceptively lists an incorrect inventorship, the patent will be unenforceable.

Under U.S. law the inventor or **inventors are those who** *conceive* **of the invention**, *i.e.*, "think it up". It is recognition of the "mental part" of the inventive process. Those who "do all the work" may not be inventors if they only contributed mental activity of a routine na-

ture. This strikes many as unfair, but that's the law. Similarly, giving useful advice does not automatically make one a co-inventor. The act of invention often involves selecting among alternative suggestions and prior art material and techniques. Also, setting a desirable goal is not an inventive conception.

Note that inventorship and authorship are judged by two very different standards. Often a person who is properly a co-author of a paper is not an inventor and sometimes a co-inventor is not properly a co-author.

Duty of Disclosure

All persons involved in the prosecution of a patent application, *e.g.*, inventors and patent counsel, have an affirmative and active duty to disclose information known to them which a reasonable patent examiner would find material to determining patentability. This duty extends throughout prosecution, ending only when a patent actually issues. The inventors and all patent counsel carry the greatest duty of disclosure burdens. The Patent Office and the courts are very strict on this matter. Failure to meet one's duty of disclosure can result in an unenforceable patent and a disbarred patent attorney.

Prior art can be a printed publication, *e.g.*, a sales brochure or another patent, a public use, *e.g.*, a demonstration at a trade show, or a sale or even an offer for sale. Prior art carries the date it became public, *e.g.*, the first day a journal article or advertisement was received in the mail. Disclosures after the priority date of the application are usually irrelevant. Information concerning whether the invention really works as stated in the application may be relevant up to the date of issue.

Patenting in Foreign Countries

Usually patent application is filed in its country of origin and then in foreign countries within one year. If this is done **the filing date of the first country is recognized as the filing date in most countries** for substantive matters. There is a second route which can be followed which involves use of the **Patent Cooperation Treaty** (PCT) which allows one to delay payment of fees somewhat. European countries have a combined patent system which simplifies filing and reduces prosecution costs.

Probably the biggest difference between U.S. patent law and foreign law, is that under foreign statutes there may be essentially **no prior public revelation of the invention prior to filing** an application. This adds to the pressure to file early and to carefully control dissemination of potentially patentable intellectual property. The fact that under foreign patent laws the first to file is considered the first to invent further encourages prompt filing. For companies in the international market, the American grace period is usually irrelevant; one always tries to file before disclosure.

Patents in **underdeveloped countries** are seldom worth the expense. The markets may be small (as measured in dollars) and local courts are not sympathetic to patents owned by Americans.

Costs

Patents are expensive. Many of the procedures done during prosecution require payment of fees to the Patent Office and all generate legal bills and consume internal institutional resources. Even after allowance of the claims one pays an issue fee and periodic maintenance fees during the life of the patent. Therefore, one should usually not patent inventions where the monopoly granted is not reasonably expected to pay the costs either by increasing market share, increasing profit margin, or generating licensing fees.

Right now (May, 1993), the absolute minimum amount that one can pay in Patent Office fees alone to obtain a U.S. patent is $1920, and that figure can easily double. Maintaining a U.S. patent for the full 17 years costs a further $5620. (Small entities having less than 500 employees and non-profit institutions like universities get about half price, $980 and $2810, respectively, unless the technology is licensed to a large entity.) Add to that the cost of lawyers, generally billing $100-

$200 an hour, costs of any foreign patent applications, and the inventor's time, and perhaps that of his colleagues. Figure $10,000-12,000 to draft and prosecute a U.S. patent, $100,000 if someone else is trying to patent the same invention and the Patent Office has to decide who invented first, and $1,000,000 if it is ever litigated. If an application is filed in Europe or Japan, add another $12,000-15,000 each. After grant of a European patent, national patents must be obtained in each country, which can exceed $30,000 for all of Europe.

Recent Changes in Patent Law

Compared to the changes in technology, the changes in the law have been relatively minor. The main change is one of interpretation, where, in the case of *Diamond v. Chakrabarty*, the Supreme Court ruled that *any* subject matter can be patented as long as it is new, useful, and unobvious. Specifically, "animal, vegetable, or mineral?" is not an appropriate question when considering whether an invention is statutory subject matter. Note that the Court's decision interpreted the statute, not the Constitution. Congress is free to restrict or eliminate patenting of life forms or any other subject matter.

Arguably, the effect of the *Chakrabarty* case may be just a bit overblown. Biological inventions had been protected by patents long before this decision, even if the claims did not cover the organism *per se*. *Chakrabarty* made the job of the patent counsel easier, but many of the changes we are seeing would be happening without it. Its prime effect may have been psychological. It made it easy for non-lawyers to understand how biological technology is protected, thereby attracting investment, especially venture capital.

Some of the greatest substantive effects of changes to the law may be areas where microorganisms are released into the environment, such as in pest management. The old way of claiming, which is still valid, was to claim "a biologically pure culture". In may products involving environmental dispersal, the microorganisms are not "biologically pure". So here the ability to claim the organism *per se* is especially important. Arguably, even without *Chakrabarty*, the Patent Office could have allowed claims of the form "A DNA molecule comprising [elements and their relationships to each other], wherein the DNA is contained by a cell." Here, one claims a chemical composition, the cell being a *further limitation*. When one sells a seed or inoculum, one is also selling the claimed DNA construction, itself an infringing act. Indeed, a claim on the construction covers the organism even if it does not mention that organism. Methods using an organism have long been patentable.

U.S. patents have been issued claiming higher eucaryotes *per se* since Hibberd *et al.*, *Tryptophan Overproducer Mutants of Cereal Crops*, issued as patent 4,581,847 in 1986 and *Transgenic Non-Human Mammals* (the "Harvard mouse") by Leder & Stewart, patent 4,736,866, issued in 1988.

Effects on the Public Sector

On the plus side, the public sector will better be able to protect its intellectual property rights. All of the legal tools available to the private companies will be available to non-profit organizations provided that they are willing to pay the bills. DNA and other genetic identification technologies will allow detection of violations by unlicensed organizations which sell protected public germplasm as proprietary. As scientists develop transgenic plants, improved inbred corn lines, and new potato varieties, they will be able to protect them with utility patents and to license them to industry.

On the minus side, some germplasm will be off-limits to the public sector breeder. Until now, a breeder had a legal right to introduce any legally obtained germplasm into his or her breeding program. Arguably, this has changed. A utility patent allows the owner to forbid others to make, use, or sell the claimed invention. This potential exclusion follows the claimed material, and even covers independently generated material.

Of course, a buyer of a patented article has an implied license to use or resell an article bought from the patentee. What this may mean for living entities, where reproduction is an essential property, is arguable.

"But I'm just doing research!" says the university scientist. Until Congress legislates otherwise, that does not give permission to trespass. Geologists and archaeologists must have permission to take samples or excavate on private property; a historian needs permission of the owner of a document to see it; and a scientist needs a license from the intellectual property's owner before using an invention.

The "experimental use" defense is legally uncertain, having not been litigated frequently. It is certain to apply only to philosophical exercises. Pest management is not a philosophical exercise. It has a commercial goal: to benefit the farmer or forester. Both the institutions and the people who do this research receive remuneration for the materials produced. Even when the materials developed are given away *gratis*, the institution's appropriation, the researcher's salary, and the reputations of both, are dependent on getting results. Public sector research and development is not done on weekends by rich eccentrics in remote corners of their estates; it is a very serious *business*. The courts may not ignore patent infringement which financially damages the patentee just because the infringing institution is considered non-profit in the tax law.

Whether companies will try to enforce their patent rights against public sector institutions is another matter. Suing a university makes it difficult to hire graduating students or to get consultants from that institution. However, when it comes to protecting the return on a multi-million dollar investment, other considerations may have more weight.

Other effects will concern how the university scientist does his or her daily work. Signed and witnessed research records in bound notebooks are essential when two parties apply for the same patent and the Patent Office must decide who invented first. They are also useful during normal prosecution to overcome a prior art rejection. Public disclosure before filing kills all foreign patent rights. This pressures scientists not to disclose the latest results at scientific meetings. Even informal nonconfidential discussions with colleagues can prove to be a bar under some circumstances.

A Private Industry Approach: Development of GlioGard™ for Disease Control in Horticulture

Angel S. Mintz and James F. Walter, Biochemical Engineering, W. R. Grace & Co.-Conn., 7379 Route 32, Columbia, Maryland 21044.

GlioGard™ is a formulation of the fungus, *Gliocladium virens* for controlling *Rhizoctonia solani* and *Pythium ultimum*. The product is targeted at bedding plants and ornamentals in the horticulture market. GlioGard™ was developed by W. R. Grace & Co.-Conn. in cooperation with the Biocontrol of Plant Diseases Laboratory at the USDA in Beltsville, Maryland, and was the first fungus to be registered by the Environmental Protection Agency as a soil fungicide for the U. S.

The procedures for product development can be divided into seven major components; Discovery, Product Concept, Formulation, Process Development, Efficacy, Environmental Protection Agency (EPA) Registration, and Test Marketing. These procedures are not unique to the development of GlioGard™ or to the development of biocontrol agents in general, but are different factors that industry should consider in the development of any new product.

Discovery

Two separate discoveries are necessary for product development. The first is the actual discovery or identification of the organism or technology. The second is the discovery or recognition by industry that the first exists. Upon identification of the potential product by industry, economic importance and possible financial commitments must be assessed. The discovery of the *Gliocladium virens* strain used in GlioGard™, GL-21, was the result of an extensive screening program of the Biocontrol of Plant Diseases Laboratory (BPDL) at the United States Department of Agriculture (USDA) (Lumsden and Locke 1989). Concurrent with this screening program, the BPDL was investigating various delivery systems for the application of biocontrol fungi to several ecosystems (soils and soilless mixes) (Lumsden and Lewis 1989). One such system

involved the formulation of beneficial antagonistic fungi into alginate prill granules. After the formulation technology was patented by the USDA and efficacy of the GL-21 strain against *Rhizoctonia* and *Pythium* was demonstrated in greenhouse trials (Lumsden and Locke 1989), W. R. Grace and Co.-Conn. entered into a cooperative research and development agreement (CRADA) with the USDA.

Financial profit is only one aspect from which companies can benefit from product development. This makes definition of concrete dollar numbers difficult to determine for decisions on whether or not a product should be developed. Commercial planners at W. R. Grace and Co.-Conn. originally targeted marketing of GlioGard™ for greenhouse crops worth $146,000,000 (wholesale) annually. Since biopesticides are a new marketing area for W. R. Grace and Co.-Conn., introduction of a biocontrol product on the market could help establish company interest and reputation in a new area with the possibility of expansion into markets other than the greenhouse industry which could eventually increase profitability. These conclusions led to the next step in product development which was product concept.

Product Concept

Product concept is a critical part of the

initial stages of product development. Product concept can be easily ignored because it is a "concept" and is not essential for the material product. The concept issue, however, should be addressed early in development to avoid problems that are likely to arise later. Too often researchers choose to evaluate microbial agents for minor problems that are financially too small to justify the cost of development. Conversely, researchers are also guilty of trying to solve problems that are so expansive and universal that it is impossible to clearly define what the product must do to be commercially acceptable. Recognizing the risks of entering a relatively new area on the market (biocontrol agents), W. R. Grace and Co.-Conn. wanted a project that could be clearly defined and easily targeted.

Rhizoctonia and *Pythium* damping-off were selected as the target diseases. The risks of introduction and cost of development of a new product can be high. Therefore, to recover potential costs, bedding plants and ornamentals were chosen as target plants, since they are high-value crops for protection. Greenhouses were chosen as the site-of-use for two primary reasons; limited control of environmental parameters would exist and soilless potting mixes could be used. Soilless medium was attractive as a potting mix because it provided a well-defined ecosystem where the disease may occur and because Grace-Sierra, a major contributor in the development of GlioGard™, manufactured these media.

Selection of damping-off diseases without consideration of the target plants or site-of-use as part of the product concept, would have included such markets as turf industry and field crops. With the inclusion of several major markets, the propensity to become distracted would have been much greater and EPA registration approval would have taken considerably longer. Thus, by selection of the targeted disease, plants, and site-of-use, a multitude of variables were eliminated from the beginning.

A major component of product concept is the product's characteristics. With the end-use of GL-21 being in plant protection, it obviously could not be phytotoxic. For worker and user safety concerns, the product needed to be non-toxic and non-dusting. A product that would be easy to handle and to package were also important considerations. Finally, the product had to be cost-effective, as efficacious as products on the current market (i.e., chemical fungicides), and it required a shelf-life of at least one year.

Formulation

Formulation is often the most difficult aspect of product development and can mean the difference in success or failure, since it is the key to delivery, shelf-life and stability.

Several steps must be considered in the formulation process of a biological control agent other than the end-product formulation. Seed inoculum preparation and media for mass-producing the organism are important considerations. With GL-21 fermentations, a medium had to be developed that would allow for maximum chlamydospore production and minimal production of the biologically active metabolites typically associated with *Gliocladium* species (Lumsden, et al. 1992). Neither viridin nor viridiol have been detected in the fermentation broth, the manufactured product or the end-use product (unpublished results). The level of gliotoxin detected has not exceeded 10 ppm in fermentation, therefore, the formulated product contains only a trace of this antibiotic (a maximum of 1.2 ppm). Consequently, the amounts of these compounds in the final product are minute and of no toxicological significance.

In addition to alginate prill as the end-product, vermiculite powders, spray-dried powders and extruded alginate were also examined as GL-21 formulations. The alginate prill most appropriately fit the description originally defined in the product concept and with chlamydospores as the active ingredient, a satisfactory shelf life was achieved.

Quality control is essential for formulation development. A simple, but well defined quality control program should be in place for comparison of different formulations and should examine, among other factors; viability, stability and efficacy. A quality control program should be initiated early in development, since it can take months to establish protocols and standardizations that ensure product satisfaction. If a quality control program is not in place during formulation research, product development can be severely delayed, which will ultimately impact profitability. Fortunately, in the exclusive licensing of GL-21 by W. R. Grace and Co.-Conn., a quality control program that had been established at the BPDL was also acquired.

Process Development

After successful formulation at the laboratory bench, a scaleable and reproducible process must be developed. Since this was the first fungus to be marketed as a soil biocontrol agent, not only were the considerations of a new product necessary, but also the development of new technology. The use of existing equipment was important for the project for cost-effective calculations, but also because W. R. Grace and Co.-Conn. did not want to simultaneously deal with a new product, new technology and new equipment that might also require additional training or outside expertise.

During fermentations for production of biomass; aeration, agitation, nutrient depletion, etc. can be drastically different for different systems, so the ability to manipulate the organism regardless of the system is critical. The ability to produce GL-21 in 250 milliliter erlenmeyer flasks had to also be achieved in various sizes of fermentors ranging from 10 liters to 4,000 liters. The process should also be reproducible and transferable to a production facility. Again, an established quality control program is essential for the evaluation of the effects of process changes on the end-product.

Efficacy

Preliminary efficacy of GL-21 was demonstrated through a disease control bioassay, but with successful mass-production, more elaborate efficacy tests were needed.

Efficacy trials began with collaborative tests with university researchers who were given a specific protocol to follow (Lumsden et al. 1990). With favorable results from these tests, trials were expanded to include several additional cooperators to ensure that results could be replicated. When similar results were obtained, trials were conducted with several bedding plant grower cooperators using different protocols. With the exception of artificial infestation with pathogens, which was used in the earlier trials, researchers were at liberty to use whatever plants they desired and their own application schedules. The design of the trials in a step arrangement allowed for less control in the way the product was handled with each phase of testing. By the time commercial growers were included in the trials, natural infestation was used to determine efficacy, and the grower was simply given the product with a basic set of instructions similar to those now provided on the product label. Efficacy of GL-21 was demonstrated at all levels of testing and only at this point was W. R. Grace and Co.-Conn. willing to apply for EPA registration.

EPA Registration

The process for registration of GL-21 as a biocontrol treatment was exclusively the preview of the EPA. The EPA reviews and regulates microorganisms used as disease control agents if they are engineered, non-indigenous, or field tested on more than ten acres (4.05 ha) of land or one acre surface of water (Betz et al. 1987). An Experimental Use Permit (EUP) is required for these conditions. The legislation charging the EPA with this responsibility is the Federal Insecticide, Fungicide and Rodenticide Act (FIFRA).

According to subdivision M of the EPA Pesticide Testing Guidelines (1989) (Federal Register 1989), microbial agents for the control of plant pests are in many ways treated similar to chemical pesticides and companies applying for registration must provide extensive information for approval to use microbial products commercially. Product testing is set up in a tier system which recognizes the inherent risks and degrees of exposure associated with different uses of pesticides. In addition to production and taxonomic data, long and short-term effects on a variety of organisms including non-target organisms may be necessary. According to the regulations, studies may be required for effects that are toxicological, mutagenic, carcinogenic, fetotoxic, teratogenic and oncogenic depending on the envisioned use pattern.

Detailed data requirements may include: 1) acute oral, dermal, respiratory, eye irritation, dermal irritation, dermal sensitization and acute delayed neurotoxicity; 2) subchronic studies to include 90 day feeding studies, 21 day repeat dermal tests, 9 day dermal toxicity, 9 day inhalation and 90 day neurotoxicity tests; 3) chronic and longer term feeding and oncogenicity studies, teratogenicity tests and reproduction studies; 4) mutagenicity studies to evaluate gene mutations, chromosomal aberrations and chemotoxic effects; 5) general metabolism studies; 6) dermal penetration studies and: 7) domestic animal safety determinations.

These requirements were originally designed for evaluation of synthetic chemicals and not living organisms, but some latitude is recognized for differences. The testing is done on a multi-tier system so that microorganisms passing certain requirements at the Tier 1 level need not be progressively tested at a more stringent, long-term Tier 2 level. The EPA is willing and able to waive or minimize the impact of certain requirements of the legislated regulations when appropriate. Each situation is judged individually.

In the case of GL-21, only the trials outlined in Table I were necessary (also see Lumsden et al. 1991). In view of these considerations, GL-21 was evaluated and reviewed in an equitable but thorough manner for its safety as the active component of an agricultural product intended for use to control damping-off diseases.

Two formulations of GL-21 were approved.

WRC-GL-21 is a manufacturing-use product for use in formulation of biocontrol products. GlioGard™ is an end-use granular product containing calcium alginate, wheat bran and proprietary additives to prolong shelf life. The granular material is mixed with soil or soilless plant growing media at least one day prior to planting or incorporated into the medium surface in plant beds prior to or at planting. The formulation is used at the rate of $1\ 1/2\ lb/yd^3$ (approximately 1 g/l) of media when mixed, or at the rate of $3/4$-1 ounce/ft^2 (approximately $0.1 g/cm^2$) when applied to the bed surface. With the receipt of registration for GL-21 from the EPA it can now be sold as a product.

Test Marketing

In order to properly test market GlioGard™, it was necessary to move the product through normal sales and distribution channels, but with feedback from each grower. To maintain close contact with growers, visits and questionnaires were employed and distribution, through Grace-Sierra, was limited to four states; Florida, Ohio, Michigan and Texas. A total of seventy-six growers used GlioGard™ on a variety of plants. Despite the great variability of plant types and media used, there were no reports of phytotoxicity and only two reports of less than expected performance. Overall, growers reported excellent disease control, uniform growth and in some cases, growth enhancement (Lumsden et al. 1990).

Unlike efficacy trials, test marketing employs no disease assessment nor statistical analysis; everything is based on perception. This is probably the most important aspect of test marketing, since perception determines whether or not a grower will make future purchases and what type of information will be passed on to other growers. As expected from the results of the efficacy trials of GL-21, the overall perception from the test market was very positive (Knauss 1992).

Table 1. Toxicology Tests of GL-21

STUDY	RESULTS
Acute oral toxicity/pathogenicity	No acute toxicity, pathogenic effects or infections detected. Microbe expelled in feces.
Acute pulmonary toxicity/pathogenicity	No acute toxicity, pathogenic effects or infection detected. Spore takes 2-3 weeks to clear lungs.
Acute intravenous toxicity/pathogenicity	No apparent acute toxicity pathogenic effects or infections detected. Mycelium is trapped in lungs, liver and spleen and cleared after 14 days.
Acute dermal/primal dermal	Waved due to product form.
Primary eye	Waved due to product form.

OTHER TOXICOLOGICAL INFORMATION

1. Study shows GL-21 does not grow at or near body temperature of mammals or birds.
2. Quantification of antibiotic products shown not to be of toxicological significance.
3. Documentation of worker exposure to GL-21 production over 2-3 years shows no toxicological response.
4. Documentation of methods to validate freedom from contaminating microorganism.
5. Confirmation of exempt status of inert ingredients generally regarded as safe material (food grade).
6. Studies showing extent of persistence of GL-21 in environment and extent of spread in soil.

Conclusions

The development of GlioGard™ took many years of collaboration and a diversity of skills and expertise among plant pathologists, microbiologists, biochemical engineers, commercial planners and others. The close proximity of the USDA and W. R. Grace and Co.-Conn., and the ability of the scientists at both facilities to work together played a major role in the successful development of GlioGard™. The importance of close collaborations and continued communications between project participants for successful technology transfer to occur is emphasized, since technology transfer is often the limiting factor in commercialization of biocontrol products.

In the initial stages of product development, a product concept was defined which targeted specific pathogens, *Rhizoctonia* and *Pythium*, in ornamental and bedding plants. This conservative approach was elected, in part, because of the unchartered waters that laid ahead in using a fungus as a soil fungicide. Now that W. R. Grace and Co.-Conn. is confident of the efficacy of GlioGard™ in the horticulture market, other markets such as turf and field crops can be investigated for future development of this and other biocontrol agents.

References

Betz, F.; Rispin, A.; Schneider, W. "Biotechnology Products Related to Agriculture. Overview of Regulatory Decisions at the U.S. Environmental Protection Agency." *ACS Symposium Series 334: American Chemical Society.* Washington, D.C. 1987, pp 316-327.

Federal Register "Data Requirements for Pesticide Registration; Final Rule." **1984**, 49:42856-42905.

Knauss, J.F. "*Gliocladium virens*, a New Microbial for Control of *Pythium* and *Rhizoctonia.*" *Florida Foliage.* **1992**, 18:6-7.

Lumsden, R.D.; Lewis, J.A. "Selection, Production, Formulation and Commercial Use of Plant Disease Biocontrol Fungi, Problems and Progress." In *Biotechnology of Fungi for Improving Plant Growth.* Whipps, J.M. and Lumsden, R.D. (Eds.). Cambridge University Press, Cambridge, UK. **1989**, pp 171-190.

Lumsden, R.D.; Locke, J.C. "Biological Control of Damping-off Caused by *Pythium ultimum* and *Rhizoctonia solani* with *Gliocladium virens* in Soilless Mix." *Phytopathology.* **1989**, 79:361-366.

Lumsden, R.D.; Locke, J.C.; Walter, J.F. "Approval of *Gliocladium virens* by the U. S. Environmental Protection Agency for Biological Control of *Pythium* and *Rhizoctonia* Damping-off." *Petria.* **1991**, 1:138.

Lumsden, R.D.; Locke, J.C.; Lewis, J.A.; Johnston, S.A.; Peterson, J.L.; Ristaino, J.B. "Evaluation of *Gliocladium virens* for Biocontrol of *Pythium* and *Rhizoctonia* Damping-off of Bedding Plants at Four Greenhouse Locations." *Biol. Cult. Control Tests.* **1990**, 5:90.

Lumsden, R.D.; Ridout, C.J.; Vendemia, M.E.; Harrison, D.J.; Waters, R.M.; Walter, J.F. "Characterization of Major Secondary Metabolites Produced in Soilless Mix by a Formulated Strain of the Biocontrol Fungus *Gliocladium virens.*" Canadian Journal of Microbiology. 1992, 38:1274-1280.

Biologically Based Regulatory Pest Management

Charles P. Schwalbe, U.S Department of Agriculture, Animal and Plant Health Inspection Service, Plant Protection and Quarantine, 6505 Belcrest Road, Hyattsville, MD 20740

The role of regulatory plant protection in preventing the establishment or mitigating the impacts of invading species is discussed. Examples of programs of exclusion, detection and control of quarantine significant pests utilizing an array of biologically based technologies are given. To an increasing extent, semiochemicals, biological control agents, sterile insect techniques and growth regulators form the technological basis on which these programs are conducted.

The importance of regulatory support to the world's production and distribution of agricultural products is as significant today as it ever has been in our history. Many of the serious pests that are the subjects of intensive research in this country have been accidentally introduced and the scale of the impacts of these introductions is exemplified by the notoriety of the pests themselves: cotton boll weevil, oriental fruit moth, gypsy moth, pink bollworm, Japanese beetle, European corn borer. The complexity of the impacts of these invading species on agriculture defies full comprehension; losses from major introduced plant pests amount to about $28.8 billion per year, and expenditures for their prevention and control are about $3.2 billion per year (USDA, 1993). Methods for their satisfactory management are still imperfect.

The main objective of regulatory programs is to minimize the risk of spread of agricultural pests in commerce through quarantines and prescribed conditions for the importation of foreign agricultural products. Essentially all countries of the world with significant agricultural interest have regulations designed to prevent the introduction of new and potentially damaging pests. These countries communicate regulatory concepts and programs through regional plant protection organizations which are coordinated by the Food and Agriculture Organization of the United Nations.

Exclusion

Pest exclusion efforts are stimulated by a variety of practical considerations. It is difficult to predict the potential pest impacts of organisms introduced into a new environment, especially when they may be free of their co-evolved natural enemies. Infestations may cause direct economic losses to production agriculture or indirect damage by displacing native species and upsetting natural ecological systems. Damage from invading pest species may compromise availability and cost of agricultural products through lessened efficiency of agricultural production. Obviously, increased occurrence of outbreaks generally leads to heightened pesticide use with concomitant environmental implications. Freedom from quarantine significant pests also affords countries with foreign marketing advantages. Agricultural commodities factor high in importance in international trade and there is intense competition for export markets. The presence of quarantine significant pests is a serious competitive disadvantage when meeting the stringent phytosanitary requirements of importing countries. Major additional requirements for survey, inspection and certification can be imposed on a large production area due to an introduced pest of limited distribution.

It has become essential to found exclusion efforts on credible pest risk analyses encompassing risk assessment, risk management, decision-making and risk communication; the view of total protection through inspection and quarantine is not congruent with todays rapid and complex movement of agricultural products. Improved understanding of the behavior and ecology of pests enables more accurate prediction of impacts foreign pests might have on American agriculture. Such information can be used to focus exclusion resources on critical entry pathways of the most serious pests and facilitate surveillance for new introductions. Basic information supports programs to minimize impacts of newly introduced pests on agricultural production, export markets and the environment. Accurate processes for determining risk and suitable methods for managing and communicating that risk have to be founded on solid and contemporary scientific principles and knowledge. This is a chief motivation for maintaining productive working relationships between action and research agencies.

Exclusion programs have traditionally relied on the inspection of products at ports of entry. Past inspection records indicate pathways with the highest incidence of interception and this is used to focus ongoing inspection efforts. Visual examination of baggage is enhanced with passenger profiling techniques and x-ray systems and in recent years APHIS has pioneered the use of specially trained dogs to locate baggage carrying prohibited fruits, vegetables and animal products.

An increasingly important technique for certifying commercial shipments of fruits and vegetables safe for entry into importing countries involves production in pest free areas or production fields. Certification of commodities can be based on the execution of complementary safeguards (pest management practices, survey and quarantine) in such a way to demonstrate and maintain pest freedom. Pheromones and other attractants are fundamental to the success of this approach and provide the means by which we can state that specific pests do not occur in designated production areas or fields. As a specific example, Florida citrus exported to Japan is certified free of Caribbean fruit fly, *Anastrepha suspensa*, based on specific management protocols. Data to support official certifications of pest freedom are obtained from very specific trapping protocols; without sensitive survey technology, the industry would rely on less cost effective and environmentally acceptable approaches to meet Japan's regulatory requirements. APHIS supports this approach and

recognizes a variety of specific fruits and vegetables produced in designated areas in Australia, Bahamas, Belize, Brazil, Canada, Chile, Costa Rica, Ecuador, Israel, Japan and Mexico as free of certain quarantine significant pests.

This pest free area or regionalization concept is an important innovation in the regulatory field; free trade and other General Agreement on Tariffs and Trade (GATT) related agreements ensure increasing use of the approach for certifying products free of quarantine significant pests. For reliable designation of pest free areas, accurate and sensitive surveys using pheromones and attractants need to be usefully integrated with other components of the pest free system for certifying local or regional pest freedom in the US. Currently codling moth, *Cydia pomonella* and the apple maggot, *Rhagoletis pomonella* are examples of pests of concern to our international trading partners. For maximum effectiveness, such programs should be approached on an area-wide or regional basis. Area-wide pest management systems can be developed and implemented that promise enormous benefits to production agriculture, the nation's export interests and the environment. Biologically based technologies (biological control, sterile insect, mating disruption, trapping) have been tested and are available for use against many of these pests. Our challenge is to strengthen partnerships with industry and discover ways in which these technologies can be integrated into area-wide management programs that provide advantages beyond those realized by controlling pests at the local or grower level.

Exclusion interests can be advanced by facilitating pest management programs in neighboring countries. The classic example of this idea is the current international effort to eradicate New World screwworm, *Cochliomyia hominovorax* from North America and Central America. The US was declared officially free of screwworm in 1976, and it is clearly in the interest of US agriculture to engage in cooperative programs which reduce the likelihood of reintroduction of this pest. The sterile insect technique (SIT) is as fundamental to the program in Mexico as it was to the original eradication of screwworm from the US.

Similarly, the SIT is in use in southern Mexico to eradicate the medfly, *Ceratitis capitata* from that region. Left unchecked in the hemisphere, this pest would likely spread to the north, increasing the risk of artificial and natural entry into the US. The Mexican fruitfly, *Anastrepha ludens* is endemic in Mexico and occasionally moves into southern Texas and California. Cooperative management programs utilizing SIT are conducted along the US-Mexico border region to reduce the risk of spread into citrus production areas of the US. Research on augmentation of the parasite, *Diachasmimorpha longicaudatus* (Knipling, 1992) may lead to its use in managing Mexfly for that purpose.

The brown citrus aphid, *Toxoptera citricida*, is an efficient vector of citrus tristezia virus, a devastating disease of citrus in many parts of the world. This aphid was first introduced into Argentina in the 1930's, probably through infested planting material from South Africa. It has since been found in Puerto Rico and Dominican Republic in the Caribbean region, and as far north as Nicaragua in Central America. Regulatory actions to prevent its artificial entry into the US have been taken, but are considered unlikely to protect US citrus producing areas in the long term. Therefore, a research and action plan is being prepared to minimize the impact of the aphids eventual entry into the US. Prominent in the plan is the discovery and possible establishment and manipulation of natural enemies to minimize the impact of this pest .

Detection

Perhaps the greatest application of natural products in the regulatory arena involves the ongoing effort to detect the introduction of exotic species. Recognizing that exclusion programs can never be completely foolproof, APHIS cooperates with states in detection survey programs for quarantine significant species. Since a pilot survey was initiated in 1985 there has been an ongoing detection survey program using pheromone baited traps for exotic pests. These surveys are statistically designed to provide credible evidence for the presence or absence of pests in certain areas. In 1991 and 1992 approximately 25,000 traps were placed in 45 states for 17 exotic species. Fortunately, none of these species has yet been detected in this country.

Pheromone-baited traps have been particularly useful in determining the distribution of invading species and monitoring their spread. In recent years the cherry bark tortrix, *Enarmonia formosana*, and apple ermine moth, *Yponomeuta malinellus,* have entered the Pacific Northwest. Only limited areas have been placed under quarantine to prevent their further artificial spread because accurate distribution data are available through systematic trapping programs. Pheromones also figure prominently in national programs against the gypsy moth, *Lymantria dispar*, and pink bollworm, *Pectinophora gossypiella.*

Natural dispersal of the gypsy moth occurs by larval ballooning and since the female is flightless natural spread is rather slow (ca. 10-15 miles/year). Given these circumstances, regulatory action to prevent the long distance artificial spread of this insect is cost effective. Since the publication of the discovery of female-produced sex attractant (disparlure) (Bierl et al.,1970) this powerful pheromone has been crucial to the success of the regulatory program. Each year approximately 300,000 disparlure baited traps are deployed throughout susceptible areas of the United States, leading to the annual discovery of 20-30 isolated infestations. While some infestations have gone undetected for several years and have been difficult to contain and eradicate, usually newly detected populations are extremely localized, occupying only a few acres. Under this program, hundreds of isolated infestations in many states have been detected, delimited and eradicated. Had not detection and delimitation of infestations been enabled by disparlure, large areas of the United States would now likely be infested by this pest.

The pink bollworm is a serious pest of cotton and a cooperative program is conducted to prevent its spread to uninfested cotton growing areas. Trapping is conducted to detect new infestations; and to protect the uninfested San Joaquin Valley of California using pheromone systems, sterile moth release and short-season cultural practices. Since 1967, APHIS has cooperated with California cotton growers and the California Department of Food and

Agriculture in a successful sterile moth release program to prevent pink bollworm from establishing in cotton in the San Joaquin Valley. The tactic is to negate the reproductive potential of moths that annually migrate there from the Colorado River Basin, The program involves systematic trapping at the rate of 1 trap/60 acres over 1,166,134 acres (in 1992) of cotton to monitor the occurrence of native moths. Fields with native moth finds are trapped with 1 trap/20 acres and sterile pink bollworm moths are released from May through October 16. Traps are used to monitor wild moth densities in order to adjust release numbers to maintain the desired overflooding ratios. While not yet perfected, mating disruption is being developed as an additional tactic for use in this program.

Susceptible fruit and vegetable producing areas are under constant surveillance for introductions of fruitflies. In 1992 82,577 traps of various designs were maintained by state and federal agencies for detection of exotic tephritids. Trap densities are maintained at 1-16 traps per square mile, and in large portions of high risk areas of California 10 traps per square mile are deployed. Recognizing that the probability of capturing flies increases when traps are close to a population and that flies tend not to fly far from host plants, trap locations are rotated on a 1-6 week interval (depending on lure formulation and time of year) to increase the likelihood of finding newly established populations. We believe that this program is effective in locating infestations shortly after colonization. The frequent discovery of medfly in California in recent years has prompted refinements in the response to introductions of that pest. Immediately following detection, highly efficient yellow sticky panels baited with trimedlure are deployed at the rate of 1000 traps per square mile. While the actual control value of this high density trapping is still under investigation, such concentrations of traps provide very sharp detail to the boundaries of the infestation. With this precise delimitation information, control operations can be scaled to the size of the infestation and we are finding that larger scale aerial malathion bait sprays can sometimes be effectively replaced with treatments applied with ground equipment. The key to the success of this strategy hinges on precise knowledge of population distribution derived from attractant baited traps.

Control

An interesting application of natural products in a regulatory program concerns the witchweed, *Striga asiatica*. This weed is parasitic on maize, sorghum, sugarcane, crabgrass, rice, millet, tobacco and legumes and is a devastating pest of corn world wide. Since it's detection in the US in 1956 an aggressive regulatory and eradication program has been underway. This effort has been highly successful and eradication is expected to be completed by the end of the 1996 growing season. Herbicides are used on infested fields and an extensive menu of treatment options for various crops and field conditions has been developed. Elimination of the microscopic seeds in the soil is critical to program success. Seed populations naturally diminish or, as they sprout, will be controlled by the herbicide program. Since witchweed seed can persist in the soil for up to 10 years, efficient eradication requires accelerating the natural rate of germination. Ethylene gas is injected into the soil under proper environmental conditions, stimulating seed germination. Sprouting witchweeds are deprived of suitable hosts through crop management and grass control, and the seedlings die. This technique is in general use in the witchweed eradication program: in 1992 10,330 acres in 1273 fields were treated with ethylene.

When pest populations are of limited distribution, host-plant resistance can be used effectively for eradication. This technique has largely supplanted chemical treatments for the eradication of golden nematode, *Globodera rostochiensis*, a pest of potato, from Long Island and other places it may be detected. A crop management sequence of resistant varieties and non-host crops is prescribed in the regulated area. This biological approach is environmentally sound and highly effective in eliminating golden nematode from the region.

The state-federal program to prevent the long distance spread of the gypsy moth is almost entirely based on biological technologies. As mentioned earlier, disparlure is used extensively to detect and delimit dozens of isolated infestations remote from the generally infested area every year. These infestations are discovered so early in development and delimited so precisely that non-chemical methods of eradication have proven effective. The typically localized and sparse infestations are often treated with *Bacillus thuringiensis* with ground equipment to reduce larval populations. Later in the season, mass trapping at the rate of ca. 9 traps/acre prevents mating of the insects that survive to the adult stage. This approach is now widely used in many states as an economical and environmentally sensitive method for eradicating isolated infestations of this insect. Much research has been done and great progress made in advancing mating disruption as a viable option for eliminating localized gypsy moth infestations. Recent field tests in Virginia with an improved controlled release formulation of disparlure have encouraged prospects for employing this method of eradication of isolated infestations. The nucleopolyhedrosis virus of gypsy moth, registered with EPA as Gypchek, is relied on to an increasing degree for suppression of gypsy moth populations in environmentally sensitive areas. Difficulty in obtaining commercial supplies is currently the only major impediment to Gypchecks wide use for management.

Summary

Competition for agricultural export markets in a global economy, free trade initiatives and increasing awareness of environmental concerns present real challenges to programs thwarting the artificial movement of agricultural pests. Concern over the economic and environmental impacts of introduced pests has prompted countries around the world to protect their agricultural resources through regulatory pest management programs. The prospect of large potential damage resulting from ineffective exclusion actions tends to shape regulatory programs as conservative and risk-averse. None-the-less, the continued reliance on chemical pesticide solutions is short-sighted and unrealistic. Methyl bromide has been listed as an ozone depleting chemical and is scheduled to be banned by the year 2000. Treatment with this highly effective fumigant is required by our trading partners for certification of many US export products. Preliminary economic analyses reveal that in the period October, 1991 to September, 1992 over $200 million in US exports of cotton, oak logs, cherries, peaches, nectarines and walnuts depended upon such treatments (USDA, 1993). It is unlikely that foreign requirements of quarantine security will be relaxed and acceptable chemical or non-chemical alternatives to methyl bromide fumigations for many products are not available. Thus, there is an intense sense of urgency to prevent introduction of additional quarantine significant pests that could limit US agricultural production and marketing. Similarly, the loss of methyl bromide as a quarantine treatment could result in an import ban of many economically important commodities. Estimates of net loss in the US from such an import ban range from $973.0 to $990.1 million over five years for 9 selected commodities (USDA, 1991). Serious disruptions in the domestic market would occur seasonally as imports form the basis of US supplies of fresh commodities during the winter months. The facts clearly dictate that we must adjust to this changing economic and environmental climate by developing and implementing modern solutions to our pest problems. Quarantine approaches that are based on biologically sound estimates of risk will have to be adopted. Regionalization and biologically-based regulatory pest management will be the norm of the future. Effective programs employing strategies consistent with this new order have been implemented and this prompts optimism that future challenges can be met successfully.

References

Bierl, B. A.; Beroza, M.; Collier, C. W. Potent sex attractant of the gypsy moth: Its isolation, identification and synthesis." Science, **1970**, 170, 87-89.

Knipling, E. F. "Principles of Insect Parasitism Analyzed From New Perspectives." *U. S. Department of Agriculture, Agriculture Handbook 693*, 1992, 337 pp.

USDA, Policy Analysis and Development,

Policy and Program Development, Animal and Plant Health Inspection Service. Economic impact of losing methyl bromide as a quarantine treatmentfor nine selected commodities. USDA, 1991, 31 pp.

USDA, Policy Analysis and Development, Policy and Program Development, Animal and Plant Health Inspection Service, The Costs of Invading Pests to U.S. Agriculture. March, 1993, 15 pp.

Pest Management and Biologically Based Technologies: A Look To The Future

Waldemar Klassen, Joint FAO/IAEA Division of Nuclear Techniques in Food and Agriculture, Vienna, Austria

The paper summarizes the Symposium highlights. The Symposium suggests a shift in the paradigms concerning the most fruitful and dominant pathways to effectively and durably cope with pests harmful in agriculture and to public health. The dominant trend for the foreseeable future is to seek solutions through an in depth understanding of biological mechanisms and relationships. Several speakers discussed methods within the context of strategies that guard against their rapid obsolescence (e.g., the development of resistance), and/or assure the greatest economic benefit. This paper also elaborates on such strategies. The development and commercial application of these methods is strongly influenced by legal provisions and mechanisms protecting intellectual property. This issue is of overriding importance in the transfer of these technologies among developing countries. The trend in developed countries toward greater private sector involvement in partnerships with public sector institutions for commercial development of biological inventions has increased the disadvantages of developing countries. The Convention on Biological Diversity, signed by more than 150 countries, provides a framework for resolving many of these difficulties. It is likely that progress will be facilitated by the Uruguay round of GATT. Finally, the implications of data pertaining to the growth in markets and the use of biologically-based pest control methods are discussed.

This Symposium has provided a banquet of exciting developments and seminal concepts concerning innovations in the quest for effective and ecologically benign methods of staving off the ravages of organisms harmful to agriculture and public health. We have been stirred by the enthusiasm, the resolute commitment and the strong sense of mission of those advancing the science and practice in this multifaceted field. Biologically-based pest control technologies have been and are being developed with unprecedented rapidity that will increasingly help to improve the environment, increase food and fiber production, combat disease, and underpin sustained development.

During the past five years this field has "taken off". Spectacular advances have been made in both the industrialized and the developing countries (Kumar, 1992). Perhaps the greatest achievement in classical biological control of the Twentieth Century occurred recently in Africa. Under the leadership of the International Institute of Tropical Agriculture, the ravages of the cassava mealybug, *Phenacoccus manihoti*, were brought under control by the mass production and distribution of the host specific encyrtid wasp, *Epidinocarsis lopezi,* throughout the cassava growing zone, which extends over 34 countries of sub-Saharan Africa (Herren; Neuenschwander, 1991). In Indonesia, control of the brown plant hopper and other pests of rice has been restored by fostering the resurgence of natural enemies through the elimination of excessive insecticide use (Oka, 1991). By means of the sterile insect technique the melon fly, *Bactrocera cucurbitae*, was eradicated from Japan (Yamagishi *et al.* in IAEA, in press.), the Mediterranean fruit fly, *Ceratitis capitata* was eradicated from Mexico, and the screwworm, *Cochliomyia hominivorax* was eradicated from Mexico and Libya [Food and Agriculture Organization (FAO), 1992].

World-wide, more than twenty microbial or nematode agents are being used in managing arthropods. Most prominent are bacteria that express one or more of the δ-endotoxins of *Bacillus thuringiensis*; some of these δ-endotoxins have been improved through genetic engineering technology. Lack of stability of the δ-endotoxin under field conditions and the significant cost of production have been overcome by encapsulating the protein crystal in killed recombinant *Pseudomonas fluorescens* (Anonymous, 1991a, 1991b; Commandeur; Komen, 1992), or through inserting the δ-endotoxin gene into plants.. Also sun screens and other adjuvants have improved the performance of the δ-endotoxin (Bateman, 1992). This endotoxin can provide systemic protection to various crops because it has been engineered into the endophyte, *Clavibacter xyli*. The use in Brazil of a NPV against the soybean looper on more than one million hectares per year is a shining example of what determined people in a developing country can do even with rudimentary technology (Jaffe; Rojas, 1992). In recent years the use of entomophagous nematodes has grown

2726–8/93/0410$06.00/0 © 1993 American Chemical Society

dramatically, especially in China and Australia (Curran, 1991, 1992; Wang, 1992).

Microbial agents are used against plant parasitic nematodes. For example in Peru *Paecilomyces* is used against potato cyst nematodes. Also *Hirsutella rhossiliensis* is used against *Crionella xeneplax* on peach trees.

The use of nonvirulent and recombinant *Agrobacterium* against bacterial diseases has had a seminal effect on this entire field (Ream, 1989). The development of *Pseudomonas syringae* to combat epiphytes that increase frost damage and *Erwinia amylovora*, the causative agent of fire blight, (Lindow; Loper, this Symposium; Wilson *et al.*, 1990) and bacteriophages for combating foliar bacterial pathogens on melons, *E. ananas*, and potato, *E. carotovora*, (Earye, this Symposium) would be landmark achievements.

Kaper (this Symposium) has pioneered the use of satellite RNA as a special form of biological control of viruses (Gallitelli *et al.*, 1991; Montasser *et al.*, 1991). Satellite RNA is used in Asia on a practical scale to inoculate plants against cucumber mosaic virus.

A number of microbial products are in use for combating fungi. *Gliocladium virens* is the most recent to enter the market place. An array of agents are under development against soil-borne fungi, postharvest pathogens and foliar pathogens. The work against the chestnut blight fungus may bring back one of the finest native American trees.

Fittingly, in 1992 the World Food Prize was awarded to Drs. E.F. Knipling and R.C. Bushland, who made many contributions to biologically based methods of pest management.

Cook (this Symposium) and Cook *et al.* (1987), have asserted that "biological control should become the primary method used in the United States to assure the health and productivity of important plant and animal species." The extent to which pests should be managed by biological versus chemical means has been a public policy issue since about 1950 (Klassen, 1988). The balance will change depending on the need for food production, environmental protection, availability of superior technologies, and the quality of people in leadership positions. Thus one can safely assume that importance of biological controls as components of pest management systems is likely to progressively increase. Moreover they will be increasingly deployed in a defensive rather than a curative mode, and according to strategies that maximize their effectiveness and durability.

Even now biologically based technologies, especially resistant crop varieties (Table 1), command a significant share of the market along with synthetic chemical pesticides. Currently the world market in chemical pesticides is US $26 billion (Table 2). By contrast the world market for crop seeds and propagules, virtually all of which have genetic resistance to at least several harmful organisms is about US $6 billion (van den Eltzen, 1993), but this does not include the seeds of resistant crops that farmers produce for their own needs. Further the aggregate of the niche markets for bacterial, fungal, viral and pheromonal pesticides although still quite small is growing rapidly (Commandeur; Komen, 1992; Lisansky, 1989). Moreover an impressive number of additional biopesticides are entering the market or are in "the pipeline". Further, a report in Genetic Technology News (Anonymous, 1991a) projects that by 2000 AD the share of the U.S.A. insecticide market captured by recombinant microbial insecticides will be 25 percent for field crops and 50 percent for vegetable crops to reach US $350 million. However penetration of the market by biopesticides will be constrained eventually by competition with resistant crop varieties developed both by conventional breeding and by genetic engineering.

SHIFT IN PARADIGMS

As implied by the title of the Symposium, <u>Pest Management: Biologically based technologies</u>, a widespread shift in paradigms appears to have occurred from seeking solutions to pest problems through largely empirical approaches to approaches based on in-depth understanding of biological mechanisms and relationships. Hypothesis driven research and objectivity are the major pillars of this paradigm, but some empirical work will always be needed.

Pest management has been defined as the selection, integration and implementation of pest control methods based on predicted economic, ecological and sociological consequences (Council on Environmental Quality, 1972). Thus our efforts against pests must be guided by objective criteria provided by economics, ecology and sociology. Moreover in order to be widely adopted biologically based methods must deliver predictable and consistent results at competitive costs (Mackauer, 1991). In many cases this requires detailed knowledge of the processes by which control is achieved, as well as the ready availability of superior technologies for the mass production, formulation, storage and application of biological materials.

IMPERATIVES FOR AN ERA OF UNPRECEDENTED OPPORTUNITIES

I propose an agenda for this era of auspicious opportunities for advancing the development of

Crop	Acreage x 10^6	Disease or Insect Pest	Percent of U.S. Acreage Seeded to Resistant Varieties	States Primarily Affected
Wheat	70	Yellow Dwarf	10	East U.S.
		Stem rust	30	ND, SD, MN, NB, MT
		Leaf rust	30	IN, IL, MO, KS, NC, OH, ND, SC, GA, AR, LA
		Soil virus	15	East one-half of U.S.
		Streak mosaic	20	KS, NB, CO
		Bunt (smut)	15	All
		Hessian fly	30	KS, IN, IL, MO, GA, AL, NC, SC, FL
		Sawfly	3	MT, ND
		Septoria	5	East U.S.
Corn	75	Stalk rots	60	Cornbelt and south
		Leaf blights	80	Cornbelt and south
		Corn earworm	50	All
		European corn borer	90	East one-half of U.S.
		Corn rootworm	10	Cornbelt and south
Barley	12	Yellow dwarf	20	CA, SD, MN
		Greenbug	15	OK, TX, KS, CA
		Smut	20	All
		Mildew	40	East U.S.
Oats	13	Yellow dwarf	30	All
		Crown rust	80	East U.S.
		Stem rust	75	East U.S.
		Soil virus	20	East U.S.
Sorghum	18	Chinch bug	40	OK, KS, ND
		Greenbug	30	TX, OK, KS
Rice	3	Blast	60	TX, LA, AR
		White tip	60	TX, LA, AR
Soybean	57	Phytophthora Root rot	80	All Midwest, Midsouth Southeast
		Soybean Cyst nematode	10	South, Southeast, Some in Midwest
		Root knot nematode	5	Southeast
		Caterpillars and various foliar feeding insects	2	Mississippi Delta
Cotton	10	Bacterial blight	70	Southwest, West
		Fusarium wilt	15	Southeast, Midsouth
		Lygus spp.	15	Midsouth
Alfalfa	26	Bacterial wilt	80	MN, WI, NY, PA, MI, ND, SD, WA, ID, MT, OR
		Anthracnose	50	NC, VA, MD, PA, NY, MI, IL, ID, OH, KS, OK
		Verticillium wilt	30	WA, ID, OR, MT, MN, WI, MI, NY, PA
		Phytophthora RR	50	NC, PA, NY, MN, WI, MI, OH
		Aphids	90	CA, NV OR, ID, WA, UT
		Stem nematode	20	WA, NV, ID, OR,NE, SD, ND, MT, WY

Source: Pers. Commun. by J.H. Elgin, Jr., P.A. Miller, and C.F. Murphy, National Program Staff, ARS, USDA, U.S.A.

Table 2. Approximate world market for chemical pesticides, 1991

Class	Percent	US Dollars (billions)
Herbicides	44	$11.44
Insecticides and acaricides	29	$7.54
Fungicides	21	$5.46
Nematicides and plant growth regulators	6	$1.56
Total	100	$26

Sources: Beyer 1991; Evans 1992; Finney 1991

biologically based technologies, and for exploiting their advantages in increasing the efficiency and sustainability of production and marketing, and for enhancing both income and employment.

1. Achieving Greater Private Sector Involvement and Greater Collaboration Between the Public and Private Sectors. Progress in the biological sciences is very dependent on strong programs in the universities and government laboratories. However most of the added acceleration of the development of biologically based methods during the past decade was derived from legislation that encouraged closer cooperation of the public and private sectors (such as the Technology Transfer Act of 1986 in the U.S.A.), harmonization of guidelines for the introduction of beneficial organisms (FAO, 1992) and for the release of organisms (UN Industrial Development Organization, 1991), and changes in patent law that created new opportunities especially for the private sector (Panetta, 1992). During the past decade we have witnessed an explosion in the commercial activity relative to tissue culture based techniques (Maluszynski, 1993) and a surge in use of microbial products (Commandeur; Komen, 1992; Lisansky, 1989) and of beneficial insects (King, this Symposium; Olkowski *et al.*, 1992; van Lenteren, this Symposium). Coming on line are transgenic pest tolerant plants (Casper; Landesmann, 1992), antagonists against soilborne pathogens, and entomophagous nematodes. Steady progress is being made toward bringing to the market biological agents against weeds, nematodes and pest molluscs. During the past decade this field was entered by many new companies and by large agrichemical firms.

Some agrichemical firms have acquired seed companies and are now accelerating the development of resistant cultivars including transgenic cultivars with resistance to insects (Huttner *et al.*, 1992), plant pathogens (Cubitt, 1991; Day, 1992) and herbicides (Bijman, 1992; Giaquinta, in NRC, 1986).

However, there is much more that should be done to engage the private sector. For example the sterile insect technique, which has been in practical use since 1957, remains the exclusive activity of governments with one exception of the small private program against the onion fly in The Netherlands. A decade ago a strong case could be made to keep all aspects of sterile insect technique programs firmly within governmental bureaucracies on the basis that specifications for quality of mass reared insects could not be adequately defined and that mass rearing was not a robust technology. However these arguments are no longer valid. Moreover it seems clear that without greater involvement of the private sector, programs that involve the mass rearing of beneficial insects for area wide use will not become common place. Likewise the biological control of cropland weeds is unlikely to become a significant practice without greater engagement of the private sector.

Specific changes in policies and legislation that would further enhance the symbiosis between the public and private sectors may be considered by the Board on Agriculture of the United States National Research Council, since it is currently engaged in a study on pest and pathogen control through management of biological control agents.

2. Conservation of Biological Diversity, Global Establishment of Intellectual Property Rights, and Increased Transfers of Technologies Between Developed and Developing Countries.
Commercial development and application of several biologically based methods is strongly influenced by the protection of intellectual property. Most industrialized countries have effective schemes for protection of intellectual property, but these are lacking or inadequate in many developing countries (Straus; Moufang, 1990). Even the Union on Protection of Plant Varieties (UPOV) has been joined by less than 30 countries. Consequently many governments and firms in the industrialized countries are reluctant to provide developing countries with access to advanced technologies. Industrialized nations are attempting to incorporate provisions in the General Agreement on Tariffs and Trade which would penalize countries that do not vigorously enforce intellectual property laws (Anonymous, 1992).

Differences in attitudes regarding intellectual property rights that exist between the industrialized and the developing countries need to be minimized (Swaminathan, 1992). The attitude of some developing countries was explained by the Chairman of the Government of India's Recombinant DNA Committee, S. Varadarajan (1991), as a reaction against colonialism. He stated that patents are not granted by some developing countries on many categories of inventions because this allows them to manufacture products developed by industrialized nations or import them without payment of royalties. This policy has enabled developing countries to foster indigenous capabilities in industrial production and achieve wide use of some advanced technologies.

During the past decade public sector scientists in industrialized countries increasingly have been required to patent their inventions. This pervasive trend has worked to the profound disadvantage of developing countries. In various aspects, the role of the private sector has come to exceed that of the public sector in developing technologies useful in agriculture and public health. In many industrialized countries public research institutions now are required to raise substantial portions of their research budgets via contractual research, cooperative research and development agreements, licenses and royalties. Many new biotechnology/biocontrol firms have been created through venture capital, and giant corporations have entered this arena. Much current information is being shared through privileged channels. These developments have combined to create a new pattern of research activity in which open interactions and communications have become restricted, and in which intellectual property rights influence the extent both of private and public sector investment (Brumby, 1991; Meeusen, 1992). For example some of the difficulties facing developing countries with respect to the use of genetic engineering to improve crops and biological control agents are summarized in Table 3 and Table 4, respectively.

Furthermore, successful importation and use of many new technologies into developing countries is critically dependent upon the effectiveness of their public sectors in conducting relevant problem solving research. Only a few developing countries have the research infrastructure needed to import and adapt technology from elsewhere and to enter into collaborative research agreements for its further development (Franzen; Wolpers, in Casper; Landsmann, 1992). Strengthening their capability in basic and applied research is essential and should be given far greater emphasis in bilateral aid programs and by international agencies. Also, collaborative agreements that facilitate technology transfer between the private sectors

of industrialized nations and the poorer countries are needed badly.

Swaminathan (1991) has urged that developing countries should not view themselves merely as receivers of technologies, but, also as technology innovators and providers, that each country should have intellectual property rights appropriate to its research initiatives, and which can be harmonized into regional patent conventions, and for the creation of common markets. The emergence of large common markets would accelerate economic development including advanced research and development pertaining to biologically based methods of pest management. The World Intellectual Property Organization has drafted relevant model laws.

I firmly believe that the Convention on Biological Diversity, already signed by 162 countries since the Earth Summit (UN Environment Programme, 1992), has created a propitious opportunity to mitigate this essentially North-South confrontation. It requires all Parties to the Convention to enact intellectual property laws and to safeguard intellectual property rights (Frye, 1992). Burhenne (1992) noted that the Convention addresses the genetic level of biodiversity in a legally binding manner. The Convention requires that all parties have in place legislation that will permit access to their genetic resources, and that will allow exchanges of materials and technologies to take place on mutually agreed terms. Of course such legislation will function only if the developed countries play fairly, because collection can be done surreptitiously. Therefore, developed countries should enact legislation that penalizes unlawful collection and that requires disclosure of the origin of genetic material used (Burhenne, 1992). An Intergovernmental Committee on the Convention on Biological Diversity is being established by UNEP, and it is expected to establish modalities of a protocol for transfer of living organisms and technologies resulting from them. On an interim basis, the Convention probably will be funded through the Global Environmental Facility of the World Bank.

3. Symbiosis of Biosystematics and Pest Management. Knutson (1981) stated:"The association of taxonomists and biological research has been mutually beneficial from the days of Charles Valentine Riley to the present."

The quest for better biological agents and genes involves the full diversity of living organisms on earth (Waage, 1991). Yet only a fraction of the world's fauna and flora has been described. Clearly there is an urgent need to more strongly engage and to reinforce biosystematics (FAO Council, 1992).

Table 3. Some pros and cons of developing resistant cultivars via genetic engineering versus conventional breeding

CONSIDERATION	GENETIC ENGINEERING	CONVENTIONAL
Overall costs	High--need large market	Modest--can target niche market
Regulatory requirements	Extensive	Modest or none
Access to genes	Almost unlimited	Limited to those in crop, its relatives or new mutations
Legal protection	Gene patents in industrialized countries only	UPOV; but less than 30 countries have signed convention
World trade	Problematical	Few problems
Can be developed by least developed countries?	Need extensive donor support	Need modest donor support

Table 4. Some pros and cons of improving biological control agents via genetic engineering versus conventional means.

CONSIDERATION	GENETIC ENGINEERING	CONVENTIONAL
Overall costs	High--need large market	Modest-can target niche markets
Regulatory requirements	Extensive	Modest
Access to genes	Almost unlimited	Limited to those in agent, its relatives (perhaps) or new mutations
Legal protection	Gene patents in industrialized countries only	Utility patents in industrialized countries only
World trade	Problematical	Few problems
Can be developed by least developed countries?	Need extensive donor support	Need modest donor support

4. *Vastly Improving Bioprocess and Mass Production Technology for Biological Agents and Natural Compounds.* Inadequate technology for the mass production of biological agents for use in pest management has been recognized as a major constraint for several decades. The production of most biological agents is still in the cottage industry phase. Indeed in 1992 the National Research Council conducted a study in which it noted that even though the U.S. has nurtured the discovery phase of biotechnology, it has not been aggressive in developing bioprocess engineering. According to this study inadequacies in the industrial bioprocessing will be a major impediment to the growth of the biotechnology industry, which nevertheless is expected to expand ten-fold within a decade (Borman, 1993).

5. *Vastly Improving Formulations and Shelf-life of Biological Agents and of Natural Compounds.* Phenomenal progress has been made in improving the performance in the field and the shelf-life of a number of biological agents and natural compounds (Shapiro; Dougherty this Symposium). For example certain entomopathogenic nematode products have been developed with a shelf-life of up to 12 months that are compatible with standard application equipment. Consequently nematodes can outcompete chemical treatments in controlling the carpenterworm, *Holocerus insularis* in Australia, and the major apple pest, *Carposina nipponensis*, in China (Curran, 1991, 1992; Ishibashi, 1992; Wang, 1992). In several cases remarkable advances have been achieved in improving the shelf-lives and effectiveness in the field of a number of products based on bacteria, fungi and viruses (Bateman, 1992). Such advances are essential for gaining widespread acceptance of biological agents. However essentially no headway has been made in developing practical methods of stockpiling mass reared beneficial insects (Leopold, in IAEA, in press). This should be one of the major goals of applied entomology.

6. *Discovering Biological Agents with Broader Host Ranges and Other Desirable Traits, or Inducing Such Traits in Known Biological Agents.* The excessively narrow host specificity of biological agents has been a major impediment to their commercial development. Doubtless, many agents with a suitably broad host range exist in nature, and we need to search for them. On the other hand, several reports in this Symposium indicated that it is sometimes possible to broaden the host range of biological agents by means of adjuvants, conventional genetic selection, and mutagenesis. The use of genetic engineering for modifying bacteria, fungi, viruses and crops has proven to be a very useful approach (Beachy *et al.*, 1990; Benedict *et al.*, 1993; Bonning *et al.*, 1991; Broglie *et al.*, 1991; Cornelissen; Melchers, 1993; Harman *et al.*, this Symposium; Huybrechts *et al.*, in IAEA, in press; Jenkins et al, 1993; Jordan; Hammond, this volume; Oakeshott *et al.*, in IAEA, in press).

In some instances it may be unwise to broaden the host range of a beneficial agent. For example it is likely that a host specific arthropod parasite or predator searches more intensively for a given pest at low pest densities than does a generalist with other available host resources (Knipling, 1979, 1992).

The mechanisms of pest and disease resistance need to be clarified in order to discover and develop additional cloned genes for developing resistant cultivars and livestock. Tobacco genes for chitinase and glucanase when transferred into tomato provide resistance to *Fusarium* (Cornelissen; Melchers, 1993; van den Eltzen, 1993). Perhaps cloned genes encoding some of the endocrine and neuroendocrine products will prove useful in developing improved biological agents against arthropods and nematodes (Bueds *et al.* in IAEA in press; or for developing resistant cultivars. Genes encoding monoclonal antibodies against important enzymes of harmful entities may be useful in developing resistant cultivars (Bakker *et al.*, *in* Casper; Landsmann, 1992) and cattle breeds (Oakeshott *et al.* in IAEA, in press). Vaccines against a number of multicellular ecto and endo parasites need to be developed.

For the autocidal control of insect pests, the development of genetic sexing strains has distinct advantages. For example in applying the sterile insect technique against tropical fruit flies, the released females, although sexually sterile, make oviposition punctures in fruit through which rot microorganisms gain entry. Therefore a strain of the Mediterranean fruit fly, *Ceratitis capitata*, has been developed in which a temperature sensitive lethal gene can be easily activated in the egg stage to remove all females but spare the males. This strain's advantages are that no sterile females are released, the cost of mass rearing is reduced, and releases of only sterile males increases their effectiveness (Franz; Kerremans; Hendrichs *et al.*, both in IAEA, in press).

As reported by Hoy (this Symposium) considerable benefits have been derived from the genetic improvement of beneficial insects. However the technology for the transformation of the germline of economically important insects needs considerable improvement (Atkinson; Crampton; Handler, all in IAEA, in press).

Inadequate fundamental knowledge of the biochemistry, physiology and ecology of parasites and their hosts continues to be a serious impediment to the discovery of useful principles and technologies.

7. Discovering and Characterizing Additional Biologically Active Natural Compounds.
Powerful attractants have been discovered for many of the most important arthropod pests in industrialized countries, but for only a small number in developing countries. In addition semiochemicals are needed for most nematodes and beneficial insects. As indicated by Tumlinson and Lewis at this Symposium, the prospects are favorable for developing innovative ways to use semiochemicals in causing natural enemies to perform more effectively. Additional compounds such as azadirachtin from the neem tree (Khanna, 1992) are needed to serve as models for new crop protection chemicals.

8. Guarding Biologically Based Technologies Against Obsolescence.
The effectiveness of living biological control agents, viruses and classical crop rotations (e.g. grasses followed by legumes), generally, appears to be highly durable. However the formidable ability of many harmful organisms to overcome most other methods of control has been demonstrated (Georghiou, 1990; Klassen, 1990). Thus the concern is valid that as the use of the δ-endotoxin of *Bacillus thuringiensis* increases, additional toxin-resistant strains will surely arise (Feitelson; Payne; Kim, 1992; McGaughey; Whalon, 1992). Moreover this hazard will increase as transgenic plants that express the δ-endotoxin are widely grown. Devising strategies to guard such technologies against obsolescence is receiving well deserved attention. A variety of practices are being considered such as crop rotation, revised field sanitation schedules, use of refugia to preserve susceptible pests, use of mixed seed lines some of which are susceptible to the pest, and integration with other biological agents (Moffat, 1992).

Much can be learned from strategies that have been implemented to preserve the usefulness of insecticides and fungicides. Experience has shown that it is essential to implement the strategy at the very outset of the commercial use of the new technology (Brent, in NRC, 1986). Further monitoring changes in the frequencies of genes for resistance in field populations is essential (Brent, Delp, Keiding, Frisbie *et al.*, all in NRC, 1986). Until recently, this required tests of susceptibility of whole organisms and in a few cases assays for the mechanisms of resistance. Now DNA probes and monoclonal antibodies are some of the technologies that can greatly facilitate this process (Hardy in NRC, 1986; Heckel, 1993).

The usefulness of benomyl in protecting bananas against Sigatoka disease (caused by *Mycosphaerella* spp.) in Central America, has been prolonged by postponing the use of benomyl so long as the proportion of resistant ascospores exceeds 5 percent (Brent in NRC, 1986). In Denmark, insecticides needed on livestock farms have been shielded from the development of resistance to them by livestock insects. This was facilitated by The Danish Act on Chemical Compounds and Products, which empowers the Danish Ministry of Environment to require, before registration, experimental data on cross-resistance and on the potential for the development of resistance. If resistance is likely to develop against a new product, or if its use is likely to result in resistance to other useful products, then registration may be refused (Keiding in NRC, 1986). Also registration may be withdrawn if resistance develops.

An insecticide resistance management (IRM) scheme in Australia has allowed Cox and Forrester (1992) to conclude that its economic benefits have been very substantial. They underscore the very important point - that previously had been made by Mironowski and Carlson (in NRC, 1986) - that susceptibility of harmful organisms to control measures may be viewed as a non-renewable resource that cannot be allocated effectively by the market price mechanism. Thus, resistance management must be achieved as a social technology.

In cases in which the harmful organism disperses widely, resistance management cannot be left to the individual farmer. A public resistance management scheme may be implemented if the benefits outweigh the costs. Mironowski and Carlson (in NRC, 1986) indicated that such a scheme might include a regional pest-management cooperative, pesticide application restrictions, education programs, etc. In countries with strong antitrust laws, and in which the prolongation of product life is not consistent with the objectives of regulatory agencies, it seems that resistance management schemes will have to be developed and led by the users (Dover; Croft; Johnson; Hawkins, all in NRC, 1986).

9. Instituting Systems Approaches for the Sustainable Management of Pests during the Production and Postharvest Phases of Agriculture.
As indicated in Agenda 21 (UN, 1992), a long term, iterative, world-wide effort must be devoted to designing and continuously improving sustainable systems of production and marketing. It is during the design stages that we must aim for enhanced economic and social benefits, and for the avoidance of unacceptable pesticide residues and degradation of natural resources.

Also, it is not enough to select good component technologies in designing a system. Other factors to be considered include the degree of uncertainty in decision making, the economics of scale and the overall strategy within which the system and its components will be deployed (Reichelderfer et al., 1984; Ridgway in IAEA, in press).

Indeed, the economic returns from some pest management systems are profoundly influenced by the scale of application. This is so in part because of the elimination of externalities and the reduction in numbers of "free riders". For example experience in the conservation of natural enemies to suppress the brown plant hopper in rice fields in Indonesia has shown that the use of natural enemies is most effective if there is a high degree of participation by growers on a regional basis (Reichelderfer et al., 1984; Oka, 1991). The movements of biological control agents and of the brown rice hopper are not affected by boundaries of farms. Thus, if a farmer or a small group of farmers in an agricultural area attempts to implement biological control, then the farmers who have not contributed to the effort are likely to receive some benefit. The non-participating farmers receive a free ride. This is a positive external effect. But the density of biological control agents will diminish as they disperse over the fields of non-participating farmers. Worse yet, these non-participating farmers may apply a broad spectrum insecticide which will drift onto the fields of participating growers and diminish their benefits. Clearly the higher the degree of participation, the greater are the total benefits realized in the region.

Rigorous adoption of an area-wide strategy has proven to be absolutely essential in attempting to eradicate pest populations such as screwworm and tropical fruit flies by means of the sterile insect technique (Klassen, 1989). Eradication can be realized only if the system of suppression is applied thoroughly in all ecological niches of the pest throughout an ecologically isolated area. Resistance against the suppressive system has rarely developed in such programs, perhaps because eradication has been achieved usually within the time span of only a few generations and because refugia have not been left in which genes for resistance could be assembled.

The avoidance of untreated pockets of the pest population is a very important consideration in the application of those biological controls against which resistance is unlikely to develop. For example, Knipling (1979) showed that "part of the population not subjected to treatment, even though it may be relatively small in percentage of the total, can make a vast difference in the number of survivors after a few generations". For insect pests such as the codling moth, *Laspeyresia pomomella* and the boll weevil, *Anthonomus grandis*, Knipling calculated that 90 percent control applied for four generations to total populations of these pests in an ecosystem would decimate the absolute size of the population, whereas the failure to control them on only ten percent of area would result in a dense population.

10. Developing Technologies for the Resource-Poor and Reaching the Unreached. A traditional concern in politics throughout the world is whether agricultural technologies developed at taxpayers' expense are both scale-neutral and resource-neutral. Recently, I visited a developing country in Africa in which the country's 9,000 large-scale farmers produce more food than the 2 million resource-poor subsistence farmers. The large-scale farmers are able to use modern technologies. Technologies such as genetically engineered disease resistant crops are scale-neutral, but they are not resource-neutral. However, many of the biologically based technologies for pest management are both scale-neutral and resource neutral. We need to establish institutional mechanisms to assure that ways will be found for adequately assisting resource poor farmers in adopting such technologies.

Currently, some social scientists (Hahn, 1993) are charging that the Green Revolution bypassed both women and poor farmers. While we must always welcome the critical hindsight of social scientists, we also must demand that they explain how we can succeed from the outset in reaching resource poor farmers. This means that such scientists and agricultural economists need to work in a timely manner with research scientists and with extension specialists.

Finally, it is absolutely essential that scientists from developing countries be included in research teams with scientists from developed countries. Developing country scientists need to work within their own countries, yet meet their cooperators both in developed and developing countries no less than every 18 months. Coordinated Research Programmes using this approach has been conducted effectively by the International Atomic Energy Agency for 30 years (Benson-Wiltschegg, 1991), and a somewhat similar approach has been adopted by the Science and Technology Directorate of the European Community.

11. Strengthening Education and Engaging the Mass Media. In order to gain a better insight into the need for improved education and information flow in this rapidly moving field, let us consider the challenge facing the extension services. An extension agent is expected to serve as a reliable source of information to

418

farmers, entrepreneurs, political leaders and the general public. Thus the extension agent must distill the essential facts from a large number of scientific and trade publications. The extension agent needs to perform objective evaluations of biologically based "products" that flow from private and public sector laboratories. Since product evaluation necessarily engenders a healthy antagonism with those who provide the new products, it must be conducted in a professionally impeccable manner. In order for the extension services to carry out their responsibilities, it is necessary that a national network of extension specialists be maintained technologically up to date, and with the means to do its work (Knutson; Richardson, 1988).

The conduct of agricultural fairs dealing largely with biologically based technologies for pest management is one device for keeping current that would likely prove helpful to extension agents, farmers, entrepreneurs, political leaders and the general public. Perhaps the leadership for periodically organizing such fairs could be provided by consortia of relevant scientific societies. Such consortia might organize distance learning courses for teachers in elementary schools. The mass media in developed countries receive materials from universities, governmental agencies and industry. Greater and more systematic efforts need to be made to develop materials that might be used by Voice of America, BBC World Service, Deutsche Wellen and others that focus on global audiences. Ministries and Departments of Agriculture need to develop special information policies on what have been referred to as "frontier technologies" (Swaminathan, 1991).

12. *Promoting National and International Biofutures Committees.* The Seminar on Public Policy Implications of Biotechnology for Asian Agriculture (1989, New Delhi) urged that every country should have a Biofutures Committee or Board in order to clearly delineate goals, targets and investment needs, and to help set priorities in bridging the gap between the current state of knowledge and technology and its applications in practice (Swaminathan, 1991). This Committee could serve as an institutional mechanism for implementing UN Agenda 21, which places strong emphasis on biologically based technologies. Biofutures Committees could advance Agenda 21 by periodically bringing together six key groups into a symbiotic relationship: financial institutions, industry, mass media, universities, scientific academies and government agencies.

Both the Brundtland Report, "Our Common Future", and the Rio Earth Summit (UN Conference on Environment and Development) emphasized that ecological concerns and technological development are causing inexorably the futures of all peoples and countries to be more closely interrelated (Frye, 1992, World Commission on Environment and Development, 1987). While the gap between rich and poor may be narrowing with regard to some countries with dynamic economies in Asia, the gap is surely growing in most countries of Africa, parts of Latin America and parts of the former Soviet Union. Biologically based methods of pest management and biotechnology provide many resource neutral and scale neutral opportunities to promote sustainable development throughout the world.

The urgency of the problems of food, agriculture and of the global environment do not permit us time to relax. The number of malnourished is growing, land per capita is shrinking, and abiotic and biotic stresses are mounting. Therefore let us take advantage of the precious opportunities that we have to make a significant difference through developing biologically based methods of pest management.

Acknowledgments

I am grateful for valuable inputs from S. Barbosa, L.R. Batra, V. Campbell, G. Franz, J. Komen, J. Hendrichs, D.A. Lindquist, M. Maluszynski, A. Papasolomontis, C. Thottakara and N. Van Der Graaff.

References

Anonymous. "Recombinant microbial insecticides: tough competition from genetically engineered plants." *Genetic Technology News.* **1991a,** *11,* 8 and 11.

Anonymous. "US approval for Mycogen's genetically-engineered biopesticides." *Biotechnology Business News.* **1991b,** *1,* 1 and 7.

Anonymous. "United States: USTR names India, Taiwan, Thailand "Special 301" priority offenders." *World Intellectual Property Report.* **1992,** *162,* 161-162.

Bateman, R. "Techniques for enhancing the efficacy of entomopathogen sprays against insect pests." *Proc. XIX Internat. Cong. Entomol.,* Beijing, **1992,** 301, (Abstract).

Beachy, R.N.; Loesch-Fries, S.; and Tumer, N.E. "Coat protein-mediated resistance against virus infection." *Ann. Rev. Phytopathol.,* **1990,** *28,* 451-474.

Benedict, J.H.; Sachs, E.S.; Altman, D.W.; Ring, D.R.; Stone, T.B.and Sims, S.R. "Impact of δ-endotoxin-producing transgenic cotton on insect-plant interactions with *Heliothis virescens* and *Helicoverpa zea* (Lepidoptera: Noctuidae)." *Environ. Entomol.,* **1993,** *22,* 1-9.

Benson-Wiltschegg, T. "Development through science: the IAEA Research Contract Programme." *IAEA/PI/A31E, 91-01874. International Atomic Energy Agency,* Vienna, Austria, **1991**, 62.

Beyer, E.M. Crop Protection - Meeting the Challenge. *Brighton Crop Protection Conference - Weeds*; British Crop Protection Council: Farnham, U.K., 1991, pp 3-22.

Bijman, J. "Will herbicide-tolerant crops be tolerated?" *Biotechnology and Development Monitor,* **1992**, *13*, 4-5.

Bonning, B.C.; Merryweather, A.T.; Possee, R.D. "Genetically engineered baculovirus insecticides." *Agbiotech News and Information,* **1991**, *3*, 29-31.

Borman, S. "Bioprocess engineering: Lack of U.S. support for field assailed." *C&EN,* **1993**, *70*, 4-5.

Broglie, K.; Chet, I.; Holliday, M.; Cressman, R.; Biddle, P.; *et al.* "Transgenic plants with enhanced resistance to the fungal pathogen *Rhizoctonia solani.*" *Science,* **1991**, *254*, 1194-1197.

Brumby, P. An international perspective on agricultural biotechnology in Asia. In *Biotechnology for Asian Agriculture.*; Getubig, Jr., I.P.; Chopra, V.L.; Swaminathan, M.S., Eds.; Asian and Pacific Development Centre: Kuala Lumpur, Malaysia, 1991, pp 307-321.

Burhenne, W.E. "Biodiversity - the legal aspects." *Environ. Policy & Law,* **1992**, *22*, 324 -326.

Casper, R.; Landsmann, J., Eds. *The Biosafety Results of Field tests of genetically Modified Plants and Microorganisms*; Goslar, Germany, 11-14 May, **1992**, 296 pp.

Commandeur, P.; Komen, J. "Biopesticides: options for biological pest control increase." *Biotechnology and Development Monitor*, **1992**, *13*, 6-8.

Cook, R.J.; Andres, L;, deZoeton, G.A.; Doane, C.; Gwadz, R.W.; Hardy, R.; Hemming, B.; Kuc, J.; Mankau, R.; Miller, D.; Ryan, Jr., C.A.; Smith, S. *Report of the Research Briefing Panel on Biological Control in Managed Ecosystems.* National Academy Press: Washington, D.C., 1987, 12 pp.

Cornelissen, B.J.C.; Melchers, L.S. "Strategies for control of fungal diseases with transgenic plants." *Plant Physiol.,* **1993**, *101*, 709-712.

Council on Environmental Quality. *Integrated Pest Management*; U.S. Govt. Printing Office: Washington, D.C., 1972, pp 41.

Cox, P.G.; Forrester, N.W. "Economics of insecticide resistance management in *Heliothis armigera* (Lepidoptera: Noctuidae) in Australia." *J. Econ. Entomol.* **1992**, *85*, 1539-1550.

Cubitt, I.R. "The commercial application of biotechnology to plant breeding." *Plant Breeding Abstracts,* **1991**, *61*, 151-158.

Curran, J. "Biological and economic feasibility of using entomopathogenic nematodes to control insect pests." *Proc. XII Internat. Cong. Plant Pro.*, Rio de Janeiro, 11-16 August **1991**, 138 (Abstract).

Curran, J. "Post-application biology of entomopathogenic nematodes in soil." *Proc. XIX Internat. Cong. Entomol.*, Beijing, 28 June - 4 July **1992**, 302 (Abstract).

Day, P.R. "Plant pathology and biotechnology: choosing your weapons." *Ann. Rev. Phytopathol,* **1992**, *30*, 1-13.

Evans, D.A. "Designing more efficient herbicides." *Proc. First Internat. Weed Control Congr.*, Melbourne, Australia, Weed Science Soc. Victoria Inc. **1992**, 34-42.

Feitelson, J.S.; Payne, J.; Kim. L. "*Bacillus thuringiensis*: insects and beyond." *Biotechnology,* **1992**, *10*, 271-275.

Finney, J.R. Where do we stand - where do we go? In *Pesticide Chemistry*; Frehse, H., Ed.; VCH: Weinheim, Germany, 1991, pp 555-575

FAO. *New World Screwworm Eradication Programme: North Africa 1988-1992.* FAO, Rome, Italy. **1992**.

FAO Council. *Expert Consultation on Guidelines for the Introduction of Biological Control Agent. Meeting Report,* **1992b**, AGP/1992/ M/3. pp 21.

Frye, R.S. "Uncle Sam at UNCED." *Environ. Policy and Law,* **1992**, *22*, 340-346.

Gallitelli, D.; Volas, C.; Martelli, G.; Montasser, M.S.; Tousignant, M.E.; Kaper, J.M. "Satellite-mediated protection of tomato against cucumber mosaic virus. II. Field test under natural epidemic conditions in southern Italy." *Plant Disease,* **1991**, *75*, 93-95.

Georghiou, G.P.; Green, M.B.; LeBaron, H.M.; Moberg, W.K., Eds.; *Managing Resistance to Agrichemicals: From Fundamental Research to Practical Strategies*; ACS Series No. 421.; American Chemical Society: Washington, D.C, 1990, 18 pp.

Hahn, N.D. "Victims of the green revolution." *Ceres,* **1993**, *25*, 41-42.

Heckel, D. "Comparative genetic linkage mapping in insects." *Ann. Rev. Entomol.,* *1993*, 381-408.

Herren, H.; Neuenschwander, P. "Biological control of cassava pests in Africa." *Ann. Rev. Entomol.,* **1991**, *36*, 257-283.

Huttner, S.L.; Arntzen, C.; Beachy, R.; Breuning, G.; Nester, E; Qualset, Q.; Vidaver, A. "Revising oversight of genetically modified plants." *BioTechnol.,* **1992**, *10*, 967-971.

IAEA [International Atomic Energy Agency]. *Management of Insect Pests: Nuclear, Molecular and Genetic Techniques.* Proceedings of a FAO/IAEA Symposium, Vienna, Austria, 1992 (in press).

Ishibashi, N. "Integrated control of soil pests by beneficial nematodes." *Proc. XIX Internat. Cong. Entomol. Abstracts*, Beijing, 28 June - 4 July **1992**, 302.

Jaffe, W.; Rojas, M. "Brazilian and Mexican efforts to protect plants biologically." *Biotechnology and Development Monitor*, **1992**, *13*, 10.

Jenkins, J.N.; Parott, W.L.; McCarty, Jr., J.C.; Callahan, F.E.; Berberich, S.A.; Deaton, W.R. "Growth and survival of *Heliothis virescens* (Lepidoptera: Noctuidae) on transgenic cotton containing a truncated form of delta endotoxin gene from *Bacillus thuringiensis*." *J. Econ. Entomol.*, **1993**, *86*, 181-185.

Khanna, A. "Neem compounds commercialized." *Biotechnology and Development Monitor*, **1992**, *13*, 12.

Klassen, W. "Biological pest control: needs and opportunities." *Amer. J. Alternative Agric.*, **1988**, *2* and *3*, 117-122.

Klassen, W. "Eradication of arthropods: theory and historical practice." *Miscellaneous Publication No. 73*, **1989**, Entomol. Soc. Amer., Lanham, Maryland, 29 pp.

Klassen, W. Insects and agriculture. In *Progress and Perspectives for the 21st Century*; Menn, J.J.; Steinhauer, A.L., Eds.; Entomol. Soc. Amer. Centennial Symposium: Lanham, Maryland, 1990, pp 44-68.

Knipling, E.F. "The basic principles of insect population suppression and management." *Agriculture Handbook No. 512*; USDA, Washington, D.C., **1979**, 659 pp.

Knipling, E.F. "Principles of insect parasitism analyzed from new perspectives: practical implications for regulating insect populations by biological means." *Agriculture Handbook No. 693*; USDA, Washington, D.C., **1992**; 337 pp.

Knutson, L. "Symbiosis of biosystematics and biological control." *Beltsville Symposia in Agricultural Research*, **1981**, *4*, 61-78.

Knutson, R.D.; Richardson, J.W. "Implications of biotechnology for agricultural and food policy." *The cotton Gin and Oil Mill* , 13 August **1988**, 6-9.

Kumar, N. "Biological pest control becomes established technique in Asia." *Biotechnology and Development Monitor*, **1992**, *13*, 9.

Lisansky, S.G. "Biopesticides." *AgBiotech News and Infor.*, **1989**, *1*, 349-353.

Mackauer, M. "Advances and future trends of biological control for plant protection." *Proc. XII Internat. Congress of Plant Protection*, Rio de Janeiro, Brazil, 11-16 August **1991**, 132 pp.

Maluszynski, M. "Applications of biotechnologies in plant breeding." Working Paper for FAO Expert Consultation on Plant Biotechnology, Joint FAO/IAEA Division, Vienna, Austria, **1993**, 10 pp.

McGaughey, W.H.; Whalon, M.E. "Managing insect resistance to *Bacillus thuringiensis* toxins." *Science*, **1992**, *258*, 1451-1455.

Meeusen, R. "Biotechnology and development issues it presents to the industrial research community." *North American Plant Protection Bulletin No. 8*, **1992**, 31-35.

Moffat, A.S. "Firms revamp Bt strategies to overcome resistance problems." *Gen. Eng. News*, **1992**, *12*, 1, 16 & 17.

Montasser, M.S.; Tousignant, M.E.; Kaper, J.M. "Satellite-mediated protection of tomato against cucumber mosaic virus: I. Greenhouse experiments and simulated epidemic conditions in the field." *Plant Disease*, **1991**, *75*, 86-91.

N.R.C. [National Research Council U.S.A., Board of Agriculture] Ed.; *Technology and Agricultural Policy: Proceedings of a symposium*; National Academy Press: Washington, D.C., 1986, 257 pp.

Oka, I.N. "Success and challenges of the Indonesia National Integrated Pest Management Program in rice-based cropping system." *Crop Protection*, **1991**, *10*, 163-165.

Olkowski, W.; Dietrick, E; Olkowski, H. "The biological control industry in the United States, Part I." *IPM Practitioner*, **1992**, *14*, 1-4.

Panetta, J.D. "Biological pest control: an industry perspective of a streamlined regulatory process." *BFE*, **1992**, *9*, 134-137.

Ream, W. "Agrobacterium tumefaciens and interkingdom genetic exchange." *Ann. Rev. Phytophathol.*, **1989**, 27, 583-618.

Reichelderfer, K.H.; Carlson, G.A.; Norton, G.A. "Economic guidelines for crop pest control." *FAO Plant Production and Protection Paper*, **1984**, *58*, 1-93.

Straus, J.; Moufang, R. "Legal aspects of acquiring, holding and utilizing patents with reference to the activities of the International Centre for Genetic Engineering and Biotechnology (ICGEB)." *Max Planck Institute of Foreign and International Patent, Copyright, and Competition Law*, Munich, Germany, **1990**, 85 pp.

Swaminathan, M.S. Biotechnology and a better common present: a synthesis. In *Biotechnology for Asian Agriculture.*; Getubig, Jr., I.P.; Chopra, V.L.; Swaminathan, M.S., Eds.; Asian and Pacific Development Centre, Kuala Lumpur, Malaysia, 1991, pp 1-9.

Swaminathan, M.S. "Contribution of biotechnology to sustainable development within the framework of the United Nations System."

United Nations Development Organization. IPCT.148/Rev.1 (SPEC.), **1992**, 10 pp.

UN (United Nations) Agenda 21: "United Nations conference on environment and development." Rio de Janeiro, 3-14 June **1992**, Parts I, 207pp; II, 275 pp.; III, 244 pp.; & IV, 65 1992. UN Administrative Coordinating Committee, New York.

UN Environment Programme. "Convention on biologica diversity." 5 June **1992**, UNEP document No. 92-7807, 24 pp.

UN Industrial Development Organization. "Voluntary code of conduct for the release of organisms into the environment." UNIDO/UNEP/WHO/ FAO Working Group on Biosafety. 1991, 7 pp.

van den Elzen, P. "Mogen becomes first to produce genetically engineered plants capable of surviving fungal attack." Mogen, Leiden, the Netherlands, 1993, 2 pp.

Varadarajan, S. Policy implications in intellectual property rights. In *Biotechnology for Asian Agriculture*; Getubig, Jr., I.P.; Chopra, V.L.; Swaminathan, M.S., Eds.; Asian and Pacific Development Centre, Kuala Lumpur, Malaysia, 1991, pp 123-130.

Waage, J.K. Biodiversity as a resource for biological control. In *The Biodiversity of Microorganisms and Invertebrates: Its Role in Sustainable Agriculture*; Hawkesworth, D.H., Ed.; CAB International, London, 1991, pp 149-163.

Wang, J-X. "Control of the peach fruit moth, *Carposina nipponensis* using entomopathogenic nematodes." *Proc. XIX Internat. Cong. Entomol.*, Beijing, 28 June-4 July, **1992**, 302 (Abstract).

Wilson, M.; Epton, H.A.S.; Sigee, D.C. "Biolgical control of fire blight of hawthorn." *Acta Hortic. Wageningen*, **1990**, *272*, 363-365.

World Commission on Environment and Development. *Our Common Future*; Oxford University Press, N.Y., 1987, 383 pp.

422

INDEX

Index

425

427

Microbial metabolites with biological activity against
    plant pathogens
  antibiotic(s), role in ecological competence, 176
  antibiotic isolation from natural habitats, 175–176
  biotic and abiotic factors affecting production,
      176–177
  2,4-diacetylphloroglucinol, 174–175
  enzymes, 176
  genetic engineering, 177–178
  genetic strategy, 174
  hydrogen cyanide, 174–175
  phenazines, 174
  recombinant DNA technology, role, 173–174
Microbial pesticides, description, 14
Microorganisms, use in biological control, 15–16
Modification of behavior, semiochemical-based
    strategy against forest insect pests,
    268–269t,273–274
Molecular genetic manipulations of beneficial
    arthropods, goal, 357
Molecular genetics, importance, 338
Monitoring, semiochemical-based strategy against
    forest insect pests, 265–271
Monitoring of insects, use of pheromones, 280–281
Monocerin, 292
Multiple-agent approach to biological control,
    rationale, 12–13
Muscoid flies
  bacterial control in livestock operations, 47–52
  need for bacteria, 47
Mycoherbicides, biological control agents in
    combination with wheat flour granules, 238–239
Mycoparasites, fungal, *See* Fungal mycoparasites
Mycostop, bacterial biological control agent, 144
Myzus persicae, resistance mediated by plant
    hormones, 378,379f,t

N

Nagilactones, 296–297
Natural communities of beneficial organisms, 13–14
Natural control
  definition, 70
  regulators of insect and mite populations, 101
  semiochemical-based enhancement strategy against
      forest insect pests, 269t,274–275
Natural enemies
  conservation in agriculture, 102–105
  evaluation and selection, 75–77
  suppression of arthropod pests
    benefits and efficacy, 85–88
    economic analysis, 86
    history and trends, 82–85
    risks, 88
  understanding behavior variability, 77–78
Natural products for weed control, 290–291
Naturally occurring disease-suppressive soils
  biological control of soilborne plant pathogens,
      206–207
  cultural practices, 205
  mechanisms, 204
  properties, 205
  suppression of *Fusarium* wilts, 205–206
Nematicides, groundwater contamination, 214
Nematodes
  biological control agents in combination with
      wheat flour granules, 238–240f

Nematodes—*Continued*
  biological control organism research, 214
  fungal biological control, 214–216
  fungus–bioregulator combination biological
      control, 217–219
  management methods, 214
  pheromone biological control, 216–217
Neochetina spp., use in biological control, 244
Neohydronomous affinis, use in biological control, 244
Nicotiana, sucrose ester identification, 285–286
Nicotiana plumbaginifolia, plant hormone mediated
    insect resistance, 378,379f,t
Nitrogen fertilizers, control of *Sclerotium rolfsii*,
    211–213
Nonoccluded viruses, description, 313–314
North American, semiochemical-based strategies and
    tactics used against forest insect pests, 265–275
Novel associations, descriptions, 22
Nuclear polyhedrosis virus
  description, 314
  use as microbial agent, 40

O

Oomycin A, alteration of regulation for improved
    biological control property, 340–341
opd gene, use for genetic manipulation of beneficial
    arthropods, 363
Operational problems to be solved for biologically
    based pest management
  interdisciplinary relationships and interfunctional
      activities, 7
  methods, information, and systems, 9
  risk investment, 8–9
Ophiobolins, 293,294
Optical brighteners, 40–42
Opuntia spp., biological control, 231
Orchard(s), control of arthropod pests with
    augmentative release of parasites and predators,
    93–94
Orchard pests, pheromones for monitoring and
    mating disruption, 280
Organisms, beneficial, management of natural
    communities, 13–14
Organization, *Trichoderma* and *Gliocladium*
    genomes, 348–349
Ostrinia spp., control using *Trichogramma* spp., 92–93
Oxysporone, 294,295

P

P-element vectors, germ line transformation
    methods, 358
Paecilomyces lilacinus, use in biological control, 15
Parabiological control
  definitions, 23,26
  examples, 25
  relationships and components, 23–25
Parapheromones, definition, 217
Parasites
  augmentative release for suppression of arthropod
      pests, 90–97
  role in biological control, 68–72,78–79
Parasitic rhabditids, insect, *See* Insect parasitic
    rhabditids for control of arthropod pests
Parasitoids, conservation of arthropod natural
    enemies in agriculture, 103–105

431

434

Production: Margaret J. Brown
Indexing: Deborah H. Steiner
Acquisition: Anne Wilson
Cover design: Amy Meyer Phifer

Printed and bound by Maple Press, York, PA

435